Introduction to Applied Quantum Chemistry

S. P. McGLYNN, *Louisiana State University*

L. G. VANQUICKENBORNE, *University of Louvain*

M. KINOSHITA, *University of Tokyo*

D. G. CARROLL, *University College, Dublin*

HOLT, RINEHART and WINSTON, INC.
New York   Chicago   San Francisco   Atlanta
Dallas   Montreal   Toronto   London   Sydney

QD
462
I58

Preface

Quantum mechanics has penetrated into the undergraduate curriculum of the chemistry, biochemistry, and biophysics major. Cataloguing in chemistry usually occurs under the title "Introductory Quantum Chemistry" and the content often approximates a diluted version of the analogous physics course. Indeed, many such courses confront the "real world" of the average chemist and biologist only insofar as they provide discussion of the hydrogen molecule or the Hückel theory for ethylene! The advanced graduate-level quantum chemistry and biology curricula fare little better: the undergraduate material is presented again at a deeper level and the practice of actual chemical computations is often relegated to a terminal three-week period during which the instructor attempts, often haphazardly and always hurriedly, to achieve relevance.

This book represents one attempt to eliminate some of the aforementioned deficiencies. It does not pretend to teach quantum mechanics. It merely tries to illustrate the manner in which quantum mechanics is used in calculating molecular properties which can be compared with experiment. In other words, this is a "how to" book—perhaps even a "Fanny Farmer approach" to simple quantum chemistry. It is not particularly concerned with the correlation of experimental and computed data: such correlation lies beyond the interfacial character we have tried to achieve between quantum mechanics (that is, theory) on the one hand and experiment on the other.

The book is written in a way that minimizes the need for simultaneous classroom instruction. Indeed, experience acquired at Louisiana State University, with students who had prior exposure to an introductory undergraduate course in quantum chemistry, indicates that the material can be adequately covered in eight two-hour sessions spaced at weekly intervals. Furthermore, the end result of these lectures was invariably rewarding: a marked increase in student ability to handle abstract quantum mechanical ideas, to approach experiment critically and with insight, and to feel "at home" in the current journal literature. Thus, although we do not visualize the material of this text supplanting current course work in quantum chemistry, we can advise its inclusion as supplementary material. It makes quantum chemistry real and useful and it converts research into an imaginative interplay of experimental and calculated data and correla-

iii

tions. In particular, for physical-chemistry majors we can advise the insertion of a short course such as this between the undergraduate and graduate courses in quantum chemistry; for non-physical-chemistry majors, we can advise this same short course as a terminal quantum chemistry effort.

The material contained in this text reflects our prejudices: we are all molecular spectroscopists. We had planned to include other material (for example, molecular electronic absorption-band shapes, nuclear quadrupole moments, advanced group theory, and so forth) but were forced to curtail our plans because of book size. The discussion of many-electron theories has also been abbreviated for similar reasons. We make no apology for these decisions; we thought it wiser to produce a readable book of modest pretensions.

This book was planned and to a large degree written while all of us were in residence at Louisiana State University. Our stay there was supported in its entirety by the United States Atomic Energy Commission—Biology Branch. We are grateful to LSU and the USAEC for the many courtesies shown us during our stay in Louisiana.

Professors W. Flygare (Illinois), C. Ballhausen (Copenhagen–LSU), N. Kestner (LSU), and C. Cusachs (Tulane) read and criticized the book in its entirety; we are deeply grateful for the time they invested and the concerns they showed. The book was also reviewed by many graduate students, past and present, at LSU; it is a pleasure to acknowledge the many suggestions they made. The initial and final typed drafts of this manuscript and the preparation of the author and subject indices were works of Judy Brignac; we are certain that little of this work was pleasant and we must record a high indebtedness to her for patience, skill, and composure under fire. Finally, we wish to thank Dr. Peter Hochmann (LSU) and Dr. John Maria (LSU) for their assistance at various stages during the course of this work.

Baton Rouge, Louisiana

S. P. MCGLYNN

L. G. VANQUICKENBORNE

M. KINOSHITA

D. G. CARROLL

Contents

‖ SPIN-ORBIT COUPLING, 339

⫿ SPIN-SPIN COUPLING, 393

Introduction to Applied Quantum Chemistry

CHAPTER 1

Atomic Orbitals

The concept of electronic shell structure is basic to our present understanding of the Periodic Table. Indeed, the whole subject of chemistry is thoroughly suffused by the concept of shells and orbitals. The high-school student speaks of valence shells, d orbitals, orbital degeneracies, etc., emphasizing, at once, both the ubiquity and the viability of this concept.

And yet the orbital concept attains precise definition only in the case of a one-electron atom such as H, He^+, Li^{2+}, etc. It is merely an approximation to the proper description of electron behavior in multielectron atoms and, in many instances, it is not even a very good approximation. Why, then, is the concept so useful? Because, goes the retort, it is the only simple concept we have; because, were we to abandon it, we should become completely reliant on computers; and because we have learned by experience how to handle the orbital concept so as to make it yield reasonable answers.

The purpose of the present chapter is mainly to discuss atomic orbitals as starting points for molecular considerations. We intend to discuss the orbital concept, to point out its deficiencies, and to catalogue some of the various types of functional representations of atomic orbitals (i.e., Morse, Slater, self-consistent field, etc.) which have found use. This, plus a knowledge of where to go to get atomic-orbital functions for molecular purposes, is our sole aim.

3

Precision indicates a few word conventions. The term *center*, as we use it, refers to an atom or nucleus. Thus, a *one-center wave function* is one for which the confining nuclear potential is supplied by some one atom. An *orbital* is a one-electron space wave function. An *atomic orbital* is a one-center one-electron space wave function, whereas a *molecular orbital* is a many-center one-electron space wave function; these are abbreviated to AO and MO, respectively. The term *spin orbital* is used to denote a one-electron wave function which is dependent on both the space and spin coordinates of the electron; hence, the terms *atomic spin orbital* and *molecular spin orbital*.

1. ONE-ELECTRON ATOMS

The simplest type of atom consists of one electron moving in the field of a single nucleus of charge $+Ze$. This is the only type of atom for which it is possible to obtain explicit closed analytic expressions for the energies, ϵ_r, and eigenfunctions, χ_r, of the stationary states. The polar coordinate frame appropriate to such an atom is shown in Fig. 1.1. The Schrödinger equation[1] is given by

$$-\frac{\hbar^2}{2\mu r^2}\left(r\frac{\partial^2}{\partial r^2}r + \frac{1}{\sin\theta}\frac{\partial}{\partial\theta}\sin\theta\frac{\partial}{\partial\theta} + \frac{1}{\sin^2\theta}\frac{\partial^2}{\partial\phi^2}\right)\chi - \left(\epsilon + \frac{Ze^2}{r}\right)\chi = 0 \quad (1.1)$$

where μ is the reduced mass of the electron-nucleus system. Separability of this partial differential equation into three differential equations, each one functionally dependent on only one of the variables r, θ, or ϕ, is assured; thus, a solution of the form $R(r)\Theta(\theta)\Phi(\phi)$ is sought. If we now insert the

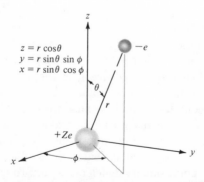

$z = r\cos\theta$
$y = r\sin\theta\,\sin\phi$
$x = r\sin\theta\,\cos\phi$

FIG. 1.1.   Polar coordinates for a one-electron atom. The origin of coordinates is at the center of mass of the nucleus-electron system. The reduced mass of the system is given by $\mu = mM/(m + M)$, where m is electron mass and M is nucleus mass.

equality $\chi(r, \theta, \phi) = R(r)\Theta(\theta)\Phi(\phi)$ into Eq. (1.1) and divide the result by χ/r^2, we find

$$-\frac{\hbar^2 r}{2\mu R}\frac{\partial^2(rR)}{\partial r^2} - \frac{\hbar^2}{2\mu\Theta \sin\theta}\frac{\partial}{\partial\theta}\left(\sin\theta \frac{\partial\Theta}{\partial\theta}\right)$$

$$-\frac{\hbar^2}{2\mu\Phi \sin^2\theta}\frac{\partial^2\Phi}{\partial\phi^2} - (\epsilon r^2 + Ze^2 r) = 0 \quad (1.2)$$

In Eq. (1.2), the first and fourth terms depend on r but not on either θ or ϕ; on the other hand, the second and third terms depend only on θ and ϕ. Consequently, we may write

$$-\frac{\hbar^2 r}{2\mu R}\frac{d^2(rR)}{dr^2} - (\epsilon r^2 + Ze^2 r) = \lambda \qquad (1.3)$$

where λ is a constant independent of r, θ, and ϕ. The remaining partial differential equation in θ and ϕ is readily separated into two parts to yield

$$-\frac{\hbar^2}{2\mu\Theta \sin\theta}\frac{d}{d\theta}\left(\sin\theta \frac{d\Theta}{d\theta}\right) - \frac{\hbar^2\gamma}{2\mu \sin^2\theta} = -\lambda \qquad (1.4)$$

$$\frac{1}{\Phi}\frac{d^2\Phi}{d\phi^2} = \gamma \qquad (1.5)$$

where γ is also a constant.

Equation (1.5) integrates to

$$\Phi(\phi) = e^{\pm im\phi}, \qquad (m = 0, 1, 2, \ldots) \qquad (1.6)$$

where $\gamma = -m^2$, and where the single valuedness of Φ elicits the integer nature of m. Replacement of γ in Eq. (1.4) and imposition of the dual requirements: that Θ be single valued and that it possesses no singularities, ascertains the fact that the only solutions to Eq. (1.4) are

$$\Theta_m = P_l{}^m(\cos\theta), \qquad (l \geq m) \qquad (1.7)$$

where the $P_l{}^m(\cos\theta)$ are associated Legendre polynomials[1] and where λ is given by

$$\lambda = -(\hbar^2/2\mu)l(l+1), \qquad (l = 0, 1, 2, \ldots) \qquad (1.8)$$

The functions Φ and Θ_m are not normalized. When normalized, they are denoted Φ_m and Θ_{lm}, respectively, and a new function, the angular wave function $Y_{lm}(\theta, \phi)$, is defined as

$$Y_{lm} = \Theta_{lm}\Phi_m \qquad (1.9)$$

It may be shown that the functions Y_{lm} are eigenfunctions of the angular momentum operators[1] with eigenvalues specified as follows:

$$\hat{l}^2 Y_{lm} = l(l+1)\hbar^2 Y_{lm} \tag{1.10}$$

$$\hat{l}_z Y_{lm} = m\hbar Y_{lm} \tag{1.11}$$

Upon substitution for λ from Eq. (1.8), Eq. (1.3) becomes

$$-\frac{\hbar^2 r}{2\mu}\frac{d^2(rR)}{dr^2} - Ze^2 rR - \epsilon r^2 R = -\frac{\hbar^2}{2\mu}l(l+1)R \tag{1.12}$$

The boundary conditions require that $R \to 0$ as $r \to \infty$ and that R be finite at $r = 0$. Thus, if we make the substitution $R' = rR$, we find

$$-\frac{\hbar^2}{2\mu}\left(\frac{d^2 R'}{dr^2} - \frac{l(l+1)R'}{r^2}\right) - \frac{Ze^2 R'}{r} - \epsilon R' = 0 \tag{1.13}$$

with $R' \to 0$ both as $r \to \infty$ and $r \to 0$. Consequently, at large r Eq. (1.13) abbreviates to

$$-\frac{\hbar^2}{2\mu}\frac{d^2 R'}{dr^2} = \epsilon R' \tag{1.14}$$

The solutions to this limiting equation are of the form $e^{-\zeta r}$; thus, it is implied that Eq. (1.13) might possess solutions of the form

$$R'(r) = f(r)e^{-\zeta r} \tag{1.15}$$

where $f(r)$ is a power series in r of the form

$$f(r) = \sum_{t=0}^{\infty} a_t r^t \tag{1.16}$$

Insertion of the trial function of Eq. (1.15) into Eq. (1.13) and imposition of the boundary conditions, provides a general solution as a product of a polynomial in r and an exponential in r. Some of the normalized solutions R_{nl}', as well as some of the normalized Θ and Φ functions, are tabulated in Appendix A.

2. ORBITALS FOR THE ONE-ELECTRON ATOM

The solutions to the hydrogen atom problem are well known.[1] Thus, we merely enumerate a few of their more important characteristics:

(i)   The probability of finding a 1s electron within a spherical shell with radii r and $r + dr$ is given by

$$(R')^2 dr = 4(Z/a_0)^3 e^{-(2Z/a_0)r} r^2 dr \tag{1.17}$$

Differentiation of this function with respect to r yields a maximum at $r = a_0/Z$. At this distance, the amplitude of the wave function is largest and the probability density of the electron is greatest. For a hydrogen $1s$ electron, $r_{max} = a_0$, where $a_0 = \hbar^2/\mu e^2 = 0.529$ Å; the distance a_0 is known as the Bohr radius. The distance a_0 serves as a convenient unit of measurement for atomic distances in molecules and for electron-nuclear *separations* in atoms and molecules; thus, a_0 may be written as 1 *bohr*, 1 *atomic unit of distance* or, simply, 1 a.u.

(ii)   All of the radial solutions R_{nl} for bound states have certain properties in common:

They all go to zero exponentially as $r \rightarrow \infty$.

The function of lowest energy for any given l is nodeless. Each successive higher-energy function with the same l has one more node than the one prior to it.

Thus, the radial probability density of a $2s$ electron exhibits two maxima, etc.

The functions are orthonormal.

Each function $(R')^2$ associated with a given l achieves maximum amplitude in its outermost lobe and each successively higher-energy function of that same l has its outer lobe at larger r than the one immediately prior to it.

The characteristics enumerated are general for all potentials of the form $V(r) = -Ar^{-q}$ where $q = 1, 2, 3, \ldots$. The radial functions of a number of hydrogen orbitals are diagrammed in Fig. 1.2.

(iii)   The general expression for the energy is given by

$$\epsilon_n = -Z^2e^2/2n^2a_0 \qquad (1.18)$$

The energy is a function of the principal quantum number n only. Thus, all orbitals of different l and m, but the same n, are degenerate. The energy unit e^2/a_0 is termed *an atomic unit of energy*, 1 a.u., or 1 hartree. It equals approximately 27.2 eV.

(iv)   The angular functions are the well-known spherical harmonics— the forms taken by standing waves in any spherical problem (e.g., standing waves on a flooded planet[2]). They possess the following general properties:

Each $Y_{lm}(\theta, \phi)$ is a polynomial of degree l in $\sin \theta$ and $\cos \theta$.

Functions of even l have even parity and those of odd l have odd parity. Parity refers to the operation of coordinate inversion in the origin.

The number of angular nodes increases with l and, as it does, the wave functions become more directional or pointed. Very high l functions with large numbers of angular nodes describe electrons that are nearly classical

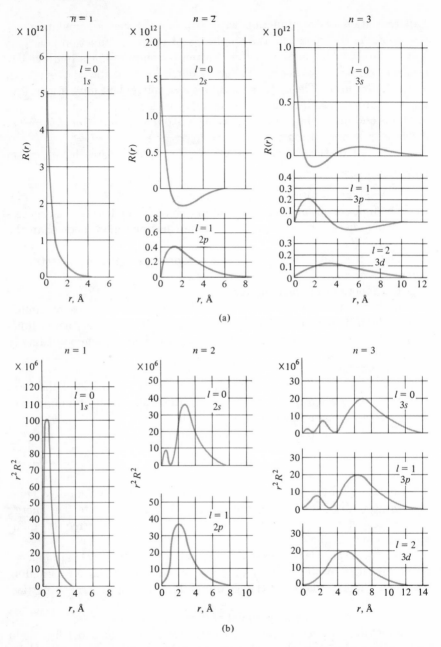

FIG. 1.2.   (a) Radial wave functions $R(r)$ versus r for a hydrogen atom. (b) Radial distribution functions $r^2 R^2(r)$ versus r for a hydrogen atom. Note that $R'(r) \equiv rR(r)$. [This diagram is taken from *Atomic Spectra and Atomic Structure* by G. Herzberg, Dover Publications, Inc., New York, 1944. Reprinted through permission of the publisher.]

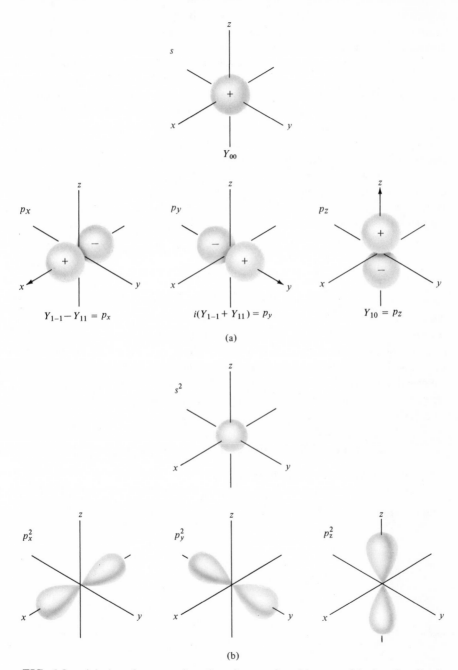

FIG. 1.3. (a) Angular wave functions for *s* and *p* electrons. (b) Angular distributions for *s* and *p* electrons.

in their orbital rotations. This follows from the Bohr Correspondence Principle which states that when the wavelength of an electron wave is small compared with the dimensions of the volume in which the electron moves, the particle ceases to show wave character.

In contrast, the *radial* part of one-electron bound-state wave functions retain quantum character, even for large n, because the outermost lobes are always the largest and most spread out. The inner parts of highly excited states, the parts at low r values, do have rapid oscillations. Consequently, in states of high n, electrons behave almost classically in their radial coordinates when they come near the nucleus, but as waves when they move to large r and their momentum becomes small.

(v)   Pictures of the functions $Y_{lm}(\theta, \phi)$ may not easily be visualized because of the imaginary exponents contained in $\Phi_m(\phi)$ for all $m \neq 0$. However, because of the degeneracy of the Y_{lm} functions associated with a given l, it is permissible to take linear combinations of the Y_{lm}. Thus,

$$-Y_{11} + Y_{1-1} = (3/2\pi)^{1/2} \sin\theta \cos\phi \qquad (1.19a)$$

$$i(Y_{11} + Y_{1-1}) = (3/2\pi)^{1/2} \sin\theta \sin\phi \qquad (1.19b)$$

$$Y_{10} = (3/4\pi)^{1/2} \cos\theta \qquad (1.19c)$$

All of these functions are real. Furthermore, it is clear from Fig. 1.1 that $\sin\theta \cos\phi$ transforms as the atomic orbital p_x, $\sin\theta \sin\phi$ transforms as p_y, and $\cos\theta$ transforms as p_z. Hence, Eqs. (1.19a)–(1.19c) may be labeled p_x, p_y, and p_z, respectively. It is in this manner that the real forms of the spherical harmonics and their labels are derived. Some such functions, after renormalization, are collected in Appendix A. They are diagrammed in Fig. 1.3.

3. TWO-ELECTRON ATOM

The simplest two-electron atom is helium. Relevant coordinates for such an atom are shown in Fig. 1.4. The Schrödinger equation may be

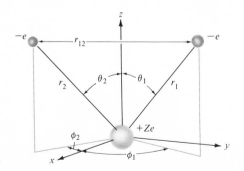

FIG. 1.4.   Polar coordinates for a two-electron atom.

written[1]

$$(\hat{H} - E)\Phi = -(\nabla_1{}^2 + \nabla_2{}^2)\Phi - (2\mu/\hbar^2)(E + Ze^2/r_1 + Ze^2/r_2 - e^2/r_{12})\Phi$$
$$= 0 \qquad (1.20)$$

where $\hat{H}$ is the Hamiltonian operator for the system, where $-(\hbar^2/2\mu)\nabla_1{}^2$ is the kinetic-energy operator for electron 1, and where $\Phi = \Phi(1, 2)$ is a two-electron space wave function.[1,2] Let us suppose that the e^2/r_{12} term can be set equal to zero or, equivalently, that there is no interelectronic repulsion. With this condition, the variables in Eq. (1.20) are separable into two sets which we may write in operator form

$$\hat{H} = \hat{H}_1 + \hat{H}_2 \qquad (1.21)$$

The Hamiltonian is now a sum of *two* one-electron Hamiltonians. Given this crude assumption, it follows that the eigenfunctions of $\hat{H}$ are simple product wave functions of the form

$$\Phi(1, 2) = \chi_r(1)\chi_s(2) \qquad (1.22)$$

where, in the specific case of helium, χ_r (or χ_s) may be any one of the hydrogenlike wave functions for He+. The normalized space part of the ground state of helium is then given by

$$\Phi_G(1, 2) = \chi_{1s}(1)\chi_{1s}(2) = (8/a_0{}^3\pi)e^{-2r_1/a_0}e^{-2r_2/a_0} \qquad (1.23)$$

Thus, in this instance, the utility of the atomic-orbital concept is validated. However, the supposition that the interelectronic repulsion $\langle \Phi(1, 2) \,|\, e^2/r_{12} \,|\, \Phi(1, 2) \rangle \simeq 0$, amounts to complete neglect of the effects of one electron on the other—and this is clearly an oversimplification.

(A) Screening

We might attempt to rescue the situation by introducing the concept of screening: The supposition that the effect of any one electron is merely to shield the other from the full effects of the nuclear field. Thus, we introduce an effective nuclear charge $Z = Z - \sigma$, where σ is the *screening constant*. A number of prescriptions are available for the estimation of *best* σ; if we use that provided by Slater,[3] we find for helium

$$\Phi_G = (1/\pi)(1.7/a_0)^3 e^{-1.7r_1/a_0}e^{-1.7r_2/a_0} \qquad (1.24)$$

---

[1] Note that Φ is a many-electron space wave function, $\Phi(r_1, r_2, \theta_1, \theta_2, \phi_1, \phi_2)$. In Secs. 1 and 2, Φ was used to denote the ϕ-dependent part of the one-electron hydrogenic space wave function.

[2] The validity of Eq. (1.20) hinges on the assumption that nuclear and electronic motions are rigorously separable. This is not true in a two-electron system. One usually makes the assumption that the nucleus, in a relative sense, is infinitely heavy—whereupon, Eq. (1.20) follows.

The wave function of Eq. (1.24) is probably better than that of Eq. (1.23). However, it is certainly not exact. If the many-electron Schrödinger equation could be solved directly, the result would not be a set of orbitals to be combined as in Eqs. (1.23) and (1.24); instead, for the ground state it would be a more complex many-electron wave function which would be simultaneously descriptive of all electrons and which would not be a simple product of two one-electron wave functions. In order to illustrate this, and the deficiencies of the orbital concept, we now proceed to a discussion of one of the most accurate approximation methods presently available.

(B)   Hylleraas Wave Functions for Helium

Hylleraas[4] proposed a number of wave functions for the helium ground state, one of which (unnormalized) is[3]

$$\Phi = e^{-1.82r_1}e^{-1.82r_2}f(r_1, r_2, r_{12})$$

$$f(r_1, r_2, r_{12}) = 1 - 0.1008(r_1 + r_2) + 0.0331(r_1 + r_2)^2$$

$$+ 0.1285(r_1 - r_2)^2 + 0.354r_{12} - 0.0318r_{12}{}^2 \qquad (1.25)$$

where all coordinate units are quoted in bohrs. Other more complex functions of the same type[4] are available; they provide excellent agreement with experiment. All of these functions contain an explicit dependence on r_{12}, and this dependence is, in fact, the source of their excellence. Equation (1.25) does not admit of the orbital concept. However, further investigation of Eq. (1.25) might indicate the conditions under which a limited orbital concept is valid.

The exponential factors in Eq. (1.25) may be considered to pertain to $1s$ AO's for a helium nucleus of charge $Z = 1.82$, $\sigma = 0.18$. However, the reason for success of Eq. (1.25) is the power series $f(r_1, r_2, r_{12})$ which introduces *electron correlation*.[5] The effect of electron correlation is to make the wave function explicitly dependent on r_{12} so as to decrease $|\Phi|$ when the electrons are close together and to increase $|\Phi|$ when the electrons are relatively far from each other. To illustrate this, define $\Phi_{coincident}$ for $r_1 =$

---

[3] The wave function of Eq. (1.25) may be supposed to result from the following considerations. Assume a wave function of the form

$$\Phi(1, 2) = e^{-\zeta(r_1+r_2)}[1 + a(r_1 + r_2) + b(r_1 + r_2)^2 + c(r_1 - r_2)^2 + dr_{12} + er_{12}{}^2]$$

Then, the exponent ζ and the coefficients a, b, c, d, and e are to be obtained from the variation theorem as those which minimize the energy $E = \langle \Phi | \hat{H} | \Phi \rangle$. It is possible to assume many different types of wave function; all of those used by Hylleraas contained an explicit dependence on $r_{12} = |\mathbf{r}_1 - \mathbf{r}_2|$.

[4] See footnote 3.

[5] See Sec. 10.

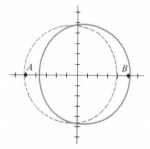

FIG. 1.5. Contours of $\Phi(1, 2)$ for a Hylleraas-like helium wave function. Solid line: electron 1 fixed at point A where $y = -0.5$ bohrs. Dashed line: electron 1 fixed at point B where $y = 0.5$ bohrs. The value of r_1 is 0.5 bohrs in both cases. The contour lines are the cross sections of the spheroidal surface on which $\Phi(1, 2)$ equals 0.2 of its maximum value. These contours were computed by Cohen and Bustard (Ref. [14]) from the 39-term Hylleraas-like wave function of Kinoshita (Ref. [25]).

$r_2 = r$ and $r_{12} = 0$, and Φ_{opposite} for $r_1 = r_2 = r$ and $r_{12} = 2r$. Then, the ratio

$$\frac{\Phi_{\text{opposite}}}{\Phi_{\text{coincident}}} = \frac{1 + 0.506r + 0.0052r^2}{1 - 0.2016r + 0.1324r^2} \qquad (1.26)$$

indicates that, at least at small r, the amplitude is larger when the electrons are opposite each other than when they are coincident.[6]

The significance of the above is simply that the probability distribution of one electron can *not* be defined in such a way as to be independent of the other electron. Or, in other words, the concept of an atomic orbital is inadequate except in a one-electron atom. To further illustrate this, Fig. 1.5 diagrams two contours of a wave function such as that of Eq. (1.25) when one of the electrons is held fixed at two different locations. A complete representation of the two-electron atom would show a set of contours for $\Phi(1, 2)$ for every possible position of one of the electrons and would also indicate the relative probabilities of the different contours. Now, we may simplify this by drawing a single contour which is an average of all the possible one-electron contours, weighted according to the relative probabilities. Such a diagram omits the detailed structure due to electron correlation but *it is the best possible representation* of the atom *in terms of independent one-electron wave functions* (i.e., orbitals).

---

[6] This conclusion breaks down at large r. However, this is not disastrous because the amplitude of Φ is small when r is large. Nonetheless, this result is unsatisfactory and denotes deficiencies in Φ of Eq. (1.25). A better Φ would retain consistency with the conclusion of Eq. (1.26) out to larger r.

Clearly, a one-electron function can, at least in principle, be a completely accurate average. What, then, is wrong with it? The error is in its implications for the distribution of two or more electrons, for if we assign the two helium electrons, for example, to any specific orbital, we are postulating that the distribution of each electron is independent of the other, that the ratio in Eq. (1.26) is identically one, or that the two electrons are as likely to be found close together as far apart. An orbital, then, is useful only if we obtain it in some way which averages the effects of other electrons, and if we remember that it shows the time-averaged distribution of an electron and does not at all indicate electron correlation.[5]

Although Hylleraas functions rank among the most accurate wave functions yet devised, their use is limited; they are difficult to obtain even for helium and they have not yet been prepared for atoms of higher complexity.

4. MANY-ELECTRON ATOM

The Hamiltonian for a many-electron atom may be written in the form

$$\hat{H} = \sum_{i=1}^{N} \hat{H}^c(i) + \sum_{\substack{i,j=1 \\ i<j}}^{N} e^2/r_{ij} \tag{1.27}$$

where $\hat{H}^c(i)$, the so-called core Hamiltonian for electron i, consists of the kinetic-energy and nuclear-attraction terms for electron i. The core Hamiltonian $\hat{H}^c(i)$ is a one-electron operator whereas $1/r_{ij}$, being dependent on the interelectron distance, is a two-electron operator. The energy of the N-electron system is given by $E = \int \Psi^* \hat{H} \Psi d\tau_1 \cdots d\tau_N$ and our problem consists of the determination[7] of both Ψ and E.

It is concerning Ψ that we must make our first arbitrary decision. We choose to employ the orbital concept for two reasons: Firstly, it is convenient to be able to enumerate what the individual electrons are doing (on an average, of course); and, secondly, it makes computations on atoms larger than beryllium more tractable. Thus, Ψ is chosen as a product of one-electron functions. However, Ψ must also take electron spin into account

[7] Note that Ψ is a many-electron spin-orbital wave function. For a two-electron system, we may write

$$\Psi(1, 2) = \Phi(x_1, y_1, \cdots, z_2)\Theta(\omega_1, \omega_2)$$

where Θ is now a two-electron spin wave function and ω_r is the spin coordinate of electron r. Note that $\Psi = \Phi\Theta$ implies no interaction of the spin and orbital motions—as is required by the nonrelativistic Hamiltonian of Eq. (1.27). For a many electron system, even in the absence of spin-orbit coupling, we cannot generally write $\Psi = \Phi\Theta$; separability of Φ and Θ is frustrated, in this instance, by simple antisymmetry requirements and the best we can do is write

$$\Psi(1, 2) = \Psi(x_1, y_1, \cdots, z_n, \omega_1, \cdots, \omega_n)$$

and it must be written in such a way that it is antisymmetric with respect to interchange of electron coordinates (spin coordinates plus space coordinates).[8] The easiest way to achieve this is to write the N-electron wave function for a particular electronic state as a linear combination of Slater determinants, a single such determinant being of the form

$$| \chi_1\alpha\chi_1\beta\chi_2\alpha \cdots \chi_r\alpha\chi_s\alpha \cdots \chi_w\alpha |$$

$$= \left(\frac{1}{N!}\right)^{1/2} \begin{vmatrix} \chi_1\alpha(1) & \chi_1\beta(1) & \cdots & \chi_r\alpha(1) & \chi_s\alpha(1) & \cdots & \chi_w\alpha(1) \\ \chi_1\alpha(2) & \chi_1\beta(2) & \cdots & \chi_r\alpha(2) & \chi_s\alpha(2) & \cdots & \chi_w\alpha(2) \\ \vdots & \vdots & & \vdots & \vdots & & \vdots \\ \chi_1\alpha(N) & \chi_1\beta(N) & \cdots & \chi_r\alpha(N) & \chi_s\alpha(N) & \cdots & \chi_w\alpha(N) \end{vmatrix}$$

$$(1.28)$$

where α and β denote spin functions with spin quantum numbers $m_s = \frac{1}{2}$ and $-\frac{1}{2}$, respectively, and where the atomic spin orbital is assumed to be a simple product of space and spin functions. Thus,

$$\chi_1\alpha(2) = \chi_1(x_2, y_2, z_2)\alpha(\omega_2)$$
$$\chi_1\beta(2) = \chi_1(x_2, y_2, z_2)\beta(\omega_2) \qquad (1.29)$$

where ω is the spin coordinate. Let us further limit ourselves to consideration of a totally symmetric ground state composed of a *closed shell* of electrons, so that we may represent the wave function by a single determinant

$$\Psi_0 = | \chi_1\alpha(1)\chi_1\beta(2) \cdots \chi_r\alpha(2r-1)\chi_r\beta(2r)$$
$$\times \chi_s\alpha(2r+1)\chi_s\beta(2s) \cdots \chi_w\alpha(N-1)\chi_w\beta(N) | \quad (1.30)$$

where N, the number of electrons, equals $2w$.

For the ground-state energy we obtain, by combining Eqs. (1.27) and (1.30)

$$E_0 = 2\sum_{r=1}^{w} I_r + \sum_{r,s=1}^{w} (2J_{rs} - K_{rs}) \qquad (1.31)$$

where

$$I_r = \langle \chi_r | \hat{H}^c | \chi_r \rangle = \int \chi_r^*(1)\hat{H}^c(1)\chi_r(1)\,d\tau_1 \qquad (1.32)$$

$$J_{rs} = \langle \chi_r\chi_s | \chi_r\chi_s \rangle = \iint \chi_r^*(1)\chi_s^*(2)\frac{e^2}{r_{12}}\chi_r(1)\chi_s(2)\,d\tau_1 d\tau_2 \quad (1.33)$$

$$K_{rs} = \langle \chi_r\chi_s | \chi_s\chi_r \rangle = \iint \chi_r^*(1)\chi_s^*(2)\frac{e^2}{r_{12}}\chi_s(1)\chi_r(2)\,d\tau_1 d\tau_2 \quad (1.34)$$

---

[8] A similar requirement should have been imposed on the eigenfunctions of Sec. 3(B)—but was not because of the preliminary nature of our discussion in Sec. 3.

The integral J_{rs} is known as the *Coulomb integral*; it represents the electrostatic interaction between a charge density $\chi_r^*(1)\chi_r(1)$ and another charge density $\chi_s^*(2)\chi_s(2)$. The integral K_{rs} is known as the *exchange integral*; it represents the electrostatic interaction of two overlap charge densities $\chi_r^*(1)\chi_s(1)$ and $\chi_s^*(2)\chi_r(2)$; this integral possesses no classical counterpart since it arises from the antisymmetry requirement imposed on all electronic wave functions Ψ. It is clear from the definitions that

$$J_{rr} = K_{rr} \tag{1.35}$$

As yet, however, we have not devised any procedure for determining the AO's $\chi_1, \chi_2, \ldots, \chi_r, \ldots \chi_w$ themselves. According to the variation principle they must be determined in such a way that E_0 is minimized. Furthermore, this minimization is restrictive in the sense that certain accessory conditions must be fulfilled. For example, we might choose the $(N/8)(N-2)$ conditions of orthogonality

$$\langle \chi_r \mid \chi_s \rangle = \int \chi_r^*(1)\chi_s(1)d\tau_1 = 0, \quad \text{for} \quad r \neq s \tag{1.36}$$

and the $N/2$ conditions of normality

$$\langle \chi_r \mid \chi_r \rangle = \int \chi_r^*(1)\chi_r(1)d\tau_1 = 1 \tag{1.37}$$

The standard mathematical technique for dealing with a restricted extreme-value problem is the Lagrange method of undetermined multipliers. Straightforward application of this method leads to the Hartree–Fock equations.[6],[7]

5.   HARTREE–FOCK METHOD

Let us now define the *pseudo-one-electron* Coulomb and exchange operators

$$\hat{J}_r\chi_s(1) = \left(\int \chi_r^*(2) \frac{e^2}{r_{12}} \chi_r(2) d\tau_2 \right)\chi_s(1) \tag{1.38}$$

$$\hat{K}_r\chi_s(1) = \left(\int \chi_r^*(2) \frac{e^2}{r_{12}} \chi_s(2) d\tau_2 \right)\chi_r(1) \tag{1.39}$$

It follows from these definitions that the Coulomb and exchange integrals may be expressed as pseudo-one-electron *integrals*

$$J_{rs} = \int \chi_r^* \hat{J}_s \chi_r d\tau = \int \chi_s^* \hat{J}_r \chi_s d\tau \qquad (1.40)$$

$$K_{rs} = \int \chi_r^* \hat{K}_s \chi_r d\tau = \int \chi_s^* \hat{K}_r \chi_s d\tau \qquad (1.41)$$

and that

$$\hat{J}_r \chi_r(1) = \hat{K}_r \chi_r(1) \qquad (1.42)$$

Consequently, we may rewrite Eq. (1.31) in the forms

$$E_0 = \sum_r \left[2I_r + \sum_s \left(2 \int \chi_r^* \hat{J}_s \chi_r d\tau - \int \chi_r^* \hat{K}_s \chi_r d\tau \right) \right] \qquad (1.43)$$

$$E_0 = \sum_r \int \chi_r^* \{ 2\hat{I}_r + \sum_s (2\hat{J}_s - \hat{K}_s) \} \chi_r d\tau \qquad (1.44)$$

where $\hat{I}_r$ is merely $\hat{H}^c$ confined to operation on the orbital χ_r. The minimization process satisfying the restrictions of Eqs. (1.36) and (1.37) leads to the Hartree–Fock equations[8] for the orbitals χ_r

$$\{ \hat{H}^{\text{core}}(1) + \sum_{s=1}^{w} (2\hat{J}_s - \hat{K}_s) \} \chi_r(1) = \epsilon_r \chi_r(1) \qquad (1.45)$$

where the leftmost quantity in curly brackets is defined as the Hartree–Fock Hamiltonian $\hat{H}^{\text{HF}}$ so that

$$\hat{H}^{\text{HF}} \chi_r(1) = \epsilon_r \chi_r(1) \qquad (1.46)$$

The Hartree–Fock operator is dependent on its own eigenfunctions because the index s of Eq. (1.45) runs over all *occupied* orbitals. Thus, we cannot know the operator without knowing its eigenfunctions. The approach to solution of Eq. (1.45) must be iterative. A set of functions $\chi_1, \chi_2, \ldots,$ $\chi_r, \ldots \chi_w$ must be guessed and $\hat{H}^{\text{HF}}$ evaluated. Knowing $\hat{H}^{\text{HF}}$ we may now evaluate a new χ_1 and an ϵ_1 from Eq. (1.45); this new χ_1 is termed a *first improved* χ_1 and replaces the *guessed* χ_1 in the original set. We now use the set χ_1 (*first improved*), $\chi_2, \chi_3, \ldots, \chi_w$ and evaluate a first-improved χ_2; cycling is continued until a set of *first-improved orbitals* $\chi_1, \ldots, \chi_w$ is available. This set of first-improved orbitals is used to generate a set of *second-improved orbitals,* and so on until self-consistency is achieved. That set of orbitals which, when used to evaluate $\hat{H}^{\text{HF}}$, generates itself via Eqs. (1.45) and (1.46) is termed the *self-consistent Hartree–Fock set.* Clearly, once the self-consistent Hartree–Fock operator is known, we may also generate wave functions for all the unoccupied orbitals; these latter ones are known as *virtual* or *excited* orbitals.

The ϵ_r originate[9] as undetermined multipliers in the restricted extremum problem based on E_0. They may be termed *orbital energies* because, to the extent that $\hat{H}^{\mathrm{HF}}$ is one electron in nature, they are one-electron energies.

Two following characteristics of the orbital energies ϵ_r must be emphasized:

(i)   The sum of occupied orbital energies does not equal E_0(minimum); in other words,

$$E_0(\text{minimum}) \neq 2 \sum_r \epsilon_r \qquad (1.47)$$

Certain electron interaction energies must be included in order to achieve equality to E_0(minimum). Comparison of Eqs. (1.44) and (1.45) indicates that

$$E_0(\text{minimum}) = \sum_{r=1}^{w} (I_r + \epsilon_r)$$

$$= \sum_{r=1}^{w} \left\{ 2\epsilon_r - \sum_{s=1}^{w} \int \chi_r^* (2\hat{J}_s - \hat{K}_s) \chi_r d\tau \right\}$$

$$= \sum_{r}^{w} \sum_{s}^{w} (2\epsilon_r - 2J_{rs} + K_{rs}) \qquad (1.48)$$

(ii)   The energy difference of two states does not equal the appropriate difference of occupied orbital energies. The energies of an orbital ϵ_r which is occupied in the ground state and of a virtual orbital ϵ_k are given by

$$\epsilon_r = I_r + \sum_s (2J_{rs} - K_{rs}); \qquad r = 1, 2, \ldots \text{ or } w \qquad (1.49)$$

$$\epsilon_k = I_k + \sum_s (2J_{ks} - K_{ks}); \qquad k > w \qquad (1.50)$$

A cancellation of terms $J_{rr} - K_{rr}$ always occurs in Eq. (1.49); no such cancellation is possible in Eq. (1.50). This difference between filled and virtual orbitals leads to the result that the energies of the singlet and triplet excited states deriving from the single orbital excitation $k \leftarrow r$ does not equal $\epsilon_k - \epsilon_r$ but is, instead, given[2] by

$$E(^1\Psi_{k\leftarrow r}) - E_0(\text{min}) = \epsilon_k - \epsilon_r - J_{rk} - 2K_{rk} \qquad (1.51)$$

$$E(^3\Psi_{k\leftarrow r}) - E_0(\text{min}) = \epsilon_k - \epsilon_r - J_{rk} \qquad (1.52)$$

The terms J_{rk} and K_{rk} are usually of the order of 1 eV; they may not be neglected in the evaluation of excitation energies.

(A) Hartree Method

The Hartree method[6] was prelude to the Hartree–Fock method.[7] It was based on an assumed many-electron wave function of the form $\chi_1\alpha\chi_1\beta\ldots\chi_r\alpha\chi_s\alpha\ldots\chi_w\alpha$; in other words, it neglects antisymmetry requirements. As a result, no exchange terms appeared in the Hartree operator. Now, the effect of the exchange term is to reduce the Coulombic repulsions between electrons with the same spin. Physically, its effects may be simulated by surrounding each electron with a small spherical volume within which other electrons of the same spin may not intrude; this excluded volume is often referred to as the *Fermi hole*. This mutual avoidance of electrons with the same spin permits the space orbitals to be closer to the atomic nucleus and therefore more strongly bound in these instances.[10]

The Hartree–Fock method takes the required antisymmetry of the total wave function into account. In the case of partly filled shells, it enables us to obtain acceptable wave functions for each of the different states arising from the given electron configuration. In specific, for the p^2 configuration which yields ^3P, ^1S and ^1D states, the Hartree–Fock method provides a reasonable means of finding the individual state wave functions.

6. APPROXIMATE HARTREE–FOCK SELF-CONSISTENT FIELD WAVE FUNCTIONS

The Hartree–Fock self-consistent field (HF-SCF) wave functions are solutions of a one-electron wave equation [Eq. (1.46)] obtained by replacing the electron interaction by an effective potential of spherical symmetry, namely, $\sum_s(2\hat{J}_s - \hat{K}_s)$. Consequently, the angular wave functions which result are identical to those of the one-electron atom; they are the Y_{lm} of Appendix A. The radial wave functions have been computed numerically to high accuracy for some atoms, and approximately for the entire Periodic Table. However, while it is computationally simple to provide a numerical solution to the Hartree–Fock equation, the result is often inconvenient for further computations.

(A) Slater-Type Function

For a neutral atom, or for an ion of charge q, the total potential seen by an electron must approach $-e^2/r$, or $-(q+1)e^2/r$, as that electron is removed to large distances from the nucleus. Therefore, at least at large values of r, the radial solutions should be hydrogenlike and contain a term $r^{n-1}e^{-\zeta r}$; when normalized and designated as $R(r)$, this term becomes

$$R(r) = (2\zeta)^{n+1/2}(1/2n!)^{1/2}r^{n-1}e^{-\zeta r} \tag{1.53}$$

A careful analysis of some of the initial HF-SCF calculations led Slater to suggest[3] the general use of wave functions of the above form as a first approximation to the results of the HF-SCF computations. Functions of the type of Eq. (1.53) are known as Slater-type functions (STF). The parameter n is the effective principal quantum number, and ζ is given by $\zeta = (Z - \sigma)/n$. The earliest calculations, and most recent ones, have assumed that n equals the principal quantum number of the orbital in question, and ζ has been varied in order to minimize the total energy. However, Slater considered both n and ζ as variables and he provided a set of rules by means of which they might be evaluated. These *Slater rules* are given in Appendix B.

In the last 35 years, the increase in the available computational facilities has made more replete calculations feasible. Thus, Clementi and Raimondi[11] have determined, by true optimization procedures, the value of ζ for a single STF which leads to the minimum electronic energy; on this basis, they have proposed an alternative set of rules for determination of ζ. Another approach assumes the availability of HF-SCF functions or numerical tables thereof, and seeks that single STF which best reproduces some important quantity calculated directly from the HF-SCF AO—for example, the average value of r: $\langle r \rangle$, or the average value of $1/r$: $\langle 1/r \rangle$. In this spirit, Burns has proposed[12] another set of rules for ζ. A comparison of ζ values for single STF representations of HF-SCF wave functions is given in Table 1.1.

Finally, a fourth type of single STF representation of HF-SCF orbitals has been proposed.[13] In contrast to the three other types (which, respectively, seek to reproduce the HF-SCF energy, to minimize the energy of a

TABLE 1.1.   *Exponents for single STF.*[a]

Atom	Orbital	Slater rules	Clementi–Raimondi rules	Burns rules	OMAO
C	$2s$	1.625	1.608	1.550	1.60
C	$2p$	1.625	1.568	1.325	1.43
S	$3s$	2.883	2.122	1.97	1.53[b]
S	$3p$	2.883	1.827	1.52	1.21[b]
Cu	$3d$	2.62	4.40	3.70	$\cdots$

[a] This table is from Cusachs, Carroll, Trus, and McGlynn (Ref. [13]).
[b] This number is the exponent of an STF with $n=2$; in other words, an STF with $n = 2$ is used to represent an AO of principal quantum number 3.

single STF, or to reproduce various atomic moments), this fourth type is concerned with interatomic affairs and attempts to reproduce the HF-SCF overlap over the range of chemically significant interatomic distances; this type of STF is said to be an overlap-matched AO and is termed OMAO. The ζ values of a few OMAO's are also given in Table 1.1.

(B)   Validity of Single Slater-Type Orbital Representation

A single STF representation is *qualitatively* improper. The STF is node-less; thus, a single STF representation of a $4p$ AO must necessarily move all radial nodes of this orbital to the origin. If the single STF representation is so gross in its qualitative characteristics, it is difficult to see how it might be *quantitatively* satisfactory. Thus, it is well to know what STF representations of AO's can not do; hence, we enumerate the following:

(i)   Quantities that depend on the gradient of the wave function, or on its nodal properties are not realistically evaluable in a single STF representation (i.e., spin-spin or spin-orbit coupling interactions, nuclear quadrupole coupling, etc.).

(ii)   Inspection of Table 1.1 indicates significant variation of the exponents from method to method. Now, kinetic energy in a single STF basis is proportional to ξ^2, whereas potential terms of $1/r$ nature are proportional to ξ. In the case of the copper $3d$ AO of Table 1.1, it is clear that kinetic- and potential-energy variations are very gross indeed. Thus the query: What method shall we follow?

(iii)   Since STF's do not contain radial nodes, it follows that functions with identical values of both the l and m quantum numbers, but different values of n, are not orthogonal. A Schmidt orthogonalization process[9] can remove the difficulty—but it does so by mixing different AO functions, and by changing AO energies. If there be any content in the HF-SCF procedure, this is clearly not an allowable tactic.

(iv)   It simply is not possible to represent most HF-SCF orbitals by a single STF. A case in point is the $3d$ AO of the first transition-metal series where a minimum of two STF's is required for even a gross representation.

(v)   Significant difficulties exist with virtual orbitals of high principal quantum number (for example, the $4s$ AO of manganese or of sulfur). These orbitals are so diffuse in the case of manganese that they extend into and over the ligands, and into the surrounding medium in which the manganese complex is embedded. To a large extent, it is the surroundings which determine these AO's.

Thus, one might ask, why bother with single STF representations? The answer is both pragmatic and sensible. It is pragmatic because the

---

[9] See Sec. 9(A).

use of numeric HF-SCF AO's or multi-STF AO's severely limits the size of molecules which can be studied—and we do want to get on with the business of chemistry (which does include some large molecules). It is sensible because we do not know what constitutes a HF-SCF AO for an atom *in situ* in a molecule. All we know are the HF-SCF AO's for isolated atoms; to insist that these should be the same for the atom in a molecule is, at best, ridiculous and, at worst, hypocrisy. However, it is equally hypocritical to proceed with *the business of chemistry* by doing calculations with a set of wave functions which are poor representations of the proper AO sets. Thus, in view of the fact that the single STF represents a given AO best on the outer fringes of that AO (i.e., at large r), it is best to limit the use of single STF representations to calculations of overlap. However, even overlap calculations, as are discussed in Chapter 2, must be processed warily.

(C)   Hydrogen-Like Functions

It is assumed that the form of $R(r)$ is identical to that for the corresponding orbital in hydrogen. Thus, this type of orbital differs from an STF only in that it contains a polynomial in r instead of a monomial in r. Otherwise, the SCF procedure is straightforward: one considers Z as a variable which is subject to SCF optimization. The best orbitals for carbon were obtained[14] with $Z = 4$ for the $2s$ AO and 3.1 for the $2p$ AO.

(D)   Morse Functions

All H-like orbitals are a linear combination of monomials in r multiplied by an exponential in r. In *this sense*, all H-like orbitals are simply linear combinations of STF's. For example, we may write the $2s$ AO of a hydrogen-like atom (unnormalized) as

$$\chi_{2s} = e^{-Zr/2} - \tfrac{1}{2}Zre^{-Zr/2} \qquad (1.54)$$

Morse rewrote[15] the above AO as

$$\chi_{2s} = A\,(Z_{2s}{}^3/8\pi)^{1/2}[Be^{-Z_{2s}'r/2} - (Z_{2s}r/2)\,e^{-Z_{2s}r/2}] \qquad (1.55)$$

where the constants A and B maintain normalization. Morse then subjected both Z_{2s} and Z_{2s}' to HF-SCF optimization procedures. Procedures such as this have produced results for the first-row atoms which are of equivalent accuracy, at least from the point of view of energy, to numerical HF-SCF procedures. It is clear that a Morse AO representation is identical to the H-like AO representation for all AO's with $l = l_{\max}$. For carbon, Tubis found[16] the following effective nuclear charges: $Z_{2s} = 3.31$; $Z_{2s}' = 10.24$.

The interpretation of Eq. (1.55) is suggestive. It may be considered as the sum of two curves where the inner behaves according to the Z_{2s}'

exponent and the outer according to Z_{2s}. The inner curve is determinative of electron behavior at low r, the outer at large r. Such a dissection is shown in Fig. 1.6.

(E) Multi-STF SCF Wave Functions

The results of Morse suggest that a more exact AO representation might be obtained by expanding all AO's into the form

$$\chi_{jlm}(r, \theta, \phi) = (\sum_i a_{ij}\text{STF}_{ij})\, Y_{lm}(\theta, \phi)$$

$$= (\sum_i a_{ij} r^{n_{ij}-1} e^{-\zeta_{ij}r})\, Y_{lm}(\theta, \phi) \qquad (1.56)$$

where the principal quantum number of the orbital of interest is $n \equiv j$, and where the expansion occurs into a linear combination of STF's with variable quantities n_{ij} and ζ_{ij}. The quantity n_{ij} is treated as a fictitious quantum number but is retained as an integer. In order to obtain the HF-SCF function based on Eqs. (1.46) and (1.56), one makes use of the variational method for the coefficients a_{ij} and of *good sense* in the choice of the STF basis set (i.e., namely, in the choice of n_{ij} and ζ_{ij}). Unfortunately, good sense does not go too far and a process of trial and error in the choice of the STF basis set is required. This process, namely, orbital exponent optimization, is carried out by computer in a way which compromises between efficiency, accuracy, and complete generality of the programmatic effort.

Examples of the type of multi-STF wave functions (i.e., multi-ζ WF's) which are available are given in Table 1.2. This table presents such wave functions for the three different electronic states of the $1s^2 2s^2 2p^2$ configura-

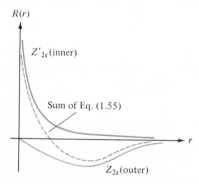

FIG. 1.6. A diagrammatic dissection (Ref. [14]) of Eq. (1.55). The 2s AO may be considered to consist of two parts, an outer and an inner part for which the **Z**'s (or ζ's) are different.

TABLE 1.2.   *Multi-STF SCF wave functions for 1s, 2s, and 2p AO's of carbon.*[a]

(i)   Configuration: $1s^2 2s^2 2p^2$; State: ^3P; SCF Energy: -37.688611 a.u.

C^0 SPECIES

s Basis Set STF type	Exponent	Coefficients 1s AO	2s AO	p Basis Set STF type	Exponent	Coefficients 2p AO
1s	5.41250	0.92695	−0.20786	2p	0.95540	0.24756
1s	9.28630	0.07665	−0.01175	2p	1.42090	0.57774
2s	1.03110	0.00073	0.06494	2p	2.58730	0.23563
2s	1.50200	−0.00167	0.74109	2p	6.34380	0.01090
2s	2.58975	0.00539	0.34626			
2s	4.25950	0.00210	−0.13208			

(ii)   Configuration: $1s^2 2s^2 2p^2$; State: ^1D; SCF Energy: -37.631317 a.u.

C^0 SPECIES

s Basis Set STF type	Exponent	Coefficients 1s AO	2s AO	p Basis set STF type	Exponent	Coefficients 2p AO
1s	5.43000	0.91965	−0.20727	2p	0.93720	0.31917
1s	9.15000	0.07951	−0.01235	2p	1.41470	0.50063
2s	1.22550	−0.00090	0.23681	2p	2.55450	0.25045
2s	1.61420	0.00220	0.61238	2p	6.30210	0.01097
2s	2.69900	−0.00043	0.31624			
2s	4.21310	0.01067	−0.14709			

(iii)   Configuration: $1s^2 2s^2 2p^2$; State: ^1S; SCF Energy: -37.549535 a.u.

C^0 SPECIES

s Basis Set STF type	Exponent	Coefficients 1s AO	2s AO	p Basis Set STF type	Exponent	Coefficients 2p AO
1s	5.38420	0.91962	−0.20999	2p	1.10600	0.68350
1s	9.06000	0.08620	−0.01333	2p	0.50740	0.05446
2s	1.21000	0.00068	0.19885	2p	2.35900	0.35041
2s	1.59290	−0.00137	0.61835	2p	6.20000	0.01237
2s	2.59640	0.00466	0.33657			
2s	4.25000	−0.00019	−0.13361			

TABLE 1.2.   (*Continued*)

(iv)   Configuration: $1s^22s^22p^1$; State: ^2P; SCF Energy: -37.292213 a.u.

C^{+1} SPECIES

s Basis Set STF type	Exponent	Coefficients 1*s* AO	2*s* AO	*p* Basis Set STF type	Exponent	Coefficients 2*p* AO
1*s*	5.55187	0.92143	-0.21033	2*p*	1.85882	0.44498
1*s*	9.66378	0.05979	-0.01400	2*p*	3.04800	0.12186
2*s*	2.13087	0.00173	0.50835	2*p*	1.36571	0.46744
2*s*	4.93221	0.03035	-0.09950	2*p*	6.80717	0.00883
2*s*	1.55797	-0.00049	0.58179			

(v)   Configuration: $1s^22s^2$; State: ^1S

C^{+2} SPECIES

s Basis Set STF type	Exponent	Coefficients 1*s* AO	2*s* AO
1*s*	5.42393	0.93181	-0.23221
1*s*	9.32901	0.07348	-0.01730
2*s*	2.05374	0.00162	0.86317
2*s*	4.68904	0.00188	-0.12048
2*s*	1.67424	-0.00093	0.24404

[a] These SCF AO's are taken from a compilation by E. Clementi [*Tables of Atomic Functions* (IBM Corp., San Jose, Calif., 1965)]

tion of neutral carbon (namely, ^3P, ^1D, and ^1S) as well as those for the ground states of the ions C^{+1} and C^{+2}. In order to facilitate reading of these tables, we note that the $2p$ AO radial wave function in the ^3P state of C^0 is given by

$$R(2p) = 0.24756\chi(2p, 0.9554) + 0.57774\chi(2p, 1.4209)$$

$$= +0.23563\chi(2p, 2.5873) + 0.01090\chi(2p, 6.3438) \quad (1.57)$$

where the number in parenthesis specifies ζ for the STF (which is also specified in the parenthesis). A compilation of sources of HF-SCF AO's is given in Appendix B.

Granted the availability of AO's of multi-STF nature for much of the Periodic Table, it now remains to query their value for molecular computations. We itemize the following:

(i)   Inspection of Table 1.2 indicates differences between the wave functions of the $2p$ AO in the ^3P, ^1D, and ^1S states of neutral carbon, with the difference $2p(^3P) - 2p(^1S)$ being considerable. Suppose we wish to perform a computation on CF_4—which of the three available wave functions: $2p(^3P)$, $2p(^1D)$, $2p(^1S)$, should be used in this calculation? Many authors would choose $2p(^3P)$ because this happens to be the ground state of neutral carbon. However, the rejection of $2p(^1D)$ and $2p(^1S)$ on the basis of energetics is invalid because the energy differences of the three states in question span only $\sim$0.15 a.u. of energy.

(ii)   One might evade item (i) if, instead of AO's deriving from different states, one used AO's from different configurations. For example, the $1s^1 2s^1$ configuration of He yields two states: ^1S and ^3S. If, instead of minimizing the energy of the individual states, we ask for the $1s$ AO and $2s$ AO which minimize the average energy of both states, namely, $\frac{1}{4}[1E(^1S) + 3E(^3S)]$, we should obtain *HF-SCF average of configuration wave functions*. However, this attitude also runs into difficulties fully comparable to those of item (i), namely, many atoms and ions possess more than one configuration of comparable energy, and one must inquire as to which configuration it is that should yield the AO's of choice. In a similar vein, one might use AO's from different *valence states*. The chemist considers the $1s^2 2s^2 2p^2$ configuration of C^0 to yield two valence states, $1s^2 2s^2 2p_x^2$ and $1s^2 2s^2 2p_x 2p_y$, and three spectroscopic states, ^3P, ^1D, and ^1S. The energies of these are related by[10]

$$E(\ldots p_x^2) = \tfrac{1}{3}[2E'(^1D) + E'(^1S)] \tag{1.58}$$

$$E(\ldots p_x p_y) = \tfrac{1}{4}[3E'(^3P) + E'(^1D)] \tag{1.59}$$

where, for example,

$$E'(^3P) = [5E(^3P_2) + 3E(^3P_1) + 1E(^3P_0)]/9 \tag{1.60}$$

In other words, $E'(^3P)$ is the average energy of all the terms in ^3P each weighted by the factor $2J + 1$—a procedure which removes the spin-orbit coupling energy and makes energies of the type E' compatible with our usage of a nonrelativistic Hamiltonian. If we now ask for those AO's which minimize the energy $E(\ldots p_x^2)$, or $E(\ldots p_x p_y)$, we should obtain *SCF average of valence-state wave functions*. While these may be intuitively more

[10] See Chapter 4.

reasonable than average of configuration wave functions, they also suffer from defects to which we refer later [see item (iii) below and Sec. 6(G)].

(iii) The charge on carbon in CF_4 is not zero; it is given by $C^{\delta+}$ where δ is probably less than 1. Now, HF-SCF AO's are available only for atoms of integer charge; consequently, some interpolative device must be engaged and the questions arise: Do we interpolate coefficients and/or exponents between corresponding states of C^0 and C^+, and how do we weight the interpolation process? No definitive answer can be given to any of these questions.

Although the situation is by no means as black as we have painted it, it is clear that the results of computation based on multi-ζ orbitals must be viewed as a basis for the asking of further relevant experimental questions, which must then be pursued experimentally; in other words, the purpose of the computation should be to goad experiment toward definitive ends. Secondly, computations of *a priori* pretensions should not use such orbitals as input data. Instead, the MO's for the molecule should themselves be generated by molecular SCF procedures. Unfortunately, such a computation severely restricts the size and complexity of molecules which can be studied and, besides, the SCF molecular problem is itself beset by a number of difficulties of a fundamental nature.

(F)   Multi-Gaussian Wave Functions

In the Gaussian representation we seek an orbital of the form[17]

$$\chi_{jlm}(r, \theta, \phi) = (\sum_i a_{ij}\mathrm{GTF}_{ij}) Y_{lm}(\theta, \phi)$$

$$= (\sum_i a_{ij} r^{n_{ij}-1} e^{-\zeta_{ij} r^2}) Y_{lm}(\theta, \phi) \qquad (1.61)$$

where the abbreviation GTF signifies a Gaussian-type function and where the coefficients a_{ij} and the exponents ζ_{ij} are to be determined by SCF procedures identical to those used in the STF representation. Gaussian functions have the advantage that they lead to simpler integrals than do exponentials in r but, unfortunately, more of them are required to reach a given level of accuracy.

The advantage of Gaussians is that the product of two of them, G_A and G_B centered on a and b, is a new Gaussian centered on e. Thus, the four-center problem, $\langle G_A G_B \mid G_C G_D \rangle = \langle G_E \mid G_F \rangle$, has reduced to a two-center one and can be computed a few hundred times faster than the corresponding integral $\langle S_A S_B \mid S_C S_D \rangle$ in a Slater basis set. On the other hand, one needs a much larger basis set in order to attain, with equal accuracy, a given SCF energy by using Gaussian as opposed to Slater functions. The need of a larger basis set offsets part or nearly all of the speed advantages.

The necessity for a larger basis set is exemplified very clearly in Table 1.3. This table enumerates the energies obtained with various basis sets of different sizes. The relevant criterion is that the energy be minimal. It is clear, using this criterion, that a basis set of 6 STF's is significantly and consistently better than a basis set of 14 GTF's.

(G) Hartree–Fock–Slater Method

In the lighter atoms, the HF equations can be handled directly. However, the process is tedious for larger atoms. Thus, there is need for an approximate treatment. Consider the HF expression of Eq. (1.45). The Coulomb term is a potential seen by an electron; it arises from the average charge distribution in the system and is the same for all orbitals, occupied or unoccupied. However, the exchange potential is different for each occupied orbital. Furthermore, it does not provide, in the case of a virtual orbital, for removal of self-interaction.[19] Consequently, HF AO's are probably not the best AO's for discussions pertaining to excited atomic states.

TABLE 1.3. *Comparison of Gaussian and Slater basis sets using* HF-SCF *energies (energies in a.u.)*[a]

Element and State	Gaussian set[b] $9(s), 5(p)$	Slater Set 1[c] $4(s), 2(p)$	Slater Set 2[d] $5(s), 4(p)$
Li(^2S)	-7.432279	-7.432718	-7.432726
Be(^1S)	-14.57207	-14.57237	-14.57301
B(^2P)	-24.52713	-24.52789	-24.52905
C(^3P)	-37.68525	-37.68668	-37.68858
N(^4S)	-54.39534	-54.39787	-54.40090
O(^3P)	-74.80029	-74.80476	-74.80935
F(^2P)	-99.39559	-99.40116	-99.40921
Ne(^1S)	-128.5267	-128.53480	-128.5470

[a] From Clementi (Ref. [18]).

[b] The notation $9(s), 5(p)$ gives the size and type of the basis set. In particular, the Slater set 1 basis set consists of four STF's with $l = 0$ and two STF's with $l = 1$.

[c] Slater set 1 is the so-called double-ζ representation and consists of two $1s$ STF's, two $2s$ STF's, and two $2p$ STF's.

[d] Slater set 2 is the most replete SCF set of this type available.

Now the exchange part of $\hat{H}^{HF}\chi_r(1)$ may be written as [see Eqs. (1.39) and (1.45)]

$$-\sum_s \hat{K}_s\chi_r(1) = -\sum_s \int \chi_s{}^*(2)\frac{e^2}{r_{12}}\chi_r(2)d\tau_2\chi_s(1)$$

$$= -\left[\sum_s \int \chi_s{}^*(2)\chi_r{}^*(1)\frac{e^2}{r_{12}}\chi_s(1)\chi_r(2)d\tau_2 \Big/ \chi_r{}^*(1)\chi_r(1)\right]\chi_r(1)$$

$$(1.62)$$

We now *average* this latter quantity to find

$$-\sum_s \hat{K}_s\chi_r \simeq -\left[\sum_r \sum_s \int \chi_s{}^*(2)\chi_r{}^*(1)\frac{e^2}{r_{12}}\chi_s(1)\chi_r(2)d\tau_2 \Big/ \sum_r \chi_r{}^*(1)\chi_r(1)\right]\chi_r(1)$$

$$(1.63)$$

This apparent desimplification is not drastically improper, and it does result in an exchange potential which is identical for all orbitals and which, at the same time, removes electron self-interaction. It provides significant computational advantages over the HF Hamiltonian and it is ideally suited to SCF average of configuration (or valence-state) wave-function determinations.

7. LOCALIZED SELF-CONSISTENT FIELD ATOMIC ORBITALS

The closed-shell ground-state determinantal wave function Ψ is uniquely determinable within the limits of the HF-SCF procedure. However, the individual AO's χ out of which Ψ is composed are not uniquely determinable. This may be seen as follows: Let the transformation

$$\chi_r' = \sum_{s=1}^w t_{rs}\chi_s \qquad (1.64)$$

define a new set of AO's χ_1', χ_2',..., χ_w'. Then the new determinantal function

$$\Psi' = |\chi_1'\alpha(1)\chi_1'\beta(2)\chi_2'\alpha(3)\ldots\chi_w'\beta(N)| \qquad (1.65)$$

is related to the previous one by

$$\Psi = \Psi'\cdot(\det t)^2 \qquad (1.66)$$

since each AO occurs twice in the determinantal wave function. If the transformation is orthogonal then, by definition

$$(t) \cdot \widetilde{(t)} = (1); \qquad (\det t)^2 = 1 \qquad (1.67)$$

and

$$\Psi' = \Psi \qquad (1.68)$$

Thus, Ψ is invariant to orthogonal transformations of the real AO's χ. Thus, uniqueness of Ψ does not imply uniqueness of the AO's χ.

It may also be shown[9] that the Hartree–Fock operator is itself invariant to orthogonal transformations of the orbitals on which it is dependent. Furthermore, it transpires that the off-diagonal elements of the matrix of undetermined multipliers, (ϵ_{rs}), can be chosen arbitrarily within certain limits. Since there are $(N/8)(N-2)$ such elements, it follows that we may impose a maximum of $(N/8)(N-2)$ additional restrictive conditions on the energy optimization process.[20] In the fashion in which we have obtained the Hartree–Fock energies of Eq. (1.46), these restrictions have been a requirement that all off-diagonal elements ϵ_{rs} be zero or, more specifically, that

$$\epsilon_{rs} = \epsilon_r \delta_{rs} \qquad (1.69)$$

where δ is the Kronecker δ. The resultant equations are known as the canonical Hartree–Fock equations and are given by Eq. (1.46).

However, we might specify other conditions. For example, we might ask for those SCF AO's which are as spatially separate as possible, without the prior specification of any location in space for them. The resultant orbitals in this instance are known as *localized* or *directed orbitals*. In the specific instance of neon, the ground-state configurational wave function is given by

$$\Psi = | \ 1s\alpha(1)\,1s\beta(2)\,2s\alpha(3)\,2s\beta(4)\,2p_x\alpha(5)\,2p_x\beta(6)$$

$$\times \ 2p_y\alpha(7)\,2p_y\beta(8)\,2p_z\alpha(9)\,2p_z\beta(10 \ | \qquad (1.70)$$

where the constituent AO's are *canonical AO's*, or by

$$\Psi = | \ 1s\alpha(1)\,1s\beta(2)\,t_1\alpha(3)\,t_1\beta(4)\,t_2\alpha(5)\,t_2\beta(6)\,t_3\alpha(7)\,t_3\beta(8)\,t_4\alpha(9)\,t_4\beta(10) \ |$$

$$(1.71)$$

where the constituent AO's are *localized AO's*. The t_k turn out to be the standard tetrahedrally hybridized orbitals.

The relevance of the ambiguity intrinsic to the HF-SCF AO set is not so evident in the case of atoms, although it does justify to some extent the ordinary chemical notions about hybridization. It is for molecules that this ambiguity leads to its most exciting results. In the case of benzene, for example, the several sets of HF-SCF localized MO's turn out to correspond, in a one to one way, to the Kékulé structures.[20] Thus, the old classical concept acquires an *ab initio* quantum-mechanical justification!

8. CHARGE AND CONFIGURATION OF AN ATOM IN A MOLECULE

The importance of atom charge and atom configuration with regard to the proper choice of an SCF atomic orbital has been mentioned previously.[11] We now propose to discuss the methods which enable us to determine the charge and configuration of *an atom in a molecule*.

Molecular orbitals are usually constructed from a linear combination of the atomic orbitals available in the molecule (LCAO approximation). The reliability of the MO results is dependent not only on the approximations intrinsic to the particular MO method itself, but also on the radial functions of the AO basis set which is used. It would seem that these radial wave functions should refer to the atom *in situ* in the molecule. Therefore, a method which converts the charge densities which result from an MO calculation into a sum of atomic charge densities is required in order to define not only the charge but also the configuration of each atom in the molecular ground state. If fractional charges and fractional atomic-orbital populations result from such a computation—as is invariably the case—one presumably should pick the atom or ion configuration closest in nature to that suggested by the population analysis and hope that radial SCF functions are available for it. For example, MO calculations on ferrocene suggest[21] that the iron configuration is $3d^{7.5437}4s^04p^0$. Now, average of configuration wave functions for the $3d$ AO are available for Fe^0: $3d^84s^04p^0$ and Fe^{+1}: $3d^74s^04p^0$—but which of these should be used to describe the $3d$ AO of $3d^{7.5437}4s^04p^0$? The answer is certainly not obvious. In the same computation, the carbon configuration turns out to be $2s^{1.1435}2p^{2.9229}$ for which a good parent configuration might be $2s^12p^3$. In any case, some quite arbitrary decisions must be made.

Let us now turn to a discussion of the population analysis methods. Consider a diatomic molecule for which one of the MO's, φ_r, is given by

$$\varphi_r = c_{r\mu}\chi_\mu + c_{r\nu}\chi_\nu \tag{1.72}$$

where χ_μ is an AO on center μ and χ_ν is on the center ν. The charge density associated with this MO, if doubly occupied, is

$$2\varphi_r\varphi_r = 2c_{r\mu}^2\chi_\mu\chi_\mu + 4c_{r\mu}c_{r\nu}\chi_\mu\chi_\nu + 2c_{r\nu}^2\chi_\nu\chi_\nu \tag{1.73}$$

---

[11] See Sec. 6(E) (i)–(iii).

which upon integration yields

$$2 = 2c_{r\mu}^2 + 4c_{r\mu}c_{r\nu}S_{\mu\nu} + 2c_{r\nu}^2 \tag{1.74}$$

where $S_{\mu\nu}$ is the overlap integral $\langle \chi_\mu(1) \mid \chi_\nu(1) \rangle$. This charge may now be divided as follows: A fraction $2c_{r\mu}^2$ of the charge of the two electrons is on atomic center μ, a fraction $2c_{r\nu}^2$ is on atomic center ν, and the remainder, namely, $4c_{r\mu}c_{r\nu}S_{\mu\nu}$, is an *overlap charge*. But how do we divide the overlap charge among the atomic centers? It is at this juncture that the population analysis methodologies become unsatisfactory.

Mulliken suggests[22] that the overlap charge be divided equally between the center μ and ν—a suggestion which forms the basis of the *Mulliken population analysis*. This approach appears to be reasonable if the center of gravity of the charge distribution $\chi_\mu(1)\chi_\nu(2)$ is located roughly half-way between centers μ and ν. This situation prevails if the two orbitals in question are very similar in spatial extent about their respective centers.

If, however, one of the AO's χ_μ or χ_ν is very diffuse spatially whereas the other is not, the Mulliken approximation does not make sense. For example, consider the overlap charge distribution between a $4s$ AO of manganese and a $2p_\sigma$ AO of oxygen in MnO_4^-. The $4s$ AO is very diffuse spatially, and is almost constant in that region of the Mn-O bond where the $2p_\sigma$ AO of oxygen has most amplitude. As a result, the overlap charge distribution resembles the $2p_\sigma$ AO and is dominantly centered on the oxygen. In this instance, it makes physical sense to give the whole overlap charge density to oxygen.

9.   ORTHOGONALIZED ORBITALS

The determination of the AO coefficients in an LCAO MO requires, as a computational intermediate, a transformation to a basis set of orthogonal AO's. The problem is solved in the orthogonal basis set, and the AO coefficients in the original basis set are found by transforming back from the orthogonal basis set to the original set. In addition to the above, we have already mentioned in Sec. 6(B) the problem of nonorthogonality of single-STF representations of AO's.

A number of orthogonalization processes are feasible. We limit our discussion to the Löwdin cyclic orthogonalization process and the asymmetric Schmidt orthogonalization process.

(A)   Schmidt Orthogonalization

Let the nonorthogonal set be a, b, c, ... Choose a as the first member of the orthogonal set a', b', c' . . . ; in other words let $a = a'$. Then $b' =$

$b - aS_{ab}$ and $c' = c - aS_{ac} - cS_{bc}$, etc. The process may be repeated until the whole set is orthogonalized and should be terminated by appropriate renormalization of b', c', d',.... The final result is dependent upon which member of the set is chosen as the initial member of the orthogonal set.

(B) Löwdin Orthogonalization

Let the original nonorthogonal set be represented in the row-matrix form $\{a, b, c, ...\}$. Let the orthogonal set be similarly designated $\{a'', b'', c'', ...\}$. Then[23]

$$\{a'', b'', c'', ...\} = \{a, b, c, ...\}(S)^{-1/2} \qquad (1.75)$$

where $(S)^{-1/2}$ is the inverse square root of the matrix of overlap integrals (S). All of the difficulty in the Löwdin orthogonalization resides in determination of $(S)^{-1/2}$; however, standard computer programs are available[24] for this purpose. If (S) may be written

$$(S) = (1) + (R) \qquad (1.76)$$

where (1) is the unit matrix, and the matrix of residuals, (R), contains elements less than 0.5, then we may expand to find the convergent series

$$(S)^{-1/2} = (1) - \tfrac{1}{2}(R) + \tfrac{3}{4}(R)^2 ... \qquad (1.77)$$

This series converges more rapidly the smaller the elements of the residual matrix.

10. CORRELATION ENERGY

The correlation energy for a given state is commonly defined as the difference between the Hartree–Fock energy and the exact nonrelativistic energy limit. Now, there are several HF-SCF energies, since these depend on a given author's choice of basis set and the precision of his optimization routines. Finally, the exact nonrelativistic limit is rather difficult to assess since not too many computations of the Hyellaraas[4] or Kinoshita[25] types are available. Despite this, fair estimates of the correlation energy are available; a compilation is given in Fig. 1.7.

The correlation energy is very large. In the ^3P state of carbon it amounts to 0.158 a.u. Thus, in benzene the correlation energy must be of the order of 1 a.u., which is a great amount of energy indeed. Nonetheless, correlation would not worry us too much if it remained constant for an atom in all environments—which, of course, it does not. Moreover, correlation is the dominant factor involved in dispersion forces between molecules; it is, consequently, the origin of the cohesive forces in molecular crystals. Clearly, the correlation problem is most important.[26]

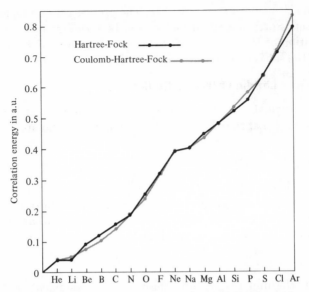

FIG. 1.7. Correlation energy obtained from HF multi-STF consideration is given by the solid curve. The correlation energy evaluated by simple Coulomb hole considerations is given by the dashed curve. The agreement is amazingly good. This diagram is adapted from Clementi (Ref. [18]).

(A)   Coulomb Holes, Configuration Interaction, and Cluster Expansions

The tendency of one electron to repel another and to force the wave functions of the two electrons to have low amplitudes when the electrons are near gives rise to the concept of the *correlation hole*. This hole takes the form of a region around each electron where no other electron is likely to be. The Pauli Exclusion Principle establishes this hole moderately well for electrons of the same spin,[12] but it has no effect on the spatial distribution of electrons with opposite spins, so does not help to introduce any correlation effect in this instance. In this latter case, the correlation hole must be a *pure Coulomb hole*. Clementi has taken up this suggestion[18] and, by making certain assumptions about the magnitude of the *diameter* of the Coulomb hole surrounding the electron in question, he has been able to compute correlation energies. In particular, Clementi apparently replaces integrals of the form

$$\int_{-\infty}^{\infty} \int_{-\infty}^{\infty} f\,(r_1, r_2)\,dr_1 dr_2$$

---

[12] This hole is known as the *Fermi hole*.

by an integral of type

$$\underbrace{\int_{r_1+\epsilon}^{\infty} \int_{-\infty}^{r_1-\epsilon}}_{\substack{\text{electron} \\ 1}} \underbrace{\int_{-\infty}^{\infty}}_{\substack{\text{electron} \\ 2}} f(r_1, r_2)\, dr_1 dr_2$$

where 2ϵ is the hole diameter. Despite some misgivings about the propriety of this sort of procedure, it must be admitted that Clementi achieved notable success. This success is apparent in Fig. 1.7.

The most extensively explored of the standard approaches to the correlation problem[27]–[30] is that of configuration interaction (CI).[10] As its name suggests, this approach consists of mixing all HF-SCF states of the same total symmetry, so that the ground state is represented as a linear combination of HF-SCF state wave functions. The mixing coefficients are determined by minimization of E for the given basis set with respect to the complete Hamiltonian. This process is very slowly convergent and quite lengthy, but it should provide a very accurate description of the ground state.

The other general approach has been developed by Sinanoğlu.[10],[31] It is based on the idea of a *cluster expansion*. In other words, deviations from the HF-SCF description may be evaluated from considerations of simple pair interactions (namely, binary interactions of two electrons in the HF field of the remainder), triad interactions, etc. This approach appears to work well for both small and large atoms.

The configuration interaction and the cluster expansion attitudes force us to abandon the simple idea that a given state can be characterized by any single set of atomic orbitals.

EXERCISES

1.   Use the Slater rules of Appendix B to find the following:

(a) the screening constant, σ, for the $1s$, $2s$, $2p$, $3s$, $3p$, $4s$, and $3d$ AO's of neutral iron;

(b) the effective nuclear charge Z for the $6s$ electron of gadolinium;

(c) the exponent ζ for the $2p$ AO of nitrogen.

Answer: (a) $\sigma = 0.30, 4.15, 4.15, 10.25, 10.25, 23.25, 19.75$ in the order quoted;

(b) $Z = 3.0$;

(c) $\zeta = 1.95$.

2.   (a) Transform the imaginary f orbitals of Table 1.3,

Appendix A into the real forms also shown. Show that

$$f_{xz^2} = (21/16\pi)^{1/2} \frac{x}{r} \left(\frac{5z^2}{r^2} - 1 \right)$$

$$f_{z(x^2-y^2)} = (105/8\pi)^{1/2} \frac{z}{r} \left(\frac{x^2}{r^2} - \frac{y^2}{r^2} \right) \tag{1.78}$$

Hint: If m is *even* use the combinations

$$i(Y_{lm} - Y_{l,-m}) \quad \text{and} \quad (Y_{lm} + Y_{l,-m}) \tag{1.79a}$$

If m is *odd* use the combinations

$$(-Y_{lm} + Y_{l,-m}) \quad \text{and} \quad i(Y_{lm} + Y_{l,-m}) \tag{1.79b}$$

For axes and trigonometric identities, see Fig. 1.1.

3. (a) Find the normalization constant N_{1s} of the Slater AO:
$\chi_{1s} = N_{1s}e^{-\zeta r}$;
(b) Find the normalization constant N_{2p_σ} of the Slater AO:
$\chi_{2p_\sigma} = N_{2p_\sigma} r \cos \theta e^{-\zeta r}$;
(c) Use your results from (a) and (b) to generate the general
form of the normalized STF representation of an s AO: $\chi = N r^{n-1} e^{-\zeta r} Y_{lm}(\theta, \phi)$.

Answer: (a) $N_{1s} = \zeta^{3/2}/\pi^{1/2}$;
(b) $N_{2p_\sigma} = \zeta^{5/2}/\pi^{1/2}$;
(c) $N = \zeta^{(2n+1)/2}/\pi^{1/2}$.

Hint: The normalization of an AO requires

$$N^2 \int_0^\infty \int_0^{2\pi} \int_0^\pi |\chi|^2 r^2 \sin\theta \, d\theta d\phi dr = 1 \tag{1.80}$$

$$\uparrow \quad \uparrow \quad \uparrow$$
$$r \quad \phi \quad \theta$$

In the case of the $1s$ AO, the only difficult integral is the radial part

$\int_0^\infty r^2 e^{-2\zeta r} dr$ and this is readily done by parts ($r^2 = u$; $e^{-2\zeta r} dr = dv$).

4. (a) Normalize the radial part $r^{n-1}e^{-\zeta r}$ of the Slater AO and
verify the general form given by Eq. (1.53) of text;
(b) Evaluate the expectation value of r, $\langle r \rangle$;
(c) Evaluate the expectation value of $1/r$, $\langle r^{-1} \rangle$;

(d) Use the results of (b) and (c) to generalize and evaluate $\langle r^q \rangle$.

Answer: (a) $R(r) = (2\zeta)^{n+1/2}(1/2n!)^{1/2}r^{n-1}e^{-\zeta r}$;

(b) $\langle r \rangle = (n + \frac{1}{2})/\zeta$;

(c) $\langle r^{-1} \rangle = \zeta/n$;

(d) $\langle r^q \rangle = (2n + q)!/(2n)!(2\zeta)^q$.

5. Consider a $1s$ AO of hydrogen atom described by a hydrogenic wave function. Evaluate $\langle r \rangle$, r_{mp}, and $\langle r^2 \rangle$, where r_{mp} is the most probable value of r. In what way do the results differ from those of the previous question?

Answer: $\langle r \rangle = 3a_0/2$; $r_{mp} = a_0$; $\langle r^2 \rangle = 3a_0^2$. Note that $a_0 \equiv h^2/4\pi^2\mu e^2$. The results are identical to those of the previous exercise and remain so for any AO of hydrogen for which $l = l_{max}$.

6. Make plots of the $2p$ AO of carbon using STF's obtained from Table 1.1 (Slater, Clementi-Raimondi, Burns, and OMAO). Make a similar plot using the SCF $2p$ AO of the ^3P state of carbon given in Table 1.2(i). Can you rationalize the differences in these plots in terms of the methods used to generate the various functions?

7. (a) The radial part of the kinetic-energy operator is given by

$$\hat{T}_r = -\frac{h^2}{8\pi^2 m}\frac{1}{r^2}\frac{d}{dr}\left(r^2\frac{dR}{dr}\right) \qquad (1.81)$$

Derive a general expression for $\langle \hat{T}_r \rangle$ in the Slater basis.

(b) The angular part of the kinetic-energy operator is given by

$$\hat{T}_{\theta;\phi} = \frac{h^2}{8\pi^2 m}\frac{l(l+1)}{r^2} \qquad (1.82)$$

Derive a general expression for $\langle \hat{T}_{\theta;\phi} \rangle$ in the Slater basis.

Answer: (a) $\langle \hat{T}_r \rangle = \zeta^2/(2n-1)$ a.u.;

(b) $\langle \hat{T}_{\theta;\phi} \rangle = \zeta^2 l(l+1)/n(n-\frac{1}{2})$ a.u.

Note: The proportionality of $\langle \hat{T} \rangle$ to ζ^2 or Z^2 specified in the text is quite evident.

8. In the determinantal wave function of Eq. (1.65), the number of electrons is N and the number of different AO's is $N/2 = w$. Hence, show that the number of different off-diagonal elements in the energy matrix is $\frac{1}{8}N(N-2) = \frac{1}{2}w(w-1)$.

BIBLIOGRAPHY

[1] H. Eyring, J. Walter, and G. E. Kimball, *Quantum Chemistry* (John Wiley & Sons, Inc., New York, 1954).

[2] W. Kauzmann, *Quantum Chemistry* (Academic Press Inc., New York, 1957).

[3] J. C. Slater, *Phys. Rev.* **36,** 57 (1930).

[4] E. A. Hylleraas, *Z. Physik* **54,** 347 (1939). [See p. 36 of Slater, Ref. [19] below.)

[5] O. Sinanoğlu, *Proc. Natl. Acad. Sci. U.S.* **47,** 1217 (1961).

[6] D. R. Hartree, *Proc. Cambridge Phil. Soc.* **24,** 89 (1927); **24,** 111 (1927). [See D. R. Hartree, *The Calculation of Atomic Structures* (John Wiley & Sons, Inc., New York, 1957.)

[7] V. Fock, *Z. Physik* **61,** 126 (1930); **62,** 795 (1930). (See pp. 1–30 of Slater, Ref. [19] below.)

[8] See C. C. J. Roothaan [*Rev. Mod. Phys.* **23,** 69 (1951)] for a closed shell; see C. C. J. Roothaan [*ibid.* **32,** 179 (1960)] for one open shell; see S. Huzinaga [*Phys. Rev.* **122,** 131 (1961)] for two open shells; see C. C. J. Roothaan and P. Bagus [*Methods in Computational Physics* (Academic Press Inc., New York, 1963)], Vol. 2.

[9] K. Ruedenberg, in *Modern Quantum Chemistry*, edited by O. Sinanoğlu (Academic Press Inc., New York, 1965), Vol. I, p. 85.

[10] R. S. Berry, *J. Chem. Ed.*, **43,** 283 (1966).

[11] E. Clementi and D. L. Raimondi, *J. Chem. Phys.* **38,** 2686 (1963).

[12] G. Burns, *J. Chem. Phys.* **41,** 1521 (1964).

[13] L. C. Cusachs, D. G. Carroll, B. Trus, and S. P. McGlynn, *Int. J. Quantum Chem.*, Slater Symposium Issue, 1967; F. A. Cotton and C. B. Harris, *Inorg. Chem.*, **6,** 369 (1967).

[14] I. Cohen and T. Bustard, *J. Chem. Ed.* **43,** 187 (1966).

[15] P. M. Morse, L. A. Young, and E. S. Haurwitz, *Phys. Rev.* **48,** 948 (1935).

[16] A. Tubis, *Phys. Rev.* **102,** 1049 (1956).

[17] S. F. Boys, *Proc. Roy. Soc. (London)* **A200,** 542 (1950); **A258,** 402 (1960). [See also F. E. Harris, *Rev. Mod. Phys.* **35,** 558 (1963); M. Krauss, *J. Res. Natl. Bur. Std.* **68B,** 35 (1964).]

[18] E. Clementi, *IBM J. Res. Develop.* **9,** 2 (1965).

[19] J. C. Slater, *Quantum Theory of Atomic Structure* (McGraw-Hill Book Co., Inc., New York, 1960), Vol. II.

[20] C. Edmiston and K. Ruedenberg, *Rev. Mod. Phys.* **35,** 457 (1963).

[21] A. T. Armstrong, D. G. Carroll, and S. P. McGlynn, *J. Chem. Phys.* **47,** 1104 (1967).

[22] R. S. Mulliken, *J. Chem. Phys.* **23,** 1833 (1955).

[23] P.-O. Löwdin, *Svensk. Kem. Tidskr.* **67,** 380 (1955).

[24] QCPE-1.1 (Prosser); QCPE-2 (Prosser); QCPE-26 (Schaad and Joy); QCPE-28 (Schaad and Joy); Quantum Chemistry Program Exchange, Chemistry Department, Room 204, Indiana University, Bloomington, Ind. 47401.

[25] T. Kinoshita, *Phys. Rev.* **105,** 1490 (1957); C. L. Pekeris, *ibid.*, **112,** 1649 (1958).

[26] *Modern Quantum Chemistry*, edited by O. Sinanoğlu (Academic Press Inc., New York, 1965), Vol. II.
[27] J. E. Lennard-Jones and J. A. Pople, *Phil. Mag.* **43,** 581 (1952).
[28] J. E. Lennard-Jones, *Proc. Natl. Acad. Sci. U.S.* **38,** 496 (1952).
[29] P. G. Dickens and J. W. Linnett, *Quart. Rev.* **11,** 291 (1957).
[30] J. W. Linnett, *The Electronic Structure of Molecules. A New Approach*, (John Wiley & Sons, Inc., New York, 1964).
[31] O. Sinanoğlu, *Proc. Natl. Acad. Sci. U.S.* **47,** 1217 (1961).

General References

1. E. U. Condon and G. H. Shortley, *Theory of Atomic Spectra* (Cambridge University Press, Cambridge, England, 1953).
2. J. C. Slater, *Quantum Theory of Atomic Structure* (McGraw-Hill Book Co., Inc., New York, 1960), Vols. I and II.
3. J. S. Griffith, *The Theory of Transition-Metal Ions* (Cambridge University Press, Cambridge, England, 1961).
4. C. J. Ballhausen, *Introduction to Ligand Field Theory*, (McGraw-Hill Book Co., Inc., New York, 1962).
5. *Rev. Mod. Phys.* **35,** (1963) (Proceedings of The International Symposium on Atomic and Molecular Quantum Mechanics, Sanibel Island, Florida, January, 1963).
6. *J. Chem. Phys.*, R. S. Mulliken Issue, Nov. 15, 1965.
7. R. M. Hochstrasser, *Behavior of Electrons in Atoms* (W. A. Benjamin, Inc., New York, 1964).
8. P. A. M. Dirac, *The Principles of Quantum Mechanics* (Oxford University Press, New York, 1947).
9. W. Kauzmann, *Quantum Chemistry* (Academic Press Inc., New York, 1957).
10. C. W. Sherwin, *Introduction to Quantum Mechanics* (Henry Holt and Co., New York, 1959).
11. F. L. Pilar, *Elementary Quantum Chemistry* (McGraw-Hill Book Co., Inc., New York, 1968).

CHAPTER 2

Overlap Integrals

The overlap integral between two atomic orbitals, χ_A and χ_B, is denoted S_{AB} or $S(\chi_A, \chi_B)$ and is defined as

$$S_{AB} = \int \chi_A{}^*\chi_B d\tau = \langle \chi_A \mid \chi_B \rangle \qquad (2.1)$$

The overlap integral is the only integral which is exactly evaluated in almost all semiempirical theories. Indeed, there also exists a widely used supposition[1]–[4] termed *The Principle of Maximum Overlapping* which presumes a direct relation of the magnitude of the overlap between two centers to the strength of the chemical bond between those two same centers. Finally, the overlap integral is necessary in the evaluation of many other important quantities: normalization factors for MO's, dipole moments, spin-orbit mixing, etc.

The purpose of this chapter is to discuss the evaluation of S_{AB}. To that end, we proceed as follows: First, we evaluate a specific atomic overlap integral and generalize this to all cases of $\sigma\sigma$, $\pi\pi$, $\delta\delta$, etc., overlaps. Second, we discuss the influence of AO selection (i.e., whether Slater, Clementi, etc.) and of atomic charge on the magnitude of the overlap. Third, we show how to transform an atomic overlap integral which is not specifically $\sigma\sigma$,

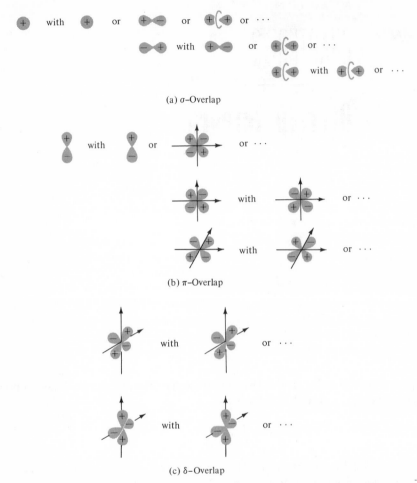

(a) σ–Overlap

(b) π–Overlap

(c) δ–Overlap

FIG. 2.1.　Schematic illustration of σ, π, and δ overlaps involving $1s$, $2p$, and $3d$ AO's. Orbital phasing is arbitrary, but is always arranged so that the overlap integral $S \geq 0$.

$\pi\pi$, $\delta\delta$, etc., into the proper linear combination of known $\sigma\sigma$, $\pi\pi$, $\delta\delta$, etc., overlaps. Fourth, we treat hybrid AO overlaps and group overlaps.

　　The discussion to follow is entirely concerned with central-field atomic orbitals, classified in the usual way: ns, np, nd, nf,... with $n \geq l + 1$. However, since we are concerned with the evaluation of the overlap of two AO's, each on different atomic centers, and since each atomic orbital is influenced by the presence of the cylindrically symmetrical field due to the two nuclear centers, it is convenient to introduce a subclassification defined by the values of the diatomic quantum number λ, where $\lambda = |m_l|$. It is in this manner that we obtain the commonly used nomenclature

$$ns, np\sigma, nd\sigma \quad \text{for} \quad \lambda = 0$$

$$np\pi, nd\pi \quad \text{for} \quad \lambda = 1$$

$$nd\delta \quad \text{for} \quad \lambda = 2$$

etc.

1. OVERLAP TYPES

The conventions for overlaps are identical to those for bonds: a σ overlap requires the participation of two σ AO's, one from each of the two centers; a π overlap requires the participation of two π AO's, one from each of the two centers; and so on. Mixed overlaps (for example, a σ, π overlap) which involve two atomic orbitals of different λ, one from each center, are zero for symmetry reasons and need not be discussed further. Thus, we need only define σ overlaps of the type shown in Fig. 2.1(a), π overlaps of the type shown in Fig. 2.1(b), δ overlaps of the type shown in Fig. 2.1(c), etc. A great many other types of overlap involving the AO's of Fig. 2.1 may be diagrammed; one such overlap is shown[1] in Fig. 2.2. However, all overlaps may be decomposed into linear combinations of overlaps of the σ, π, δ, etc., types which have already been defined in Fig. 2.1. For example, as is shown in Sec. 6(D), the overlap depicted in Fig. 2.2 equals 25% $S(d\sigma, d\sigma)$ plus 75% $S(d\delta, d\delta)$. Indeed, the set of overlaps: $S(nd\sigma, nd\sigma)$,

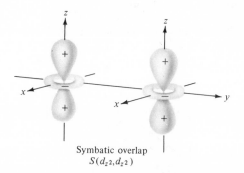

Symbatic overlap
$S(d_{z^2}, d_{z^2})$

FIG. 2.2.   Illustration of an overlap which is not of σ, π, δ, etc., type. This type of overlap has been termed σ *symbatic* by D. P. Craig, A. Maccoll, R. S. Nyholm, L. E. Orgel, and L. E. Sutton [*J. Chem. Soc.* **1954,** 332].

[1] Note that the AO's of Fig. 2.2 do not have a defined value of λ (i.e., they are not irreducible representations of the cylinder point group).

$S(nd\pi, nd\pi)$, and $S(nd\delta, nd\delta)$, constitutes a complete set of nd overlaps. Therefore, the only types of overlap which need be evaluated are those of the σ, π, δ, etc., types.

2.   A SAMPLE CALCULATION: $S(1s_A, 2p\sigma_B)$

(A)   Axes and Axes Transformations

The Cartesian coordinates affixed to the two atomic centers, A and B, are illustrated in Fig. 2.3. The coordinate frame on A is a left-handed set whereas that on B is a right-handed set; the positive directions of the z axes are opposed (i.e., pointed toward the opposite atomic center). The spherical coordinates of a point q relative to both the A and B centers are also shown in Fig. 2.3; the angle ϕ is the angle between the plane AqB and the plane xz. The natural symmetry of the overlap problem, however, is of a cylindrical nature; hence, it might be expected that the computation of an overlap would be more readily carried out in a coordinate system which possesses such symmetry. The use of a spheroidal coordinate set appears to provide the greatest simplification. The transformation from spherical to spheroidal coordinates is given by[5]

$$\mu = (r_A + r_B)/R; \qquad \nu = (r_A - r_B)/R; \qquad \phi = \phi \qquad (2.2a)$$

where the coordinate limits are

$$1 \leq \mu \leq \infty; \qquad -1 \leq \nu \leq 1; \qquad 0 \leq \phi \leq 2\pi \qquad (2.2b)$$

and the differential volume element is

$$d\tau = (R/2)^3(\mu^2 - \nu^2)\,d\mu\,d\nu\,d\phi \qquad (2.3)$$

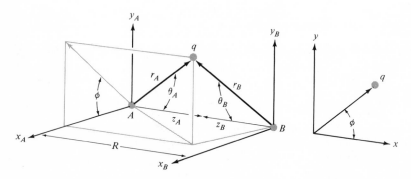

FIG. 2.3.   Cartesian coordinate systems.

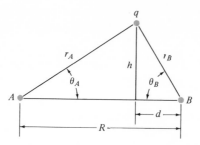

FIG. 2.4. Construct for Eq. (2.5).

The inverse transformations are also readily found. Simple addition and subtraction of μ and ν yields

$$r_A = (R/2)(\mu + \nu); \qquad r_B = (R/2)(\mu - \nu) \qquad (2.4)$$

The trigonometric functions of ϕ are the same in both sets of coordinates. Those of θ may be found from the construct of Fig. 2.4. For example, it is clear that the elaboration of $\cos\theta_A = (R - d)/r_A$ as a function $f(\mu, \nu)$ requires expansion of d as a function of r_A, r_B, and R. In view of the identity

$$h^2 = r_B{}^2 - d^2 = r_A{}^2 - (R - d)^2 \qquad (2.5)$$

it follows that

$$d = (r_B{}^2 - r_A{}^2 + R^2)/2R \qquad (2.6)$$

Therefore, using Eq. (2.4) for r_A and r_B, it is found that

$$\cos\theta_A = (R^2 - r_B{}^2 + r_A{}^2)/2Rr_A$$
$$= (1 + \mu\nu)/(\mu + \nu) \qquad (2.7)$$

Similarly,

$$\cos\theta_B = d/r_B = (1 - \mu\nu)/(\mu - \nu) \qquad (2.8)$$

(B) Evaluation of Overlap

The two atomic orbitals in question are represented by single STF's of the form

$$\chi_{1s,A} = (\zeta_A{}^{3/2}/\pi^{1/2})\,e^{-\zeta_A r_A} \qquad (2.9)$$

$$\chi_{2p\sigma,B} = (\zeta_B{}^{5/2}/\pi^{1/2})\,r_B e^{-\zeta_B r_B}\cos\theta_B \qquad (2.10)$$

with r specified in atomic units. The overlap region is depicted in Fig. 2.5. The overlap integral is given by

$$S(1s_A, 2p\sigma_B) = (\zeta_A{}^{3/2}\zeta_B{}^{5/2}/\pi)\int_0^\infty\int_0^\pi\int_0^{2\pi} r_B e^{-(\zeta_A r_A + \zeta_B r_B)}\cos\theta_B\,d\tau \qquad (2.11)$$

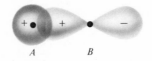

FIG. 2.5. Overlap $S(1s_A, 2p\sigma_B)$ pictorialized. Phasing of AO's is arbitrary.

This integral is readily transformed to spheroidal coordinates. However, the exponential term becomes somewhat unwieldy unless we introduce a few abbreviations. Thus, from the definitions of r_A and r_B of Eq. (2.4), it follows that

$$-\zeta_A r_A - \zeta_B r_B = -(\zeta_A + \zeta_B)(R/2)[\mu + \nu(\zeta_A - \zeta_B)/(\zeta_A + \zeta_B)]$$

$$= -p(\mu + \nu t) \tag{2.12}$$

where p and t are defined as

$$p \equiv (\zeta_A + \zeta_B)(R/2) \tag{2.13}$$

$$t \equiv (\zeta_A - \zeta_B)/(\zeta_A + \zeta_B) \tag{2.14}$$

Consequently, substituting for $\cos\theta_B$, r_B, $d\tau$, and the exponent, and changing the limits of integration, we find

$$S(1s_A, 2p\sigma_B) = (\zeta_A{}^{3/2}\zeta_B{}^{5/2}/\pi) \int_1^\infty \int_{-1}^1 \int_0^{2\pi} (R/2)(\mu - \nu)e^{-p(\mu+\nu t)}$$

$$\times [(1 - \mu\nu)/(\mu - \nu)](R/2)^3(\mu^2 - \nu^2)\,d\mu\,d\nu\,d\phi \tag{2.15}$$

$$= (\zeta_A{}^{3/2}\zeta_B{}^{5/2}R^4/2^3) \int_1^\infty \int_{-1}^1 (\mu^2 - \nu^2)(1 - \mu\nu)e^{-p(\mu+\nu t)}d\mu\,d\nu$$

$$\tag{2.16}$$

This integral is a sum of products of the form

$$\int_1^\infty \int_{-1}^1 \mu^K \nu^J e^{-p(\mu+\nu t)}d\mu\,d\nu = \int_1^\infty \mu^K e^{-p\mu}d\mu \int_{-1}^1 \nu^J e^{-p\nu t}d\nu \tag{2.17}$$

$$= A_K(p)B_J(pt) \tag{2.18}$$

where

$$A_K(p) \equiv \int_1^\infty \mu^K e^{-p\mu}d\mu \tag{2.19}$$

$$B_J(pt) \equiv \int_{-1}^1 \nu^J e^{-p t\nu}d\nu \tag{2.20}$$

Therefore, we finally find

$$S(1s_A, 2p\sigma_B) = (\zeta_A{}^{3/2}\zeta_B{}^{5/2}R^4/2^3)(A_2B_0 - A_0B_2 - A_3B_1 + A_1B_3) \quad (2.21)$$

However, it is clear from Eqs. (2.13) and (2.14) that ζ_A and ζ_B are also functions of p and t. Indeed, it may be shown that

$$\zeta_A = (1 + t)(\zeta_A + \zeta_B)/2 = (1 + t)p/R \quad (2.22)$$

$$\zeta_B = (1 - t)(\zeta_A + \zeta_B)/2 = (1 - t)p/R \quad (2.23)$$

Therefore, it follows that

$$\zeta_A{}^{3/2}\zeta_B{}^{5/2}R^4/8 = (1 + t)^{3/2}(1 - t)^{5/2}p^4/8 \quad (2.24)$$

and

$$S(1s_A, 2p\sigma_B) = 8^{-1}(1 + t)^{3/2}(1 - t)^{5/2}p^4(A_2B_0 - A_0B_2 - A_3B_1 + A_1B_3)$$

$$(2.25)$$

3. GENERAL OVERLAP INTEGRAL

Consider the overlap of two atomic orbitals χ_A and χ_B located at centers A and B, respectively, of Fig. 2.3. The orbital χ_A is given in Slater form as

$$\chi_A(n_A, l_A, m_A) = \frac{(2\zeta_A)^{n_A+1/2}}{[(2n_A)!]^{1/2}} e^{-\zeta_A r_A} r_A{}^{n_A-1} \Theta_{l_A}{}^{m_A}(\theta_A) \Phi_{m_A}(\phi) \quad (2.26)$$

where the principal quantum number n_A is a positive integer. Evaluation of the overlap in spheroidal coordinates[6] yields

$$S(\chi_A, \chi_B) = [p(1 + t)]^{n_A+1/2}[p(1 - t)]^{N_B+1/2}[(2n_A)!(2n_B)!]^{1/2}$$

$$\times \left\{ \int_1^\infty \int_{-1}^1 (\mu + \nu)^{n_A}(\mu - \nu)^{n_B} e^{-p(\mu+t\nu)} \Theta_{l_A}{}^{m_A}(\mu, \nu) \Theta_{l_B}{}^{m_B}(\mu, \nu)\, d\mu\, d\nu \right\}$$

$$(2.27)$$

The integral part, which we have put in curly brackets for convenience, is not qualitatively different[2] from that of Eqs. (2.11), (2.15), or (2.16). Hence, it may be expressed in terms of the auxiliary functions, $A_K(p)$ and $B_J(pt)$. Thus, the overlap evaluation devolves on a rewriting of the integral in terms of the A_K and B_J auxiliary functions and an evaluation of these latter functions. The expansions in terms of A_K and B_J functions are avail-

---

[2] Providing $n_A \geq l_A$ and $n_B \geq l_B$; if $n < l$, it is not possible to rewrite the integral in terms of A_K and B_J solely.

able in a number of sources (see, for example, Mulliken *et al.*[7] and Lofthus[8]). The A_K and B_J functions are discussed in Sec. 3(A).

(A) Auxiliary Functions

The integrals A_K and B_J are readily evaluated. They are given by[7]

$$A_K(p) = e^{-p} \sum_{i=1}^{K+1} [K!/p^i(K-i+1)!] \tag{2.28}$$

$$B_K(pt) = -e^{-pt} \sum_{i=1}^{K+1} [K!/(pt)^i(K-i+1)!]$$

$$- e^{+pt} \sum_{i=1}^{K+1} [(-1)^{K-i}K!/(pt)^i(K-i+1)!] \tag{2.29}$$

$$B_K(pt = 0) = 2/(K+1) \qquad K \text{ even}$$

$$= 0 \qquad K \text{ odd} \tag{2.30}$$

The equation for $A_K(p)$ breaks down as $p \to 0$ (i.e., as $R \to 0$). In this instance, both AO's are centered on the same nucleus and it is natural to perform the overlap integration in spherical coordinates. However, we now note that both of the z axes can point in the same direction (i.e., they need not be opposed). Furthermore, orthogonality of $\Theta_l{}^m$ and Φ_m angular functions applies. Consequently,

$$S(n_1, l_1, m_1; n_2, l_2, m_2)$$

$$= \{(1+t)^{n_1+1/2}(1-t)^{n_2+1/2}(n_1+n_2)!/[(2n_1)!(2n_2)!]^{1/2}\}\delta_{l_1,l_2}\delta_{m_1,m_2} \tag{2.31}$$

where δ is the Kronecker δ. This type of integral is, of course, the non-orthogonality integral of two Slater functions on the same center; it equals unity when $n_1 = n_2$ and $t = 0$.

The situation $pt = 0$ occurs when certain overlaps for which $t = (\zeta_A - \zeta_B)/(\zeta_A + \zeta_B) = 0$ are being evaluated between orbitals on two identical atoms. This case includes not only identical orbital overlap [for example, $S(ns, ns)$, $S(nd\pi, nd\pi)$, etc.] but also such overlaps as $S(ns, np\sigma)$ for which $\zeta_A = \zeta_B$ (when center A is identical to center B).

The parameter t always lies between -1 and 1. The parameter p covers the region 0 to ∞, but it is clear that much of this extent lies outside of the chemically relevant range. In any case, tables of A_K and B_J functions are available for $-1 \leq t \leq 1$ and for relevant ranges of p.[9]−[11] Use of these tables simplifies the evaluation of overlap integrals.

Our discussion of the A_K and B_J integrals applies only when the principal quantum number n is a positive integer. The situation of positive noninteger n does occur[3] and has been discussed.[8],[12],[13]

4. OVERLAP TABLES

In view of the fact that expansions of the overlap integral are available in terms of the A_K and B_J auxiliary functions, and that extensive tables of values of these latter integrals exist, it is but a small step to combine both and generate overlap tables. Such overlap tables are now available for almost all relevant overlap integrals. A compilation of references to such tables is given in Appendix C; a typical overlap table is reproduced in Table 2.1. Thus, to evaluate a given overlap, we calculate the appropriate

TABLE 2.1. $S(2s_A, 2p\sigma_B)$ overlap table

p	$t = 0.0$	$t = 0.1$	$t = 0.2$	$t = 0.3$	$t = 0.4$	$t = 0.5$	$t = 0.6$
0.0	0.000	0.000	0.000	0.000	0.000	0.000	0.000
0.5	0.143	0.167	0.180	0.180	0.166	0.139	0.103
1.0	0.276	0.319	0.343	0.342	0.315	0.265	0.198
1.5	0.386	0.443	0.473	0.472	0.437	0.369	0.278
2.0	0.464	0.526	0.561	0.560	0.521	0.446	0.340
2.5	0.504	0.567	0.603	0.602	0.569	0.493	0.383
3.0	0.509	0.570	0.606	0.612	0.582	0.513	0.407
3.2	0.503	0.562	0.598	0.606	0.580	0.515	0.412
3.4	0.492	0.549	0.586	0.596	0.573	0.513	0.415
3.6	0.479	0.534	0.570	0.582	0.563	0.509	0.416
3.8	0.463	0.515	0.551	0.565	0.550	0.502	0.415
4.0	0.444	0.494	0.529	0.545	0.535	0.492	0.412
4.2	0.425	0.471	0.505	0.523	0.518	0.481	0.407
4.4	0.402	0.446	0.480	0.500	0.499	0.469	0.401
4.6	0.380	0.421	0.455	0.476	0.479	0.454	0.394
4.8	0.357	0.396	0.428	0.451	0.458	0.439	0.386
5.0	0.334	0.370	0.402	0.426	0.436	0.423	0.377
5.5	0.278	0.308	0.337	0.364	0.382	0.382	0.351
6.0	0.226	0.250	0.277	0.305	0.328	0.339	0.334
6.5	0.180	0.200	0.224	0.251	0.278	0.297	0.295
7.0	0.141	0.157	0.178	0.205	0.233	0.258	0.266
7.5	0.109	0.122	0.140	0.165	0.193	0.222	0.238
8.0	0.083	0.093	0.109	0.131	0.159	0.189	0.212

[3] The effective principal quantum number for the case $n = 4$ is 3.7 and for the case $n = 6$ is 4.2 (see Appendix B).

values of p and t for the pair of atomic orbitals in question and look up the magnitude of S in the overlap tables.

Since there is arbitrariness associated with the definition of t, the following conventions have been adopted:

(i)   In an overlap between two AO's of unequal n, the orbital with the smaller n is identified with center A and is written first in the expression $S(\chi_A, \chi_B)$. Thus, for example, we write $S(2s, 3p\sigma)$, but not $S(3p\sigma, 2s)$.

(ii)   When $n_A = n_B$, the AO with the larger value of ζ is identified with the orbital on center A and is written first in the expression $S(\chi_A, \chi_B)$.

We now provide an example of the use of the overlap tables.

(A) $S(2s_O, 2p\sigma_C)$ at $R = 2.324$ a.u.

We are interested in the overlap of a $2s$ oxygen AO and $2p\sigma$ carbon AO at an interatomic distance of 2.324 a.u. Appendix B yields

$$\zeta(2s_O) = 2.275; \qquad \zeta(2p_C) = 1.625 \tag{2.32}$$

Consequently, we find

$$p = 1/2(\zeta_A + \zeta_B)R = 4.532 \tag{2.33}$$

$$t = (\zeta_A - \zeta_B)/(\zeta_A + \zeta_B) = 0.165 \tag{2.34}$$

From the table of overlaps (Table 2.1) for $S(2s, 2p\sigma)$ we find

t	0.1	0.2
p		
4.4	0.446	0.480
4.6	0.421	0.455

Simple interpolation for the values of p and t quoted yields $S = 0.453$.

As is indicated in the example given, it is usually necessary to interpolate to obtain the overlap required. In most instances, this is a satisfactory procedure.

5.   DEPENDENCE OF OVERLAP ON WAVE FUNCTION

We have been engrossed, so far, with the evaluation of overlaps between single-ζ Slater-type wave functions engineered according to the recipes of Appendix B. However, there are many different types of Slater-type wave functions available—ranging all the way from the multi-ζ SCF type to the single-ζ overlap-matched type which was briefly discussed in Chap. 1, Sec. 6(A). The question of the overlap dependence on the choice of wave functions made for the two overlapping AO's arises immediately. It is to this question that we now turn our attention. Prior to doing so, however, we demonstrate how to evaluate the overlap of two multi-ζ SCF-type wave functions.

(A) Overlap Integral of Multi-ζ SCF Atomic Orbitals

It is necessary to use a linear combination of Slater functions in order to obtain any reasonably accurate analytical representation of the Hartree–Fock SCF functions in terms of STF's. The following are examples of this type of function for oxygen and carbon:

$$\chi^{\text{SCF}}(2p_\text{C}) = 0.629925\chi(2p, 1.0789) + 0.42868\chi(2p, 2.1444)$$

$$+ 0.021553\chi(2p, 5.9216) \quad (2.36a)$$

$$\chi^{\text{SCF}}(2p_\text{O}) = 0.53086\chi(2p, 1.3632) + 0.505716\chi(2p, 2.7487)$$

$$+ 0.062116\chi(2p, 5.9169) \quad (2.37a)$$

where the numbers in parentheses define ζ for the type of STF (also

TABLE 2.2. *Evaluation of* $S(2p\pi_\text{C}^{\text{SCF}}, 2p\pi_\text{O}^{\text{SCF}})$ *[wave functions of Eqs. (2.36a) and (2.37a)].*[a]

Term C_iO_j	p	t	$S_{ij}(\pi, \pi)$	$c_io_jS_{ij}$
C_1O_1	2.7921	0.1164	0.500	0.167
C_1O_2	4.3761	0.4363	0.189	0.060
C_1O_3	7.9983	0.6916	0.036	0.001
C_2O_1	4.0702	0.2227	0.270	0.061
C_2O_2	5.5943	0.1235	0.115	0.025
C_2O_3	9.2165	0.4680	0.020	0.001
C_3O_1	8.3287	0.6257	0.034	0.000
C_3O_2	9.9128	0.3660	0.011	0.000
C_3O_3	13.5350	0.0004	0.000	0.000

$$S(2p\pi_\text{C}^{\text{SCF}}, 2p\pi_\text{O}^{\text{SCF}}) = \sum_{i,j=1}^{3} c_io_jS_{ij} = 0.315$$

[a] For purposes of tabulation, Eqs. (2.36a) and (2.37a) are rewritten

$$\chi^{\text{SCF}}(2p_\text{C}) = c_1C_1 + c_2C_2 + c_3C_3 \quad (2.36b)$$

$$\chi^{\text{SCF}}(2p_\text{O}) = o_1O_1 + o_2O_2 + o_3O_3 \quad (2.37b)$$

where the lower-case letters represent coefficients in the χ^{SCF}.

specified in the parentheses). An overlap integral between these two functions involves the sum of nine single-ζ-type overlaps multiplied by the proper coefficients. The manner in which $S(2p\pi_C{}^{SCF}, 2p\pi_O{}^{SCF})$ is evaluated at $R = 2.2866$ a.u. is detailed in Table 2.2. The SCF overlap is 0.315.

(B) Overlap Integral of Single-ζ Slater-type Functions[4]

Using single-ζ wave functions formulated by Slater's rules,[5] we find $\zeta(2p_C) = 1.625$, $\zeta(2p_O) = 2.275$; this yields $S(2p\pi_C, 2p\pi_O) = 0.220$. The difference in the two values of S is significant:

$$S^{SCF} - S^{Slater} = 0.095 \cong 45\% \ S^{Slater} \tag{2.38}$$

Thus, comparison of the two overlaps indicates very severe discrepancy. This observation is more or less general. It is further illustrated in Fig. 2.6.

(*i*) *Clementi–Raimondi Wave Functions:*   The Clementi–Raimondi[14] set of single-ζ STF's are those which minimize the total electronic energy. Since Slater's recipes were set up with precisely the same purpose in mind, it is not surprising that the Clementi–Raimondi and Slater sets of AO's are not very different in the case of the lighter atoms. It is also not surprising that they reproduce the Hartree–Fock SCF energy

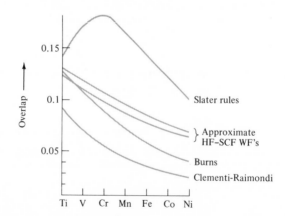

FIG. 2.6.   Overlap integrals $S(2p\sigma_C, 3d\sigma_M)$ for first-row transition metals obtained using the $3d$ wave functions specified. The internuclear distance used was 2.1 Å. The $2p$ wave function was a carbon HF-SCF function of multi-ζ STF type. This diagram is adapted from the work of Brown and Fitzpatrick [D. A. Brown and N. J. Fitzpatrick, *J. Chem. Soc.* **1966A,** 941].

[4] See Chap. 1, Secs. 6(A) and 6(B).
[5] See Appendix B.

quite closely: The HF-SCF, Clementi–Raimondi, and Slater energies for the ^3P state of neutral carbon are 37.68861, 37.622389, and 37.61855 a.u., respectively. However, since the energy of a $2p$ electron of $C^0(^3P$ state) is only 0.43334 a.u., it is clear that the major contribution to the total energy arises in the closed shells where the charge density is relatively high, and that the description given by the Clementi–Raimondi and Slater sets of AO's are prejudiced in favor of these inner shells. Indeed, one might expect that the outer or chemically important orbitals are poorly represented and not ideally suited for the evaluation of properties which are heavily dependent on the shape of χ at large R—for example, the overlap S. This, as is clear in Fig. 2.6, is indeed the case.

(*ii*) **Burns Wave Functions:** The Burns[15] set of single-ζ STF's were obtained by matching various moments of r with the SCF wave functions. Since the higher moments are less sensitively dependent than energy on inner-shell affairs, it might be expected that the overlap integrals estimated using the Burns wave functions would correspond better with S^{SCF} than would those based on either the Clementi–Raimondi or Slater sets. This, indeed, is usually the case; this claim is also supported by the results of Fig. 2.6.

(*iii*) **Overlap-Matched Atomic Orbitals:** The overlap-matched atomic orbitals (OMAO's) are chosen so as to mimic the overlap properties of the SCF wave functions over a range of representative bond distances.[16]–[18] A comparison of ζ's is given in Table 2.3 (see also Table 1.1). It is clear that, insofar as the first-row transition metals are concerned, the OMAO's represent a compromise between the Slater set, for which

TABLE 2.3. *Orbital exponents for some transition-metal 3d orbitals.*

Atom and Configuration	Clementi–Raimondi[a]	Burns[b]	Slater[c]	Overlap Matched[d]
$Cr^{+1}, 3d^5$	3.11	2.70	1.53	2.10
$Fe^{+1}, 3d^7$	3.60	3.13	1.97	2.35
$Co^{+1}, 3d^8$	3.84	3.35	2.18	2.42
$Ni^0, 3d^{10}$	4.00	3.45	2.28	2.35
$Cu^0, 3d^{10}4s^1$	4.40	3.70	2.62	2.70

[a] See Ref. [14].
[b] See Ref. [15].
[c] See Appendix B.
[d] See Ref. [16].

overlaps are usually too large, and the Clementi set, for which the overlaps are usually too small. Extension of this comparison to some of the more common elements indicates that this observation—namely, that the OMAO and Burns STF's are the best single-ζ STF's for overlap calculations—is reasonably general.

The matching process employed in the generation of the OMAO set of STF's referred to an overlap of two identical AO's on identical centers. The transferability of the ζ's so obtained to overlap between different AO's on different centers is validated by the results of Table 2.4, wherein it is seen that HF-SCF and OMAO overlaps correspond fairly closely.

TABLE 2.4. *A comparison of overlap integrals from SCF and overlap-matched wave functions.*[a]

Orbitals			
Chromium	Carbon	S (SCF)[b]	S (Overlap Matched)[c]
$4s$	$2s$	0.330	0.338
$4s$	$2p\sigma$	0.236	0.241
$3d\sigma$	$2s$	0.125	0.112
$3d\sigma$	$2p\sigma$	0.130	0.148
$3d\pi$	$2p\pi$	0.094	0.090
$4p\sigma$	$2s$	0.436	0.446
$4p\sigma$	$2p\sigma$	-0.003	0.007
$4p\pi$	$2p\pi$	0.207	0.195

[a] At $R = 4.2$ a.u.

[b] The chromium orbital wave functions are taken from Richardson *et al.* [J. W. Richardson, W. C. Nieuwpoort, R. R. Powell, and W. F. Edgell, J. Chem. Phys. **36**, 1057 (1962); J. W. Richardson, R. R. Powell, and W. C. Nieuwpoort, *ibid.* **38**, 796 (1963)] and were for the following configurations:

$$3d: \text{Cr}^{+1} (3d^5)$$

$$4s: \text{Cr}^0 (3d^4 4s^2)$$

$$4p: \text{Cr}^0 (3d^5 4p^1).$$

The carbon $2s$ and $2p$ wave functions were those of Clementi [IBM J. Res. Develop. Suppl. **9**, 2 (1965), see Table 45-01].

[c] Orbital exponents used were: chromium, $\zeta (3d) = 2.17$, $\zeta (4s) = 1.31$, $\zeta (4p) = 0.77$; carbon, $\zeta (2p) = 1.42$. For the carbon $2s$ orbital a $1s$ wave function with orbital exponent $\zeta (1s) = 1.02$ was used.

(C) Effect of Charge on Overlap Values

In simple calculations, the wave functions used are those for neutral atoms even though in the molecular environment the atoms may be far from neutral. It is true that in some calculations on transition-metal complexes, where it was suspected that the metal would carry a positive charge, some workers have used metal wave functions corresponding to a charge of $+1$. The point that concerns us is whether, among all the other approximations implicit in simple calculations, it is worthwhile adding a refinement to take account of the variation of the wave function with charge. The answer, in most cases, is no.[18],[19] Table 2.5 provides comparison of some $C^0 - O^0$ and $C^+ - O^-$ overlaps; the variations, relatively speaking, are seen to be small.

(D) Consensus

At this stage, the reader might ask why one should bother with approximate single exponential functions when accurate HF-SCF functions are available. The reason, of course, is because of their convenience. In many semiempirical MO calculations the only property of the wave function of real concern is its overlap integral with other functions. The Hamiltonian matrix elements are not calculated from the wave functions but are inferred or approximated from experimental data. Therefore, if a single exponential function can be found which reproduces reasonably well the overlap characteristics of an SCF function, its use can represent a considerable saving in the time and effort required to carry out such semiempirical calculations on large molecules.

6. ANGULAR CORRECTIONS

All overlaps which we have considered have been of $\sigma, \pi, \delta, \phi, \ldots$ type. In the few instances where this was not the case, the overlap integral was

TABLE 2.5. *Effect of atom charge on overlap*[a]

Overlap	C^0, O^0	C^+, O^-
$S(2s_O, 2s_C)$	0.368	0.371
$S(2s_O, 2p\sigma_C)$	0.453	0.420
$S(2p\pi_O, 2p\pi_C)$	0.210	0.214

[a] Using Slater's Rules: $\zeta_{C^+}(2s, 2p) = 1.80$ and $\zeta_{O^-}(2s, 2p) = 2.10$.

partitioned into parts which were separately of σ, π, δ,... type. In view of the primacy of these types of overlap we will, henceforth, refer to σ, π, δ,... overlaps as *pure overlaps* or *overlaps of pure type*.

In the case of diatomic molecules, it is a simple matter to arrange the coordinate systems on the individual atoms so that all overlap integrals are of pure type. However, such is not the case in more complex molecules. The case of formaldehyde is shown in Fig. 2.7. While it is possible to arrange axes on any two centers to meet the requirements of pure overlap, it is not possible to arrange the axes on all centers so that all overlaps are of pure type. Therefore, the following problem must be solved: Given two sets of AO's, one set on center A defined with respect to Cartesian axes x_A, y_A, and z_A, and the other set on center B defined with respect to Cartesian axes x_B, y_B, and z_B, find the set of overlap integrals S_{AB} when the sets of axes on A and on B are spatially arbitrary. In order to use the tables of overlap integrals, it follows that we must seek a way of resolving any one of the set of AB overlaps into a linear combination of overlaps of the pure type.

The obvious way to solve this problem and stay within the restrictions imposed by the available overlap tables is to use the Euler transformation scheme outlined in Sec. 6(A). Prior to doing so, however, we wish to emphasize two points:

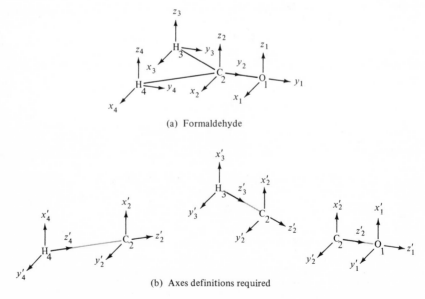

(a) Formaldehyde

(b) Axes definitions required

FIG. 2.7.   (a) Right-handed axes in H_2CO and (b) the axes transformations required for pure overlap characteristics for three different *diatomic* groupings present in H_2CO.

(i)   The directions of the transformed z_B and z_A axes should be opposed in order to conform to the conventions of the overlap tables. However, such opposition merely changes the sign of the overlap relative to the case in which the z axes are codirectional. On the other hand, the codirectional usage of Fig. 2.7 enables us to use right-handed coordinate systems, whereas the antiparallel situation (see Fig. 2.3) requires simultaneous use of both left- and right-handed coordinate systems. We use right-handed coordinate systems consistently in the following. This may require a few abrupt sign changes here and there but forewarning should evade any confusion.

(ii)   Rotation of the coordinate frame does not affect the radial part of the wave function; it alters only the angular part. Consequently, the procedures used are the same whether one is concerned with hydrogenlike, Slater-type, or Gaussian-type atomic orbitals.

(A)   Euler Angles and Transformations[6]

A convenient way to describe a general rotation is to make use of the three Euler angles a, b, and c. These angles are such that when one rotates the original coordinate system over these angles in some very specific way, one produces the new desired coordinate system.[22] The chosen rotations are illustrated and defined in Fig. 2.8. First one rotates the original coor-

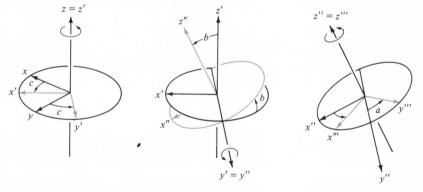

1. Rotate positively about z by angle c.
$(x,y,z) \xrightarrow[c_z]{} (x',y',z'=z)$

2. Rotate positively about y' by angle b.
$(x',y',z') \xrightarrow[b_{y'}]{} (x'',y''=y',z'')$

3. Rotate positively about z'' by angle a.
$(x'',y'',z'') \xrightarrow[a_{z''}]{} (x''',y''',z'''=z'')$

FIG. 2.8. Euler angles a, b, and c defined. A positive rotation carries $x \to y$, $y \to z$, or $z \to x$ when the axis of rotation is z, x, y, respectively.

[6] See Ballhausen and Gray (Ref. [20]) and Yeranos (Ref. [21]) for alternate procedures.

dinate system in a positive direction over an angle c around the z axis; this yields $(x', y', z' = z)$. Then one rotates the (x', y', z') system in a positive direction over an angle b around the y' axis, producing $(x'', y'' = y', z'')$. Finally, one rotates (x'', y'', z'') over an angle a around the z'' axis, resulting in the desired coordinate system $(x''', y''', z''' = z'')$.

A positive rotation of the (x, y, z) coordinate system around the z axis by an angle c yields the new set

$$\begin{pmatrix} x' \\ y' \\ z' \end{pmatrix} = \begin{pmatrix} \cos c & \sin c & 0 \\ -\sin c & \cos c & 0 \\ 0 & 0 & 1 \end{pmatrix} \begin{pmatrix} x \\ y \\ z \end{pmatrix} \qquad (2.39)$$

Similarly, the second rotation of Fig. 2.8 yields

$$\begin{pmatrix} x'' \\ y'' \\ z'' \end{pmatrix} = \begin{pmatrix} \cos b & 0 & -\sin b \\ 0 & 1 & 0 \\ \sin b & 0 & \cos b \end{pmatrix} \begin{pmatrix} x' \\ y' \\ z' \end{pmatrix} \qquad (2.40)$$

and the third rotation of Fig. 2.8 yields

$$\begin{pmatrix} x''' \\ y''' \\ z''' \end{pmatrix} = \begin{pmatrix} \cos a & \sin a & 0 \\ -\sin a & \cos a & 0 \\ 0 & 0 & 1 \end{pmatrix} \begin{pmatrix} x'' \\ y'' \\ z'' \end{pmatrix} \qquad (2.41)$$

Applying the three rotations consecutively, one obtains by matrix multiplication

$$\begin{pmatrix} x''' \\ y''' \\ z''' \end{pmatrix} = \begin{pmatrix} & \\ & A & \\ & \end{pmatrix} \begin{pmatrix} x \\ y \\ z \end{pmatrix}$$

where (A) is the matrix

$$\begin{pmatrix} \cos a \cos b \cos c - \sin a \sin c & \cos a \cos b \sin c + \sin a \cos c & -\cos a \sin b \\ -\sin a \cos b \cos c - \cos a \sin c & -\sin a \cos b \sin c + \cos a \cos c & \sin a \sin b \\ \sin b \cos c & \sin b \sin c & \cos b \end{pmatrix}$$

$$(2.42)$$

The inverse operation is represented by

$$
\begin{pmatrix} x \\ y \\ z \end{pmatrix} = \begin{pmatrix} \\ A \\ \end{pmatrix}^{-1} \begin{pmatrix} x''' \\ y''' \\ z''' \end{pmatrix}
$$

where the inverse matrix $(A)^{-1}$ is

$$
\begin{pmatrix}
\cos a \cos b \cos c - \sin a \sin c & -\sin a \cos b \cos c - \cos a \sin c & \sin b \cos c \\
\cos a \cos b \sin c + \sin a \cos c & -\sin a \cos b \sin c + \cos a \cos c & \sin b \sin c \\
-\cos a \sin b & \sin a \sin b & \cos b
\end{pmatrix}
$$

$$(2.43)$$

(B) $p\text{-}p$ Overlap

Consider the evaluation of the orbital overlap shown in Fig. 2.9—where it is to be emphasized that the planes defined by any two rotation axes, one for each of the two p AO's, do not coincide with any of the coordinate planes xy, xz, yz.

The z''' axes are fixed in the usual way.[7] The positions of x''' and y''' may be chosen so as to keep the transformation as simple as possible. The most obvious way to proceed is to give the Euler angles a, b, c the following

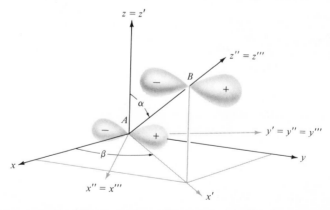

FIG. 2.9. $p\text{-}p$ overlap situation where the atoms A and B are not both in one coordinate plane. Successive rotations over the three Euler angles to generate the suitable coordinate system (x''', y''', z''') are illustrated.

[7] Along the AB bond.

values:

$$c = \beta, \qquad b = \alpha, \qquad a = 0$$

The resulting coordinate systems, indicated by one, two, and three primes, respectively, are shown on Fig. 2.9. The transformation matrix of Eq. (2.43) simplifies to

$$\begin{pmatrix} x \\ y \\ z \end{pmatrix} = \begin{pmatrix} \cos\alpha\cos\beta & -\sin\beta & \sin\alpha\cos\beta \\ \cos\alpha\sin\beta & \cos\beta & \sin\alpha\sin\beta \\ -\sin\alpha & 0 & \cos\alpha \end{pmatrix} \begin{pmatrix} x''' \\ y''' \\ z''' \end{pmatrix} \qquad (2.44)$$

Therefore, a p_y orbital is expressed in the new coordinate system as

$$p_y = (\cos\alpha\sin\beta)(p_x''') + (\cos\beta)(p_y''') + (\sin\alpha\sin\beta)(p_z''') \qquad (2.45)$$

and the overlap integral between the original orbitals, $(p_y)_A$ and $(p_y)_B$, is given by

$$\begin{aligned} S[(p_y)_A, (p_y)_B] &= (\cos^2\alpha\sin^2\beta)\, S[(p_x''')_A, (p_x''')_B] \\ &\quad + (\cos^2\beta)\, S[(p_y''')_A, (p_y''')_B] \\ &\quad + (\sin^2\alpha\sin^2\beta)\, S[(p_z''')_A, (p_z''')_B] \\ &= (\cos^2\alpha\sin^2\beta + \cos^2\beta)\, S(p\pi, p\pi) \\ &\quad - (\sin^2\alpha\sin^2\beta)\, S(p\sigma, p\sigma) \qquad (2.46) \end{aligned}$$

This equation reduces to

$$S[(p_y)_A, (p_y)_B] = S(p\pi, p\pi) \qquad (2.47)$$

for $\alpha = 0$ or $\beta = 0$; it reduces to

$$(\cos^2\alpha)\, S(p\pi, p\pi) - (\sin^2\alpha)\, S(p\sigma, p\sigma) \qquad (2.48)$$

for $\alpha = \pi/2$ or $\beta = \pi/2$. The minus sign of the $S(p\sigma, p\sigma)$ contribution arises because of the head-to-tail position of $(p_z''')_A$ and $(p_z''')_B$ while, as always, the published tables refer to the head-to-head (or bonding) position.

(C) *s-d* Overlap

A pure σ-type overlap results when the s and d orbital have a common axis with complete rotational symmetry; the d orbital, consequently, has to be a d_{z^2} orbital ($\alpha = 0$ in Fig. 2.10). If $\alpha \neq 0$, one obtains a partly σ bonded, partly nonbonded atom pair.

Consider the example of Fig. 2.10 where a d_{z^2} orbital is situated on center A and an s orbital is situated on center B. The appropriate transformation is shown on Fig. 2.10 and is expressed again by Eq. (2.44). Now, since the angular part of the $3d_{z^2}$ orbital is given by

$$(3d_{z^2})_{\text{ang}} = \left(\frac{5}{16\pi}\right)^{1/2} (3\cos^2\theta - 1) = \left(\frac{5}{16\pi}\right)^{1/2} \frac{3z^2 - r^2}{r^2} \quad (2.49)$$

and since, in view of Eq. (2.44), we have

$$z = -(\sin\alpha)(x''') + (\cos\alpha)(z''')$$

we obtain

$$\frac{3z^2 - r^2}{r^2} = \frac{3[(\sin^2\alpha)(x''')^2 + (\cos^2\alpha)(z''')^2 - 2\sin\alpha\cos\alpha(x'''z''')] - (r''')^2}{(r''')^2}$$

$$= \left[\frac{3(x''')^2 - (r''')^2}{(r''')^2}\right]\sin^2\alpha + \left[\frac{3(z''')^2 - (r''')^2}{(r''')^2}\right]\cos^2\alpha$$

$$- \frac{6(x'''z''')}{(r''')^2}\sin\alpha\cos\alpha \quad (2.50)$$

The normalized angular part of a $3d_{xz}'''$ orbital is

$$(3d_{xz}''')_{\text{ang}} = \left(\frac{15}{4\pi}\right)^{1/2} \frac{x'''z'''}{(r''')^2} \quad (2.51)$$

The first term of Eq. (2.50) gives rise to a $3d_{x^2}$ orbital, functionally equiva-

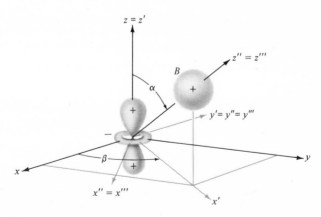

FIG. 2.10. *s-d* overlap; the original coordinate system is (x, y, z); the final coordinate system is (x''', y''', z'''). The intermediate coordinate systems, obtained by applying successively the three Euler rotations, are indicated by one and two primes.

lent with $3d_{z^2}$ but oriented along the x axis. Such an orbital can be expressed as a linear combination of the five usual d orbitals:

$$3x^2 - r^2 = \tfrac{3}{2}(x^2 - y^2) - \tfrac{1}{2}(3z^2 - r^2) \tag{2.52}$$

The normalized part of a $3d_{x^2-y^2}{}'''$ orbital is given by

$$(3d_{x^2-y^2}{}''')_{\text{ang}} = \left(\frac{15}{16\pi}\right)^{1/2} \frac{x^2 - y^2}{r^2} \tag{2.53}$$

Combining Eqs. (2.49)–(2.53), we find

$$3d_{z^2} = (\cos^2 \alpha - \tfrac{1}{2}\sin^2 \alpha)(3d_{z^2}{}''') + (\tfrac{3}{4})^{1/2}(\sin^2 \alpha)(3d_{x^2-y^2}{}''')$$
$$- (3)^{1/2}(\sin \alpha \cos \alpha)(3d_{xz}{}''') \tag{2.54}$$

Since only the first term contributes to the value of the overlap integral in question, we find

$$S(s, d_{z^2}) = (\cos^2 \alpha - \tfrac{1}{2}\sin^2 \alpha)[S(s, d\sigma)] \tag{2.55}$$

where $S(s, d\sigma)$ represents a pure σ-overlap integral.

(D) *d-d* Overlap

Besides the three simple pure-type overlaps between d AO's, shown in Fig. 2.1, there are some other interesting situations. Two of these are diagrammed in Fig. 2.11 (see also Fig. 2.2).

(i) $d_{z^2} - d_{z^2}$: The appropriate coordinate transformation is obviously a rotation over 90° about the x axis; if the rotation is taken clockwise, the transformation is described by

$$y = z', \qquad z = -y', \quad \text{and} \quad x = x' \tag{2.56}$$

where the primes indicate the new coordinate system. Since

$$3z^2 - r^2 = 3(y')^2 - (r')^2 = -\tfrac{1}{2}[3(z')^2 - (r')^2] - \tfrac{3}{2}[(x')^2 - (y')^2]$$

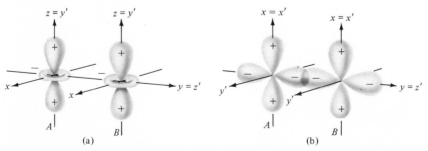

FIG. 2.11. Two less-simple *d-d* overlap situations.

we obtain

$$d_{z^2} = -\tfrac{1}{2}(d_{z^2}') - \tfrac{1}{2}(3)^{1/2}(d_{x^2-y^2}') \tag{2.57}$$

and

$$S(d_{z^2}, d_{z^2}) = \tfrac{1}{4}S(d\sigma, d\sigma) + \tfrac{3}{4}(d\delta, d\delta) \tag{2.58}$$

Indeed, $S[(d_{z^2}')_A, (d_{x^2-y^2}')_B] = S[(d_{z^2}')_B, (d_{x^2-y^2}')_A] = 0$; therefore, the $d_{z^2} - d_{z^2}$ overlap is 25% σ and 75% δ in nature.

(ii) $d_{x^2-y^2} - d_{x^2-y^2}$: The same transformation as in (i) is adequate; Eqs. (2.56) are the substitution equations and lead to

$$x^2 - y^2 = (x')^2 - (z')^2 = \tfrac{1}{2}[(x')^2 - (y')^2] - \tfrac{1}{2}[3(z')^2 - (r')^2]$$

Therefore,

$$d_{x^2-y^2} = \tfrac{1}{2}(d_{x^2-y^2}') - \tfrac{1}{2}(3)^{1/2}(d_{z^2}') \tag{2.59}$$

and

$$S(d_{x^2-y^2}, d_{x^2-y^2}) = \tfrac{1}{4}S(d\delta, d\delta) + \tfrac{3}{4}S(d\sigma, d\sigma) \tag{2.60}$$

7. HYBRIDIZED ORBITALS

No new principles are introduced in the evaluation of overlap integrals involving a hybridized atomic orbital. Hence, we content ourselves with an example calculation. The example we choose is the overlap of a $1s$ AO of hydrogen with a tetrahedrally hybridized sp^3 AO of carbon.

The methane molecule is diagrammed in Fig. 2.12. It is enclosed in a cube, the length of whose side is two units. The hydrogen atoms are positioned at points 1, 2, 3, and 4. The sp^3 hybrids directed toward these hydrogen atoms are labeled t_1, t_2, t_3, and t_4, respectively. The carbon atom is located at the origin of coordinates [i.e., at $(0, 0, 0)$]. The coordinates of the H-atom locations are also specified in Fig. 2.12. Hence, it is an easy matter to write down the hybridized carbon AO's. An orbital directed toward 1, must contain equal amounts of $2p_x$, $2p_y$, and $2p_z$. It may contain

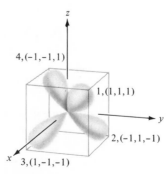

FIG. 2.12. Methane molecule.

any amount of $2s$. However, distributing $2s$ equally among all four sp^3 orbitals, we find by simple coordinate enumeration:

$$t_1 = \tfrac{1}{2} 2s + \tfrac{1}{2}(2p_x + 2p_y + 2p_z)$$

$$t_2 = \tfrac{1}{2} 2s + \tfrac{1}{2}(-2p_x + 2p_y - 2p_z)$$

$$t_3 = \tfrac{1}{2} 2s + \tfrac{1}{2}(2p_x - 2p_y - 2p_z)$$

$$t_4 = \tfrac{1}{2} 2s + \tfrac{1}{2}(-2p_x - 2p_y + 2p_z) \tag{2.61}$$

These hybrid AO's are all normalized; they are also mutually orthogonal.

Consider now the overlap of t_3 with $1s_1$, where the s_1 AO pertains to the H atom at corner 1. Then

$$S(t_3, 1s_1) = \tfrac{1}{2} S(2s + 2p_x - 2p_y - 2p_z, 1s_1)$$

$$= \tfrac{1}{2} S(2s, 1s_1) + \tfrac{1}{2} S(2p_x, 1s_1) - \tfrac{1}{2} S(2p_y, 1s_1) - \tfrac{1}{2} S(2p_z, 1s_1) \tag{2.62}$$

Now the symmetry of the problem dictates that

$$S(2p_x, 1s_1) = S(2p_z, 1s_1) = S(2p_y, 1s_1) \tag{2.63}$$

Hence

$$S(t_3, 1s_1) = \tfrac{1}{2} S(2s, 1s_1) - \tfrac{1}{2} S(2p_z, 1s_1) \tag{2.64}$$

Resolving $2p_z$ into components along and perpendicular to the line joining $(0, 0, 0)$ and $(1, 1, 1)$, we find

$$S(t_3, 1s_1) = \tfrac{1}{2} S(2s, 1s) - \tfrac{1}{2}(3)^{-1/2} S(2p\sigma, 1s_1) \tag{2.65}$$

The formalism of the overlap tables necessitates that this result be written

$$S(t_3, 1s_1) = \tfrac{1}{2} S(1s_\text{H}, 2s_\text{C}) - [2(3)^{1/2}]^{-1} S(1s_\text{H}, 2p\sigma_\text{C}) \tag{2.66}$$

Thus, it is seen that overlap evaluations involving hybridized AO's are a relatively straightforward matter.

8.   GROUP-OVERLAP INTEGRALS[8]

Group-orbital considerations reduce the labor involved in quantum-chemical computations to a degree which increases with increasing symmetry of the molecule under consideration.[23] The subject of group overlaps has been discussed by Ballhausen and Gray[20] in a very admirable fashion; these same authors also provide tables of group-orbital overlaps for some of the more symmetrical point groups. Hence, we content ourselves with one simple example of the evaluation of a group-overlap integral.

---

[8] A group orbital is a molecular orbital of a multicenter nature. It is composed of AO's from those (identical) centers in the molecule which are transformed into each other by symmetry operations of the point group to which the molecule belongs. The group orbital is completely determined by symmetry.

The molecule of interest may be supposed to be the tetrahedral sulfate ion, $SO_4^=$. Furthermore, we may suppose ourselves to be interested in the involvement of sulfur d orbitals in the bonding to the oxygens. Thus, we might wish to evaluate the overlap of the sulfur d_{yz} AO with one of the group orbitals constituted of oxygen $2s$ AO's. Now d_{yz} transforms as one of the components of the t_2 representation in the tetrahedral T_d point group. The group orbital based on $2s$ AO's which transforms similarly in T_d is

$$\varphi_{2s} = \tfrac{1}{2}(2s_1 - 2s_2 + 2s_3 - 2s_4) \tag{2.67}$$

where the normalization factor of $\tfrac{1}{2}$ is correct only in the absence of $2s_i - 2s_j$ ($i \neq j = 1, 2, 3, 4$) overlap. The two orbitals, d_{yz} and φ_{2s} are diagrammed in Fig. 2.13. If the S_{ij} overlap, $S(2s_i, 2s_j)$, be non-negligible, the normalized group orbital is given by

$$\varphi_{2s} = \{1/[2(1 - S_{ij})^{1/2}]\}(2s_1 - 2s_2 + 2s_3 - 2s_4) \tag{2.68}$$

and it is with this latter form of φ_{2s} that we are concerned.

The overlap integral of relevance is

$$S(d_{yz}, \varphi_{2s}) = \{1/[2(1 - S_{ij})^{1/2}]\}S(d_{yz}, [2s_1 - 2s_2 + 2s_3 - 2s_4]) \tag{2.69}$$

Inspection of Fig. 2.13 from the point of view of the symmetry of the problem indicates that

$$S(d_{yz}, 2s_1) = S(d_{yz}, -2s_2) = S(d_{yz}, 2s_3) = S(d_{yz}, -2s_4) \tag{2.70}$$

Hence, we find

$$S(d_{yz}, 2s_1) = [2/(1 - S_{ij})^{1/2}]S(d_{yz}, 2s_1) \tag{2.71}$$

The evaluation of the group-overlap integral reduces to calculation of $S(2s_i, 2s_j)$, which is trivial, and $S(d_{yz}, 2s_1)$ which is given as an exercise for the reader (see Exercise 10).

Other group-orbital overlaps might be more complex, but they introduce no new principles above and beyond those elaborated here.

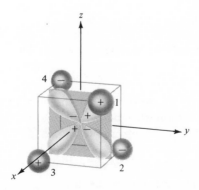

FIG. 2.13. d_{yz} AO and φ_{2s} GO diagrammed.

EXERCISES

1. If the AO's on centers A and B of Fig. 2.3 are given by

$$1s_A = (1/\pi^{1/2})e^{-r_A} \quad \text{and} \quad 1s_B = (1/\pi^{1/2})e^{-r_B} \qquad (2.72)$$

where $\zeta_A = \zeta_B = 1$, show that

$$\langle 1s_A \mid 1s_B \rangle = S(1s_A, 1s_B) = e^{-R}(1 + R + R^3/3) \qquad (2.73)$$

Hint: Use spheroidal coordinates and extended integration by parts.

2. If $1s_A$ and $1s_B$ are defined as in Exercise 1, show that

$$\langle 1s_A \mid 1/r_B \mid 1s_B \rangle = \int \frac{1s_A 1s_B}{r_B}\, d\tau = e^{-R}(R + 1) \qquad (2.74)$$

3. (a) Derive a general equation for the overlap of a $1s$ AO with orbital exponent ζ and a $2s$ AO with orbital exponent ζ' located on the same center. The result obtained should be

$$S = \iiint \chi_{1s}\chi_{2s}r^2 \sin\theta d\theta d\phi dr = \frac{\zeta^{3/2}(\zeta')^{5/2}\, 2^2 6}{(\zeta + \zeta')^4\, 3^{1/2}} \qquad (2.75)$$

when both AO's are single-ζ STF's.
(b) Show the identity of this result to that obtained by insertion in Eq. (2.31).
(c) $\zeta = 5.6727$ and $\zeta' = 1.6083$ are reasonable values for the $1s$ AO and $2s$ AO, respectively, of carbon. Hence, show that $S = 0.2185$.

4. Evaluate $S(1s_H, 1s_H)$ at $R = 3.452$ a.u. Use Table IV of Mulliken, Rieke, Orloff, and Orloff [*J. Chem. Phys.* **17,** 1248 (1949)].

Answer: $p = 3.452;\quad t = 0;\quad S = 0.267$ \qquad (2.76)

5. Evaluate $S(2s_C, 2s_O)$ at $R = 1.229$ Å.

Answer: $p = 4.53;\quad t = 0.167;\quad S = 0.374$ \qquad (2.77)

6. Derive an expression for the overlap integral between a p_z and a p_y orbital when both orbitals are situated in the (yz) plane. This situation is depicted in Fig. 2.14.

Answer: $S = -\sin\alpha \cos\alpha[S(p\pi, p\pi) + S(p\sigma, p\sigma)]$ \qquad (2.78)

where α is defined as in Fig. 2.14.

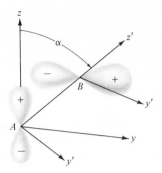

FIG. 2.14. Overlap to be evaluated in Exercise 6.

7.   Find the angular dependence of a $(d_{yz} - p_y)$ overlap integral provided both atoms are situated in the (yz) plane as depicted in Fig. 2.15.

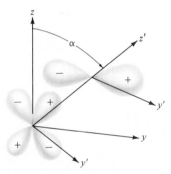

FIG. 2.15. Overlap to be evaluated in Exercise 7.

Answer: $S = (\cos \alpha \cos 2\alpha)[S(p\pi, d\pi)] - \frac{1}{2}(3)^{1/2}(\sin \alpha \sin 2\alpha)[S(p\sigma, d\sigma)]$

$$(2.79)$$

8.   Jaffé [*J. Chem. Phys.* **21,** 258 (1953)] has published overlap tables of $S(3d\delta, 3d\delta)$. Roberts and Jaffé [*J. Chem. Phys.* **27,** 883 (1957)] have published tables for $S(3d\sigma, 3d\sigma)$. Craig *et al.* [*J. Chem. Soc.* **1954,** 354] give tables for $S(3d_{z^2}, 3d_{z^2})$. Check the relationship between the values of these three types of overlap integrals using Eq. (2.58) of text.

9.   Consider a rotation by $2\pi/3$ about the axis connecting the points $(0, 0, 0)$ and $(1, 1, 1)$ in the right-handed coordinate system of Fig. 2.12. Show, by simple symmetry considerations,

that the new set of axes is given by

$$
\begin{pmatrix} x' \\ y' \\ z' \end{pmatrix} = \begin{pmatrix} 0 & 1 & 0 \\ 0 & 0 & 1 \\ 1 & 0 & 0 \end{pmatrix} \begin{pmatrix} x \\ y \\ z \end{pmatrix} \tag{2.80}
$$

Determine the set of Euler angles (a, b, c). Insert these into Eq. (2.42) of text and verify the symmetry determined transformation of Eq. (2.80).

Answer: $c = 0;\quad b = \pi/2;\quad a = \pi/2$ \hfill (2.81)

10. We wish to evaluate the overlap integral of a d_{yz} AO situated at $(0, 0, 0)$ and a $1s$ AO situated at $(1, 1, 1)$—see Figs. 2.12 and 2.13. Consequently, we must transform to a new coordinate system where z''' is the line connecting points $(0, 0, 0)$ and $(1, 1, 1)$. Choose y''' to be the line connecting points $(0, 0, 0)$ and $(-1, 1, 0)$; choose x''' to be the line connecting $(0, 0, 0)$ and $(1, 1, -1)$. Find the relationship between the sets of axes (x, y, z) and (x''', y''', z''').

Answer: $\cos a = 1;\quad \sin a = 0;\quad \cos b = (\tfrac{1}{3})^{1/2};\quad \sin b = (\tfrac{2}{3})^{1/2};$

$\cos c = (\tfrac{1}{2})^{1/2};\quad \sin c = (\tfrac{1}{2})^{1/2}$ \hfill (2.82)

$$
\begin{pmatrix} x \\ y \\ z \end{pmatrix} = \begin{pmatrix} (\tfrac{1}{6})^{1/2} & -(\tfrac{1}{2})^{1/2} & (\tfrac{1}{3})^{1/2} \\ (\tfrac{1}{6})^{1/2} & (\tfrac{1}{2})^{1/2} & (\tfrac{1}{3})^{1/2} \\ -(\tfrac{2}{3})^{1/2} & 0 & (\tfrac{1}{3})^{1/2} \end{pmatrix} \begin{pmatrix} x''' \\ y''' \\ z''' \end{pmatrix} \tag{2.83}
$$

11. The d_{yz} AO is given by

$$
d_{yz} = (15/4\pi)^{1/2} \frac{yz}{r^2} \tag{2.84}
$$

Substitute for y and z from Eq. (2.83) and collect terms to find

$$
d_{yz} = (1/3)^{1/2} d_{z'''^2} - (1/3) d_{x'''^2-y'''^2} - (1/18)^{1/2} d_{x'''z'''}
$$
$$
- (1/3)^{1/2} d_{x'''y'''} + (1/6)^{1/2} d_{y'''z'''} \tag{2.85}
$$

Hence, show that

$$
S(d_{yz}, s) = (\tfrac{1}{3})^{1/2} S(d\sigma, s) \tag{2.86}
$$

Hint: See Ref. [20], pp. 111–116.

12. Normalize the group orbital $\varphi_s = (s_1 - s_2 + s_3 - s_4)$ of Eq. (2.67) when the $S(s_i, s_j)$ overlap is non-negligible.

Answer: See Eq. (2.68).

BIBLIOGRAPHY

[1] See, for example, H. Suzuki, *Electronic Absorption Spectra and Geometry of Molecules* (Academic Press Inc., New York, 1967).

[2] J. C. Slater, *Phys. Rev.* **37,** 481 (1931); **38,** 325 (1931); **38,** 1109 (1931).

[3] R. S. Mulliken, *Phys. Rev.* **41,** 67 (1932).

[4] L. C. Pauling, *J. Am. Chem. Soc.* **53,** 1367 (1937).

[5] H. Margenau and G. M. Murphy, *The Mathematics of Physics and Chemistry* (D. van Nostrand Co., New York, 1961), 2nd ed., Chap. 5.

[6] H. Preuss, *Integraltafeln zur Quantenchemie* (Springer-Verlag, Berlin, 1956).

[7] R. S. Mulliken, C. A. Rieke, D. Orloff, and H. Orloff, *J. Chem. Phys.* **17,** 1248 (1949).

[8] A. Lofthus, *Mol. Phys.* **5,** 105 (1962).

[9] J. Miller, J. M. Gerhauser, and F. A. Matsen, *Quantum Chemistry Integrals and Tables* (University of Texas Press, Austin, Tex. 1959).

[10] M. Kotani, A. Amemiya, and S. Simose, *Proc. Phys. Math. Soc. Japan,* **20,** 1 (1938); **22,** 1 (1940).

[11] N. Rosen, *Phys. Rev.* **38,** 255 (1931).

[12] D. A. Brown, *J. Chem. Phys.* **29,** 1086 (1958).

[13] L. Leifer, F. A. Cotton, and J. R. Leto, *J. Chem. Phys.* **28,** 364 (1958); **28,** 1253 (1958).

[14] E. Clementi and D. Raimondi, *J. Chem. Phys.* **38,** 2686 (1963).

[15] G. Burns, *J. Chem. Phys.* **41,** 1521 (1964).

[16] L. C. Cusachs, D. G. Carroll, B. L. Trus, and S. P. McGlynn, *Intl. J. Quantum Chem.* **1,** 423 (1967).

[17] L. C. Cusachs and J. H. Corrington, in *Sigma Molecular Orbital Theory,* [edited by O. Sinanoğlu and K. Wiberg] (Yale University Press, New Haven, Conn., 1969).

[18] F. A. Cotton and C. B. Harris, *Inorg. Chem.* **6,** 369 (1967).

[19] A. T. Armstrong, D. G. Carroll, and S. P. McGlynn, *J. Chem. Phys.* **47,** 1104 (1967).

[20] C. J. Ballhausen and H. B. Gray, *Molecular Orbital Theory* (W. H. Benjamin, Inc., New York, 1965).

[21] W. A. Yeranos, *Inorg. Chem.* **5,** 2070 (1966).

[22] J. C. Slater, *Quantum Theory of Molecules and Solids* (McGraw-Hill Book Co., New York, 1965), Vol. 2.

[23] S. F. A. Kettle, *J. Chem. Ed.* **43,** 21 (1966); *Inorg. Chem.* **4,** 1821 (1965).

General References

1. C. J. Ballhausen and H. B. Gray, *Molecular Orbital Theory* (W. H. Benjamin, Inc., New York, 1965).

2. R. S. Mulliken, C. A. Rieke, D. Orloff, and H. Orloff, *J. Chem. Phys.* **17,** 1248 (1949).

3. A. Lofthus, *Mol. Phys.* **5,** 105 (1962).

CHAPTER 3

Hückel Molecular-Orbital Method

The Hückel molecular-orbital method (or HMO method) is probably the most widely used of all quantum-chemical computational schemes.[1] It is based on a series of assumptions, of which many are arbitrary. It is patterned on quantum mechanics, but it is really no more than a useful empirical scheme which rationalizes much chemical experience concerning π-electron molecules.

The HMO method consists of the following:

(i)   Molecular orbitals are to be formed. These are assumed to be eigenfunctions of an effective one-electron Hamiltonian $\hat{H}^h$, appropriate to the equilibrium nuclear configuration of the molecule under consideration. This eigenequation may be written

$$\hat{H}^h \varphi_r = \epsilon_r \varphi_r \tag{3.1}$$

where φ_r is the rth MO and ϵ_r is the corresponding MO energy.

(ii)   The molecular orbitals are assumed to be given as linear combinations of atomic orbitals[1]—usually one AO from each of the various

---

[1] This is usually denoted the *LCAO approximation*.

atomic centers encompassed by the MO

$$\varphi_r = \sum_\mu c_{r\mu} \chi_\mu \tag{3.2}$$

(iii) The Variation Theorem[2] is used to deduce the coefficients $c_{r\mu}$ which yield the optimal (i.e., lowest energy) set of MO's; this ultimately yields a secular determinant

$$| H_{\mu\nu} - \epsilon S_{\mu\nu} | = 0 \tag{3.3}$$

where $H_{\mu\nu} \equiv \langle \chi_\mu | \hat{H}^h | \chi_\nu \rangle$ and $S_{\mu\nu} \equiv \langle \chi_\mu | \chi_\nu \rangle$.

(iv) The introduction of certain empirical rules simplify the evaluation of the various elements in the secular determinant and, at the same time, effectively limit the HMO method to considerations of planar unsaturated compounds (i.e., to π-electron systems).

1. BASIS OF THE METHOD

(A) Effective Hamiltonian

We anticipate the restriction of item (iv) above to π electrons. We assume that the interaction between the π and the σ electrons is small and that the π electrons can be treated as a separate loosely held group of electrons. All nuclei and σ electrons are thought of as a *core* which merely provides a potential field in which the π electrons move. Thus, the Hamiltonian for the π electrons is written as

$$\hat{H}^\pi = \sum_i (\hat{T}_i + \hat{V}_i{}^{\text{core}}) + \sum\sum_{i<j} e^2/r_{ij} \tag{3.4}$$

where i and j are π-electron indices; $\hat{T}_i$ is the kinetic-energy operator for electron i; $\hat{V}_i{}^{\text{core}}$ is the potential energy of electron i in the field of the core; and e^2/r_{ij} is the repulsion of electrons i and j. We now rewrite Eq. (3.4) as

$$\hat{H}^\pi = \sum_i (\hat{H}_i{}^{\text{core}} + \hat{V}_i{}^\pi) \tag{3.5}$$

where $\hat{V}_i{}^\pi$, the effective potential acting on electron i due to all other π electrons, is defined so that

$$\sum_i \hat{V}_i{}^\pi = \sum\sum_{i<j} e^2/r_{ij} \tag{3.6}$$

Thus, granted the feasibility of Eq. (3.5), we have

$$\hat{H}^\pi = \sum_i \hat{H}_i{}^h \tag{3.7}$$

where

$$\hat{H}_i{}^h = \hat{H}_i{}^{\text{core}} + \hat{V}_i{}^\pi \tag{3.8}$$

In the specific case of ethylene, for example, we can illustrate the above by writing

$$\hat{H}^\pi = \hat{T}_1 + \hat{T}_2 + \hat{V}_1{}^{core} + \hat{V}_2{}^{core} + e^2/r_{12}$$

$$= \hat{H}_1{}^h + \hat{H}_2{}^h \tag{3.9}$$

where

$$\hat{H}_1{}^h = \hat{T}_1 + \hat{V}_1{}^{core} + \hat{V}_1{}^\pi \tag{3.10}$$

The crux of the whole matter rests on the assumed separability of π and σ electrons and the feasibility of obtaining a $\hat{V}_i{}^\pi$ such that Eq. (3.6) be satisfied.

(B) LCAO Approximation

The LCAO approximation is chosen because of its mathematical simplicity and the ease with which it allows molecular considerations to be referred back to atomic origins.

We justify the LCAO approximation in the following primitive manner. Consider $H_2{}^+$, a molecule ion containing a one-electron bond. The wave function φ of the electron is a solution of the equation

$$\hat{T}\varphi - \left(\frac{e^2}{r_A} + \frac{e^2}{r_B}\right)\varphi = \epsilon\varphi \tag{3.11}$$

where r_A and r_B are distances of the electron from protons A and B, respectively. If the electron be close to A, then

$$\frac{e^2}{r_A} \gg \frac{e^2}{r_B} \tag{3.12}$$

and

$$\hat{T}\varphi - (e^2/r_A)\varphi \simeq \epsilon\varphi \tag{3.13}$$

Since Eq. (3.13) is a hydrogenic Schrödinger equation, the solutions are the hydrogen AO's, χ_A, on center A. Similarly, if the electron be close to B, the hydrogen AO's, χ_B, on center B are good approximate solutions. Thus, a reasonable first assumption concerning the form of φ is that

$$\varphi = c_A\chi_A + c_B\chi_B \tag{3.14}$$

where c_A and c_B are to be determined. The dihedral cylindrical $(D_{\infty h})$ geometry of $H_2{}^+$ imposes certain symmetry restrictions on the relative values of c_A and c_B (namely, $c_A = \pm c_B$).

In the general run of molecule, we may assume an MO of the form of Eq. (3.2). Symmetry may be of some use in generating the coefficients $c_{r\mu}$, but it is not usually sufficient. Consequently, a more general approach is required.

(C) Secular Equation

We rewrite Eqs. (3.1) and (3.2) as

$$\varphi = \sum_{\nu} c_{\nu} \chi_{\nu} \qquad (3.15)$$

$$\hat{H}^{h} \varphi = \epsilon \varphi \qquad (3.16)$$

The functions φ (or, equivalently, the coefficients c_{ν}) and ϵ are to be determined. The usual procedure is to determine that set of coefficients which leads to the minimum one-electron energy ϵ; this set of coefficients is guaranteed by the Variation Theorem[2] to be the best possible set. The usual mode of application of the variation technique[2] leads to the secular determinant of Eq. (3.3) which possesses a number of solutions equal to the dimension of the determinant. Only the lowest-energy solution meets the requirements of the Variation Theorem. It may be shown, however, that the other solutions are approximations to the excited states of the system. Thus, in a sense, the secular routine generates the whole manifold of *best one-electron energies*. However, the secular determinant of Eq. (3.3) is obtainable by a simple projection technique without any recourse whatever to the Variation Theorem. We now reproduce this latter derivation.

Combination of Eqs. (3.15) and (3.16) yields

$$\sum_{\nu} c_{\nu} (\hat{H}^{h} - \epsilon) \chi_{\nu} = 0 \qquad (3.17)$$

We now multiply Eq. (3.17) on the left-hand side by one of the AO's, say, χ_{μ} (or χ_{μ}^{*} if we are dealing with complex functions), and integrate to find

$$\sum_{\nu} c_{\nu} \left(\int \chi_{\mu} \hat{H}^{h} \chi_{\nu} d\tau - \epsilon \int \chi_{\mu} \chi_{\nu} d\tau \right) = 0 \qquad (3.18a)$$

or, equivalently,

$$\sum_{\nu} c_{\nu} (H_{\mu\nu} - \epsilon S_{\mu\nu}) = 0 \qquad (3.18b)$$

There is one such equation for every χ_{μ} which may be used as a left-hand multiplier [i.e., one for every χ_{μ} in Eq. (3.15)]. Thus, if there be N AO's in the LCAO MO of Eq. (3.15), we obtain N equations of the form of Eq. (3.18) containing the N unknown coefficients c_{ν}. These equations only have a nontrivial solution[3] for the c_{ν} if the $N \times N$ secular determinant is identically zero [i.e., if Eq. (3.3) applies]. Expansion of the secular determinant yields a polynomial of the Nth degree in ϵ. The roots of this poly-

---

[2] See Exercise 1.
[3] The trivial solution is: Every coefficient $c_{\nu} = 0$.

nomial are the energies of the N linearly independent MO's which can be formed as linear combinations of N AO's.

Suppose that ϵ_r is one specific solution. To find $\varphi_r = \sum_\mu c_{r\mu}\chi_\mu$ with energy ϵ_r, we substitute ϵ_r back into the secular equations and solve to obtain the ratios of the coefficients. The absolute values of the coefficients are obtained by use of the normalization condition

$$\int \varphi_r{}^*\varphi_r d\tau = 1 \tag{3.19}$$

(D) Simplification of Secular Determinant

A number of approximations which simplify the secular determinant are now made. These, as stated for a system of aromatic carbon atoms, are as follows:

(i) Zero differential overlap is assumed, even between neighbors. This, stated mathematically, is

$$\langle \chi_\mu \mid \chi_\nu \rangle \equiv S_{\mu\nu} = \delta_{\mu,\nu} \tag{3.20}$$

where δ is the Kronecker δ.

(ii) The diagonal elements $H_{\mu\mu}$, termed *Coulomb integrals*, are assumed to be the same for all carbon atoms in the ring. This, stated mathematically, is

$$H_{\mu\mu} \equiv \langle \chi_\mu \mid \hat{H}^h \mid \chi_\mu \rangle = \alpha \tag{3.21}$$

(iii) The off-diagonal elements $H_{\mu\nu}$ termed *resonance integrals*, are set equal to β

$$H_{\mu\nu} \equiv \langle \chi_\mu \mid \hat{H}^h \mid \chi_\nu \rangle = \beta \tag{3.22}$$

if atoms μ and ν are nearest neighbors.

(iv) The off-diagonal elements are set equal to zero

$$H_{\mu\nu} = 0 \tag{3.23}$$

if atoms μ and ν are not nearest neighbors.

The elements α and β may not be computed in any exact way because the Hamiltonian $\hat{H}^h$ is not well defined. However, values appropriate to α and β may be guessed or approximated by suitable means.

2. EXAMPLE CALCULATIONS

(A) Allyl Radical

The allyl radical is shown in Fig. 3.1. This molecule contains three π electrons and the MO's encompass three carbon centers. The secular determinant of Eq. (3.3) is

$$
\begin{vmatrix}
H_{11} - \epsilon S_{11} & H_{12} - \epsilon S_{12} & H_{13} - \epsilon S_{13} \\
H_{21} - \epsilon S_{21} & H_{22} - \epsilon S_{22} & H_{23} - \epsilon S_{23} \\
H_{31} - \epsilon S_{31} & H_{32} - \epsilon S_{32} & H_{33} - \epsilon S_{33}
\end{vmatrix} = 0
\qquad (3.24)
$$

Introduction of the Hückel approximations of Eqs. (3.20)–(3.23), leads to the simplified determinant

$$
\begin{vmatrix}
\alpha - \epsilon & \beta & 0 \\
\beta & \alpha - \epsilon & \beta \\
0 & \beta & \alpha - \epsilon
\end{vmatrix} = 0
\qquad (3.25)
$$

If we now divide each determinantal element by β and set $(\alpha - \epsilon)/\beta = x$, we can rewrite Eq. (3.25) as

$$
\begin{vmatrix}
x & 1 & 0 \\
1 & x & 1 \\
0 & 1 & x
\end{vmatrix} = 0
\qquad (3.26)
$$

At this point, it is appropriate to note that the Hückel determinant, in the form of Eq. (3.26), can be written for any molecule by simple inspection of that molecule. The prescription is: *For any planar system of unsaturated carbon atoms, take a determinant of order equal to the number of atoms, write x along the diagonal, write 1 for off-diagonal elements between atoms which are nearest neighbors and put a zero everywhere else.*

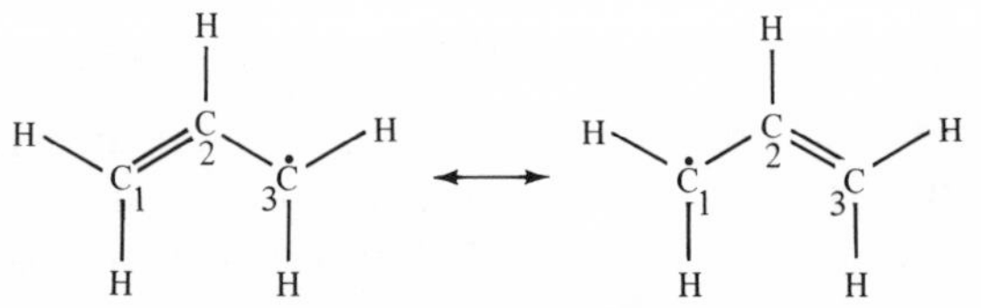

Allyl radical

FIG. 3.1.   Two resonance forms of allyl radical. The two C—C bonds are of identical length in this molecule. Atom positions are numbered 1, 2, 3; these same indices serve to identify the loci of the three $2p\pi$ AO's. [The two standard valence bond formulas are reproduced here only because we need these in Sec. 3(B) when we discuss MO delocalization energies.]

(B)   MO Energies of Allyl Radical

Expansion of Eq. (3.26) yields the polynomial

$$x^3 - 2x = 0 \qquad (3.27)$$

the roots of which are $x = 0, \pm 1.414$. The corresponding energies are

$$\epsilon_3 = \alpha - 1.414\beta; \qquad \text{(from } x_3 = +1.414)$$

$$\epsilon_2 = \alpha; \qquad \text{(from } x_2 = 0)$$

$$\epsilon_1 = \alpha + 1.414\beta; \qquad \text{(from } x_1 = -1.414) \qquad (3.28)$$

Indeed, the general form of the Hückel MO energy is

$$\epsilon_r = \alpha + w_r\beta; \qquad \text{(from } x_r = -w_r) \qquad (3.29)$$

Since α and β are both negative quantities,[4] it follows that φ_1 is the MO of lowest energy.

All MO energies are of the form $\epsilon_r = \alpha + w_r\beta$. Thus, all MO energy differences are specifiable in a natural unit of β which must be obtained from comparison with experiment. When $w_r = 0$, φ_r is said to be *nonbonding*; when $w_r > 0$ the MO φ_r is said to be *bonding*; when $w_r < 0$ the molecular orbital φ_r is said to be *antibonding*.

(C)   MO's of Allyl Radical

To obtain the coefficients, we substitute, one by one, the values of x obtained above into the secular equation. Thus, substituting x_r into the secular equations, we find

$$\begin{aligned} c_{r1}x_r + c_{r2} \qquad\qquad &= 0 \\ c_{r1} \quad + c_{r2}x_r + c_{r3} &= 0 \\ c_{r2} \quad + c_{r3}x_r &= 0 \end{aligned} \qquad (3.30)$$

If we take the simplest case, namely, $x_r = x_2$ we find upon solution

$$c_{22} = 0; \qquad c_{21} + c_{23} = 0 \qquad (3.31)$$

The normalization condition of Eq. (3.19) requires that

$$c_{21}^2 + c_{22}^2 + c_{23}^2 = 1 \qquad (3.32)$$

where

$$c_{21} = (2)^{-1/2}; \qquad c_{23} = -(2)^{-1/2}; \qquad c_{22} = 0 \qquad (3.33)$$

Repetition of this procedure for x_1 and x_3 yields the full set of MO's shown in Table 3.1. These MO's are illustrated schematically in Fig. 3.2.

[4] See Sec. 4.

TABLE 3.1.   *Hückel MO's for the allyl radical.*[a]

	MO		
AO	φ_1	φ_2	φ_3
χ_1	0.500	0.707	0.500
χ_2	0.707	0.0	−0.707
χ_3	0.500	−0.707	0.500

[a] This table is read, for example, as

$$\varphi_3 = c_{31}\chi_1 + c_{32}\chi_2 + c_{33}\chi_3$$
$$= 0.500\chi_1 - 0.707\chi_2 + 0.500\chi_3$$

For a molecule of large size, the polynomial to be solved becomes quite large and the determination of the coefficients $c_{r\mu}$ may involve a great deal of tedious mathematical manipulation. At this stage, it is usual to introduce group theory and to use it to block the secular determinant for large symmetric molecules into a set of smaller determinants which are more readily handled. Such chemical applications of group theory are discussed in several texts.[3]−[5] However, the availability of efficient computer programs[6] renders the use of group theory in Hückel problems almost unnecessary.

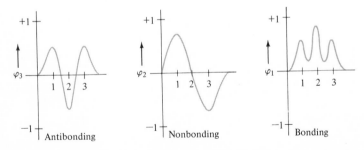

FIG. 3.2.   Schematic MO's of the allyl radical. The numbers 1, 2, and 3 along the abscissa indicate atom positions defined in Fig. 3.1. These MO's are drawn as follows: In the schematization of φ_r, mark the values of c_{r1}, c_{r2} and c_{r3} along the vertical lines that go through the points 1, 2, and 3, respectively; connect these points by a smooth curve and assume that the amplitude is zero one-half bond length to the left of C_1 and to the right of C_3.

(D) Trimethylene Methane

The central carbon of trimethylene methane is equally π bonded to three others. Carbon 4 is a maximally π-bonded carbon, hence its use as a standard in the definition of the free valence index [see Sec. 3(F)]. The three dominant resonance forms of trimethylene methane are illustrated in Fig. 3.3.

The Hückel determinant is

$$\begin{vmatrix} x & 0 & 0 & 1 \\ 0 & x & 0 & 1 \\ 0 & 0 & x & 1 \\ 1 & 1 & 1 & x \end{vmatrix} = 0 \qquad (3.34)$$

Expansion, followed by solution of the polynomial, yields the four eigenvalues

$$\epsilon_4 = \alpha - (3)^{1/2}\beta; \qquad [\text{from } x_4 = (3)^{1/2}]$$

$$\epsilon_3 = \alpha; \qquad (\text{from } x_3 = 0)$$

$$\epsilon_2 = \alpha; \qquad (\text{from } x_2 = 0)$$

$$\epsilon_1 = \alpha + (3)^{1/2}\beta; \qquad [\text{from } x_1 = -(3)^{1/2}] \qquad (3.35)$$

3. ILLUSTRATIVE USES OF HMO RESULTS

(A) Configuration Energy

The energy of a π-electron-containing molecule is the sum of the core energy and the π-electron energies. The core contribution is unknown, but it is assumed to be constant and independent of the π energy for any given

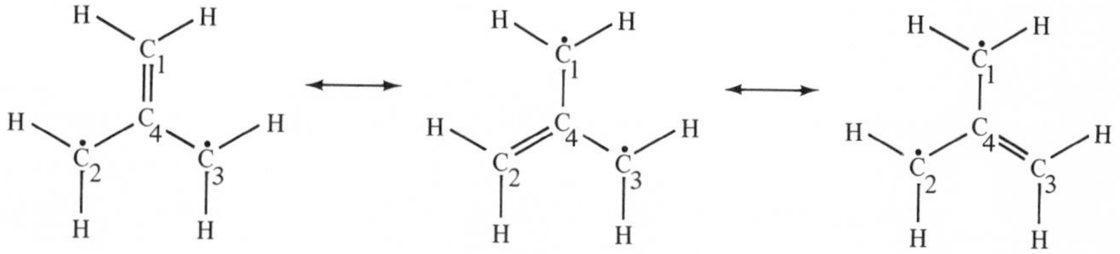

Trimethylene methane

FIG. 3.3. The trimethylene methane molecule.

molecule. Consequently, we can define a π-electron energy as

$$E_\pi = \sum_r n_r \epsilon_r = \sum_r n_r(\alpha + w_r\beta) \qquad (3.36)$$

where n_r is the number of electrons in the rth MO for the electron configuration in question.

The lowest-energy π-electron configuration of allyl radical is depicted in Fig. 3.4(a). The corresponding π-electron energy is given by

$$E_\pi = 2(\alpha + 1.414\beta) + 1(\alpha) = 3\alpha + 2.828\beta \qquad (3.37)$$

A number of excited configurations of allyl radical are also depicted in Figs. 3.4(b)–3.4(d).

(B)   Delocalization Energy

It is assumed that the lowest-energy configuration is fully representative of the ground state of the molecule. The energy difference between the ground state so defined and that calculated on the assumption that each pair of π-bonded carbons in the structural formula is ethylenelike, is known as the π-delocalization energy. Now the ground state of ethylene[5] has a π-electron energy of $2\alpha + 2\beta$, whereas a non-π-bonded carbon $2p\pi$ AO has an energy of α. Thus, the delocalization energy is readily evaluated.

The allyl radical affords a good example. Each structural formula of Fig. 3.1 contains one ethylenelike bond and one electron in a carbon $2p\pi$ AO which is not π bonded. Hence, the energy associated with either of the structural formulas is (see Fig. 3.1 and Exercise 5)

$$2(\alpha + \beta) + \alpha$$

whereas the ground-state π-electron energy of the same molecule is [see Eq. (3.37)]

$$3\alpha + 2.828\beta$$

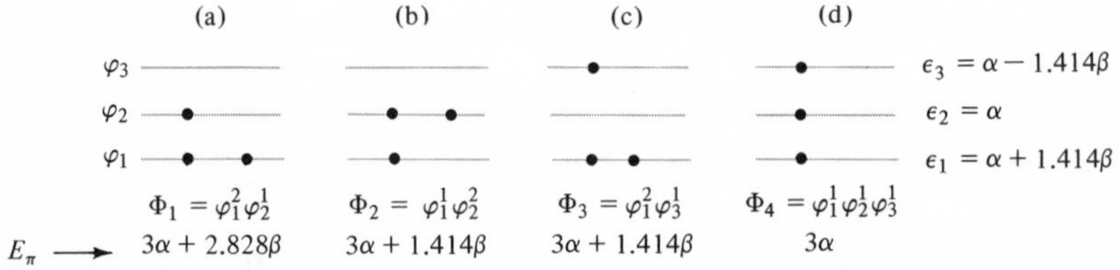

FIG. 3.4.   Various electron configurations for allyl radical. Note that Φ_2 and Φ_3 are degenerate. The total π-electron energy is denoted E_π [see Eq. (3.36)].

[5] See Exercise 3.

The π-electron delocalization energy is therefore

$$3\alpha + 2.828\beta - (2\alpha + 2\beta) - \alpha = 0.828\beta \qquad (3.38)$$

(C) Excitation Energies

It is assumed that the configurations Φ_2, Φ_3, Φ_4 of Fig. 3.4 fully represent excited states of the allyl radical. Thus, the spectrum of the allyl radical should exhibit absorption bands corresponding to the π-electron transitions

$$\Phi_2 \leftarrow \Phi_1; \qquad \Delta E = -1.414\beta;$$

$$\Phi_3 \leftarrow \Phi_1; \qquad \Delta E = -1.414\beta;$$

and

$$\Phi_4 \leftarrow \Phi_1; \qquad \Delta E = -2.828\beta \qquad (3.39)$$

(D) π-Electron Density

The π-electron density resident on the μth carbon center is denoted q_μ; it is given in atomic units of charge by

$$q_\mu = \sum_r c_{r\mu}^2 n_r \qquad (3.40)$$

where n_r is the number of electrons in the rth MO and the coefficients $c_{r\mu}$ are assumed to be real. The π-electron density obtained in this way for the ground state of allyl radical is given in Fig. 3.5. For example, it is readily seen that

$$q_1 = 2(0.500)^2 + 1(0.707)^2 = 1 \qquad (3.41)$$

(E) Mobile Bond Order

The mobile bond order is defined as

$$P_{\mu\nu} = \sum_r n_r c_{r\mu} c_{r\nu} \qquad (3.42)$$

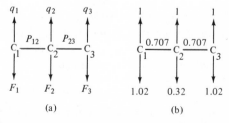

FIG. 3.5. Hückel diagram specifying the values of π-electron density (q_μ), of mobile bond order $(P_{\mu\nu})$ and of free valence index (F_μ). (a) The usual positions in which the various quantities are entered, (b) their values for allyl radical.

where μ and ν must be a directly bonded pair of carbon centers. For the allyl radical, for example, we find

$$P_{12} = 2(0.500)(0.707) + 1(0.707)(0)$$

$$P_{23} = 0.707 \tag{3.43}$$

Empirical π-bond order—π-bond length curves are available.[7] By reference of the computed values of $P_{12} = P_{23} = 0.707$ to such a curve, we estimate the two C—C bonds of allyl radical to be equal and 1.38 Å in length. The experimental value is 1.39 Å.

(F) Free Valence Index

The free valuence index is defined as

$$-F_\mu = \sum_\nu P_{\mu\nu} - \sum_\nu P_{4\nu} \text{ (trimethylene methane)} \tag{3.44}$$

where the summation over ν counts only those carbon centers directly bonded to center μ. Thus, for trimethylene methane, $\nu = 1, 2, 3$ from inspection of Fig. 3.3. The value of $\sum_\nu P_{4\nu}$ for trimethylene methane is $(3)^{1/2} = 1.73$; since C_4 of this molecule is a maximally bonded carbon center, the value of 1.73 can be considered a *maximum* π valence. Thus,

$$F_\mu = 1.73 - \sum_\nu P_{\mu\nu} \tag{3.45}$$

represents the amount of π valence unused on center μ of the molecule of interest. This index is of use in considerations of chemical reactivity. The values of F_μ for allyl radical ground state are given in Fig. 3.5.

(G) π-Dipole Moment[6]

For compounds containing π-bond networks it is convenient to dissect the total dipole moment into a σ moment and a π moment. The σ moment is estimated from analogous systems which do not contain π bonds. The π moment is taken to be the difference between the experimental total dipole moment and the estimated σ moment.

The π-electron density at each center μ is given by q_μ. Since each atom in an all-carbon system contributes an amount of nuclear charge $+1$ to the core, the net charge on any center is (in atomic units)

$$\eta_\mu = 1 - q_\mu \tag{3.46}$$

The π moment is readily computed from the η's and the known molecule geometry. For example, in allyl radical $\eta_1 = \eta_2 = \eta_3 = 0$ and the dipole moment is zero. On the other hand, in methylenecyclopropene (see Fig.

---

[6] A more sophisticated discussion of dipole moments is given in Chapter 10.

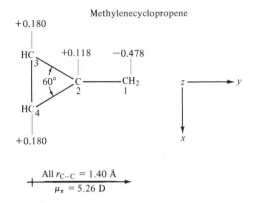

FIG. 3.6. Geometry, axes, and net charges, η_μ, for methylenecyclopropene.

3.6), while there is clearly no net moment in the x or z directions, there is one in the y direction. This moment is most readily evaluated if atom 2 is chosen as origin; it is given by

$$\mu_\pi = [2(0.180)(1.40)(\cos 30°) + (0.478)(1.40)](4.80) = 5.26 \text{ D} \quad (3.47)$$

The factor 4.80 is the electronic charge in units appropriate to a final numeric result in Debye units (D).

4. PARAMETRIZATION

(A) Cyclic Polyenes

The Hückel theory works very well if it is used for correlative purposes. For example, the experimental energy of the longest-wavelength absorption band of aromatic hydrocarbons is determinable to ± 1500 cm^{-1} using the equation

$$\Delta E = (\Delta w)\beta; \quad \beta = -2.71 \text{ eV or } -21\,900 \text{ cm}^{-1} \quad (3.48)$$

Similar good correspondence exists between prediction and the polarographic half-wave reduction potentials of aromatic hydrocarbons; in this instance, the best value of β is found to be

$$\beta = -2.37 \text{ eV} \quad (3.49)$$

Equally good predictive capacity exists for aromatic π-delocalization energies. The value of $|\beta|$ to be used here is found to be exceptionally small, namely,

$$\beta = -0.69 \text{ eV or } -16 \text{ kcal/mole} \quad (3.50)$$

This great variance in the value of β is not alarming. It results from the fact that the ill-defined Hückel Hamiltonian contains no explicit two-electron terms. Such terms enter in different ways and in different amounts into the calculation of different molecular properties—and one merely compensates for their neglect by using different values of the resonance integral.

(B)   Linear Polyenes

The polyenes differ from aromatic hydrocarbons in that they exhibit distinct bond alternation. In butadiene, for example, the 1, 2 and 3, 4 C—C bonds of Fig. 3.7(b) are considerably shorter than the 2, 3 bond. It follows that one should use different resonance integrals for these various bonds. The current approximation presumes a linear relation between the value of $|\beta|$ and the $2p\pi$-$2p\pi$ overlap integral for a given bond, namely,[7]

$$\beta_{MN} = kS_{MN}\beta_{CC} \qquad (3.51)$$

where k is assumed to be the same constant for all $2p\pi$-$2p\pi$ bonds between centers M and N, where S_{MN} is the overlap integral $S(2p\pi_M, 2p\pi_N)$ and where β_{CC} is the resonance integral appropriate to benzene. Thus, since $S_{CC} \simeq 0.25$, we find

$$\beta_{MN} = (S_{MN}/S_{CC})\beta_{CC} \simeq 4S_{MN}\beta_{CC} \qquad (3.52)$$

Application of Eq. (3.52) to the case of butadiene generates quite different β's for the two types of C—C bond present in this molecule.

(C)   Non-planar Conjugated Carbon Systems

The prototype system is biphenyl. As one phenyl group is rotated with respect to the other about the long molecular axis, the extent of conjugation between the rings must alter. That resonance integral which is most crucially affected by twisting pertains to the inter-ring C—C bond. Since the primary effect of the twisting is a simple rotation of one of the inter-ring carbon AO's with respect to the other, it follows that Eq. (3.52) reduces to

$$\beta_{CC}(\theta) = 4S_{CC}\beta_{CC}\cos\theta = \beta_{CC}\cos\theta \qquad (3.53)$$

where θ is the angle between the two benzenoid planes.

(D)   Effects of π-Electron Density on Coulomb Integral

In benzene, allyl radical, ethylene, etc., it is found, for all μ that $q_\mu = 1$. It would seem that an identical value of the Coulomb integral α is appro-

---

[7] In HMO theory there is only one $2p\pi$ AO on each center. Hence, the same notation, say, χ_μ or μth carbon, may be used to denote both the AO and the atom. We have made use of this circumstance so far in this chapter. At this point and hereafter we occasionally depart this notation; it is not adequate when the nuclear centers differ from one another.

priate for all carbons in such systems. We now ask the question: What dependence, if any, does α_μ exhibit on the value of q_μ? A case where this question is relevant is the allyl cation where $q_1 = q_3 = 0.500$ and $q_2 = 1.00$.

When q is large, Z is small and the $2p\pi$ AO in question is less tightly bound than when q is small. In other words, since α is a measure of $2p\pi$ AO energy, it follows that α is less negative when q is large. Following this reasoning, Wheland and Mann[8],[9] suggested the equation

$$\alpha_\mu = \alpha_C + (1 - q_\mu)\omega\beta_{CC}$$

$$\alpha_\mu = \alpha_C + \eta_\mu\omega\beta_{CC} \qquad (3.54)$$

where ω is a dimensionless parameter chosen to agree with experiment. The *best value*, according to Streitwieser,[9] is $\omega = 1.4$.

Because of the dependence of α_μ on q_μ, these quantities should be made self-consistent. Thus, the use of Eq. (3.54) implies an iterative procedure— which procedure is known as the ω technique. This technique is merely a simple case of attainment of charge self-consistency and is discussed further in Chap. 4. The procedure is as follows: One calculates q_μ's by simple HMO procedures; one now evaluates a new set of α_μ's from Eq. (3.54) and uses these in a new Hückel routine which eventually results in the production of a revised set of q_μ's and α_μ's. One continues the cycling until the values of α_μ's at the beginning and end of a given calculational cycle are identical within some prescribed limits.[8]

(E) Unsaturated Systems Containing Hetero-Atoms

Considerable care must be exercised in parametrizing hetero-atom systems. For example, the nitrogen atom behaves quite differently in certain instances: In pyridine, it makes a contribution of 1 electron to the π system; in pyrolle, it effectively donates 2π electrons; in pyridinium ion, while it donates only one π electron, it also possesses approximately one unit of σ charge. These effects must be taken into account in evaluating the Coulomb integral; indeed, the three types of nitrogen center which we have described are best thought of as three different *hetero-atoms*.

The great majority of hetero-atoms can be classified into two types; these are denoted Ṁ and M̈ where the number of *dots* indicates the number of electrons contributed to the π system. Examples are:

Ṁ: N in pyridine; N of C=N and N=N bonds; O of C=O bonds; S of C=S bonds; P of P=O bonds, etc.

M̈: Atoms with lone-pair electrons which are bound by single σ-bonds to adjacent atoms in the structural formula. O of furan; N of aniline; S of

---

[8] See Exercise 8.

thiophene; O of phenol; S of thiophenol; F, Cl, Br, I directly linked to unsaturated systems.

The resonance integral of an N—M π bond (where N, M, or both need not be carbon centers) is given by Eqs. (3.51) and (3.52).

The Coulomb integral of a $2p\pi$ AO on center M may be assumed to be given by a variant of Eq. (3.54), namely,

$$\alpha_M = \alpha_C + h_M \beta_{CC} \tag{3.55}$$

The constant h_M may be considered to be wholly empirical—in which case, its value is obtained by fitting HMO calculations to experiment. On the

TABLE 3.2.   *Abstraction of h from Pauling electronegativities and Slater exponents.*[a]

Atom	Z (Slater)	EN	EN − ENC	h
B (Ḃ)	2.60	2.0	−0.5	−0.5
C (Ċ)	3.25	2.5	0	0
N (Ṅ)	3.90	3.0	+0.5	+0.5[b]
N$^{\delta+}$(N̈)	...	...	...	(0.63)[c]
N$^+$ (Ṅ$^+$)	4.25	(3.27)	(+0.77)	(+0.77)[d]
O (Ȯ)	4.55	3.5	1.0	1.0
O$^{\delta+}$(Ö)	...	...	...	(+1.13)
O$^+$ (Ȯ$^+$)	4.90	(3.77)	(+1.27)	(+1.27)
F (F̈)	5.20	4.0	+1.5	+1.5

[a] From a collection by Suzuki (Ref. [12]).

[b] Use for nitrogen atom of pyridine, denoted N or N. This atom donates one π electron to the π system.

[c] Use for nitrogen atom of pyrolle, denoted N$^{\delta+}$ or N. This atom donates two π electrons to the π system. It is assumed that the charge on this center is somewhat intermediate between that on nitrogen of pyridine and that on nitrogen of pyridinium ion. Hence, the value of h quoted is $(0.5+0.77)/2$.

[d] Use for nitrogen atom of pyridinium ion, denoted N$^+$ or N$^+$. This atom donates one π electron to the π system. No value of EN is known for N$^+$. However, the value of $EN = 3.27$ may be interpolated with some assurance from the value of Z_{N^+}. The reader may care to verify that there is a good parallelism between Z and EN, and that such an interpolation is justified.

TABLE 3.3. *A compilation of hetero-atom Hückel parameters.*[a]

Atomic Center	S^c	h P^d	Y^e	S	$k^{\bullet}_{C-X}$ [b] P	Y
$\overset{\bullet}{N}$ (pyridine)	0.5	0.4	0.6	1	1	1
$\overset{..}{N}$ (pyrolle)	1.5	1	1	0.8	0.9	1
$\overset{\bullet}{N}{}^{+}$ (pyridinium)	2	2	...	...	1	...
$\overset{\bullet}{O}$	1	1.2	2	1	2	$(2)^{1/2}$
$\overset{..}{O}$	2	2	2	0.8	0.9	0.6
$\overset{\bullet}{S}$	...	0	0.9	...	1.2	1.2
$\overset{..}{S}$	...	0	0.9	...	0.6	0.5
$\overset{..}{F}$	3	...	2.1	0.7	...	1.25
$\overset{..}{Cl}$	2	...	1.8	0.4	...	0.8
$\overset{..}{Br}$	1.5	...	1.4	0.3	...	0.7
$\overset{..}{I}$	...	...	1.2	...	...	0.6
$\overset{..}{N}H_2$	...	...	0.4	...	...	0.6
$\overset{..}{O}H$	...	...	0.6	...	...	0.7
$\overset{..}{O}CH_3$	...	...	0.5	...	...	0.6
$\overset{..}{S}H$	...	...	0.55	...	...	0.6

[a] From a collection by Suzuki (Ref. [12]).

[b] Equations (3.51) and (3.52) are written in the completely empirical form

$$\beta^{\bullet}_{C-X} = k^{\bullet}_{C-X}\beta^{\bullet}_{C-C} \qquad (3.51a); \ (3.52a)$$

for the purposes of this tabulation.

[c] A. Streitwieser, Jr., Ref. [1].

[d] B. Pullman and A. Pullman, *Quantum Biochemistry* (John Wiley & Sons, Inc., New York, 1963), Chap. III, Sec. V.

[e] T. Yonezawa, C. Nagata, H. Kato, A. Imamura, and K. Morakuma, *Ryoshikagaku Nyumon* (Kagakudojin, Kyoto, Japan, 1963), Chap. 2.

other hand, it does appear reasonable that the Coulomb integral for the $2p\pi$ AO on M should equal the AO energy or, at least, that there should be some close parallel between them. Following this trend of thought, it is fairly commonly presumed[10] that

$$h_{\text{M}} = EN_{\text{M}} - EN_{\text{C}} \tag{3.56}$$

where the EN's are Pauling electronegativities.[11] Unfortunately, electronegativities are not available for all species M; hence, other ancillary relations, such as a presumed correlation between effective Slater charge Z and the Coulomb parameter h_{M}, are frequently used.[12] A compilation of h's, obtained as above, is given in Table 3.2.

Coulomb integrals are also related to valence orbital ionization potentials[9] (VOIE's)—a result which is neither surprising nor contradictory of the presumed relationship of Eq. (3.56) to Pauling electronegativity. We note here that

$$EN^M = (IP + EA)/2 \simeq IP/2 \tag{3.57}$$

where EN^M is the Mulliken electronegativity,[10] where IP is ionization potential and EA is electron affinity. Since IP is usually quite larger than EA, it is not too disastrous to neglect EA altogether—hence, Eq. (3.57). If we also note that $IP \simeq -\text{VOIE}$, the use of electronegativity data and/or VOIE data to obtain values of α_{M} seems appropriate. In addition, VOIE data enable account to be taken of hybridization, a topic which is not accessible in the electronegativity approach. However, we postpone further consideration of this topic (i.e., VOIE's) until Chap. 4, where it is discussed in some detail.

In practice, it is best to use values of h and k which have been obtained empirically. However, it is well to remember that such parameters are functions not only of the hetero-atom in question, but also of the atomic milieu of the hetero-atom and of the physical or chemical property under investigation. A compilation of *good* h and k values is given in Table 3.3. It is obvious that the various authors represented in Table 3.3 do not always concur concerning the *goodness* of the parameters.

5. ELABORATIONS ON HMO THEORY

(A) Inclusion of Overlap

The simple HMO theory neglects overlap. However, it is rather easy to include it: In this regard, we follow Wheland[13] and assume that the

---

[9] VOIE's are discussed in Chapter 4 and Appendix D.
[10] As a rough approximation, it is generally found that EN^M(Mulliken) $\simeq 3EN$(Pauling).

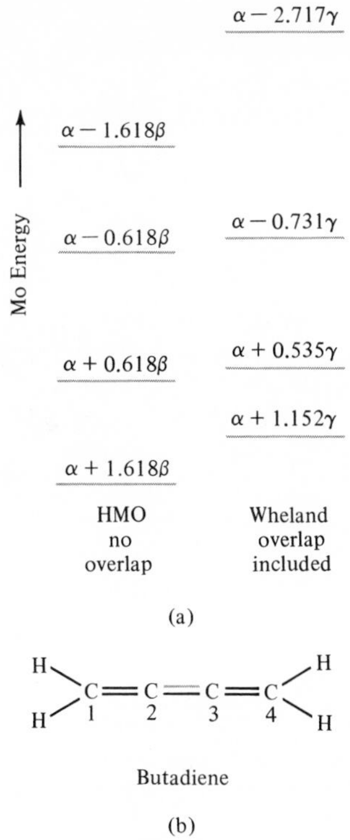

$$\underline{\alpha - 2.717\gamma}$$

(a)

Butadiene

(b)

FIG. 3.7. (a) MO diagram for butadiene with $S = 0$ is on the left; that for $S = 0.25$ is on the right. (b) Atom numbering for the butadiene molecule.

overlap between adjacent π AO's is S and between nonadjacent π AO's is 0.

The example we choose is butadiene. This molecule is diagrammed in Fig. 3.7. The secular determinant, with inclusion of overlap is

$$\begin{vmatrix} \alpha - \epsilon & \beta - \epsilon S & 0 & 0 \\ \beta - \epsilon S & \alpha - \epsilon & \beta - \epsilon S & 0 \\ 0 & \beta - \epsilon S & \alpha - \epsilon & \beta - \epsilon S \\ 0 & 0 & \beta - \epsilon S & \alpha - \epsilon \end{vmatrix} = 0 \qquad (3.58a)$$

If we divide each element by $\beta - \epsilon S$ and set

$$(\alpha - \epsilon)/(\beta - \epsilon S) = y \qquad (3.59)$$

we find

$$\begin{vmatrix} y & 1 & 0 & 0 \\ 1 & y & 1 & 0 \\ 0 & 1 & y & 1 \\ 0 & 0 & 1 & y \end{vmatrix} = 0 \qquad (3.58b)$$

This last determinant is identical to that which would be obtained if S equaled zero, except that y's replace x's. The roots of this determinant are

$$y = \pm 1.618; \qquad \pm 0.618 \qquad (3.60)$$

Thus, setting $S = 0.25$ or $S = 0$, we obtain the Wheland overlap-included MO energies or the simple HMO energies, respectively. These are diagrammed in Fig. 3.7. The observed effects are quite general: Those MO's which are antibonding expand energetically upon inclusion of overlap; those MO's which are bonding compress energetically upon inclusion of overlap.

In view of Eq. (3.59) it is clear that

$$\epsilon_r = (\alpha - y_r\beta)/(1 - y_rS) \qquad (3.61)$$

If we now define

$$\gamma = \beta - S\alpha \qquad (3.62)$$

and eliminate β between Eqs. (3.61) and (3.62), we find

$$\epsilon_r = \alpha - \frac{y_r}{1 - y_rS}\gamma = \alpha - w_r'\gamma \qquad (3.63)$$

which is of the same form as Eq. (3.29), with the reservation that γ is a resonance integral which takes overlap into account and which, in any case, must be evaluated empirically.

As far as we are aware, inclusion of overlap does not lead to considerable improvement of the simple HMO results.[14]

(B)   Hückel $4n + 2$ Rule for Cyclic Polyolefins

The secular equations, with imposition of the Hückel approximations, take the form

$$c_{\mu-1} + xc_\mu + c_{\mu+1} = 0 \qquad (3.64)$$

where $\mu - 1$, μ, and $\mu + 1$ are successive atoms in the ring of carbons. If there are N carbon atoms in the ring, any solution must satisfy the boundary conditions

$$c_\mu = c_{N+\mu} \qquad (3.65)$$

We now seek a solution which is periodic in N; as an initial attempt, we try

$$c_\mu = \sin k\mu; \qquad k \equiv 2\pi r/N \qquad (3.66)$$

where r is an integer. Substitution of Eq. (3.66) into Eq. (3.64) yields

$$\sin k(\mu - 1) + x \sin k\mu + \sin k(\mu + 1) = 0 \qquad (3.67)$$

or, equivalently,

$$2 \sin k\mu \cos k + x \sin k\mu = 0 \qquad (3.68)$$

Thus

$$x = -2 \cos k \qquad (3.69)$$

and

$$\epsilon_r = \alpha + 2\beta \cos k = \alpha + 2\beta \cos (2\pi r/N) \qquad (3.70)$$

When N is even, the possible solutions are restricted to

$$r = 0, \ldots, N/2; \qquad (N \text{ even}) \qquad (3.71)$$

When N is odd, the possible solutions are restricted to

$$r = 0, \ldots, (N - 1)/2; \qquad (N \text{ odd}) \qquad (3.72)$$

The reader may verify that

$$c_\mu = \cos k\mu \qquad (3.73)$$

is an equally satisfactory function, with solutions identical to those obtained for $c_\mu = \sin k\mu$. The reader may also verify that these (i.e., the sine and cosine functions) are the only two linearly independent solutions possible.

Thus there are two acceptable solutions to Eq. (3.64) for all values of r except[11] $r = 0$ and $r = N/2$. Thus, the highest- and lowest-energy MO's

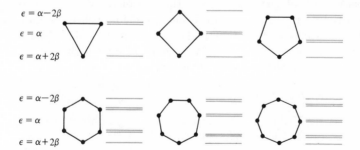

FIG. 3.8. Hückel energy levels for cyclic polyolefins. Inscribe an N-sided regular polygon in a circle of radius 2β such that one corner is the lowest point. Each remaining corner corresponds to a single Hückel orbital, the energy of which is given, in units of β, by the distance from the horizontal diameter of the circle.

---

[11] For $r = 0$ or $r = N/2$, the sine solution has all $c_\mu = 0$; these are trivial solutions.

in a cyclic polyolefin for which N is even are nondegenerate, whereas all intermediate energy MO's are doubly degenerate. On the other hand, for a cyclic polyolefin with N odd only the lowest MO is nondegenerate; the others are all doubly degenerate.

A simple mnemonic device which reproduces all the above prescriptions is illustrated in Fig. 3.8. The origin of the name $4n + 2$ *rule* is implicit in Fig. 3.8. The three-membered ring is stable with 2 π electrons (i.e., $n = 0$), the four-membered with 6 π electrons (i.e., $n = 1$), and the eight-membered with 10 π electrons (i.e., $n = 2$). In other words, a special stability is associated with cyclic systems containing $(4n + 2)$ π electrons, where n for a given cyclic molecule is some one of the integers 0, 1, 2, 3, ...

6. A CRITIQUE OF HÜCKEL THEORY

It is very easy to criticize Hückel theory. In fact, there is little in the HMO method that can be said to be *theoretically correct:* It openly neglects antisymmetrization, electron correlation, σ-π interactions, σ-charge redistribution effects, overlap influences, etc.,—the litany of theoretical deficiencies is very large! However, these deficiencies are partially compensated by the generation of a very simple computational scheme which is frankly empirical, easy to use, and which works much better than one has any right to expect of it. One must assume that the parametrization process subsumes many deficiencies and corrects for them in a wholly empirical way.

The Hückel theory provides a simple means of rationalizing and correlating experimental results for a class of related compounds. The larger the number of molecules in the class, the better are the chances that the HMO method proves successful. In the case of hetero-aromatics the size of the parameter set goes up whereas the size of the class decreases; hence, Hückel theory is only moderately successful for such molecules. In other words, the utility of HMO methods in a given class of molecules is somewhat proportionate to the size of that class; or, one might say, utility in an area is proportional to experience in that area.

EXERCISES

1. Utilize the Variation Theorem to obtain the secular equations of Eqs. (3.3) and (3.18b).

Hint: See work by Liberles.[12]

2. Show that the π-delocalization energy of trimethylene methane is $2[(3)^{1/2} - 1]\beta$. Trimethylene methane is discussed in Sec. 2(D) and diagrammed in Fig. 3.3.

---

[12] A. Liberles, *Introduction to Molecular Orbital Theory* (Holt, Rinehart and Winston, Inc., New York, 1966), p. 81ff.

3.　Write the secular determinant for ethylene and evaluate the Hückel MO energies and eigenvectors. What is the π-electron energy of the ground state of ethylene?

Answer:　Ethylene is $H_2C = CH_2$. The secular determinant is

$$\begin{vmatrix} \alpha - \epsilon & \beta \\ \beta & \alpha - \epsilon \end{vmatrix} = 0 \qquad (3.74)$$

The eigenvalues and eigenvectors are

$$\epsilon_2 = \alpha - \beta; \qquad \varphi_2 = (\tfrac{1}{2})^{1/2}(\chi_1 - \chi_2)$$
$$\epsilon_1 = \alpha + \beta; \qquad \varphi_1 = (\tfrac{1}{2})^{1/2}(\chi_1 + \chi_2) \qquad (3.75)$$

The total π-electron energy of the ground state is

$$E_\pi = 2\alpha + 2\beta \qquad (3.76)$$

4.　Assume that the same β may be used for all bonds in butadiene and calculate the energy of the lowest-energy electronic excitation.

The bond alternation in butadiene requires $\beta_{12} = \beta_{34} < \beta_{23}$. A correlation of HMO predictions with the lowest-energy electronic excitation in the linear polyenes selects $\beta_{12} = \beta_{34} = -4.00$ eV and $\beta_{23} = -2.88$ eV. Calculate the lowest-energy electronic excitation energy using these values of β.

The C—C bond distances in butadiene are $R_{12} = R_{34} = 1.34$ Å and $R_{23} = 1.48$ Å. Calculate the $2p\pi$-$2p\pi$ overlap integrals for the 1, 2 and 2, 3 bonds and, using Eq. (3.52), calculate β_{12} and β_{23}. How do these compare with the fully empirical values of -4.00 and -2.88 eV, respectively?

Answer:　With all β's identical to $\beta_{benzene}$, the first excitation energy is calculated to be 35 300 cm^{-1}. The experimental value is 46 100 cm^{-1}. The agreement is poor.

Using Eq. (3.52), it may be shown that $\beta_{12} = 1.1\beta$ and $\beta_{23} = 0.87\beta$; the ratio $\beta_{12}/\beta_{23} \simeq 1.26$, whereas $4/2.88 \simeq 1.38$. Thus, the results of Eq. (3.52) do not disagree too badly with the empirical ratio.

5.　Evaluate the HMO energy-level diagram of formaldehyde. Use the Yonezawa parameters of Table 3.3.

Answer:　The secular determinant is

$$\begin{vmatrix} \overset{\centerdot}{\alpha}_C - \epsilon & \overset{\centerdot\centerdot}{\beta}_{C-O} \\ \overset{\centerdot\centerdot}{\beta}_{C-O} & \overset{\centerdot}{\alpha}_O - \epsilon \end{vmatrix} = 0 \qquad (3.77)$$

Insertion of the Yonezawa parameters yields

$$\begin{vmatrix} \alpha - \epsilon & (2)^{1/2}\beta \\ (2)^{1/2}\beta & \alpha + 2\beta - \epsilon \end{vmatrix} = 0 \tag{3.78}$$

The eigenvalues are

$$\epsilon_2 = \alpha - 0.7\beta$$

$$\epsilon_1 = \alpha + 2.7\beta \tag{3.79}$$

The energy levels and MO's are diagrammed in Fig. 3.9.

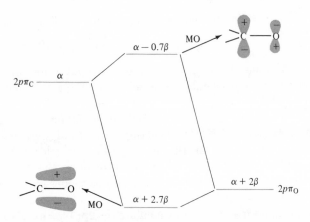

FIG. 3.9.   Relative π-AO and π-MO energy-level diagram for carbon, oxygen, and formaldehyde. The MO's of the C=O bond are also schematized.

6.   Evaluate the MO energies, π-electron densities, net charges, and resonance energy for pyrolle. Pyrolle is diagrammed in Fig. 3.10. Take $\alpha_N^{..} = \alpha + \frac{3}{2}\beta$ and $\beta_{C-N}^{..} = 1$.

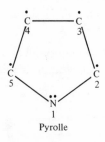

Pyrolle

FIG. 3.10.   Atom numbering in pyrolle.

Answer: The secular determinant is

$$
\begin{vmatrix}
\alpha_{\ddot{N}} - \epsilon & \beta_{C-\ddot{N}} & 0 & 0 & \beta_{C-\ddot{N}} \\
\beta_{C-\ddot{N}} & \alpha - \epsilon & \beta & 0 & 0 \\
0 & \beta & \alpha - \epsilon & \beta & 0 \\
0 & 0 & \beta & \alpha - \epsilon & \beta \\
\beta_{C-\ddot{N}} & 0 & 0 & \beta & \alpha - \epsilon
\end{vmatrix} = 0
\tag{3.80}
$$

The eigenvalues, π-electron densities, and net charges are given by

$$\epsilon_5 = \alpha - 1.62\beta; \qquad q_5 = 1.05; \qquad \eta_5 = -0.05$$

$$\epsilon_4 = \alpha - 1.20\beta; \qquad q_4 = 1.14; \qquad \eta_4 = -0.14$$

$$\epsilon_3 = \alpha + 0.618\beta; \qquad q_3 = 1.14; \qquad \eta_3 = -0.14$$

$$\epsilon_2 = \alpha + 1.15\beta; \qquad q_2 = 1.05; \qquad \eta_2 = -0.05.$$

$$\epsilon_1 = \alpha + 2.55\beta; \qquad q_1 = 1.63; \qquad \eta_1 = +0.37 \tag{3.81}$$

The delocalization energy is -1.64β.

7.   Prove that the trial solution $c_\mu = \cos k\mu$ to Eq. (3.64) leads to the same energy expressions as the sine solution. Prove that in a cyclic polyolefin all MO levels must be confined between limits $\alpha + 2\beta$ and $\alpha - 2\beta$.

8.   The HMO charge densities in the allyl cation are

$$q_1 = q_3 = 0.500; \qquad q_2 = 1.00 \tag{3.82}$$

Use the ω technique to evaluate the q's at charge self-consistency.

Answer: With $\omega = 1.4$ the results are given by Streitwieser[13] as

Number of iterations	$q_1 = q_3$	q_2
0	0.500	1.00
1	0.621	0.757
2	0.534	0.934
3	0.597	0.806
⋮	⋮	⋮
∞	0.571	0.858

$$\tag{3.83}$$

[13] See Ref. [14].

BIBLIOGRAPHY

[1]   A. Streitwieser, Jr., *Molecular Orbital Theory for Organic Chemists* (John Wiley & Sons, Inc., New York, 1961), p. vi of Foreword.

[2]   A. Streitwieser, Jr., Ref. [1], p. 33ff.

[3]   F. A. Cotton, *The Chemical Applications of Group Theory* (Wiley-Interscience Publishers, Inc., New York, 1965).

[4]   H. H. Jaffé and M. Orchin, *Symmetry in Chemistry* (John Wiley & Sons, Inc., New York, 1965).

[5]   A. Streitwieser, Jr., Ref. [1], Chap. 3.

[6]   QCPE-70 (Bloor and Gilson); QCPE-81 (Morris); Quantum Chemistry Program Exchange, Chemistry Department, Room 204, Indiana University, Bloomington, Ind. 47401.

[7]   A. Streitwieser, Jr., Ref. [1], Figs. 6.8 and 6.9, p. 168; R. Daudel, R. Lefebvre, and C. Moser, *Quantum Chemistry. Methods and Applications*, (Interscience Publishers, Inc., New York, 1959), Fig. V.5, p. 128.

[8]   G. W. Wheland and D. E. Mann, *J. Chem. Phys.* **17,** 264 (1949).

[9]   A. Streitwieser, Jr., Ref. [1], p. 115ff.

[10] A. Laforgue, *J. Chim. Phys.* **46,** 568 (1949).

[11] L. Pauling, *J. Am. Chem. Soc.* **54,** 3570 (1932).

[12] H. Suzuki, *Electronic Absorption Spectra and Geometry of Organic Molecules* (Academic Press Inc., New York, 1967), Chap. 9.

[13] G. W. Wheland, *J. Am. Chem. Soc.* **63,** 2025 (1941).

[14] A. Streitwieser, Jr., Ref. [1], p. 101ff.

General References

1.   J. N. Murrell, S. F. A. Kettle, and J. M. Tedder, *Valence Theory* (John Wiley & Sons, Inc., New York, 1965), Chap. 15.

2.   H. Suzuki, *Electronic Absorption Spectra and Geometry of Organic Molecules* (Academic Press Inc., New York, 1967) Chaps. 7–10.

3.   L. F. Phillips, *Basic Quantum Chemistry* (John Wiley & Sons, Inc., New York, 1965), Chap. 5.

4.   A. Liberles, *Introduction to Molecular Orbital Theory* (Holt, Rinehart and Winston, Inc., New York, 1966).

5.   A. Streitwieser, *Molecular Orbital Theory for Organic Chemists* (John Wiley & Sons, Inc., New York, 1961).

6.   C. Sandorfy, *Electronic Spectra and Quantum Chemistry* (Prentice-Hall, Inc., Englewood Cliffs, N.J., 1964), Chaps. 3 and 8.

7.   J. D. Roberts, *Notes on Molecular Orbital Calculations* (W. A. Benjamin, Inc., New York, 1961).

8.   R. Daudel, R. Lefebvre, and C. Moser, *Quantum Chemistry. Methods and Applications* (Interscience Publishers, Inc., New York, 1959).

CHAPTER 4

Mulliken–Wolfsberg–Helmholz Molecular-Orbital Approach

The semiempirical molecular-orbital theory discussed in this chapter is based on the 1952 work of Wolfsberg and Helmholz[1] on tetrahedral oxy-anions.

For a molecule, the Hamiltonian is of molecular extent and its eigenfunctions are molecular orbitals. It is assumed that these MO's are given in the LCAO form

$$\varphi_r = \sum_{\mu} c_{r\mu}\chi_{\mu} \tag{4.1}$$

The MO energy is given by

$$\epsilon_r = \langle \varphi_r \mid \hat{H} \mid \varphi_r \rangle = \sum_{\mu} \sum_{\nu} c_{r\mu}{}^* c_{r\nu} \langle \chi_{\mu} \mid \hat{H} \mid \chi_{\nu} \rangle \tag{4.2}$$

The MO's are orthonormalized—that is,

$$\langle \varphi_r \mid \varphi_s \rangle = \sum_{\mu} \sum_{\nu} c_{r\mu}{}^* c_{s\nu} \langle \chi_{\mu} \mid \chi_{\nu} \rangle = \delta_{r,s} \tag{4.3}$$

Minimization of energy with respect to the AO coefficients yields the secular determinant

$$\mid H_{\mu\nu} - \epsilon S_{\mu\nu} \mid = 0 \tag{4.4}$$

where

$$H_{\mu\nu} \equiv \langle \chi_\mu \mid \hat{H} \mid \chi_\nu \rangle; \qquad S_{\mu\nu} \equiv \langle \chi_\mu \mid \chi_\nu \rangle \qquad (4.5)$$

The overlap matrix has been discussed in Chapter 2. The Hamiltonian matrix, subject to very restrictive conditions (i.e., the Hückel approximations), has been discussed in Chapter 3. The purpose of the present chapter is to remove many of the Hückel approximations and to extend the method so that it can treat any type of molecule: inorganic, saturated, unstable, etc. At the same time, the semiempirical one-electron nature of the Hückel theory is retained. Hence, this chapter might be subtitled: *EXTENDED HÜCKEL THEORY*—a name first coined by Hoffmann.[2]

The Mulliken–Wolfsberg–Helmholz (MWH) method, like the Hückel method, is frankly empirical. Nonetheless, the MWH method does avoid many of the Hückel approximations. They are the following:

(i)   It does not neglect overlap.

(ii) $H_{\mu\nu}$ terms are not restricted to nearest-neighbor centers.

(iii)   None of the valence electrons are neglected.

(iv)   Inner-shell or nonvalence electrons (for example, the $1s^2$ electrons of carbon) are usually neglected; they are considered implicitly insofar as they provide a nonpolarizable core and, if so wished, they need not be neglected at all.

(v)   The theory is not restricted to planar systems.

(vi)   The $H_{\mu\mu}$ and $H_{\mu\nu}$ terms are approximated, but in much less crude fashion than in Hückel theory.

(vii)   The method can account for charge redistribution effects which occur during the formation of molecules from atoms.

The plan of this chapter is as follows: We first digress on atoms and the information available in atomic spectroscopic data. We then discuss the diagonal $H_{\mu\mu}$ terms—how they are obtained and used. Thereafter we examine the various $H_{\mu\nu}$ approximations which exist and attempt a relative evaluation of them. This is followed by a detailed presentation of two example computations: one for a highly polar but simple molecule, namely, H_2O, and the other for a transition-metal complex, namely, $TiF_6{}^{3-}$.

1.   ELECTRONS IN ATOMS

The purpose of this section is a presentation of the terminology used in the discussion of the electronic energy levels of atoms. We make no great attempt at pedagogy because this topic is excellently discussed in a great variety of places.[3]−[6]

We adopt the central-field approximation. This topic has been discussed in Chapter 1. It implies that the radial wave functions $R(r)$ are the same for all AO's with the name n and l quantum numbers. In other words, $R(r)$ is determined by n and l, and not by the spatial orientation of the AO (i.e., by m_l). The central-field approximation, of course, is not correct for poly-electron atoms; it is, however, a very convenient simplification which works well in the electronic spectroscopy of atoms. We assume the transferability of this approximation to the case of an atom in a molecule.

(A) Electronic Configurations

A specification of the (n, l)-*electron distribution describes an electron configuration.* Thus, the lowest-energy configuration of carbon is

$$C: 1s^2 2s^2 2p^2$$

where the values of n and l for all electrons are clearly defined.

Let us now inquire about the degeneracy of the $C: 1s^2 2s^2 2p^2$ configuration. In other words, let us ask: How many states arise from the $1s^2 2s^2 2p^2$ configuration? The two electrons in the $2p$ subshell can have values of $m_l = \pm 1, 0$; furthermore, if the two values of m_l are different, each electron can have $m_s = \pm \frac{1}{2}$; on the other hand, if both electrons have the same value of m_l, the Pauli principle implies that one electron has $m_s = +\frac{1}{2}$ and the other $m_s = -\frac{1}{2}$. Hence, we can tabulate all possibilities as in Table 4.1. It is clear that 15 different states emerge from the $C: 1s^2 2s^2 2p^2$ configuration.

(B) Terms

Relativistic effects are not very important in the lighter atoms. To the extent that they can be neglected, the total wave function for all electrons must correspond to a definite total orbital angular momentum $[L(L + 1)]^{1/2}\hbar$, a definite total spin angular momentum $[S(S + 1)]^{1/2}\hbar$, and a definite total angular momentum $[J(J + 1)]^{1/2}\hbar$. This situation

TABLE 4.1. *Electron arrangements within configuration* $C: 1s^2 2s^2 2p^2$.

Orbital Arrangements Within Configuration	$M_S = \sum_i m_{s_i}$	$M_L = \sum_i m_{l_i}$	Number of States
$1s^2 2s^2 2p_0 2p_1$	1, 0, 0, or -1	1	4
$2p_0 2p_{-1}$	1, 0, 0, or -1	-1	4
$2p_1 2p_{-1}$	1, 0, 0, or -1	0	4
$2p_0{}^2$	0	0	1
$2p_1{}^2$	0	2	1
$2p_{-1}{}^2$	0	-2	1
Total Number of States			15

constitutes the well-known Russell–Saunders or LS coupling scheme. *In this context, a term is specified when L and S are known.* The different values of J consistent with given L and S values are degenerate in the absence of relativistic (i.e., spin-orbit coupling) effects.

We again tabulate—but in a somewhat different manner—the different values of $M_L = \sum_i m_{l_i}$ and $M_S = \sum_i m_{s_i}$ for the $C: 1s^2 2s^2 2p^2$ configuration. This tabulation is given in Table 4.2. The symbol x of Table 4.2 indicates the presence of the (M_S, M_L) combination in question. The first column is consistent with the presence of a term with $L = 2$, $S = 0$, the second column with the presence of a term with $L = 1$, $S = 1$, and the third with the presence of a term with $L = 0$, $S = 0$. Using the Russell–Saunders term designations (i.e., ^{2S+1}L, where $2S + 1$ is the electron-spin degeneracy) the three terms in question are designated ^1D, ^3P, and ^1S, respectively. These terms possess degeneracies given by $(2J + 1)(2S + 1)$, which equal 5, 9, and 1, respectively. Thus, knowing the energies of the three terms, we can immediately define their energy barycenter. This barycenter is the average energy of the $1s^2 2s^2 2p^2$ configuration. Thus,

$$E(1s^2 2s^2 2p^2)_{av} = (1/15)[9E(^3P) + 5E(^1D) + E(^1S)] \quad (4.6)$$

The energy differences between the various terms are also expressible in terms of a small number of integrals of theoretical origin known as the Slater–Condon parameters.[5] These integrals are usually parametrized from experiment. In the present instance, they lead to the correlation of theory and experiment shown in Table 4.3. Tabulations of Slater–Condon parameters have been given by a number of authors.[7]–[9] We rarely use them, preferring instead to work with experimental information on the atomic energy levels.[10] However, if needed experimental values of atomic energy levels are unknown, the Slater–Condon parametrization provides a useful and reasonably safe way to obtain them.

TABLE 4.2. M_L and M_S values for $C: 1s^2 2s^2 2p^2$.

M_L	M_S				
	0	1	0	−1	0
2	x				
1	x	x	x	x	
0	x	x	x	x	x
−1	x	x	x	x	
−2	x				
Terms $\rightarrow$	^1D +		^3P	+	^1S

(C)   Levels

A level is specified when the values of L, S, and J are quoted. Thus, the ^1D, ^3P, and ^1S terms give rise to the levels

$$^1D_2; \quad ^3P_2, \quad ^3P_1, \quad ^3P_0; \quad ^1S_0$$
$$2J + 1 \rightarrow 5 \quad\quad 5 \quad\quad 3 \quad\quad 1 \quad\quad 1 \leftarrow \text{DEGENERACY}$$

where the subscript is the value of J in the level notation $^{2S+1}L_J$; the level degeneracy equals the number of possible values of M_J.

In the lighter elements, spin-orbit coupling is small and the level differences within a given term are more or less close to zero. As a result, the experimental pattern of energy levels is easy to follow and to correlate. In the heavier elements, however, these same level differences can exceed the energy differences between different configurations. Consequently, experimental patterns are difficult to identify and level misidentifications do occur.

Except in the case of atoms containing only s electrons, spin-orbit coupling is not zero. Consequently, the degeneracy of a specific term is, to some extent, removed. Therefore, Eq. (4.6) is invalid. Instead, we must write

$$E(1s^2 2s^2 2p^2)_{\text{av}} = (1/15)[5E(^1D_2) + 5E(^3P_2)$$

$$+ 3E(^3P_1) + E(^3P_0) + E(^1S_0)] \quad (4.7)$$

because it is only the atomic energy *levels* which are experimentally known. The validity of Eq. (4.6) may be reinstated, however, if we redefine the *term energy* at the barycenter of the levels which arise from it. For example,

TABLE 4.3.   *Correlation of theory[a] and experiment for* C: $1s^2 2s^2 2p^2$.

Term	Theoretical Energy[b]	Parametric fit of Theoretical Energy (eV)[b,c]	Experimental Energy (eV)[b]
^3P	$-3F_2$	-0.56702	-0.59857
^1D	$+3F_2$	$+0.56702$	$+0.66127$
^1S	$+12F_2$	$+2.26833$	$+2.08093$

[a] Using Slater–Condon parametrization
[b] All energies are quoted relative to $E(1s^2 2s^2 2p^2)_{\text{av}}$.
[c] The value of F_2 used is $F_2(pp)$ for the $2p$ electrons of carbon.

definitions of the type

$$E(^3P) \equiv (1/9)[5E(^3P_2) + 3E(^3P_1) + E(^3P_0)] \qquad (4.8)$$

correct any inadequacies present in Eq. (4.6).

In the following, we take an equation of the type of Eq. (4.7), which is based on energy levels (rather than terms), as our primary definition of a configuration energy.

The definition of Eq. (4.7) removes spin-orbit coupling effects from the configuration energy. This is as it should be: Considerations of spin-orbit coupling in molecules are best treated in a perturbational framework based on zero-order eigenvectors of a molecular Hamiltonian which does not contain any relativistic parts (see Chap. 11).

(D) States

A state is specified when the values of L, S, J, and M_J are quoted. All states with the same values of L, S, and J are degenerate. This degeneracy is readily removed by a magnetic field (i.e., the Zeeman effect). However, the Zeeman effect is of primary importance as a *level sorter* and it does not engross our attention.

(E) Valence Configurations

The wave functions of an electronic state of an atom are specified most simply using the exponential functions, $e^{\pm im_l\phi}$. In molecular calculations, it is expedient to avoid complex numbers. Hence, it is usual to transform from the imaginary basis to some real basis—for example, see Appendix A. The real basis which may be selected is not unique. Furthermore, the choice of a real basis is not limited by any requirement for occupancy of a given AO by an integral number of electrons. In any event, the most common real basis in the case of p AO's is the (p_x, p_y, p_z) set.

Valence configurations are usually specified only in the real basis. A valence configuration is defined when the values of n and l are given for every electron and when the occupancy of every real AO is indicated. For example, the $C: 1s^2 2s^2 2p^2$ configuration consists of two valence configuration which, in the (p_x, p_y, p_z) basis, are given by

$$1s^2 2s^2 2p_x 2p_y \qquad \text{Valence configuration 1}$$

(two electrons in different p AO's in the x, y, z basis)

$$1s^2 2s^2 2p_x{}^2 \qquad \text{Valence configuration 2}$$

(two electrons in the same p AO in the x, y, z basis)

The valence configuration $1s^2 2s^2 2p_x 2p_z$ and $1s^2 2s^2 2p_y 2p_z$ are not different from valence configuration 1 and are embraced by the definition given. Similarly, valence configuration 2 embraces both $1s^2 2s^2 2p_y{}^2$ and $1s^2 2s^2 2p_z{}^2$.

We now discuss the terms which arise from a given valence configuration. The valence configuration 2 gives rise to the (M_L, M_S) sets: $(2, 0)$, $(-2, 0)$, $(0, 0)$. These are circled in Table 4.4. The valence configuration 1 gives rise to the twelve (M_L, M_S) sets which are enclosed in rectangles in Table 4.4. Thus, from Table 4.4, we may write[1]

$$E(\text{V.C.1})_{\text{av}} = \tfrac{1}{4}[3E(^3P) + E(^1D)] \tag{4.9}$$

$$E(\text{V.C.2})_{\text{av}} = \tfrac{1}{3}[2E(^1D) + E(^1S)] \tag{4.10}$$

where the term energies used are themselves averages of the sort specified by Eq. (4.8).

(F) Resumé

All of our considerations relative to carbon are summarized in Fig. 4.1. The energies quoted in Fig. 4.1 are based on the energy levels given by Moore.[10] The equations used in constructing Fig. 4.1 are Eqs. (4.7)–(4.10).

The nomenclature proposed here is very specific. Unfortunately, it is not often used; therefore, one must be very careful concerning the meaning of the names *valence configuration, term, level, state*, etc., as used by different authors. Indeed, that to which we refer as a *valence configuration* is, more often than not, referred to as a *valence state*.

TABLE 4.4. *Relation of valence states and terms of* $C: 1s^2 2s^2 2p^2$.

M_L	M_S				
	0	1	0	−1	0
2	(x)				
1	x	x	x	x	
0	x	x	x	x	(x)
−1	x	x	x	x	
−2	(x)				
Terms →	^1D		^3P		^1S

[1] Tabulations of valence configurations based on the real orbitals of $s, p, d, \ldots$ types and their hybrids are available (Refs. [11]–[16]).

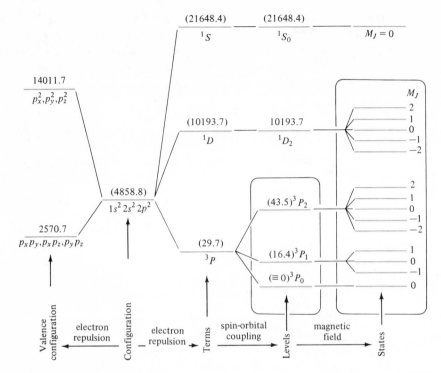

FIG. 4.1.   Energies associated with the C: $1s^2\ 2s^2\ 2p^2$ configuration. All energies are in cm^{-1}. The energy coordinate is schematic only. Enclosure of certain levels and states within frames indicates a gross expansion of the energy scale within that enclosure.

(G)   Example Calculations

We now detail some computational examples:

(i) $E(\ldots 3s^2 3p^4)_{av}$ of sulfur, S(I): The $\ldots 3s^2 3p^4$ configuration of sulfur yields three terms: ^3P, ^1D, and ^1S. The lowest energy of these is the ^3P term or, more specifically, the $^3P_2$ level. The energy difference

$$E(\ldots 3s^2 3p^4)_{av} - E(^3P_2) = P^0 \qquad (4.11)$$

is known as the promotion energy P^0. If $E(^3P_2)$ is taken as the zero of energy, then the promotion energy equals the configuration energy. The promotion energy is the weighted mean of all the levels arising from the $\ldots 3s^2 3p^4$ configuration. The weighting factor for a given level is its degeneracy $2J + 1$. The computation of P^0 proceeds as in Table 4.5.

(ii) $E(\ldots 3s^2 3p^3)_{av}$ of sulfur cation, S(II): The configuration $\ldots 3s^2 3p^3$ yields terms ^2P, ^2D, ^4S. The promotion energy P^+ is defined as

TABLE 4.5. *Promotion energy P^0 of Eq. (4.11).*

Level	Energy[a] E (cm^{-1})	Level Degeneracy $(2J+1)$	$(2J+1)E$ (cm^{-1})
$^3P_2$	0	5	0
$^3P_1$	396.8	3	1 190.4
$^3P_0$	573.6	1	573.6
$^1D_2$	9 239.0	5	46 195.0
$^1S_0$	22 181.4	1	22 181.4
		Totals → 15	70 140.4

$$P^0 = 70\ 140/15 = 4\ 676\ \text{cm}^{-1}$$

[a] From Ref. [10].

the difference

$$E(\ldots 3s^2 3p^3)_{\text{av}} - E(^4S_{1/2}) = P^+ \tag{4.12}$$

where $^4S_{1/2}$ is the ground level of S(II). The computation of P^+ proceeds as in Table 4.6. With $E(^4S_{1/2}) \equiv 0$, P^+ equals the average configuration energy.

(iii) $E(\ldots 3s^2 3p^2)_{\text{av}}$ of S(III): The computation of P^{++} is left to the reader. Its value is $P^{++} = 5921$ cm^{-1}.

TABLE 4.6. *Promotion energy P^+ of Eq. (4.12).*

Level	Energy[a] E (cm^{-1})	Level Degeneracy $(2J+1)$	$(2J+1)E$ (cm^{-1})
$^4S_{3/2}$	0	4	0
$^2D_{3/2}$	14 851.9	4	59 407.6
$^2D_{5/2}$	14 883.4	6	89 300.4
$^2P_{1/2}$	24 524.2	2	49 048.4
$^2P_{3/2}$	24 572.8	4	98 291.2
		Totals → 20	296 047.6

$$P^+ = 296\ 047.6/20 = 14\ 802.4\ \text{cm}^{-1}$$

[a] From Ref. [10].

(iv) $E(\ldots 3s^2 3p^5)_{av}$ of $S(0)$ (i.e., S^{-1}) : This configuration yields two levels: $^2P_{3/2}$ and $^2P_{1/2}$. Spectroscopic data are not available; however, the ground level is $^2P_{3/2}$. Comparison with the corresponding levels of the configuration $\ldots 3s^2 3p^3$ of $S(II)$ indicates that the level separation in $S(0)$ is certainly less than 50 cm^{-1}. Thus, we may assume $P^- \approx 0$.

2. VALENCE ORBITAL IONIZATION ENERGIES

The term *valence orbital ionization energy* (or VOIE[2]) *denotes the energy necessary to remove an electron from a given orbital of either a specific configuration or valence configuration.* This definition is clarified using $1s$ VOIE's of atoms which contain $1s$ electrons only.

The lowest ionization potential of $H(I)$ corresponds to the process

$$H : 1s^1;\ ^2S_{1/2} \rightarrow H^+$$

and that of $H(0)$ corresponds to

$$H^- : 1s^2;\ ^1S_0 \rightarrow H : 1s^1;\ ^2S_{1/2}$$

Since only one level arises from any one of these electron configurations, it is apparent that the $1s$ VOIE's are equal to the corresponding ionization potentials. These are tabulated in Table 4.7 for six different atomic species. The data presented clearly indicate that the $1s$ VOIE's are functions of the charge on the atom and the $1s$ orbital population. For example, the first

TABLE 4.7. $1s$ *VOIE's of hydrogenlike atoms.*[a]

Species	Atomic Number (Z)	Atom Charge q (in units of $\mid e \mid$)	Orbital Population[b] (No. of Electrons)	VOIE (eV)
Li^{++}	3	2	1	122.5
He$^+$	2	1	1	54.4
H	1	0	1	13.6
Li$^+$	3	1	2	75.6
He	2	0	2	24.6
H$^-$	1	-1	2	0.75[c]

[a] From Ref. [10] of text.
[b] Henceforth denoted as *POP* or n_μ.
[c] From Pekeris [C. L. Pekeris, Phys. Rev. **112**, 1649 (1948)].

[2] The abbreviations VSIE and VSIP are also quite common.

three rows of data are given by

$$1s \; \text{VOIE}(1s^1) = 13.6(q + 1)^2, \text{ in eV} \qquad (4.13)$$

according to Bohr theory for a one-electron atom, or by

$$1s \; \text{VOIE}(1s^1) = 109.84q^2 + 219.2q + 109.7, \text{ in kK} \qquad (4.14)$$

according to a *best computer fit* by Basch, Viste, and Gray.[17] On the other hand, the data of the last three rows are adequately represented by

$$1s \; \text{VOIE}(1s^2) = 13.6(q + 1.7)^2, \text{ in eV} \qquad (4.15)$$

according to Slater[18] (see Appendix D) or by

$$1s \; \text{VOIE}(1s^2) = 109.82q^2 + 301.7q + 198.4, \text{ in kK} \qquad (4.16)$$

according to a best computer fit by Basch, Viste, and Gray.[17] Carroll *et al.*[19] have parametrized the results of rows 1, 2, 3, and 6 as

$$1s \; \text{VOIE}(q) = -0.121q^3 + 13.97q^2 + 26.93q + 13.6, \text{ in eV} \quad (4.17)$$

and they have used this equation for evaluation of $1s$ VOIE's of hydrogen when the $1s$-AO population is fractional. For example, the H atoms of H_2O and H_2S have electron configurations $1s^{\sim0.85}$ and $1s^{\sim0.96}$, respectively; the corresponding VOIE's are evaluated at $q = +0.15$ and $+0.04$, respectively.

No distinction can be made between a configuration or valence configuration based solely on s AO's. However, in the general case, VOIE's based on configurations do differ from those based on valence configurations. Furthermore, somewhat different parametrization schemes have evolved from the two different types of VOIE. Consequently, we now discuss the two approaches separately.

(A) VOIE's Based on Configurations

The $3p$ VOIE of the $\ldots 3s^2 3p^4$ configuration of S(I) is the energy of the ionization process

$$\ldots 3s^2 3p^4 \text{ of S(I)} \rightarrow \ldots 3s^2 3p^3 \text{ of S(II)}$$

We already know the configuration energies of S(I) and S(II)— but we know them relative only to the ground levels of S(I) and S(II), respectively. However, the lowest ionization potential of S(I) refers to the electron-removal process which connects the ground states of S(I) and S(II). If we denote this ionization potential simply by IP, then

$$3p \; \text{VOIE}[\text{S(I)}; \ldots 3s^2 3p^4] = IP + P^0 - P^+ \qquad (4.18)$$

This equation is readily understood by reference to Fig. 4.2 and Secs. 1(G)(i), and 1(G)(ii). The ionization potential in question is 83 539.56

cm^{-1} (10.357 eV). Hence, using the P^0 and P^+ promotion energies already evaluated, we find

$$3p\ VOIE[S(I)\ ;\ \ldots 3s^2 3p^4] = 83\ 539.56 + 14\ 802.4 - 4\ 676.0$$

$$= 93\ 666\ cm^{-1}\ (11.612\ eV) \qquad (4.19)$$

The evaluation of the $3p$ VOIE$[S(II)\ ;\ \ldots 3s^2 3p^3]$ and $3p$ VOIE$[S(0)\ ;$ $\ldots 3s^2 3p^5]$ is illustrated in Fig. 4.3 and is based on the discussion of Sec. 1(G).

(B)   Charge Dependence of Configurational VOIE's

The VOIE's for sulfur are given in Table 4.8; they are very dependent on the charge on the sulfur atom. These charge dependencies are readily parametrized as

$$3s\ VOIE(\ldots 3s^2 3p^{v-2}) = 1.565q^2 + 11.225q + 20.70,\ \text{in eV} \quad (4.20)$$

$$3p\ VOIE(\ldots 3s^2 3p^{v-2}) = 0.82q^2 + 9.775q + 11.61,\ \text{in eV} \quad (4.21)$$

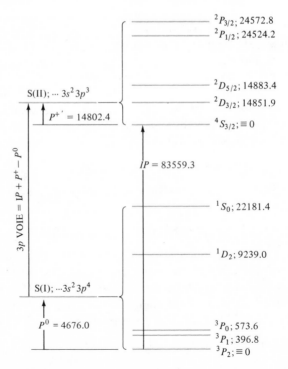

FIG. 4.2.   Evaluation of the $3p$ VOIE $[S(I)\ ;\ \ldots 3s^2 3p^4]$ of sulfur. Energies are in cm^{-1}.

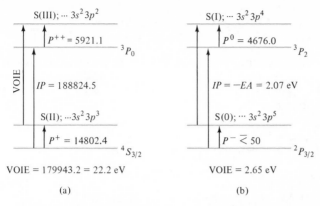

FIG. 4.3. Evaluation of (a) $3p$ VOIE [S(II); ... $3s^2 3p^3$], (b) $3p$ VOIE [S(0); ... $3s^2 3p^5$]. The electron affinity of S^0, denoted *EA*, is taken from Branscomb and Smith [L. M. Branscomb and S. J. Smith, *J. Chem. Phys.* **25,** 598 (1956)]. All numbers are in cm^{-1}, unless otherwise specified.

These equations are of a general form

$$\chi \text{ VOIE(configuration)} = Aq^2 + Bq + C \qquad (4.22)$$

Similar equations can be set up for the configurations $...3s3p^{v-1}$, $...3s^0 3p^v$, etc.; in this manner, information is generated concerning both the effect of charge and configuration on the VOIE's.

(C) VOIE in a Mixed Configuration

Suppose it is necessary to determine the $2s$ VOIE of carbon when the carbon configuration is $...2s^{1.6}2p^{1.9}$. This configuration corresponds to none of those characterized by integral occupancy of AO's. However, the known

TABLE 4.8. *Charge dependence of sulfur $3p$ and $3s$ VOIE's.*

Orbital	Configuration	VOIE (eV)[a]		
		$q = -1$	$q = 0$	$q = 1$
$3s$	$\cdots 3s^2 3p^{v-2}$	11.04	20.70	33.49
$3p$	$\cdots 3s^2 3p^{v-2}$	2.65	11.61	22.20

[a] The number of valence electrons on sulfur is denoted v. Neutral sulfur has $v = 6$. The excess charge q is given by $q \equiv 6 - v$.

$2s$ VOIE's (see Appendix D)

$$2s \text{ VOIE}(\dots 2s^2 2p^{v-2}) = 27.95q^2 + 141.6q + 156.6, \text{ in kK} \quad (4.23)$$

$$2s \text{ VOIE}(\dots 2s 2p^{v-1}) = 28.00q^2 + 141.2q + 171.0, \text{ in kK} \quad (4.24)$$

do bracket the configuration in question. We proceed as follows:
The excess charge is

$$q = 4 - (1.6 + 1.9) = +0.5 \quad (4.25)$$

At this value of q, the VOIE values are

$$2s \text{ VOIE}(\dots 2s^2 2p^{v-2}) = 234.39 \text{ kK} = 2s \text{ VOIE}(\dots 2s^2 2p^{1.5}) \quad (4.26)$$

$$2s \text{ VOIE}(\dots 2s 2p^{v-1}) = 248.60 \text{ kK} = 2s \text{ VOIE}(\dots 2s 2p^{2.5}) \quad (4.27)$$

Thus, the $\dots 2s^{1.6} 2p^{1.9}$ configuration is written as a linear combination of the known configurations, namely,

$$\dots 2s^{1.6} 2p^{1.9} = a(\dots 2s^2 2p^{1.5}) + b(\dots 2s 2p^{2.5}) \quad (4.28)$$

and a and b are obtained from the equations

$$1.6 = 2a + b \quad (4.29)$$

$$1.9 = 1.5a + 2.5b \quad (4.30)$$

Solution yields

$$a = 0.6; \quad b = 0.4 \quad (4.31)$$

Thus, the value of the $2s$ VOIE$(\dots 2s^{1.6} 2p^{1.9})$ is given by

$$0.4 \times 248.60 + 0.6 \times 234.39 = 240.07 \text{ kK} = 29.76 \text{ eV} \quad (4.32)$$

Suppose we wish to calculate the $3d$ VOIE of chromium in the configuration $\dots 3d^{5.0} 4s^{0.1} 4p^{0.2}$. The excess charge is given by

$$q = 6 - (5.0 + 0.1 + 0.2) = +0.7 \quad (4.33)$$

The $3d$ VOIE's available in Appendix D are

$$3d \text{ VOIE}(\dots 3d^v) = 14.75q^2 + 74.75q + 35.1, \text{ in kK} \quad (4.34)$$

$$3d \text{ VOIE}(\dots 3d^{v-1} 4s) = 9.75q^2 + 95.95q + 57.9, \text{ in kK} \quad (4.35)$$

$$3d \text{ VOIE}(\dots 3d^{v-1} 4p) = 9.74q^2 + 96.95q + 67.7, \text{ in kK} \quad (4.36)$$

A computation of the $3d$ VOIE at $q = +0.7$ yields

$$3d \text{ VOIE}(\dots 3d^{5.3}) = 94.65 \text{ kK} \quad (4.37)$$

$$3d \text{ VOIE}(\dots 3d^{4.3} 4s) = 129.84 \text{ kK} \quad (4.38)$$

$$3d \text{ VOIE}(\dots 3d^{4.3} 4p) = 140.34 \text{ kK} \quad (4.39)$$

We now write the $\ldots 3d^{5.0}4s^{0.1}4p^{0.2}$ configuration as a linear combination of those for which we know the VOIE's. Hence,

$$\ldots 3d^{5.0}4s^{0.1}4p^{0.2} = a(\ldots 3d^{5.3}) + b(\ldots 3d^{4.3}4s) + c(\ldots 3d^{4.3}4p) \quad (4.40)$$

and

$$5.3a + 4.3b + 4.3c = 5.0 \quad (4.41)$$

$$b = 0.1 \quad (4.42)$$

$$c = 0.2 \quad (4.43)$$

Solution yields

$$a = 0.7; \quad b = 0.1; \quad c = 0.2 \quad (4.44)$$

Therefore,

$3d$ VOIE$(\ldots 3d^{5.0}4s^{0.1}4p^{0.2})$

$$= (0.7 \times 94.65 + 0.1 \times 129.84 + 0.2 \times 140.34) \text{ kK}$$

$$= 107.31 \text{ kK} = 13.30 \text{ eV} \quad (4.45)$$

(D) VOIE's Based on Valence Configurations

Cusachs and Reynolds[20] have approached the subject of VOIE's from the point of view of valence configuration ionization energies. According to them, the utility of such valence configurations depends on the relative insensitivity of the VOIE to the detailed arrangement of other electrons in the valence configuration, providing these arrangements do not differ in any pathologic way. This point of view is illustrated in Table 4.9.

In view of the fact that the VOIE is very dependent on the occupancy

TABLE 4.9. *VOIE's of O^{2+} valence configurations.*[a]

Initial Valence Configuration (O^{2+})	Final Valence Configuration (O^{3+})	I_p (eV)
$s^2p_x^2$	s^2p_x	51.46
$sp_x^2p_y$	sp_xp_y	50.48
$p_x^2p_yp_z$	$p_xp_yp_z$	49.40
$p_x^2p_y^2$	$p_xp_y^2$	49.78
	Average =	50.28

[a] From Cusachs and Reynolds, Ref. [20] of text.

of a given orbital, Cusachs writes

$$\text{VOIE} = A'(l, POP) + B'q \tag{4.46}$$

where l, the orbital angular momentum quantum number, identifies the valence AO of the atom in question; where POP is the number of electrons in the orbital (i.e., $POP = 1$ or 2); and where q is the excess charge on the atom. Cusachs claims[20] that such VOIE's are good to ± 1.0 eV over a q range of $-1 \le q \le 1$.

Equation (4.46) can be made a continuous function of the population by modifying it to the form[19]

$$\text{VOIE} = A''(POP) + B'' + C''q \tag{4.47}$$

For a carbon $2s$ AO, the resultant equation is

$$-1.7(POP) + 22.9 + 11.9q, \text{ in eV} \tag{4.48}$$

For a net charge of $+0.5$ and a configuration of $\ldots 2s^{1.62}2p^{1.9}$, this equation gives a VOIE of 26.13 eV; this VOIE may be compared with the value of 29.76 eV of Eq. (4.32). Fortunately, the results from the two methods usually agree somewhat better than this.

VOIE's of the type of Eq. (4.47) are given in Appendix D.

(E)   Ab Initio VOIE's

Newton, Boer, and Lipscomb[21] have adapted VOIE's directly from minimal-basis-set SCF calculations on molecules similar to those being treated semiempirically. This approach has been successful with hydrocarbons, boron hydrides, and heterocyclic molecules. However, one is restricted to those types of molecules for which SCF calculations are available as models.

(F)   Reliability of VOIE Data

VOIE's are a reasonable approximation to the AO ionization potentials in an isolated atom. However, even in this instance, they are subject to some error:

(i)   The energies of atomic spectral lines can be very accurately measured—typically to seven significant figures. However, the identification of a given line is not a simple matter and, for all but a few very well investigated materials (i.e., carbon, nitrogen, oxygen, etc.), line misidentifications exist.

(ii)   In order to estimate VOIE's, we need atom ionization potentials or electron affinities. These are obtained by series-limiting processes (line convergence) or beam techniques which are frequently uncertain to at least several tenths of an eV.

(iii) It is assumed that the levels which are correlated to a given configuration refer only to that configuration. However, many effects—for example, configuration interaction—cause significant mixing of atomic levels; this, in turn, produces a backward generation of *contaminated VOIE* data.

Finally, VOIE's are generated for ultimate use in molecules. Insofar as they are a function of one-center excess charge and configuration, they do take account, to some extent, of the molecular environment. However, they neglect many other effects which may be important, for example, the effects of neighbor-atom excess charges.

Various aspects of the VOIE concept have not been touched on here. For this reason, we recommend a reading of the original literature.[12]−[17],[20],[22]−[27]

3. COULOMB INTEGRAL

The Coulomb integral $H_{\mu\mu}$ is equated to the negative value of the VOIE of the χ_μ AO. The VOIE is understood to be evaluated in terms of both an atomic charge q and a configuration appropriate for that atom when incorporated in the molecule of interest. Thus,

$$H_{\mu\mu} = E(q + 1) - E(q) = -\text{VOIE} \qquad (4.49)$$

where the electron which is removed (and which increases the excess charge to $q + 1$) is taken from the χ_μ AO. If the VOIE is given as a continuous function of charge, then we may write[28]

$$H_{\mu\mu} = \frac{\partial[\text{VOIE}(q)]}{\partial q} \qquad (4.50)$$

Equation (4.49) is the finite difference equivalent of Eq. (4.50); it is the form used by most authors.[29]

4. RESONANCE INTEGRAL

The simplest approximation to the resonance integral $H_{\mu\nu}$ is direct proportionality to the overlap:

$$H_{\mu\nu} = kS_{\mu\nu} \qquad (4.51)$$

The applicability of Eq. (4.51) is implied in simple Hückel theory by the manner in which β is varied to take account of changing bond lengths, of hetero-atom interactions, of twisting, etc. However, Eq. (4.51) is too primitive. In view of this, and the unspecified nature of the Hamiltonian $\hat{H}$, it is clear that one has considerable latitude in formulating an expression for $H_{\mu\nu}$. This liberty has been fully exercised; some (!) of the resultant forms for $H_{\mu\nu}$ are

$$H_{\mu\nu} = kS_{\mu\nu}; \qquad \text{Longuet-Higgins and Roberts}[30]-[32] \tag{4.52}$$

$$H_{\mu\nu} = kS_{\mu\nu}(H_{\mu\mu} + H_{\nu\nu})/2; \qquad \text{Wolfsberg and Helmholz}[1] \tag{4.53}$$

It is usually assumed that $1.4 < k < 2.0$ in Eq. (4.53).

$$H_{\mu\nu} = kS_{\mu\nu}(H_{\mu\mu}H_{\nu\nu})^{1/2}; \qquad \text{Ballhausen and Gray}[33]-[35] \tag{4.54}$$

$$H_{\mu\nu} = (2 - |S_{\mu\nu,\text{local}}|)S_{\mu\nu}(H_{\mu\mu} + H_{\nu\nu})/2; \qquad \text{Cusachs}[36] \tag{4.55}$$

where $|S_{\mu\nu}|$ is the absolute value of $S_{\mu\nu}$ and $S_{\mu\nu,\text{local}}$ implies that this is an overlap evaluated in the local (or diatomic) coordinate system defined by the two atomic centers on which the AO's χ_μ and χ_ν are located.

$$H_{\mu\nu} = kS_{\mu\nu}[2H_{\mu\mu}H_{\nu\nu}/(H_{\mu\mu} + H_{\nu\nu})]; \qquad \text{Yeranos}[37] \tag{4.56}$$

$$H_{\mu\nu} = S_{\mu\nu}[(H_{\mu\mu} + H_{\nu\nu})/2 + k]; \qquad \text{Morokuma}[38] \tag{4.57}$$

$$H_{\mu\nu} = T_{\mu\nu}(\text{exact}) + S_{\mu\nu}(U_{\mu\mu} + U_{\nu\nu})/2; \qquad \text{Newton, Boer and Lipscomb}[21] \tag{4.58}$$

(A) Wolfsberg–Helmholz Form

The Wolfsberg–Helmholz form [see Eq. (4.53)] is based on the Mulliken approximation. The Mulliken approximation[39] is: The expectation value $\langle \chi_\mu | \hat{f} | \chi_\nu \rangle$ of a function $\hat{f}$ which occupies the region between the coordinate origins of χ_μ and χ_ν and which is symmetrical about the midpoint between these two coordinate origins, is given approximately by

$$\langle \chi_\mu | \hat{f} | \chi_\nu \rangle \approx (S_{\mu\nu}/2)(\langle \chi_\mu | \hat{f} | \chi_\mu \rangle + \langle \chi_\nu | \hat{f} | \chi_\nu \rangle) \tag{4.59}$$

It is clear, given $\hat{f} = \hat{H}$, that Eqs. (4.53) and (4.59) are identical. Furthermore, it may be shown that Eq. (4.53), with $k = 1$, provides the lower boundary value of $H_{\mu\nu}$. However, it is doubtful that Eq. (4.59) obtains when $\hat{f} = \hat{H}$ and, thus, the above rationalization of Eq. (4.53) is dubious.

(B) Cusachs Form

The Cusachs form [see Eq. (4.55)] is based on

(i)   the decomposition $\hat{H} = \hat{T} + \hat{U}$;
(ii) the approximate validity of the virial theorem in the form

$$T_{\mu\mu} \approx -U_{\mu\mu}/2; \qquad H_{\mu\mu} \approx U_{\mu\mu}/2 \tag{4.60}$$

for a $1/r$ potential and an orbital χ_μ which is *almost* atomic;

(iii) the probability that Eq. (4.59) is inapplicable when $\hat{T} = \hat{f}$;
(iv) the empirical observation by Ruedenberg[40] that for carbon-

carbon $2p\pi$-$2p\pi$ interactions, the equation

$$T_{\mu\nu} = S_{\mu\nu}{}^2(T_{\mu\mu} + T_{\nu\nu})/2 \tag{4.61}$$

is valid within 1% over a wide range of C—C distances. The assumed generality of this equation by Cusachs[36] constitutes a weakness in the resultant expression for $H_{\mu\nu}$ [see Eq. (4.55)];

(v)   the applicability of Eq. (4.59) when $\hat{f} = \hat{U}$.

Combining items (i), (iv), and (v), it follows that

$$H_{\mu\nu} = \tfrac{1}{2}(T_{\mu\mu} + T_{\nu\nu})S_{\mu\nu}{}^2 + \tfrac{1}{2}(U_{\mu\mu} + U_{\nu\nu})S_{\mu\nu} \tag{4.62}$$

Substitution of Eq. (4.60) leads to

$$H_{\mu\nu} = (2 - S_{\mu\nu})S_{\mu\nu}(H_{\mu\mu} + H_{\nu\nu})/2 \tag{4.63}$$

In order to satisfy the upper boundary value of $H_{\mu\nu}$ it is necessary to write $(2 - S_{\mu\nu})$ as $(2 - |S_{\mu\nu}|)$; in order to satisfy rotational invariance conditions [see Sec. 4(F)], it is required that $(2 - |S_{\mu\nu}|)$ be written as $(2 - |S_{\mu\nu,\text{local}}|)$. In this way, Eq. (4.55) is obtained.

(C)   Numerical Comparison of Various $H_{\mu\nu}$

In Table 4.10, we compare the various $H_{\mu\nu}$ relationships for nearest-neighbor valence orbital interaction of σ type. Inspection of this table indicates that there is little difference between the results of Eqs. (4.53)–(4.58) but that Eq. (4.52) is inferior.

(D)   One-Center Resonance Integrals

The radial parts $R(r)$ of Slater-type wave functions are not orthogonal. Therefore, two AO's on the same atomic center which have the same l and m_l quantum numbers, but different principal quantum numbers, are not orthogonal in a single Slater basis. Thus, if the $3s$ AO and $4s$ AO of sulfur are represented by single-ζ wave functions, it follows that $S(3s, 4s) \neq 0$. By the same token, $H_{3s,4s}$ is also nonzero.

Now the $3s$ AO and $4s$ AO are orthogonal in the isolated atom, at least in the sense that they are eigenfunctions of the same one-electron Hamiltonian. It is merely our representative AO's, χ_{3s} and χ_{4s}, which are nonorthogonal. Thus, common sense dictates that we proceed in one of two following ways:

(i)   We arbitrarily define all one-center overlaps equal to zero, in which case all $H_{\mu\nu}$ of one-center nature are also zero. Thus, for example, $S(3s, 4s) \equiv 0$; $H_{3s,4s} = 0$. However, the resultant MO's now contain nonorthogonalities.

(ii)   We allow Schmidt orthogonalization of the one-center AO's in question. This produces zero overlap for all orbitals on any one center, but

TABLE 4.10. *Comparison of some semiempirical $H_{\mu\nu}$ expressions.*[a]

Orbital Interactions	$S_{\mu\nu}$	$H_{\mu\nu}$ (a.u.) (SCF)	$H_{\mu\nu}$ (a.u.) [Eq. (4.52); $k = 1.4592$ a.u. $= 39.69$ eV]	$H_{\mu\nu}$ (a.u.) [Eq. (4.53); $k = 1.441$]	$H_{\mu\nu}$ (a.u.) [Eq. (4.54); $k = -1.514$]	$H_{\mu\nu}$ (a.u.) [Eq. (4.56); $k = 1.591$]	$H_{\mu\nu}$ (a.u.) [Eq. (4.55); i.e., $(1.907 - \lvert S_{\mu\nu}\rvert)$]; $k = 1.907$	$H_{\mu\nu}$ (a.u.) [Eq. (4.58)]
				CH$_4$, Methane				
1s$_H$–2s$_C$	0.569	0.807	0.830 (2.85)	0.861 (6.69)	0.823 (2.0)	0.787 (−2.5)	0.196 (−0.9)	0.747 (−7.4)
1s$_H$–2p$_C$	0.269	0.238	0.393 (65.1)	0.196 (−17.64)	0.201 (−15.54)	0.206 (−13.44)	0.223 (−6.3)	0.240 (0.84)
				HCN, Hydrocyanic Acid				
1s$_H$–2s$_C$	0.537	0.794	0.784 (−1.25)	0.819 (3.14)	0.772 (−2.77)	0.728 (−8.32)	0.779 (−1.88)	0.756 (−4.78)
1s$_H$–2p$_C$	0.498	0.443	0.727 (64.1)	0.524 (18.28)	0.54 (21.9)	0.557 (25.77)	0.512 (15.55)	0.586 (32.27)
2s$_C$–2s$_N$	0.469	1.035	0.684 (−33.9)	1.169 (12.94)	1.220 (17.87)	1.274 (23.09)	1.167 (12.75)	0.966 (−6.66)
2s$_C$–2p$_N$	0.397	0.793	0.579 (−26.98)	0.624 (−21.31)	0.602 (−24.08)	0.580 (−26.84)	0.654 (−17.52)	0.679 (−14.4)
2p$_C$–2s$_N$	0.499	1.022	0.728 (−28.75)	1.008 (−1.36)	0.979 (−4.17)	0.952 (−6.84)	0.985 (−3.62)	0.915 (−10.44)
2p$_C$–2p$_N$	0.303	0.391	0.442 (13.08)	0.333 (−14.85)	0.346 (−11.5)	0.361 (−7.67)	0.371 (−5.11)	0.269 (−31.2)

C₂H₂, Acetylene

$1s_H$-$2s_C$	0.536	0.771	0.782 (1.42)	0.782 (1.42)	0.736 (−4.55)	0.694 (−10.0)	0.744 (−3.54)	0.731 (−5.19)
$1s_H$-$2p_C$	0.498	0.436	0.727 (66.74)	0.471 (8.02)	0.49 (12.40)	0.509 (16.74)	0.461 (5.73)	0.549 (25.9)
$2s_C$-$2s_C$	0.508	0.94	0.741 (−21.17)	1.07 (13.82)	1.123 (19.46)	1.18 (25.53)	1.037 (10.31)	0.866 (−7.87)
$2s_C$-$2p_C$	0.470	0.783	0.685 (−12.51)	0.748 (−4.46)	0.744 (−4.97)	0.74 (−5.49)	0.746 (−4.73)	0.702 (−10.34)
$2p_C$-$2p_C$	0.293	0.271	0.428 (57.9)	0.316 (16.60)	0.332 (22.5)	0.349 (28.78)	0.354 (30.62)	0.197 (−27.3)
$\bar{x}$			+11.28	+1.64	+2.19	+3.0	+2.42	−5.12
Δ			38.7	13.1	15.3	18.4	12.4	17.6

[a] The tabulated $H_{\mu\nu}$'s were calculated from SCF $H_{\mu\mu}$ data. The $H_{\mu\mu}$'s so approximated were compared with the exact SCF $H_{\mu\nu}$'s and a *best average value* of the parameter k was obtained. These values of k are listed along the top of the table. The bracketed numbers are percentage errors relative to the SCF $H_{\mu\nu}$ values. The average percent error $\bar{x}$ and the standard deviation Δ are listed along the bottom of the table. The SCF data were taken from minimal-basis-set SCF computations by Palke and Lipscomb [W. E. Palke and W. N. Lipscomb, *J. Am. Chem. Soc.* **88**, 2384 (1966)]. Further detail on this type of comparison is given by Carroll and McGlynn [D. G. Carroll and S. P. McGlynn, *J. Chem. Phys.* **45**, 3827 (1966)].

it also alters a number of the AO's. Any such alteration changes the Coulomb integrals. We cannot permit such a change because the Coulomb integrals are of an experimental nature and are not dependent on the vagaries of the chosen AO basis set. In order to avoid difficulty, we must proceed in the manner outlined for $3s$ and $4s$ AO's of sulfur: Set $H_{3s,4s} = H_{33}S(3s, 4s)$ and, using an abbreviated notation, write the 2×2 secular determinant

$$\begin{vmatrix} H_{33} - \epsilon & H_{33}S_{34} - \epsilon S_{34} \\ H_{33}S_{34} - \epsilon S_{34} & H_{44} - \epsilon \end{vmatrix} = 0 \qquad (4.64)$$

Solution provides

$$\chi_{3s}' = \chi_{3s}; \qquad \epsilon' = H_{33} \qquad (4.65)$$

$$\chi_{4s}' = (\chi_{4s} - S_{34}\chi_{3s})/(1 - S_{34}{}^2)^{1/2}; \qquad \epsilon' = (H_{44} - H_{33}S_{34}{}^2)/(1 - S_{34}{}^2) \qquad (4.66)$$

Thus, the AO's, χ_{3s} and χ_{4s}, have been orthogonalized and the Coulomb integral for the $3s$ AO has remained unchanged. However, the Coulomb integral for the $4s$ AO has been changed and must be *back corrected* to its original value. This we achieve by setting [see Eq. (4.66)]

$$H_{44} = H_{44}'(1 - S_{34}{}^2) + H_{33}S_{34}{}^2 \qquad (4.67)$$

where H_{44}' (or ϵ') is evaluated in the orthogonal basis.

(E) Coordinate Considerations

It is convenient to adopt a molecular coordinate system and to specify the positions of all atoms in the molecule relative to it. Symmetry considerations sometimes specify a convenient choice for the axes of this coordinate system. We may also fixate coordinate systems on the various atoms of the molecule; these are called the *local coordinate systems*. If the local coordinate systems on two centers are chosen to facilitate overlap computations between these centers, these are called the *diatomic coordinate systems*.

In the evaluation of the overlap matrix, the set of overlaps between the subset of (nl) AO's on one center with those on another comprises a total of $(2l + 1)^2$ entries (9 for p AO's, 25 for d AO's, etc.). However, in the diatomic coordinate system, there are only two distinct p overlaps (σ and π) and three distinct d overlaps (σ, π, and δ). Thus, it is convenient to transform from the molecule coordinate system to various diatomic coordinate systems, evaluate the overlaps in these systems, and then *back transform* to the molecule system. In other words, we write

$$\chi_\mu{}^M(n_\mu, l_\mu, m_\mu) = \sum_\nu c_{\mu\nu}\chi_\nu{}^D(n_\mu, l_\mu, m_\nu) \qquad (4.68)$$

where M and D refer to the molecule and diatomic coordinate systems, respectively. It is clear that this procedure simplifies computation only if the complete (nl) subset of AO's is available on each atomic center and if the radial parts of each AO in this subset are identical.

The overlap of $\chi_\mu{}^M$ with $\chi_\zeta{}^M$ is given by

$$S_{\mu\zeta} = \langle \chi_\mu{}^M \mid \chi_\zeta{}^M \rangle$$

$$= \langle \sum_\nu c_{\mu\nu}\chi_\nu{}^D (n_\mu, l_\mu, m_\nu) \mid \sum_\eta c_{\zeta\eta}\chi_\eta{}^D (n_\zeta, l_\zeta, m_\eta) \rangle$$

$$= \sum_\nu \sum_\eta c_{\mu\nu}{}^* c_{\zeta\eta} \langle \chi_{(\mu)\nu}{}^D \mid \chi_{(\zeta)\eta}{}^D \rangle \tag{4.69a}$$

By virtue of the orthogonality determined by $S_{\nu\eta}{}^D = S\delta_{\nu,\eta}$ in the diatomic coordinate system, it follows that

$$S_{\mu\zeta}{}^M = \sum_\nu c_{\mu\nu}{}^* c_{\zeta\nu} S_{(\mu)\nu(\zeta)\nu}{}^D \tag{4.70}$$

This is the simplification referred to in the previous paragraph. A similar simplification occurs for any operator $\hat{P}$ which is not dependent on the angular functions θ and ϕ. If, however, the operator is dependent on angle the simplification of Eq. (4.69) to Eq. (4.70) may not be effected. Worse still, the computation of the expectation value $\langle \hat{P}(\theta, \phi) \rangle$ can become rotationally noninvariant.

(F) Transformation Invariance

LCAO MO's should be invariant under any linear transformation of the basis set.[41] Within the framework of the MWH method, this means that a transformation which rotates the local coordinate systems with respect to the fixed molecular coordinate system, and which interchanges orbitals with the same n and l quantum numbers on the same atom (e.g., the three carbon $2p$ AO's or the five $3d$ AO's of a transition metal), should not affect the calculated MO energies. We can achieve this invariance by imposing two conditions:

(i)   All orbitals with the same n and l quantum numbers on the one center have the same value for $H_{\mu\mu}$ and interact with other orbitals using the same value of k in Eqs. (4.52)–(4.56).

(ii) H_{ij} is linearly related to S_{ij}.

There can be some relaxation of condition (i) for practical purposes. For instance, in planar aromatic hydrocarbons the π orbitals could be identified with the carbon $2p_z$ AO's and the σ framework then taken to lie in the xy plane. In this case, invariance is required only with respect to rotation in the xy plane and we can use different $H_{\mu\mu}$ and k values for the $\sigma(2p_x, 2p_y)$ and $\pi(2p_z)$ orbitals. Equations (4.52)–(4.54) and (4.56) obey condition (ii); Eq. (4.55) at first sight, does not. To retain rotational invariance using Eq. (4.55), we proceed as follows: We adopt Eq. (4.69a) in the form

$$S_{\mu\varsigma}{}^M = \sum_\nu \sum_\eta c_{\mu\nu}{}^* c_{\varsigma\eta} \langle \chi_{(\mu)\nu}{}^D \mid \chi_{(\varsigma)\eta}{}^D \rangle \tag{4.69b}$$

The quantity $[S(2 - \mid S \mid)]_{\mu\varsigma}$ is now computed for pairs of Slater AO's in the diatomic coordinate system and is then rotated to the molecular coordinate system, the entire factor being treated as a single term which transforms as an overlap integral. Thus

$$[S(2 - \mid S_{\text{local}} \mid)]_{\mu\varsigma}{}^M \equiv \sum_\nu \sum_\eta c_{\mu\nu}{}^* c_{\varsigma\eta} [S_{(\mu)\nu(\varsigma)\eta}{}^D (2 - \mid S_{(\mu)\nu(\varsigma)\eta}{}^D \mid)] \tag{4.71}$$

Under these conditions, the proper rotational invariance is retained. An illustration of the use of Eq. (4.71) for two p AO's on two different centers is given below:

$$S_{\mu\varsigma}{}^M = S_\sigma{}^D \cos\theta \cos\phi + S_\pi{}^D \sin\theta \sin\phi \tag{4.72}$$

and

$$[(2 - \mid S_{\text{local}} \mid) S]_{\mu\varsigma}{}^M = [(2 - S_\sigma{}^D) S_\sigma{}^D] \cos\theta \cos\phi$$
$$+ [(2 - S_\pi{}^D) S_\pi{}^D] \sin\theta \sin\phi \tag{4.73}$$

5. AN EXAMPLE CALCULATION: WATER MOLECULE

Coordinate axes and molecular structure are shown in Fig. 4.4. We use the following valence orbitals:

Oxygen: $2s, 2p_x, 2p_y, 2p_z$; $\varsigma = 2.275$

Hydrogen: $1s$; $\varsigma = 1.0$

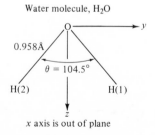

Water molecule, H_2O

x axis is out of plane

FIG. 4.4. Structure of the water molecule. Structural information on H_2O and many other molecules may be obtained in a number of sources [L. E. Sutton, *Interatomic Distances* (Special Publications 11 (1958) and 18 (1965), Burlington House, The Chemical Society, London); L. Pauling, *The Nature of the Chemical Bond* (Cornell University Press, Ithaca, N.Y., 1960); A. F. Wells, *Structural Inorganic Chemistry* (Oxford University Press, New York, 1962)].

TABLE 4.11.   *Overlap matrix for water molecule.*

	$2s$	$2p_z$	$2p_x$	$2p_y$	H(1)	H(2)
$2s$	1.0	0.0	0.0	0.0	0.509	0.509
$2p_z$	0.0	1.0	0.0	0.0	0.213	0.213
$2p_x$	0.0	0.0	1.0	0.0	0.0	0.0
$2p_y$	0.0	0.0	0.0	1.0	0.275	−0.275
H(1)	0.509	0.213	0.0	0.275	1.0	0.377
H(2)	0.509	0.213	0.0	−0.275	0.377	1.0

The total number of valence electrons is eight. The AO representation is the single STF—the orbital exponents ζ being given by Slater's rules (see Appendix B).

(A)   Overlap Matrix

The overlap matrix is given in Table 4.11.

To obtain the overlap between the $2p_z$ AO and the $1s$ AO of H(1), we first calculate $S(1s, 2p\sigma)$. The result is 0.348. The overlap desired is then obtained by multiplying $S(1s, 2p\sigma)$ by $\cos(\theta/2) = 0.612$; the result is 0.213.

(B)   Energy Matrix

The neutral atom VOIE's are taken from the promotion energies of Hinze and Jaffé.[15] They are

$$2s \text{ VOIE} = 32.30 \text{ eV}; \qquad 2p \text{ VOIE} = 14.61 \text{ eV}; \qquad 1s \text{ VOIE} = 13.6 \text{ eV}$$

$$(4.74)$$

The off-diagonal terms are obtained from the Wolfsberg–Helmholz equation [Eq. (4.53)] with $k \equiv 1.75$. The energy matrix is given in Table 4.12.

TABLE 4.12.   *Energy matrix for water molecule.*

	$2s$	$2p_z$	$2p_x$	$2p_y$	H(1)	H(2)
$2s$	−32.30	0.0	0.0	0.0	−20.443	−20.443
$2p_z$	0.0	−14.61	0.0	0.0	−5.258	−5.258
$2p_x$	0.0	0.0	−14.61	0.0	0.0	0.0
$2p_y$	0.0	0.0	0.0	−14.61	−6.788	6.788
H(1)	−20.443	−5.258	0.0	−6.788	−13.6	−8.973
H(2)	−20.443	−5.258	0.0	6.788	−8.973	−13.6

(C)   MO Energies and Eigenvectors

The secular determinant $\mid H_{\mu\nu} - \epsilon S_{\mu\nu} \mid = 0$ is constructed by addition of Tables 4.11 and 4.12. Upon diagonalization, it yields the MO energies and eigenvectors of Table 4.13.

The manner in which the various MO's transform in the C_{2v} point group is also given in Table 4.13. The oxygen $2s$ and $2p_z$ AO's transform as a_1, the $2p_x$ AO as b_1, and the $2p_y$ AO as b_2. Since no MO's in which the oxygen AO's are not involved occur in Table 4.13, the symmetries of all MO's are immediately deducible.

(D)   Population Analysis

We now perform a Mulliken population analysis as described in Sec. 8, Chapter 1. The number of electrons associated with a given AO in the rth MO is given by

$$POP_{r\mu} = n_r(c_{r\mu}{}^2 + \sum_{\nu \neq \mu} c_{r\mu}c_{r\nu}S_{\mu\nu}) \qquad (4.75)$$

when the AO coefficients are real and where n_r is the number of electrons in the rth MO. Now there are a total of eight valence electrons; in the ground-state electron configuration of water there are two electrons in each of the MO's 6, 5, 4, and 3 and no electrons in MO's 2 and 1. Thus, the AO populations in the various MO's are those given in Table 4.14. The small negative value of $POP_{6,2p_z}$ is physically nonsensical; it is caused by a deficiency in the Mulliken definition.

TABLE 4.13.   *MO energies and eigenvectors for water molecule.*

	MO					
AO	1	2	3	4	5	6
$2s$	0.9565	0	0	0.1664	0	0.8553
$2p_z$	0.4815	0	0	-0.9411	0	-0.0254
$2p_x$	0	0	0	0	0	0
$2p_y$	0	0.8450	1	0	0.7790	0
$1s$ H (1)	-0.7885	-0.9799	0	-0.1122	0.3147	0.1291
$1s$ H (2)	-0.7885	0.9799	0	-0.1122	-0.3147	0.1291
Energy (eV)	$+8.64$	$+3.16$	-14.61	-15.10	-16.44	-33.55
Transform behavior in C_{2v}	a_1	b_2	b_1	a_1	b_2	a_1

TABLE 4.14. *AO populations as a function of MO for water molecule [values of POP$_{r\mu}$ of Eq. (4.75)].*

				MO		
AO	1	2	3	4	5	6
$2s$	0	0	0	0.0174	0	1.6878
$2p_z$	0	0	0	1.8613	0	−0.0015
$2p_x$	0	0	2.0	0	0	0
$2p_y$	0	0	0	0	1.4833	0
$1s$ H(1)	0	0	0	0.0607	0.2584	0.1568
$1s$ H(2)	0	0	0	0.0607	0.2584	0.1568

The summation of POP$_{r\mu}$ over all MO's, namely,

$$POP_\mu = \sum_r POP_{r\mu} = \sum_r n_r(c_{r\mu}^2 + \sum_{\nu \neq \mu} c_{r\mu}c_{r\nu}S_{\mu\nu}) \tag{4.76}$$

provides the total electron population of the χ_μ AO. These quantities are displayed along the diagonal of Table 4.15. The quantity

$$OV\ POP_{r(\mu,\nu)} = 2n_r c_{r\mu}c_{r\nu}S_{\mu\nu}, \qquad \mu \neq \nu \tag{4.77}$$

is termed the *overlap population of AO's μ and ν in the rth MO.* The quantity

$$OV\ POP_{(\mu,\nu)} = 2\sum_r n_r c_{r\mu}c_{r\nu}S_{\mu\nu}, \qquad \mu \neq \nu \tag{4.78}$$

is termed the *total overlap population.* It relates to the total electron density induced by AO's χ_μ and χ_ν between the two centers on which they are located. This quantity (i.e., $OV\ POP_{(\mu,\nu)}$) is given by the off-diagonal elements of Table 4.15. The diagonal elements of Table 4.15, of course, are the sums of the rows of Table 4.14.

TABLE 4.15. *Total electron populations [Eq. (4.76)] and total overlap populations [Eq. (4.78)] for water molecule.*

	$2s$	$2p_z$	$2p_x$	$2p_y$	$1s$ H(1)	$1s$ H(2)
$2s$	1.7052	0	0	0	0.1867	0.1867
$2p_z$	0	1.8598	0	0	0.0872	0.0872
$2p_x$	0	0	2	0	0	0
$2p_y$	0	0	0	1.4833	0.2698	0.2698
$1s$ H(1)	0.1867	0.0872	0	0.2698	0.4758	−0.1051
$1s$ H(2)	0.1867	0.0872	0	0.2698	−0.1051	0.4758

The total electron population on oxygen is found by addition of the oxygen AO populations. It is 7.0483 electrons. Since neutral oxygen possesses six valence electrons, the net charge on the oxygen of water is $-1.048 \mid e \mid$. The hydrogen atoms each bear net positive charge $+0.524 \mid e \mid$.

(E)   Evaluation of Results

We can evaluate the worth of the results provided by the computation in two ways: correlation with the results of either experiment or *better* computations.

(*i*)   *Atomic Charges:*   A Mulliken population analysis of the results of a nonempirical H_2O calculation by Ellison and Shull[42] provides charge of $-0.35 \mid e \mid$ for oxygen; the value obtained here was $-1.048 \mid e \mid$. The discrepancy is quite large.

The VOIE's which were used as input $H_{\mu\mu}$ data referred to the neutral atomic species; however, the output eigenvector data indicated that the atoms are quite ionic. Thus, the output-input information is contradictory. It seems necessary to require consistency of both sets of data. A method for doing this (i.e., iteration to charge self-consistency) has been developed[19]; it is discussed in Sec. 7.

(*ii*)   *Total Energy and Molecule Geometry:*   The total energy of the molecule is

$$E = E' + \sum_r n_r \epsilon_r \qquad (4.79)$$

where E' is the energy associated with the inner or nonvalence electrons. Thus, for H_2O, the total energy is

$$E = E' + 2(-33.35 - 16.44 - 15.10 - 14.61)$$
$$= (E' - 159.0) \text{ eV} \qquad (4.80)$$

If E' is assumed to be the same for the isolated atoms as it is for H_2O, the total energy of the three separated neutral atoms is

$$E = E' + 2(-32.3) + 2(-13.60) + 4(-14.61)$$
$$= (E' - 150.24) \text{ eV} \qquad (4.81)$$

Consequently, the binding energy of the water molecule is 8.76 eV.

Binding energies, computed as above, do reproduce the experimental trend of the atomization energies for different series of hydrocarbons. However, the computed values are usually larger than the experimental numbers.

If E be computed as a function of molecular geometry, it follows that the minimum E should pertain to the equilibrium molecule. Many authors

TABLE 4.16.   *Orbital energies for water molecule.*

MO	Extended Huckel (eV)	SCF (eV)[a]	Experiment (eV)
a_1	-33.35	-36.19	$\cdots$
b_2	-16.44	-18.55	16.2 ± 0.3^b
a_1	-15.10	-13.2	14.5 ± 0.3^b
b_1	-14.61	-11.79	12.61^c

[a] From Ellison and Shull, Ref. [42] of text.
[b] Electron impact data of W. C. Price and T. M. Sugden, *Trans. Faraday Soc.* **44,** 108 (1948).
[c] F. H. Field and J. L. Franklin, *Electron Impact Phenomena* (Academic Press Inc., New York, 1957), p. 280.

have obtained good agreement between the computed equilibrium geometry and the experimental molecular structure data.

That any agreement of the types specified above should exist, is very surprising. The Hamiltonian does not include any nuclear repulsion terms; it takes account of interelectron repulsion using a very specious $H_{\mu\nu}$ recipé; and it pays little attention to electron-other-nucleus attractive forces. Some effort to rationalize the existence of such agreement has been made by Boer *et al.*[43]

(*iii*) *MO Energies and Ionization Potentials:* MO energies are compared in Table 4.16. The computations of Ellison and Shull[42] are in qualitative agreement with those of the extended Hückel approach, and both of these compare equally well with the experimental data.

(*iv*) *Electronic Spectra:* The lowest-energy electronic transition is predicted to occur at $\sim$17.77 eV and to be of an MO excitation type $b_2 \leftarrow b_1$. The observed transition energy is $\sim$7.75 eV. There is significant discrepancy. It is doubtful if the method is of utility for electronic spectroscopy studies.

6. AN EXAMPLE CALCULATION: HEXAFLUOROTITANIUM(III) ANION

The calculation described here is patterned after the work of Bedon, Horner, and Tyree.[44] The coordinate system and molecule geometry is specified in Fig. 4.5.

Titanium hexafluoride

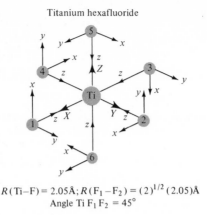

$R(\text{Ti–F}) = 2.05\text{Å}; R(\text{F}_1 - \text{F}_2) = (2)^{1/2} (2.05)\text{Å}$
Angle Ti F₁ F₂ = 45°

FIG. 4.5. Molecular coordinate system (X, Y, Z) and diatomic coordinate systems (x, y, z) for TiF_6^{3-}. The molecule is octahedral. The F positions are numbered 1–6.

The calculations reported here could be performed in the same way as those on the water molecule. However, many of the methods in the literature, while not in any way better, are considerably more occult than is the treatment of H_2O given in Sec. 5. Therefore, we now introduce a few of the stratagems that frequently find use. We do this in the hope that we clarify these stratagems; however, we also run a considerable risk of overcomplicating our discussion of TiF_6^{3-}. We introduce the following computational *tactics*:

(i)   the use of group theory (to simplify the secular equation);

(ii)   the use of hybrid AO's on the ligands (which enables us, perhaps wrongly, to truncate the basis set);

(iii)   the use of improperly normalized group overlap integrals (because they are often used);

(iv)   the use of hydride ionization potentials (which are sometimes used);

(v)   the use of multi-ζ SCF AO wave functions[45] (to familiarize the reader with them);

(vi)   we perform only the last cycle of a charge self-consistent calculation.[3] (The H_2O results were unaesthetic: The output and input electron configurations were not identical.)

---

[3] Such calculations are discussed in Sec. 7. See also Exercises 11 and 12.

(A)   Atomic Orbitals

The AO basis set consists of

$$Ti: \ 3d, \ 4s, \ 4p$$

$$F: \ 2s, \ 2p$$

These AO's are represented by multi-ζ wave functions which are given in Table 4.17.

(B)   Hybridization of Fluorine Orbitals

In order to limit the computational effort—and also because it seems physically reasonable—the fluorine basis set is truncated to two $2p\pi$ AO's and one σ AO on each fluorine. The terms σ and π are defined with respect to the diatomic coordinate system based on Ti and the F atom in question.

The σ AO which is directed toward the Ti center is given by

$$\chi_\sigma{}^+ = (\sin \rho)\chi_{2s} + (\cos \rho)\chi_{2p\sigma} \tag{4.82}$$

TABLE 4.17.   *Atomic-orbital multi-ζ wavefunctions.*[a,b]

AO	Species	Configuration	Wave function
$3d$	Ti^{+1}	$3d^3$	$0.4391\,(3d, 4.55) + 0.7397\,(3d, 1.60)$
$4s$	Ti	$3d^2 4s^2$	$-0.02231\,(1s, 21.40) + 0.7751\,(2s, 8.05)$ $-0.1985\,(3s, 3.64) + 1.0164\,(4s, 1.20)$
$4p$	Ti^{+1}	$3d^2 4p^1$	$0.04642\,(2p, 8.8) - 0.17729\,(3p, 3.31)$ $+1.01438\,(4p, 1.08)$
$2s$	F	$2s^2 2p^5$	$0.05531\,(1s, 12.187) - 0.35720\,(1s, 8.189)$ $+0.01767\,(2s, 14.296) + 0.13038\,(2s, 4.096)$ $+0.67725\,(2s, 2.767) + 0.23491\,(2s, 1.716)$ $+0.04834\,(3s, 3.75)$
$2p$	F	$2s^2 2p^5$	$0.16774\,(2p, 5.954) + 0.06352\,(2p, 3.068)$ $+0.58026\,(2p, 1.752) + 0.03579\,(2p, 0.7863)$ $+0.2753\,(3p, 4.17)$

[a] From Clementi, Ref. [45] of text. Reprinted by permission.
[b] For $0.4391\,(3d, 4.55)$ read: A $3d$ STF with exponent $\zeta = 4.55$; the weighting factor of this STF in the multi-ζ wave function is 0.4391.

The hybridization parameter ρ is chosen to minimize[33],[46] the ratio[4] $\text{VOIE}(\rho)/S(\rho)$ given by

$$\frac{\text{VOIE}(\rho)}{S(\rho)} = \frac{\sin^2 \rho (2s\ \text{VOIE}) + \cos^2 \rho (2p\sigma\ \text{VOIE})}{\sin \rho S(2s\sigma, 3d\sigma) + \cos \rho S(2p\sigma, 3d\sigma)} \tag{4.83}$$

The minimum occurs at $\rho = 0.36$ rad. Consequently, we find

$$\chi_\sigma^+ = (0.13)^{1/2}\chi_{2s} + (0.87)^{1/2}\chi_{2p\sigma} \tag{4.84}$$

$$\chi_\sigma^- = (0.87)^{1/2}\chi_{2s} - (0.13)^{1/2}\chi_{2p\sigma} \tag{4.85}$$

The χ_σ^- AO points away from the titanium center; it is assumed to remain nonbonding throughout. Each such χ_σ^- contains two electrons, for a total of twelve nonbonding electrons altogether.

(C)   Transformation Properties

The manner in which the AO's and Ligand Group Orbitals (LGO's) transform in the O_h point group is given in Table 4.18. The LGO's were found in the manner outlined by Cotton.[47] The use of group theory simplifies computational considerations a great deal. Thus, if the elements of the secular determinant refer only to the metal AO and LGO basis sets of Table 4.18, this determinant may be partially diagonalized on the basis of symmetry alone. In fact, the full 11×11 secular determinant breaks down into three 2×2 determinants (for a_{1g}, e_g, and t_{2g} cases), one 3×3 determinant (for the t_{1u} case) and two 1×1 determinants (for t_{1g} and t_{2u} cases).

(D)   Overlap Integrals

The overlap integrals between the AO's of Table 4.18, in the diatomic coordinate system, are

$$S(2p\sigma, 3d\sigma) = 0.130 \qquad S(2s\sigma,\ 3d\sigma) = 0.161$$

$$S(2p\sigma, 4s\sigma) = 0.103 \qquad S(2s\sigma,\ 4s\sigma) = 0.219$$

$$S(2p\sigma, 4p\sigma) = 0.0765 \qquad S(2s\sigma,\ 4p\sigma) = 0.35$$

$$S(2p\pi, 3d\pi) = 0.101 \qquad S(2p\pi, 4p\pi) = 0.135 \tag{4.86}$$

where the ordering convention used is $S(\text{F AO, Ti AO})$.

---

[4] The hybridization parameter should provide as large an overlap of χ_σ^+ as possible and as low a χ_σ^+ VOIE as possible. These requirements are approximated by seeking the extremum of the ratio $\text{VOIE}(\rho)/S(\rho)$. The choice of $3d\sigma$ of titanium for the overlap maximization implies an assumption that all other bonding in the molecule is prostituted to the single requirement of maximum Ti-F σ bonding. This assumption may or may not be reasonable.

The overlap integrals involving the hybridized χ_σ^+ AO of fluorine are obtained in the diatomic coordinate system using Eq. (4.84). They are

$$S(\sigma^+, 3d\sigma) = 0.180$$
$$S(\sigma^+, 4s\sigma) = 0.175$$
$$S(\sigma^+, 4p\sigma) = 0.198 \qquad (4.87)$$

TABLE 4.18. *Transformation properties of metal and ligand orbitals in the octahedral point group.*

Repre-sentation	Metal Orbital	Ligand Orbitals[a,b] σ^+	π
a_{1g}	$4s$	$(1/6)^{1/2}(\sigma_1^+ + \sigma_2^+ + \sigma_3^+ + \sigma_4^+ + \sigma_5^+ + \sigma_6^+)$	
e_g	$3d_{x^2-y^2}$	$(1/2)(\sigma_1^+ - \sigma_2^+ + \sigma_3^+ - \sigma_4^+)$	
	$3d_{z^2}$	$(1/12)(2\sigma_5^+ + 2\sigma_6^+ - \sigma_1^+ - \sigma_2^+ - \sigma_3^+ - \sigma_4^+)$	
t_{1u}	$4p_x$	$(1/2)^{1/2}(\sigma_1^+ - \sigma_3^+)$	$(1/2)(y_5 + x_2 + y_6 - x_4)$
	$4p_y$	$(1/2)^{1/2}(\sigma_2^+ - \sigma_4^+)$	$(1/2)(y_1 + x_5 + y_3 - x_6)$
	$4p_z$	$(1/2)^{1/2}(\sigma_5^+ - \sigma_6^+)$	$(1/2)(x_1 + y_2 - x_3 + y_4)$
t_{2g}	$3d_{xy}$		$(1/2)(y_1 + x_2 - y_3 + x_4)$
	$3d_{xz}$		$(1/2)(x_1 + y_5 + x_3 - y_6)$
	$3d_{yz}$		$(1/2)(x_5 + y_2 + x_6 - y_4)$
t_{1g}			$(1/2)(y_1 - x_2 - y_3 - x_4)$
			$(1/2)(x_1 - y_5 + x_3 + y_6)$
			$(1/2)(x_5 - y_2 + x_6 + y_4)$
t_{2u}			$(1/2)(y_5 - x_2 + y_6 + x_4)$
			$(1/2)(y_1 - x_5 + y_3 + x_6)$
			$(1/2)(x_1 - y_2 - x_3 - y_4)$

[a] We use the abbreviation $\chi_\sigma^+ = \sigma^+$ in this table and henceforth.
[b] The symbols y and x represent $2p_y$ and $2p_x$ AO's on a fluorine atom. The numeric subscripting is defined in Fig. 4.5.

The overlap integrals between the LGO's and the metal AO's are

$$G[a_{1g}(\sigma)] = (1/6)^{1/2}\langle(\sigma_1{}^+ + \sigma_2{}^+ + \sigma_3{}^+ + \sigma_4{}^+ + \sigma_5{}^+ + \sigma_6{}^+) \,|\, 4s\sigma\rangle$$

$$= (6)^{1/2}S(\sigma^+, 4s\sigma) = 0.429 \tag{4.88}$$

$$G[t_{1u}(\sigma)] = (1/2)^{1/2}\langle(\sigma_1{}^+ - \sigma_3{}^+) \,|\, 4p_x\rangle = (2)^{1/2}S(\sigma^+, 4p\sigma) = 0.279 \tag{4.89}$$

$$G[t_{1u}(\pi)] = (1/2)\,\langle y_5 + x_2 + y_6 - x_4 \rangle \,|\, 4p_x\rangle = 2S(2p\pi, 4p\pi) = 0.270 \tag{4.90}$$

$$G[t_{2g}(\pi)] = (1/2)\,\langle(y_1 + x_2 - y_3 + x_4) \,|\, 3d_{xy}\rangle = 2S(2p\pi, 3d\pi) = 0.201 \tag{4.91}$$

$$G[e_g(\sigma)] = \langle(1/3)^{1/2}(\sigma_5{}^+ + \sigma_6{}^+) - (1/12)^{1/2}(\sigma_1{}^+ + \sigma_2{}^+ + \sigma_3{}^+ + \sigma_4{}^+) \,|\, 3d_{z^2}\rangle$$

$$= (3)^{1/2}S(\sigma^+, 3d\sigma) = 0.311 \tag{4.92a}$$

We use the term G to distinguish between an overlap integral involving group orbitals of the ligands (i.e., a multicenter overlap integral) and a two-center overlap integral which we commonly denote S.

The only G integral which might cause difficulty is $G[e_g(\sigma)]$. Its evaluation proceeds as follows: A set of very simple transformations may be used to rotate the molecular coordinate system into each one of the diatomic coordinate systems specified by the axes designations on centers 1–6 of Fig. 4.5. These transformations are

$$\text{Set 1:} \quad x^2 \to z^2;\ y^2 \to y^2;\ z^2 \to x^2$$

$$\text{Set 2:} \quad x^2 \to x^2;\ y^2 \to z^2;\ z^2 \to y^2$$

$$\text{Set 3:} \quad x^2 \to z^2;\ y^2 \to y^2;\ z^2 \to x^2$$

$$\text{Set 4:} \quad x^2 \to x^2;\ y^2 \to z^2;\ z^2 \to y^2$$

$$\text{Set 5:} \quad x^2 \to y^2;\ y^2 \to x^2;\ z^2 \to z^2$$

$$\text{Set 6:} \quad x^2 \to y^2;\ y^2 \to x^2;\ z^2 \to z^2 \tag{4.93}$$

We now represent the angular part of $3d_{z^2}$ by $3z^2 - r^2$ and subject z^2 to the transformations of Eq. (4.93). This leads to

$$G[e_g(\sigma)] = (1/3)^{1/2}\langle\sigma_5{}^+ \,|\, (3z^2 - r^2)\rangle + (1/3)^{1/2}\langle\sigma_6{}^+ \,|\, (3z^2 - r^2)\rangle$$

$$- (1/12)^{1/2}\langle\sigma_1{}^+ \,|\, (3x^2 - r^2)\rangle - (1/12)^{1/2}\langle\sigma_2{}^+ \,|\, (3y^2 - r^2)\rangle$$

$$- (1/12)^{1/2}\langle\sigma_3{}^+ \,|\, (3x^2 - r^2)\rangle - (1/12)^{1/2}\langle\sigma_4{}^+ \,|\, (3y^2 - r^2)\rangle \tag{4.94}$$

Since the subscripts 1–6 of Eq. (4.94) have now lost their previous significance, Eq. (4.94) is written as

$$G[e_g(\sigma)] = 2(1/3)^{1/2}\langle\sigma^+ \mid 3d\sigma\rangle - (1/12)^{1/2}\langle\sigma^+ \mid 6x^2 + 6y^2 - 4r^2\rangle \quad (4.95)$$

Upon noting that $6x^2 + 6y^2 - 4r^2 = 2r^2 - 6z^2 = -2(3z^2 - r^2)$, Eq. (4.95) simplifies to

$$G[e_g(\sigma)] = 2(1/3)^{1/2}\langle\sigma^+ \mid 3d\sigma\rangle + 2(1/12)^{1/2}\langle\sigma^+ \mid 3d\sigma\rangle$$

$$= (3)^{1/2}\langle\sigma^+ \mid 3d\sigma\rangle \quad (4.92b)$$

(E) Coulomb Integrals

The VOIE's for the titanium $3d$, $4s$, and $4p$ AO's are given in Table 4.19.

Prior computations of a charge self-consistent type indicate that the final charge on Ti is $+0.5138 \mid e \mid$ and that the final electron configuration is

$$\text{Ti}^{+0.5138}: \ldots 3d^{2.60863}4s^{0.37622}4p^{0.50136} \quad (4.96)$$

This knowledge, of course, is not necessary. The assumed initial configuration might be something dear to the reader's heart, say, Ti^0: $\ldots 3d^34s^1$; convergence procedures which terminate on the charge self-consistent configuration yield the configuration of Eq. (4.96). We merely wish to perform here the final cycle of a charge self-consistent computation; hence, our use of Eq. (4.96) as an initiation point.

TABLE 4.19. *Valence orbital ionization energies (VOIE's) of titanium.[a]*

AO	Configuration	VOIE
$3d$	$3d^v$	$17.15q^2 + 60.85q + 27.4$
	$3d^{v-1}4s$	$18.45q^2 + 77.85q + 44.6$
	$3d^{v-1}4p$	$18.45q^2 + 76.75q + 55.4$
$4s$	$3d^{v-1}4s$	$9.3q^2 + 50.4q + 48.6$
	$3d^{v-2}4s^2$	$9.3q^2 + 58.5q + 57.2$
	$3d^{v-2}4s4p$	$9.3q^2 + 55.0q + 66.0$
$4p$	$3d^{v-1}4p$	$7.8q^2 + 35.6q + 26.9$
	$3d^{v-2}4p^2$	$7.8q^2 + 48.9q + 35.9$
	$3d^{v-2}4s4p$	$7.8q^2 + 48.9q + 34.4$

[a] From Table D.2, Appendix D.

At any rate, the VOIE's evaluated [as in Sec. 2(C)] at the charge self-consistent configuration of Eq. (4.96) are

$$3d \text{ VOIE}(\ldots 3d^{3.4862}) = 63.192 \text{ kK}$$

$$(\ldots 3d^{2.4862}4s) = 89.47 \text{ kK}$$

$$(\ldots 3d^{2.4862}4p) = 99.705 \text{ kK} \qquad (4.97)$$

Utilizing the relation

$$\ldots 3d^{2.60863}4s^{0.37622}4p^{0.50136} = a(\ldots 3d^{3.4862}) + b(\ldots 3d^{2.4862}4s)$$

$$+ c(\ldots 3d^{2.4862}4p) \qquad (4.98)$$

we find the values of a, b, and c to be

$$a = 0.12242; \qquad b = 0.37622; \qquad c = 0.50136 \qquad (4.99)$$

The $3d$ VOIE of relevance [i.e., $3d$ VOIE corresponding to the configuration of Eq. (4.96)] is then

$$3d \text{ VOIE} = 0.12242(63.192) + 0.37622(89.47) + 0.50136(99.705)$$

$$= 91.384 \text{ kK} \qquad (4.100)$$

The $4s$ and $4p$ VOIE's for the configuration of Eq. (4.96) are obtained in like manner, they are

$$4p \text{ VOIE} = 46.060 \text{ kK}; \qquad 4s \text{ VOIE} = 78.898 \text{ kK} \qquad (4.101)$$

The VOIE data for fluorine are obtained in a quite convoluted way. The fluorine atom in TiF_6^{3-} carries a net negative charge of

$$-(3 + 0.5138)/6 \approx -0.6 \,|\, e \,|$$

In this regard, therefore, the F atom of TiF_6^{3-} must resemble the F atom of hydrofluoric acid, HF. The experimental ionization potential[48] of the $2p\pi$ AO (which is nonbonding) of F in HF is 127.2 kK; that for the $2p\sigma$ AO (or, better, for the HF MO in which the $2p\sigma$ AO is involved) is 136.9 kK. The $2s$ VOIE is obtained on the assumption that the $2s - 2p\pi$ energy difference in HF is the same as the $2s - 2p$ difference in atomic fluorine with an excess charge $q = -0.6 \,|\, e \,|$. Now the VOIE data of Table D.1, Appendix D, at $q = -0.6 \,|\, e \,|$, yield

$$2s \text{ VOIE}(\ldots 2s^2 2p^5) = 210.285 \text{ kK}$$

$$2p \text{ VOIE}(\ldots 2s^2 2p^5) = 61.155 \text{ kK} \qquad (4.102)$$

Thus, the $2s - 2p\pi$ difference is taken to be 149.130 kK. The value used by Bedon *et al.*[44] was 145.000 kK. Since the difference of $\sim$4 kK is not significant, we use the latter value. Since we have chosen the $2p\pi$ VOIE of fluorine equal to 127.2 kK, the $2s$ VOIE equals $127.2 + 145.0 = 272.2$ kK. The VOIE for $\chi_\sigma{}^+$ is now obtained from Eq. (4.84) as

$$\sigma^+ \text{ VOIE} = 0.13(272.2) + 0.87(136.9) = 154.5 \text{ kK} \qquad (4.103)$$

(F) Resonance Integrals

We adopt the Wolfsberg–Helmholz equation [Eq. (4.53)] with $k \equiv 2$.

(G) Secular Equations

It is now possible to set up and solve the various secular determinants. We do the 2×2 a_{1g} determinant only. The secular determinant is

$$\begin{vmatrix} -78.898 - \epsilon & -100.13 - 0.429\epsilon \\ -100.13 - 0.429\epsilon & -154.5 - \epsilon \end{vmatrix} = 0 \qquad (4.104)$$

The secular polynomial is

$$0.81596\epsilon^2 + 147.486\epsilon + 2163.72 = 0 \qquad (4.105)$$

The roots are

$$\epsilon = -164.64 \text{ kK}; \qquad \epsilon = -16.10 \text{ kK} \qquad (4.106)$$

For the root $\epsilon = -164.64$ kK, we solve for the eigenvector coefficients from the normalization condition and Eq. (4.104)

$$(-78.898 + 164.646)c_{4s} + [-100.13 - 0.429(-164.646)]c_{\sigma^+, a_{1g}} = 0$$
$$\qquad (4.107)$$

to find $c_{4s} = 0.28933$ and $c_{\sigma^+, a_{1g}} = 0.84112$. The resultant eigenvector is

$$0.28933\chi_{4s} + 0.84112(1/6)^{1/2}(\sigma_1^+ + \sigma_2^+ + \sigma_3^+ + \sigma_4^+ + \sigma_5^+ + \sigma_6^+) \quad (4.108)$$

The root $\epsilon = -16.10$ kK provides another eigenvector—in this case, the antibonding $a_{1g}{}^*$ MO.

The remaining secular determinants are set up with equal facility and roots and eigenvectors extracted. The group overlap integrals, MO energies and eigenvectors are collected in Table 4.20; the MO energy levels are depicted in Fig. 4.6.

(H) Population Analysis

The total number of valence electrons in $TiF_6{}^{3-}$ is

$$\text{Ti} + \quad 6\text{F} \quad + (q = -3) - 12 \text{ in } \sigma^-$$
$$4 + 6 \times 7 + \qquad 3 \qquad - 12 = 37 \qquad (4.109)$$

TABLE 4.20. *Group overlap integrals, MO energies, and MO eigenvectors for titaniumhexafluoride.*[a]

MO Representation	G_{12}	G_{13}	Bonding and Nonbonding[b,c]				Antibonding[c]			
			$-\epsilon$ (kK)	c_1	c_2	c_3	$-\epsilon$ (kK)	$c_1{}^*$	$c_2{}^*$	$c_3{}^*$
t_{2g}	⋯	0.201	133.987	?	⋯	?	75.394	?	⋯	?
e_g	0.311	⋯	163.576	0.306	0.862	⋯	55.979	−1.007	0.604	⋯
a_{1g}	0.429	⋯	164.646	0.289	0.841	⋯	16.109	−1.069	0.720	⋯
t_{1u}	0.279	0.270	155.922	0.110	0.963	0.018	9.907	1.073	−0.395	−0.404
t_{1u}	0.279	0.270	128.676	−0.118	0.091	−0.961				
t_{1g}	⋯	⋯	127.2	⋯	⋯	1				
t_{2u}	⋯	⋯	127.2	⋯	⋯	1				

[a] The subscript 1 pertains to the Ti AO; the subscript 2 pertains to the σ^+ LGO; the subscript 3 pertains to the $2p\pi$ LGO. The reader should correlate this table with Table 4.18.

[b] All of these MO's except t_{1g} and t_{2u}, are bonding; these two latter MO's are nonbonding.

[c] The reader may wish to obtain the coefficients c_1, $c_1{}^*$, c_3, and $c_3{}^*$ for the t_{2g} MO.

The ground-state configuration is

$$(1a_{1g})^2(1e_g)^4(1t_{1u})^6(1t_{2g})^6(2t_{1u})^6(1t_{2u})^6(1t_{1g})^6(2t_{2g}*)^1; \,^2T_{2g} \quad (4.110)$$

This configuration yields only one state, namely, $^2T_{2g}$.

The population analysis proceeds in a straightforward way. The probability of finding an electron from the $1a_{1g}$ MO on the titanium center is

$$c_{4s}^2 + \tfrac{1}{2}\{2c_{4s}c_{\sigma^+,a_{1g}}G[a_{1g}(\sigma)]\} \quad (4.111a)$$

In the notation of Table 4.20 this becomes

$$c_1^2 + \tfrac{1}{2}(2c_1c_2G_{12}) = 0.18811 \quad (4.111b)$$

Since the a_{1g} MO contains two electrons, it follows that the $4s$ population of titanium is 0.376. The overlap population between the $4s$ AO of Ti and the a_{1g} LGO is given by $4c_1c_2G_{12}$; it equals 0.418.

The results of the population analysis are given in Table 4.21.

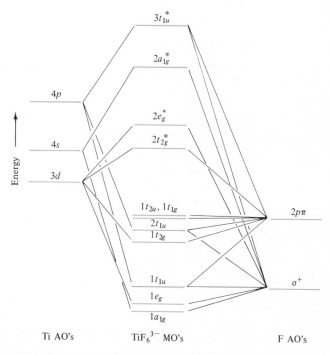

FIG. 4.6. Orbital energy-level diagram for $TiF_6{}^{3-}$. The notation $3t_{1u}*$ indicates that this is the third t_{1u} MO in order of increasing energy.

TABLE 4.21.   *Titanium orbital populations in* TiF_6^{3-}.

AO	MO	MO Occupation Number n_r	Number of Electrons on Titanium
$3d$	e_g	4	0.7012
	t_{2g}	6	1.0889
	$t_{2g}*$	1	0.8185
		$3d$ Total $= 2.6087$	
$4s$	a_{1g}	2	0.3762
		$4s$ Total $= 0.3762$	
$4p$	t_{1u}	6	0.2529
	t_{1u}	6	0.2484
		$4p$ Total $= 0.5014$	

Total number of electrons on titanium $= 3.4863$

Net charge on titanium $= +4 - 3.4863 = +0.5137 \, | \, e \, |$

(I)   Electronic Absorption Spectrum

For purposes of correlation with the experimental absorption spectrum, the important MO's are $2e_g*$ and $2t_{2g}*$ for the d-d transition; for the ligand-to-metal charge transfer band,[5] they are $1t_{2u}$ and $2t_{2g}*$. The energies of the resultant MO excitations are correlated with experiment in Table 4.22. The agreement of experiment and theory is probably fortuitous.

TABLE 4.22.   *Electronic absorption spectrum of* TiF_6^{3-}.

MO Excitation	Predicted Energy (This Work) (kK)	Predicted Energy (BHT)[a] (kK)	Experimental Energy (kK)
$2e_g*\leftarrow 2t_{2g}*$	19.415	17.51	17.50
$2t_{2g}*\leftarrow 1t_{2u}$	> 51.80	48.40	> 50.00

[a] From Bedon, Horner, and Tyree, Ref. [44] of text.

[5] Note that two other candidates for the ligand-to-metal charge transfer band are $2t_{2g}* \leftarrow 1t_{1g}$ and $2e_g* \leftarrow 1t_{1g}$. Both of these transitions are parity forbidden.

(J) Evaluation of Computational Procedure

The computations presented suffer from a number of defects. We now detail these deficiencies and provide correction and/or comment.

(i) *Neglect of Ligand-Ligand Overlap on Group Overlaps*[49]−[51]: The LGO normalization of Table 4.18 did not consider ligand-ligand (LL) overlap. These overlaps can be large; their neglect is improper.

Consider, for example, the a_{1g} LGO. In the absence of LL overlap, the normalization factor is $(1/6)^{1/2}$. Inclusion of LL overlap yields (see Chapter 2) a normalization factor

$$N = (1/6)^{1/2}[1 + \langle \sigma_1^+ \mid \sigma_3^+ \rangle_\sigma + 2\langle \sigma_1^+ \mid \sigma_2^+ \rangle_\sigma + 2\langle \sigma_1^+ \mid \sigma_2^+ \rangle_\pi]^{-1/2} \quad (4.112)$$

where the subscripts on the overlap brackets identify the type of overlap involved. Since the factor $(1/6)^{1/2}$ has already been included in the a_{1g} LGO of Table 4.18, the overlap $G[a_{1g}(\sigma)]$ of Eq. (4.88) should be corrected to

$$G'[a_{1g}(\sigma)] = G[a_{1g}(\sigma)][1 + \langle \sigma_1^+ \mid \sigma_3^+ \rangle_\sigma + 2\langle \sigma_1^+ \mid \sigma_2^+ \rangle_\sigma$$
$$+ 2\langle \sigma_1^+ \mid \sigma_2^+ \rangle_\pi]^{-1/2} \quad (4.113)$$

Tables of correction factors—the inverse square root part of Eq. (4.112), for example—are available[49],[50] for the more symmetric point groups.

(ii) *Neglect of LL Overlap on Energy Matrix:* The neglect of proper normalization also intrudes into the evaluation of the matrix elements of $\hat{H}$ in the LGO basis. In the absence of LL overlap, the Coulomb integral of the LGO φ_r equals the Coulomb integral of any one of the AO's which compose φ_r; in other words, $H_{rr} = H_{\mu\mu}{}^r$. If, on the other hand, LL overlap is included, it may be shown that

$$H_{rr} = \left(\frac{1 + 2X_r}{1 - X_r} \right) H_{\mu\mu}{}^r \quad (4.114)$$

where

$$X_r \equiv \sum_{\mu \neq \nu} \sum c_{r\mu}^* c_{r\nu} S_{\mu\nu}{}^r \quad (4.115)$$

where the superscript r on S merely indicates that the AO's μ and ν occur in φ_r. The correction factors X_r usually lie in the range $0.8 \leq X_r \leq 1.2$. It is clear that their neglect is both improper and serious.

The approximation used for the off-diagonal elements of the energy matrix is valid in the absence of LL overlap and is given by

$$H_{rs} = (H_{rr} + H_{ss})G_{rs} = (H_{\mu\mu}{}^r + H_{\nu\nu}{}^s)G_{rs} \quad (4.116)$$

If we cease neglect of LL overlap, the correct expression is

$$H_{rs} = (H_{\mu\mu}{}^r + H_{\nu\nu}{}^s)G_{rs}' \tag{4.117}$$

where G_{rs}' is a corrected LGO overlap such as given in Eq. (4.113). Since the ratio G'/G can be quite different from unity, the neglect of LL overlap is improper.

One may correct the calculations in the manner just outlined or, more directly, one may simply forget about group theory and group orbitals and perform calculations on $TiF_6{}^{3-}$ in the manner of Sec. 5 for water. The choice of either route is essentially a matter of taste.

(iii)   Assumption that $\chi_\sigma{}^-$ is Nonbonding: The $\chi_\sigma{}^-$ AO was assumed to remain nonbonding. Since $S(\chi_\sigma{}^-, 3d\sigma) = 0.103$, this non-bonding assumption is unreasonable. Explicit inclusion of $\chi_\sigma{}^-$ in the computations of Fenske[52] produced a d-d splitting for $TiF_6{}^{3-}$ which was approximately twice that when not included.

The direct procedure would include the $2s$ AO explicitly. Such computations for $TiF_6{}^{3-}$ have been made by Fenske[52] and by Basch *et al.*[53] Similar computations for $MnO_4{}^-$ and $CrF_6{}^{3-}$ are available.[50] The results are very indiscriminate: Quite often the computed quantities are better when the $2s$ AO (or, almost equivalently, $\chi_\sigma{}^-$) is neglected!

(iv)   Use of Molecular Ionization Potentials: The $2p\pi$ and $\chi_\sigma{}^+$ VOIE's were obtained from ionization potentials for HF. Such VOIE's yield different MO results than those based on atomic spectroscopic data. However, it is not *a priori* clear which is better in terms of agreement with experiment.

The use of hydride ionization potentials usually produces large co-valencies, even in the hexafluorides. This, of course, runs counter to the traditional view of almost-complete ionic bonding in these complexes. It is not clear, however, that VOIE data of atomic origins will not produce similar results.

7.   CHARGE SELF-CONSISTENCY

Considerable interest attaches to MWH calculations in which the parameters are assumed to depend on the molecular charge distribution.[19],[20],[33],[54] Such a dependence allows one to correct, in a first-order way, for interelectronic repulsion effects: For example, an electron at a relatively negative site can be assigned a lesser value of $|H_{\mu\mu}|$ than when the site is relatively positive. It is, of course, this same attitude which motivated development of the ω *technique* of simple Hückel theory[55] [see Sec. 4(D), Chapter 3].

The procedure is illustrated for a neutral molecule. One may begin the computation assuming every atom in the molecule to be electrically neutral; the secular equation is solved, and a new (or first generation) set of charges and populations is obtained. A new secular equation is now generated from these output charges and populations; it is solved and processed to provide a second generation of atomic charges and AO populations. The whole procedure is repeated until the set of atom charges and populations used as input for one cycle of the computation is identical, within some pre-scribed limit, to that obtained at termination of the cycle.

Convergence to the charge self-consistent point is not guaranteed; indeed, the computations usually diverge if the iterative procedures are not heavily damped. In other words, the input for the *jth* cycle is given in terms of that for the $(j-1)th$ cycle by[19]

$$\text{Input}_j = \text{Input}_{(j-1)} - \lambda(\text{Input}_j - \text{Input}_{(j-1)}) \qquad (4.118)$$

where the damping parameter $\lambda(0 < \lambda < 1)$ is given a value just small enough to force convergence. The *best value* of λ for a *steepest descent approach* to self-consistency has been discussed by McWeeny.[56]

The results of charge self-consistent calculations[19],[57]−[59] usually show a marked improvement over the results obtained without iteration.

8. INTERJECTION ON WOODWARD–HOFFMANN APPROACH TO CHEMICAL-REACTION MECHANISMS

The Woodward–Hoffmann approach to chemical-reaction mechanisms[60]−[62] is one of the more significant recent achievements of molecular-orbital theories. This approach provides a remarkable synthesis of the literature on chemical-reaction mechanisms[62]: It makes possible a discussion of the concertedness[6] of chemical reactions, the relative quantities of different isomeric forms among reaction products of concerted reactions, the unexpected stability of certain reactants, etc. And yet, despite the extent of its apparent success, the Woodward–Hoffmann approach is based on the very simple concept of the energy-level correlation diagram.

(A) Correlation Diagrams

Energy-level correlation diagrams of the united atom—separated atom variety for diatomic molecules are of considerable importance in chemistry.[63] Such diagrams are constructed by visualizing the approach of two

---

[6] A concerted reaction is one in which there is a degree of simultaneity between all bond-breakage and bond-formation events. In other words, a concerted reaction may not be separated into a sum of consecutive step reactions.

atoms from infinite separation until one collapses into the other to produce the *united atom*. The approach passes from the *separated atom* region (i.e., large R) through the *molecule* region (i.e., intermediate R) into the united atom region (i.e., zero R). The actual diagrammatic construction may be systematized as follows:

(i)   One obtains the atomic energy levels for the separated atoms and the united atom and classifies these with respect to the cylindrical symmetry which is maintained during the process of decreasing R.

(ii)   Levels at large R are connected with ones of similar symmetry at small R, due regard being paid to the noncrossing rule which says that levels of similar transformation properties may not intersect.

(iii)   The diagrammatic representation which emerges at intermediate R is assumed to refer to the molecule. Indeed, it was from studies of this sort that an understanding of the triplet nature of the ground state of O_2 was first obtained.

The Woodward–Hoffmann approach to a theory of concerted chemical reactions consists of the construction of energy-level correlation diagrams. For example, the separated atoms might be referred to as *reactants*, the united atom might be referred to as *product*, and the diatomic molecule region might be considered to be a *transition state* for the reaction: united atom $\rightleftarrows$ separated atoms. Cylindrical symmetry is maintained during the course of approach to and departure from the transition state and there exists no doubt that some aspects of the specified reaction could be adequately handled within the context of the diatomic correlation diagram if this reaction was, in fact, of chemical—as opposed to nuclear—nature.

(B)   Cyclo-addition Reaction

In view of (A) above, we now turn attention to a purely chemical reaction: The cyclo-addition of two olefins to yield cyclobutane. The reactants are two ethylene molecules, the product is cyclobutane, and the nature of the transition-state *molecule* is largely unknown. This situation contrasts sharply with the transition-state molecule for the united atom $\leftarrow$ separated atom reaction which, necessarily, was of cylindrical symmetry. Consequently, some assumption concerning the symmetry of the transition-state molecule must be made and, in order to simplify matters further, it must also be assumed that approach to and departure from this transition-state molecule preserves its assumed symmetry. Thus, a further restriction of compatibility is imposed on the transition-state molecule: Its geometry must derive smoothly from reactants and flow smoothly into products. This latter condition is, clearly, a consequence of the assumed concerted nature of the reaction.

The concerted cyclo-addition of two ethylene molecules appears to lead one to the assumption of a transition-state molecule of D_{2h} symmetry. This transition-state molecule is illustrated in Fig. 4.7. We now classify the levels of the reactants and products with regard to energy and to transformation properties in the D_{2h} point group.

(C)   Molecular Orbitals for Concerted Cyclo-addition

Each ethylene possesses two π MO's, one bonding and one antibonding. During the course of reaction, these MO's are converted into σ MO's of the cyclobutane molecule. Indeed, the whole process can be viewed as the loss of two π bonds and the simultaneous generation of two σ bonds per pair of ethylenic reactants, all other bonds remaining more or less unchanged. This process is diagrammed in Fig. 4.8.

(D)   Molecular-Orbital Transformation Properties

The π MO's of an individual ethylene reactant are not symmetry adapted with respect to the point group defined by the assumed transition-state molecule. Adaptation is readily achieved using methods of Chap. 7 and yields the resultant π MO's of Fig. 4.9. The bonding MO of the top

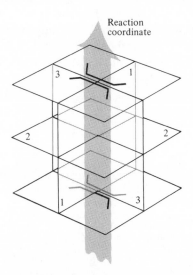

FIG. 4.7.   Concerted cyclo-addition reaction of two ethylene molecules to yield cyclobutane. The transition-state molecule is of D_{2h} symmetry. The reaction coordinate lies in plane No. 3.

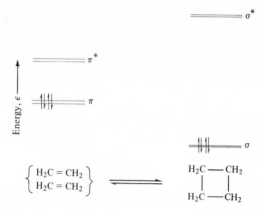

FIG. 4.8. Molecular-orbital energies for two ethylene molecules (i.e., large R) are shown on the left. The MO energies of the σ MO's of cyclobutane (i.e., small R) are shown on the right.

ethylene molecule of Fig. 4.7 is designated π_1, that of the bottom ethylene molecule is designated π_2. The final symmetry-adapted MO's of the super-molecule (i.e., transition state molecule) are classified with respect to reflection characteristics in the planes 1 and 2 defined in Fig. 4.7. These unnormalized symmetry-adapted MO's are $\pi_1 - \pi_2$, which we denote S_1S_2, and $\pi_1 + \pi_2$, which we denote S_1A_2. The appropriate symmetry-adapted linear combinations of the antibonding orbitals π_1^* and π_2^* are $\pi_1^* - \pi_2^*$, denoted A_1S_2, and $\pi_1^* + \pi_2^*$, denoted A_1A_2.

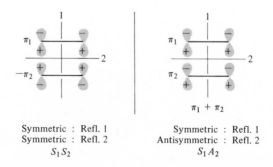

FIG. 4.9. Symmetry-adapted MO's of π parentage for the transition-state molecule. S_1 denotes symmetry with respect to reflection in plane No. 1, A_1 denotes antisymmetry with respect to reflection in plane No. 1, etc. The planes 1 and 2 are defined in Fig. 4.7.

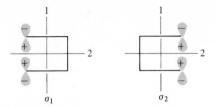

We must also process a similar description of the σ bonds in cyclobutane. If we define σ_1 and σ_2 as the bonding σ orbitals which are generated when cyclobutane is formed from two ethylene molecules disposed as in Fig. 4.7, we find that appropriate symmetry-adapted unnormalized linear combinations are $\sigma_1 + \sigma_2$, which is S_1S_2 in the previous notation, and $\sigma_1 - \sigma_2$ which is A_1S_2. The corresponding linear combinations of the antibonding σ orbitals are $\sigma_1^* + \sigma_2^*$, denoted S_1A_2, and $\sigma_1^* - \sigma_2^*$, denoted A_1A_2.

We are now in a position to redraw Fig. 4.8, and to include the newly acquired information on MO symmetry. This is done in Fig. 4.10. As a result of the symmetry information contained in Fig. 4.10, we can now complete the process of drawing the molecular-orbital correlation diagram. This is done in Fig. 4.10 also. The most striking feature of this diagram is the correlation of a bonding reactant level with an antibonding product level and vice versa.

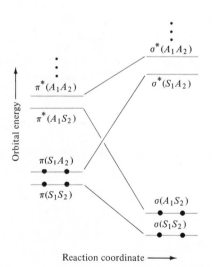

FIG. 4.10. Molecular-orbital correlation diagram for the concerted process 2 (ethylenes) $\rightleftarrows$ cyclobutane.

(E)   State Correlation Diagram

Figure 4.10 is a molecular-orbital energy-level correlation diagram. This is a necessary intermediate in the construction of a state energy-level diagram. We now use the MO correlations and the MO energies and symmetries of Fig. 4.10 to construct an approximate state correlation diagram.

The ground state of the two separated ethylene systems is $(S_1S_2)^2(S_1A_2)^2$ in the point group defined by the transition-state molecule; we may abbreviate this state as S_1S_2. The first-excited state is approximately fourfold degenerate and given by

$$(S_1S_2)^2(S_1A_2)^1(A_1S_2)^1, \text{ abbreviated as } A_1A_2$$

$$(S_1S_2)^2(S_1A_2)^1(A_1A_2)^1, \text{ abbreviated as } A_1S_2$$

$$(S_1S_2)^1(S_1A_2)^2(A_1S_2)^1, \text{ abbreviated as } A_1S_2$$

$$(S_1S_2)^1(S_1A_2)^2(A_1A_2)^1, \text{ abbreviated as } A_1A_2$$

The next excited singlet state is obtained by double-electron excitation. Four such states of designation S_1S_2 exist.

The states of the cyclobutane molecule may be obtained in a similar fashion. When correlated, by using Fig. 4.10 and the noncrossing rule, the result is Fig. 4.11.

(F)   Woodward–Hoffmann Rules for Concerted Cyclo-addition

Inspection of Fig. 4.11 provides us with a number of propositions which must hold true when the reaction coordinate is that shown in Fig. 4.7:

(i)   Two ground-state ethylene molecules cannot readily combine in a concerted reaction to yield ground-state cyclobutane. The potential energy barrier to this reaction is simply too high.

(ii)   For reasons similar to (i), cyclobutane should be rather stable with respect to a concerted thermal decomposition into two ethylene molecules.

(iii)   The concerted reaction of one ground state and one photo-excited ethylene molecule should proceed readily and yield, as an initial product, a photoexcited cyclobutane molecule. This product should decay rapidly and nonradiatively to its ground-state level.

Since it is rather difficult to conceive of any transition-state molecule for a concerted ethylene → cyclobutane reaction other than one of D_{2h}

symmetry, it follows that we must consider this reaction to be symmetry forbidden. In other words, we must expect the cyclo-addition reaction of two ethylenes not to be a concerted reaction. This conclusion agrees with experiment.

In contrast to cyclo-addition of two ethylenes, had we investigated the Diels–Alder addition of butadiene to an ethylenic bond region, we would conclude that this reaction could proceed (i.e., was allowed) *via* a concerted mechanism. This conclusion appears to accord with all the available data on Diels–Alder reactions.

(G)   General Discussion

The gist of the Woodward–Hoffmann approach lies in the construction of an energy-level correlation diagram. This diagram relates the electronic states of reactants, products, and the assumed transition-state molecule for a concerted reaction. We have constructed such a diagram, using very simplistic arguments, and have used it to predict that the concerted thermal cyclo-addition of two ethylenes is a rather improbable thermal process whereas the concerted cyclo-addition route ought to be dominant in the photochemical process. In addition, all the symmetry classification was based on two reflection planes—one was the plane of symmetry for the

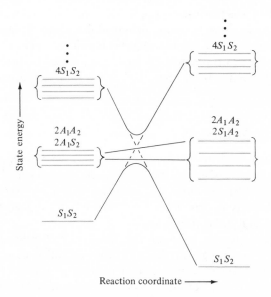

FIG. 4.11.   Correlation diagram for state energies in the concerted cyclo-addition of two ethylenes to yield cyclobutane.

newly formed σ bonds, the other was for the newly broken π bonds. This discussion had unsatisfactory aspects which derived, in their entirety, from the manner of construction of the correlation diagram. These unsatisfactory aspects could have been eliminated by performing actual computations for all molecular entities, including the transition-state entity, and classifying all states with respect to the full complement of operations of the D_{2h} point group of the *supermolecule*. Had this been done, the results would be more or less identical to those already obtained; they would, however, appear less magical in content. In other words, the Woodward–Hoffmann approach constitutes a completely formalized approach to concerted chemical-reaction mechanisms.

In order to demonstrate the power of the Woodward–Hoffmann approach, let us suppose that we are interested in the stability of some molecule. We consider all decomposition products and the various paths which a concerted reaction may take in forming such products. Having constructed the appropriate set of correlation diagrams, we may determine the most likely decomposition routes, we may ascertain ways which yield some preferred set of decomposition products, or we may derive certain intuitions concerning ways to stabilize the parent molecule. Clearly, this is very useful chemical information. The ability to generate such information constitutes the essence of the Woodward–Hoffmann approach and highlights its importance.

Many of the ideas intrinsic to the Woodward–Hoffmann approach had appeared in various guises in earlier chemical literature. This aspect of the topic of concerted chemical reactions has been well-catalogued by Dewar.[64] We refer the reader to this source[64] for a most readable discussion of the intrusion of molecular-orbital attitudes into discussions of chemical reactivity.

9. CONCLUSION

The Mulliken–Wolfsberg–Helmholz computational procedure is a very useful tool for the area of chemistry. The results of such calculation should provide ideas which can be subjected to experimental investigation. In this manner, the interplay of experiment and calculation can lead to the evolution of concepts and interpretations.

Various attempts to improve the basic MWH schema are available.[41],[65]–[71] These range all the way from calculations similar to the Pariser–Parr–Pople type, to the angular overlap method of Schäffer and Jørgensen. However, it is not clear that these more replete schemes offer a sufficient improvement to merit the extra effort and intricacies which they involve.

EXERCISES

1. Show that the terms in the right-hand column of the following tabulation are those which arise from the configurations in the left-hand column.

Configuration[a]	Terms
ss'	$^1S, ^3S$
pp'	$^1S, ^1P, ^1D; ^3S, ^3P, ^3D$
dd'	$^1S, ^1P, ^1D, ^1F, ^1G; ^3S, ^3P, ^3D, ^3F, ^3G$
s^2	^1S
p^2	$^1S, ^1D, ^3P$
d^2	$^1S, ^1D, ^1G, ^3P, ^3F$
p^3	$^2P, ^2D, ^4S$
p^4	$^1S, ^1D, ^3P$

[a] The prime indicates that the principal quantum numbers of the two AO's in question are not identical.

Hint: See Herzberg, Ref. [3] of text.

2. Calculate the $2p$ VOIE $(\ldots 2s^2 2p^2)$ of neutral carbon.

Hint: First calculate the promotion energy from the level $^3P_0$ to the $\ldots 2s^2 2p^2$ configuration of C(I). Verify your answer in terms of the data of Fig. 4.1. Next evaluate the promotion energy from $^2P_{1/2}$ to $\ldots 2s^2 2p^1$ of C(II); relevant data are given by Moore.[72]

Answer: $2p$ VOIE $= 90\,878.3 - 4\,858.8 + 42.7 = 86\,062.2 \text{ cm}^{-1}$ (4.119)

3. Calculate the $3s$ VOIE$(\ldots 3s^2 3p^4)$ of the neutral sulfur atom.

Hint: The promotion energy (i.e., P^0) for the ground $^3P_2$ level to the $\ldots 3s 3p^4$ configuration of neutral sulfur is given in Fig. 4.2. The promotion energy from the ground $^4S_{1/2}$ level to the $\ldots 3s 3p^4$ configuration of S(II) equals $88\,323.4$ cm^{-1}.[73]

Answer: $3s$ VOIE $= 83\,559.3 - 4\,676.0 + 88\,324.5 = 167\,207.8 \text{ cm}^{-1}$

$$= 20.73 \text{ eV}   (4.120)$$

4. Evaluate the $4s$ and $4p$ VOIE's appropriate to the configuration $\ldots 3d^{2.60863}4s^{0.37622}4p^{0.50136}$ of titanium carrying an excess charge $q = +0.5138 \mid e \mid$. Use the data of Table 4.19 of text.

Answer: $4p$ VOIE $= 46.060\,\mathrm{kK};$ $4s$ VOIE $= 78.898\,\mathrm{kK}$ (4.121)

5. The ionization potentials of an isoelectronic series may be written as

Configuration	$s^s p^p d^d$	$s^s p^p d^d$	$s^s p^p d^d$	$\ldots$	Equations
Excess Charge	$q = 0$	$q = 1$	$q = 2$	$\ldots$ $IP = Aq^2 + Bq + C$
Atomic Number	$Z \equiv Z_0$	$Z = Z_0 + 1$	$Z = Z_0 + 2$	$\ldots$ $IP = \alpha Z^2 + \beta Z + \gamma$

Find the relationship between the sets of parameters (A, B, C) and (α, β, γ).

Answer: $\alpha = A$ $A = \alpha$

$\beta = B - 2Z_0 A$ $B = 2Z_0\alpha + \beta$

$\gamma = C - BZ_0 + AZ_0^2$ $C = \alpha Z_0^2 + \beta Z_0 + \gamma$ (4.122)

6. Cusachs[20] has given information on VOIE's of valence configurations in the form of Eq. (4.46). The values (in eV) of the parameters in Eq. (4.46) in the case of sulfur are

$A'(s, 1)$	$A'(s, 2)$	$A'(p, 1)$	$A'(p, 2)$	B'
23.72	21.29	12.50	11.38	9.70

Use these data to generate equations of the form of Eq. (4.47) for the $3s$ and $3p$ VOIE's of S(I). Check your Eq. (4.47) against those of Table D.4, Appendix D.

Answer: $3s$ VOIE $= -2.43(POP_{3s}) + 26.15 + 9.7q$

$3p$ VOIE $= -1.12(POP_{3p}) + 13.26 + 9.7q$ (4.123)

7. The secular determinant for the water molecule can be partitioned by symmetry into one 3×3 determinant (a_1 type), one 2×2 determinant (b_2 type), and one 1×1 determinant (b_1 type). Perform the partitioning in question, and set up and solve the resultant secular equations. Verify all the results of Table 4.13.

8. Set up the secular determinants (using the LGO basis sets, improperly normalized, of Table 4.18) for the three 2×2 determinants (a_{1g}, e_g, and t_{2g}), for the one 3×3 determinant (t_{1u}), and for the two 1×1 determinants (t_{1g} and t_{2u}) of TiF_6^{3-}. Solve for MO energies and eigenvectors.

Hint: The 2×2 a_{1g} determinant is given by Eq. (4.104) of text. The data on group overlap integrals of Table 4.20 of text are useful.

Answer: See Table 4.20 of text.

9. Calculate the MO energies and eigenvalues of hydrogen, H_2. Restrict the valence set of AO's to the $1s$ orbitals and use the Cusachs approximation [Eq. (4.55) of text]. The bond length $R(H—H)$ is 0.741 Å.

Answer: Initiate the calculation with $q_{H(1)} = q_{H(2)} = 0$. Under these conditions, the $1s$ VOIE equals 13.6 eV. The exponent of the single STF is $\zeta = 1$. The bond-length matrix is

$$\begin{matrix} 1s\,H(1) \\[1em] 1s\,H(2) \end{matrix} \begin{bmatrix} 0 & 0.741 \\ 0.741 & 0 \end{bmatrix} \qquad (4.124)$$

The overlap matrix is

$$\begin{matrix} 1s\,H(1) \\[1em] 1s\,H(2) \end{matrix} \begin{bmatrix} 1.000 & 0.753 \\ 0.753 & 1.000 \end{bmatrix} \qquad (4.125)$$

Therefore, we find

$$(2 - |S_{local}|)S = 0.939 \qquad (4.126)$$

The energy matrix is (in eV)

$$\begin{matrix} 1s\,H(1) \\[1em] 1s\,H(2) \end{matrix} \begin{bmatrix} -13.600 & -12.769 \\ -12.769 & -13.600 \end{bmatrix} \qquad (4.127)$$

The MO eigenvectors and eigenvalues are

	MO_1	MO_2
$1s\,H(1)$	0.534	1.422
$1s\,H(2)$	0.534	−1.422
ϵ	−15.044	−3.362

$$(4.128)$$

The sum of MO energies is -30.088 eV. The ionization potential is 15.044 eV.

Note: This computation is already self-consistent.

10.   Calculate the MO eigenvectors and eigenvalues of LiH. Limit the AO basis set to the $1s$ AO of H and the $2s$ AO of Li and let these AO's be represented by single-ζ STF's formulated by Slater's rules. Make use of the Cusach's equation. Iterate the computation until input and output atomic charge densities differ by less than $0.005 \mid e \mid$. Use a damping factor λ [see Eq. (4.118) of text] of 0.01. The distance $R(\text{Li—H})$ is 1.595 Å.

Answer: Since H is more electronegative than Li, let us start the computation with 0.8 electrons in the $2s$ AO of Li and 1.2 electrons in the $1s$ AO of H. The Slater ζ's are: for $2s$ Li, $\zeta = 0.6500$; for $1s$ H, $\zeta = 1.00$. The bond-length matrix is

$$\begin{array}{c} 2s \text{ Li} \\ \\ 1s \text{ H} \end{array} \begin{bmatrix} 0.000 & 1.595 \\ 1.595 & 0.000 \end{bmatrix} \tag{4.129}$$

The overlap matrix is

$$\begin{array}{c} 2s \text{ Li} \\ \\ 1s \text{ H} \end{array} \begin{bmatrix} 1.000 & 0.476 \\ 0.476 & 1.000 \end{bmatrix} \tag{4.130}$$

The quantity $(2 - \mid S_{\text{local}} \mid) S$ equals 0.725. The energy matrix is (in eV)

$$\begin{array}{c} 2s \text{ Li} \\ \\ 1s \text{ H} \end{array} \begin{bmatrix} -6.480 & -5.529 \\ -5.529 & -8.774 \end{bmatrix} \tag{4.131}$$

The MO eigenvectors and eigenvalues are

	MO_1	MO_2
$2s$ Li	0.329	1.088
$1s$ H	0.801	-0.807
ϵ	-9.240	-3.676

$$\tag{4.132}$$

Thus, in the first iteration we find the output-input electron populations to be

	Input	Output	Difference	1% (Difference)	
2s Li	0.80000	0.46735	0.3327	0.0033	
1s H	1.20000	1.53265	−0.3327	−0.0033	(4.133)

As input for the second cycle we use

$$2s\ \text{Li} \qquad 0.8000 - 0.0033 = 0.79670$$
$$1s\ \text{H} \qquad 1.2000 + 0.0033 = 1.20330 \qquad (4.134)$$

We now tabulate, for various cycles of the iterative process, the results which should be obtained

		Input	Output	Difference	
Iteration 1	2s	0.80000	0.48210	0.3327	
	1s	1.20000	1.51790		
Iteration 2	2s	0.79670	0.48210	0.3146	
	1s	1.20330	1.51790		
Iteration 3	2s	0.79360	0.49631	0.2973	
	1s	1.20640	1.50369		
Iteration 20	2s	0.76160	0.66354	0.0981	
	1s	1.23840	1.33646		
Iteration 59	2s	0.74830	0.74363	0.0047	
	1s	1.25170	1.25637		(4.135)

At this point the energy matrix is

$$\begin{matrix} 2s\ \text{Li} \\ 1s\ \text{H} \end{matrix} \begin{bmatrix} -6.759 & -5.244 \\ -5.244 & -7.709 \end{bmatrix} \qquad (4.136)$$

The MO eigenvectors and eigenvalues are

	MO$_1$	MO$_2$
2s Li	0.468	1.036
1s H	0.689	−0.904
ϵ	−8.518	−3.732

$$(4.137)$$

11. Perform the same calculation as in Exercise 10, but expand the AO basis set to include the $2p_x$, $2p_y$, and $2p_z$ orbitals of Li. Limit the calculation to the final cycle of the charge self-consistent process.

Answer: At charge self-consistency, it is found that the electron configurations are

$$\text{Li:} \quad 1s^2 2s^{0.70010} 2p_x{}^0 2p_y{}^0 2p_z{}^{0.05070}$$

$$\text{H:} \quad 1s^{1.24910} \qquad\qquad (4.138)$$

The overlap matrix is

$$
\begin{array}{c}
2s \;\; \text{Li} \\
2p_x \; \text{Li} \\
2p_y \; \text{Li} \\
2p_z \; \text{Li} \\
1s \;\; \text{H}
\end{array}
\begin{bmatrix}
1.000 & 0 & 0 & 0 & 0.476 \\
(0) & 1.000 & 0 & 0 & 0 \\
(0) & (0) & 1.000 & 0 & 0 \\
(0) & (0) & (0) & 1.000 & -0.554 \\
(0.725) & (0) & (0) & (-0.801) & 1.000
\end{bmatrix}
\quad (4.139)
$$

where the bracketed quantities are the values of $(2 - |\, S_{\text{local}}\,|)\, S$. The energy matrix is

$$
\begin{array}{c}
2s \;\; \text{Li} \\
2p_x \; \text{Li} \\
2p_y \; \text{Li} \\
2p_z \; \text{Li} \\
1s \;\; \text{H}
\end{array}
\begin{bmatrix}
-6.801 & 0 & 0 & 0 & -5.278 \\
0 & -4.966 & 0 & 0 & 0 \\
0 & 0 & -4.966 & 0 & 0 \\
0 & 0 & 0 & -4.966 & 5.100 \\
-5.278 & 0 & 0 & 5.100 & -7.760
\end{bmatrix}
\quad (4.140)
$$

The eigenvectors and eigenvalues are

MO	1	2	3	4	5	
$2s$ Li	0.451	0.702	0	0	0.888	
$2p_x$ Li	0	0	1.000	0	0	
$2p_y$ Li	0	0	0	1.000	0	
$2p_z$ Li	-0.063	0.644	0	0	-1.113	
$1s$ H	0.675	-0.281	0	0	-1.268	
ϵ (eV)	-8.589	-5.790	-4.966	-4.966	2.290	(4.141)

12. Derive an expression for the energy of a charge transfer transition

$$A^\alpha - B^\beta \overset{\text{CT}}{\longrightarrow} A^{\alpha+1} - B^{\beta-1}$$

where α and β denote the charges on the atoms of the ground-state molecule.

Answer: $E_{\text{CT}} = \text{VOIE (for } A \text{ at } q = \alpha) - \text{VOIE (for } B \text{ at } q = \beta - 1)$

$$+ (\alpha - \beta + 1)e^2/R_{AB} \quad (4.142)$$

BIBLIOGRAPHY

[1] M. Wolfsberg and L. Helmholz, *J. Chem. Phys.* **20,** 837 (1952).
[2] R. Hoffmann, *J. Chem. Phys.* **39,** 1397 (1963).
[3] G. Herzberg, *Atomic Spectra and Atomic Structure* (Dover Publications, New York, 1944).
[4] C. J. Ballhausen, *Introduction to Ligand Field Theory* (McGraw-Hill Book Co., Inc., New York, 1962).
[5] J. C. Slater, *Quantum Theory of Atomic Structure* (McGraw-Hill Book Co., Inc., New York, 1960), Vols. 1 and 2.
[6] E. U. Condon and G. H. Shortley, *The Theory of Atomic Spectra* (Cambridge University Press, Cambridge, England, 1963).
[7] J. Hinze and H. H. Jaffé, *J. Chem. Phys.* **38,** 1834 (1963). [The F_k and G_k used by Hinze and Jaffé differ by a multiplicative constant from those used by Slater (F^k and G^k; Ref. [8]).]
[8] J. C. Slater, see Ref. [5], Vol. 1, Chap. 14 and Appendix 12a; Vol. 2, Appendix 22.

[9]   J. S. Griffith, *The Theory of Transition Metal Ions* (Cambridge University Press, Cambridge, England, 1961).

[10] C. E. Moore, *Atomic Energy Levels*, Circular 467 [National Bureau of Standards, Government Printing Office, Washington, D.C., 1949 (Vol. I); 1952 (Vol. II); 1958 (Vol. III)].

[11] U. Öpik, *Mol. Phys.* **4**, 505 (1961).

[12] H. A. Skinner and H. O. Pritchard, *Trans. Faraday Soc.* **49**, 1254 (1953).

[13] H. O. Pritchard and H. A. Skinner, *Chem. Rev.* **55**, 745 (1955).

[14] G. Pilcher and H. A. Skinner, *J. Inorg. Nucl. Chem.* **24**, 937 (1962).

[15] J. Hinze and H. H. Jaffé, *J. Am. Chem. Soc.* **84**, 540 (1962); *Can. J. Chem.* **41**, 1315 (1963); *J. Phys. Chem.* **67**, 1501 (1963).

[16] Palmieri P., and Zauli, C., *J. Chem. Soc.*, **1967**, A813.

[17] H. Basch, A. Viste, and H. B. Gray, *Theoret. chim. Acta (Berl.)* **3**, 458 (1965).

[18] J. C. Slater, *Phys. Rev.* **36**, 57 (1930).

[19] D. G. Carroll, A. T. Armstrong, and S. P. McGlynn, *J. Chem. Phys.* **44**, 1865 (1966).

[20] L. C. Cusachs and J. W. Reynolds, *J. Chem. Phys.* **43**, S160 (1965).

[21] M. D. Newton, F. P. Boer, and W. N. Lipscomb, *J. Am. Chem. Soc.* **88**, 2353 (1966); **88**, 2361 (1966); **88**, 2367 (1966).

[22] W. Moffitt, *Ann. Rept. Progr. Phys.* **17**, 173 (1954).

[23] A. L. Companion and F. O. Ellison, *J. Chem. Phys.* **28**, 1 (1958).

[24] J. H. Van Vleck, *J. Chem. Phys.* **2**, 20 (1934).

[25] R. S. Mulliken, *J. Chem. Phys.* **2**, 782 (1934).

[26] H. A. Skinner and F. H. Sumner, *J. Inorg. Nucl. Chem.* **4**, 245 (1955).

[27] L. C. Cusachs and J. W. Reynolds, *J. Chem. Phys.* **44**, 835 (1966).

[28] J. Hinze, M. A. Whitehead, and H. H. Jaffé, *J. Am. Chem. Soc.* **85**, 148 (1963).

[29] H. O. Pritchard, *J. Am. Chem. Soc.* **85**, 1876 (1963).

[30] H. C. Longuet-Higgins and M. de V. Roberts, *Proc. Roy. Soc. (London)* **A224**, 336 (1954); **A230**, 110 (1955).

[31] R. Hoffmann and W. N. Lipscomb, *J. Chem. Phys.* **36**, 2179, 3489 (1962); **37**, 2872 (1962).

[32] R. Hoffmann, *J. Chem. Phys.* **40**, 2745 (1964).

[33] C. J. Ballhausen and H. B. Gray, *Inorg. Chem.* **1**, 111 (1962).

[34] L. L. Lohr, Jr., and W. N. Lipscomb, *J. Chem. Phys.* **38**, 1607 (1963); *J. Am. Chem. Soc.* **85**, 240 (1963).

[35] T. Jordan, W. H. Smith, L. L. Lohr, Jr., and W. N. Lipscomb, *J. Am. Chem. Soc.* **85**, 846 (1963).

[36] L. C. Cusachs, Report of the International Symposium Atomic Molecular Quantum Theory, Sanibel Island, Fla., 1964 (unpublished) p. 36; *J. Chem. Phys.* **43**, S157 (1965).

[37] W. Yeranos, *J. Chem. Phys.* **44**, 2207 (1966).

[38] K. Fukui, in *Modern Quantum Chemistry*, edited by O. Sinanoğlu (Academic Press, Inc., New York, 1965), Vol. 1, p. 49.

[39] R. S. Mulliken, *J. Chim. Phys.* **46**, 497 (1949); **46**, 675 (1949). (An English version is available in Technical Reports, Laboratory of Molecular Structure and Spectra, Physics Dept., University of Chicago, Chicago, Ill., 1947–49.)

[40] K. Ruedenberg, *J. Chem. Phys.* **34**, 1892 (1961).

[41] J. A. Pople, D. P. Santry, and G. A. Segal, *J. Chem. Phys.* **43**, S129 (1965).

[42] F. O. Ellison and H. Shull, *J. Chem. Phys.* **23**, 2348 (1955).

[43] F. P. Boer, M. D. Newton, and W. N. Lipscomb, *Proc. Natl. Acad. Sci. U.S.* **52**, 890 (1964).

[44] H. D. Bedon, S. M. Horner, and S. Y. Tyree, Jr., *Inorg. Chem.* **3**, 647 (1964).

[45] E. Clementi, *Tables of Atomic Functions* (IBM Corp., San Jose, California, 1965) [supplement to *IBM J. Res. Develop.* **9**, 2 (1965)].

[46] J. P. Dahl and C. J. Ballhausen, *Advan. Quantum Chem.*, **4**, 180 (1968).

[47] F. A. Cotton, *Chemical Applications of Group Theory* (John Wiley & Sons, Inc., New York, 1963).

[48] J. W. C. Johns and R. F. Barrow, *Nature* **179**, 374 (1957); J. A. R. Coope, D. C. Frost, and C. A. McDowell, *ibid.* **179**, 1186 (1957).

[49] A. Viste and H. B. Gray, *Inorg. Chem.* **3**, 1113 (1964).

[50] C. J. Ballhausen and H. B. Gray, *Molecular Orbital Theory* (W. A. Benjamin, Inc., New York, 1964).

[51] A general method for the evaluation of group overlap integrals has been given by W. A. Yeranos, *Inorg. Chem.* **5**, 2070 (1966).

[52] R. F. Fenske, *Inorg. Chem.* **4**, 33 (1965).

[53] H. Basch, A. Viste, and H. B. Gray, *J. Chem. Phys.* **44**, 10 (1966).

[54] R. Rein, N. Fukuda, H. Win, G. A. Clarke, and F. E. Harris, *J. Chem. Phys.* **45**, 4743 (1966).

[55] A. Streitwieser, Jr., *Molecular Orbital Theory for Organic Chemists* (John Wiley & Sons, Inc., New York, 1961), pp. 115 and 116.

[56] R. McWeeny, *Proc. Roy. Soc.* (*London*) **A235**, 496 (1956).

[57] A. T. Armstrong, D. G. Carroll, and S. P. McGlynn, *J. Chem. Phys.* **47**, 1103 (1967).

[58] L. G. Vanquickenborne and S. P. McGlynn, *Theoret. Chim. Acta* (*Berl.*) **9**, 390 (1968).

[59] B. Bertus and S. P. McGlynn, in *Proceedings of the Yale Symposium on σ-Electron Theories*, edited by O. Sinanoğlu and K. Wiberg (W. A. Benjamin, Inc., New York, 1969).

[60] R. B. Woodward and R. Hoffmann, *J. Am. Chem. Soc.* **87**, 395 (1965); **87**, 2511 (1965); R. Hoffmann and R. B. Woodward, *J. Am. Chem. Soc.* **87**, 2046 (1965).

[61] R. B. Woodward and R. Hoffmann, *Accts. Chem. Res.* **1**, 17 (1968).

[62] R. B. Woodward and R. Hoffmann, *Angew. Chem. Intern.* **8**, 781 (1969). [See also R. B. Woodward and R. Hoffmann, *The Conservation of Orbital Symmetry* (Academic Press, Inc., New York, (1970).]

[63] See, for example, W. J. Moore, *Physical Chemistry* (Prentice-Hall, Inc., Englewood Cliffs, N.J. 1967), 3rd ed., Fig. 13.8, p. 533, and discussion relating to it.

[64] M. J. S. Dewar, *The Molecular Orbital Theory of Organic Chemistry* (McGraw-Hill Book Co., Inc., New York, 1969) [see, in particular, Chap. 8).

[65] R. G. Parr, *Quantum Theory of Molecular Electronic Structure* (W. A. Benjamin, Inc., New York, 1964).

[66] L. Oleari, G. De Michaelis, and L. Di Sipio, *Mol. Phys.* **10**, 111 (1966).

[67]   G. Berthier, P. Millie, and A. Veillard, *J. Chim. Phys.* **62,** 8 (1965); **62,** 20 (1965).

[68]   B. Roos, *Acta Chem. Scand.* **20,** 1673 (1966).

[69]   R. F. Fenske, K. G. Caulton, D. D. Radtke, and C. C. Sweeney, *Inorg. Chem.* **5,** 951, 960 (1966).

[70]   P. Ros and G. C. A. Schuit, *Theoret. Chim. Acta (Berl.)* **4,** 1 (1966).

[71]   C. E. Schäffer and C. K. Jørgensen, *Mol. Phys.* **9,** 401 (1965).

[72]   See C. E. Moore, Ref. [10], Vol. I, p. 24.

[73]   See C. E. Moore, Ref. [10], Vol. I, pp. 184 and 185.

General References

1.   C. J. Ballhausen and H. B. Gray, *Molecular Orbital Theory* (W. A. Benjamin, Inc., New York, 1964).

2.   H. B. Gray, *Electrons and Chemical Bonding* (W. A. Benjamin, Inc., New York, 1964).

3.   J. C. Slater, *Quantum Theory of Atomic Structure* (McGraw-Hill Book Co., Inc., New York, 1960), Vols. 1 and 2.

4.   H. Basch, A. Viste, and H. B. Gray, *Theoret. Chim. Acta (Berl.)* **3,** 458 (1965).

5.   L. C. Cusachs and J. W. Reynolds, *J. Chem. Phys.*, Suppl. **43,** 160 (1965); **44,** 835 (1966).

6.   F. A. Cotton, *Chemical Applications of Group Theory* (John Wiley & Sons, Inc., New York, 1963).

CHAPTER 5

Free-Electron Molecular-Orbital Theory

The free-electron molecular-orbital (FEMO) theory is a variant of the molecular-orbital approach. If interelectronic repulsion between π electrons is neglected, the FEMO and Hückel theories attain a comparable degree of sophistication. The most important groups of molecules to which FEMO theory is applicable are the linear polyene chains, many organic dyes, the polycyclic aromatics, molecules consisting partly of aromatic rings and partly of polyenic chains, porphyrins, etc. The theory is also of importance in solid-state physics; however, we confine attention to the study of individual molecules.

The FEMO theory provides a very efficient description of the π electrons in aromatics and in conjugated bond systems. A typical MO energy diagram for such a system is shown in Fig. 5.1. The σ MO's usually constitute the group of lowest-energy orbitals whereas the σ^* MO's are invariably the most antibonding set. On the other hand, the π MO's usually constitute the set of highest-energy occupied MO's whereas the π^* MO's usually correspond to the set of lowest-energy unoccupied MO's. As a result of this sort of energy-level scheme, one may conclude that:

(i) The excitation of σ electrons is largely destructive of molecular integrity.

157

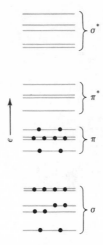

FIG. 5.1. Schematized MO energy-level scheme of a typical conjugated bond system. Orbital occupancy is indicated by circles.

    (ii)   Excitations of $\sigma^* \leftarrow \pi$ type (which are usually of very low oscillator strength) are of quite high energy.

    (iii)   Excitations of $\pi^* \leftarrow \pi$ type are almost wholly responsible for the observed spectroscopic behavior of such molecules.

The π electrons are the least-tightly-bound group of electrons. They are characterized by strong delocalization (i.e., they may be considered to move freely throughout the molecular framework). In this context, the nuclei plus σ electrons merely provide a confining potential which retains the π electrons *within* the molecule. Furthermore, the *atomic-orbital* concept is no longer needed: The π electrons traverse a bond path defined by the molecular skeleton (or, equivalently, by the confining potential). These simple assumptions constitute the basis of FEMO theory.

1.   OPEN LINEAR CHAINS[1]−[5]

    An open-chain molecule (linear polyene, dye, etc.) is treated as a one-dimensional box of length L. Neglecting electron repulsion and denoting the single coordinate in the box by x, the one-electron Schrödinger equation is given by

$$\hat{H}\varphi = \left(-\frac{\hbar^2}{2m}\frac{d^2}{dx^2} + \hat{U}\right)\varphi = \epsilon\varphi \tag{5.1}$$

where m is the electronic mass and $\hat{U}$ is the potential energy operator for the electron in the box.

(A)   Case of a Constant Potential

As indicated in Fig. 5.2, U is expected to be a periodic function of x. However, if the amplitude associated with U is small, it is a reasonable first approximation to assume U constant within the box, say, $U = 0$, and to assume $U = \infty$ outside the box. Imposition of the corresponding boundary conditions, namely, $\varphi = 0$ at $x = 0$ and L, on the solutions of Eq. (5.1), provides

$$\varphi_n = \left(\frac{2}{L}\right)^{1/2} \sin \frac{\pi n x}{L} = \left(\frac{2}{L}\right)^{1/2} \sin \frac{(2m\epsilon_n)^{1/2}}{\hbar} x \qquad (5.2)$$

$$\epsilon_n = n^2 h^2/8mL^2; \qquad n = 1, 2, 3, 4, \ldots \qquad (5.3)$$

Furthermore, the quantum number n is related to the number of nodes q_n in the wave function φ_n by

$$n = q_n - 1 \qquad (5.4)$$

Figure 5.3 displays the energies and wave functions for some of the lower values of n.

There is only one empirical parameter in Eqs. (5.2) and (5.3), namely, the length L. This length, however, does not equal the total length of the molecule—it would be unreasonable to require that the wave function falls to zero at the end atoms. Therefore, Kuhn[2] suggested an extension of the

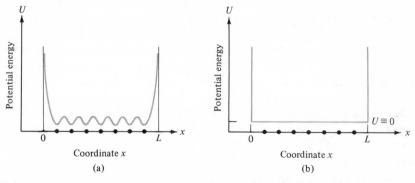

FIG. 5.2.   Atom-chain of a linear π-electron system replaced by a one-dimensional box. One of the axes of a regular Cartesian coordinate system represents the one-dimensional space defined by the bond path. The atoms are shown as points at regular intervals along this axis; (a) shows a qualitatively realistic shape of the potential U along x; (b) shows the idealized rectangular potential well used in the constant potential assumption.

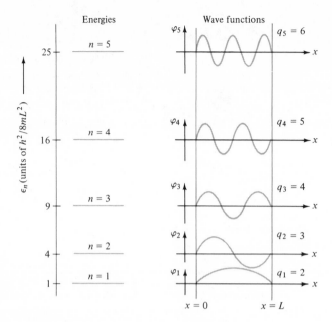

FIG. 5.3.   Five lowest-energy levels of a particle in a one-dimensional box and the corresponding wave functions for the case of a constant potential. The number of nodes q is also shown.

box by one bond length beyond each of the end atoms. Thus, $L = L' + 2R$, where L' is the total length of the molecule and R is the average bond length. Ruedenberg and Scherr[5] have advanced theoretical reasons for this choice of L.

The actual free-electron path is not likely to be the straight line postulated in Fig. 5.2. Indeed, a zig-zag path such as that provided by the all-*trans*-configuration of Fig. 5.4 would appear more realistic. The sp^2 nature of the carbon hybrids responsible for the σ bonds indicates that all angles should be $\sim 120°$. However, it is still possible to define a one-dimensional space whose variable s follows the bond path throughout all of its turns and twists.[1] The results of Eq. (5.2) remain unaltered provided x is replaced by s.

(B)   Intensity of Transitions

If the chain can be represented as a straight line (see Fig. 5.2, for example), the intensity calculations are straightforward.

---

[1] The fact that one needs two Cartesian coordinates to describe the actual molecular shape does not mean that one must cease the use of a one-dimensional box picture. The notion *one dimension* as used here merely implies that the situation can be described by one variable.

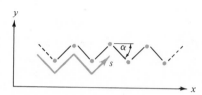

FIG. 5.4. All-*trans* geometric configuration of a chain of carbon atoms and the variable *s* which adequately describes the bond path. The angle α should equal approximately 30°.

The oscillator strength $f_{n_2 \leftarrow n_1}$ is

$$f_{n_2 \leftarrow n_1} = 2(1.085)10^{-5}\bar{\nu} \mid \mathbf{D} \mid^2 \qquad (5.5)$$

where $\bar{\nu}$ is the energy, in cm^{-1}, of the transition $n_2 \leftarrow n_1$ and $\mid \mathbf{D} \mid^2$, in Å^2, is given by

$$\mid \mathbf{D} \mid^2 = \mid \mathbf{X} \mid^2 + \mid \mathbf{Y} \mid^2 + \mid \mathbf{Z} \mid^2 \qquad (5.6)$$

where, for instance,

$$\mathbf{X} = \mathbf{i} \int \varphi_{n_2}{}^* x \varphi_{n_1} dx \qquad (5.7)$$

and **i**, **j**, and **k** are unit vectors. The factor 2 in Eq. (5.5) implies that the n_1 MO in the ground state contains two electrons.[2]

The calculation of intensity requires a three-dimensional Cartesian coordinate system. If we choose this system so that the x axis coincides with the x axis of our one-dimensional bond-path system, we have at any point in the molecule $y = z = 0$; therefore, $\mathbf{Y} = \mathbf{Z} = 0$. On the other hand,

$$\mathbf{X} = \left(\frac{2}{L}\right)\mathbf{i} \int_0^L \left(\sin \frac{\pi n_2 x}{L}\right) x \left(\sin \frac{\pi n_1 x}{L}\right) dx$$

$$= 0, \qquad \text{if } (n_2 - n_1) \text{ is even} \qquad (5.8a)$$

$$= \frac{2L}{\pi^2}\mathbf{i}\left[\frac{1}{(n_1 + n_2)^2} - \frac{1}{(n_1 - n_2)^2}\right], \qquad \text{if } (n_2 - n_1) \text{ is odd} \quad (5.8b)$$

[2] See chapter 9 for a more detailed discussion of transition probabilities.

Hence, the selection rule: Δn *is odd*; and the polarization rule: *All electric dipole-allowed transitions are polarized along the molecular axis.*

The introduction of kinks, as previously indicated, has no qualitative effect on the energy calculations. However, the presence of kinks can affect the transition probabilities in a quite marked way. The all-*trans*-configuration of Fig. 5.4 does not show considerable deviation from the polarization rule, but the f values for the allowed transitions are decreased by a factor $\cos^2 \alpha$. Reference to Fig. 5.4 indicates that

$$x = s \cos \alpha \qquad (5.9)$$

Thus **X** as calculated in Eq. (5.8b) is replaced by **X** $\cos \alpha$. The **Y** and **Z** components remain approximately zero, as can be verified by calculating the appropriate integrals in Eq. (5.6).

(C) An Example Calculation

Consider the symmetric dye cation shown in Fig. 5.5. The conjugated molecular backbone contains five carbon atoms (whence five π electrons) and two nitrogens (whence three π electrons[3]). The corresponding box of the FEMO model has a length $L = 8R$; therefore, the energies are given[4]

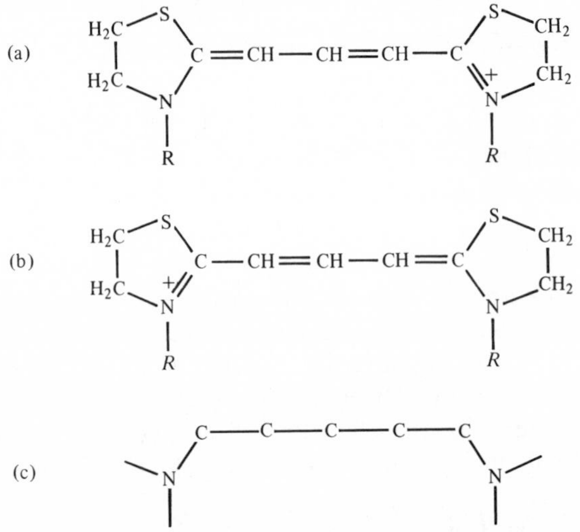

FIG. 5.5. Typical symmetric organic dye. The dye cation can be thought of as resonating between two extreme structures (a) and (b). The molecular extent of the π-electron system is shown in (c).

[3] The N center contributes two π electrons and the N^+ center contributes one π electron.
[4] It is shown in Sec. 5 that the introduction of heteroatoms in the π chain introduces no complication whatever, provided these heteroatoms are situated at the chain ends.

by Eq. (5.3) as

$$\epsilon_n = (0.3)n^2, \quad \text{in eV} \tag{5.10}$$

For the smallest transition energy $(\varphi_2 \leftarrow \varphi_1)$, we find $\Delta\epsilon = 2.7$ eV (21 800 cm^{-1}). The observed energy is 22 470 cm^{-1}.

From Eq. (5.8b), with correction made for the angle $\alpha = 30°$, the oscillator strength is $f_{2\leftarrow 1} = 1.8$; the experimental value is 1.2.

(D) Case of a Fluctuating Potential

The agreement between theory and experiment is excellent for symmetric dyes; it is much less satisfactory in the case of polyenes or unsymmetric dye cations.

It is well known that the polyenes are characterized by alternating long and short bonds. For example, the two resonance structures applicable to hexatriene are

$$CH_2{=}CH{-}CH{=}CH{-}CH{=}CH_2; \quad \text{Kekulé}$$

$$CH_2{-}CH{=}CH{-}CH{=}CH{-}CH_2; \quad \text{Thielé}$$
$$\vdots \qquad\qquad\qquad\qquad \vdots$$

The Kekulé structure is so dominant relative to the Thielé structure that the polyene may be considered to consist of an alternating series of *almost double* bonds and *almost single* bonds.

This situation contrasts with the typical symmetric organic dye cations which possess two identical extreme resonance structures [see, for example, Figs. 5.5(a) and 5.5(b)]. Since both resonance structures correspond to exactly the same energy, the actual structure should be *aromatic* in nature— at least in the sense that all bonds in the π system should be equivalent (with π-bond order = 0.5). However, if the dye cation is unsymmetric (i.e., if the two end groups are different), the two possible structures may have different energies and one of them may be dominant in an energetic sense; if this be the case, a considerable bond alternation characteristic results. Thus, the symmetric dyes are structurally more *homogeneous* than either the polyenes or the unsymmetric dyes and are more suited to the treatment outlined in Sec. 1(A).

In order to account for the electronic structure of polyenes and/or the unsymmetric dyes, Kuhn[2] introduced a periodic potential similar to that shown in Fig. 5.6. He assumed this potential to be of the form

$$U = U_0 \cos\left[2\pi s/(R_1 + R_2)\right] \tag{5.11}$$

where s has more or less arbitrary end points of the order of 1 to $\frac{1}{2}$ bond lengths beyond the terminal atoms. The resulting Schrödinger equation is exactly solvable; the solutions, however, are rather complicated functions of U_0 and $(R_1 + R_2)$. Indeed, the parameter U_0 must be adjusted to fit the position of the first intense absorption band. Rather than investigate the exact solution, it is probably more valuable to examine the qualitative changes caused by the fluctuating potential.

Let j be the number of complete U cycles in the s space under consideration. In the case of a polyene or an asymmetric dye, it is clear that $j = \Pi/2$ where Π represents the total number of π electrons. Therefore, in the energy-level diagram, groups of MO's, each group containing j levels, are drawn together; each group of j levels converges towards a limiting value[5] as $U_0 \to \infty$. This process is illustrated in Fig. 5.7. Furthermore, the highest occupied level is characterized by a quantum number $n = \Pi/2 = j$. As seen in Fig. 5.7, FEMO theory predicts, in agreement with experiment, that the absorption spectrum of an unsymmetric dye lies at higher energies than the spectrum of the corresponding symmetric dye.

The significance of this treatment is not the prediction of a spectroscopic interval (because U_0 cannot be evaluated in any *a priori* way). Our discussion merely sheds some light on why such a simple method should be so surprisingly good in a very large number of cases and so poor in so many

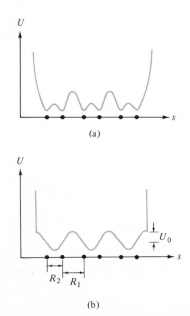

(a)

(b)

FIG. 5.6 Periodic potentials for polyenes and unsymmetric dyes.

[5] As $U_0 \to \infty$, the potential of Fig. 5.6(b) separates into three identical potential wells. If the interaction between these wells be negligible, all levels are threefold degenerate. Noting that 3(or j) $= 6/2$(or $\Pi/2$), the conclusions we derive are quite straightforward.

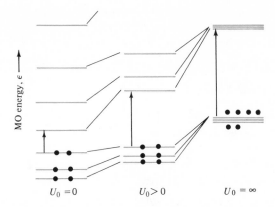

FIG. 5.7. Effect of increasing amplitude U_0 of the fluctuating potential. Note that $j \equiv 3$. The lowest-energy electronic transition is also schematized.

others. Indeed, the fluctuating potential associated with symmetric dyes [see Fig. 5.2(a)] goes through *one complete period once every atom*, while in Fig. 5.6(b), U goes through *one complete period once every two atoms*. Hence, the MO's of a symmetric dye, as may also be deduced from Fig. 5.7, draw together in groups of $2j = \Pi$. However, it is not the *inter*group but rather the *intra*group separation which now corresponds to the lowest-energy intense electronic transition (i.e., which produces color). Thus, the introduction of a fluctuating potential into considerations of symmetric dyes yields results which are not much different from those already obtained on the basis of the constant potential assumption.

2. CLOSED NONBRANCHED CHAINS

The simplest closed π system is benzene. In FEMO theory we might idealize benzene, as in Fig. 5.8, as a circular one-dimensional box of length

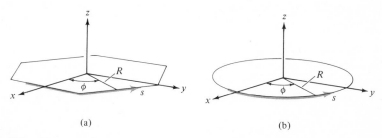

(a)   (b)

FIG. 5.8. Two one-dimensional bond-path idealizations of benzene: (a) hexagon; (b) circle.

L, in which the only coordinate, s, is the displacement along the circular bond path.[6],[7] From Sec. 1(D), we can set $U = 0$ within the box and $U = \infty$ outside the box. The resulting Schrödinger equation is identical to that for the linear case [i.e., Eq. (5.1)]. The solutions, however, are not given by Eqs. (5.2) and (5.3) because the boundary conditions are different. For a circular bond path, we require $\varphi(x) = \varphi(x + L)$ for single valuedness of the wave function.

(A)   Molecular Orbitals

The solutions are

$$\epsilon_n = \frac{n^2 h^2}{2mL^2} \tag{5.12}$$

The corresponding set of eigenfunctions is

$$\varphi_n = \left(\frac{2}{L}\right)^{1/2} \cos\,(2\pi n s/L)$$

$$\varphi_n{}' = \left(\frac{2}{L}\right)^{1/2} \sin\,(2\pi n s/L) \tag{5.13a}$$

or, alternatively,

$$\varphi_{n+} = (\tfrac{1}{2})^{1/2}(\varphi_n + i\varphi_n{}') = \left(\frac{1}{L}\right)^{1/2} e^{2\pi i n s/L}$$

$$\varphi_{n-} = (\tfrac{1}{2})^{1/2}(\varphi_n - i\varphi_n{}') = \left(\frac{1}{L}\right)^{1/2} e^{-2\pi i n s/L} \tag{5.13b}$$

where the quantum number $n = 0, 1, 2, \ldots$. Except for $n = 0$, all MO's are doubly degenerate. The solutions of Eq. (5.13b), when pertaining to the circular bond path, have the advantage that they are eigenfunctions of the angular momentum operator $\hat{l}_z$ with eigenvalues $\pm n\hbar$. From Fig. 5.8(b)

$$\hat{l}_z\varphi_{n+} = \frac{\hbar}{i}\frac{\partial}{\partial\phi}\,\varphi_{n+} = R\,\frac{\hbar}{i}\frac{\partial}{\partial s}\,\varphi_{n+} = n\hbar\varphi_{n+} \tag{5.14}$$

Thus, we might think of the electron described by Eq. (5.13b) as traveling either clockwise (φ_{n-}) or counterclockwise (φ_{n+}) around the perimeter. Alternatively, if the real wave functions are used, the quantum number n, as may be seen in Fig. 5.9, determines the number of nodes q along the perimeter,[6] namely,

$$q_n = 2n \tag{5.15}$$

---

[6] The complex functions of Eq. (5.13b) do not have any nodes.

The introduction of appropriate kinks transforms the circle into a molecular polygon (i.e., the benzene hexagon). However, the energies and the wave functions of Eqs. (5.12) and (5.13) remain unaltered provided s represents displacement along the polygon perimeter [see Fig. 5.8(a)]. In a polygon, the radius vector R is not constant; consequently, the wave functions are not, strictly speaking, eigenfunctions of $\hat{l}_z$. On the other hand, n remains a good quantum number because it still determines the number of nodes along the polygonal perimeter.

(B) State Functions

The FEMO approach for closed bond-path systems provides a qualitative account of the electron repulsion interaction—and this without any explicit consideration of interelectronic repulsion integrals.

Consider an aromatic polygon containing $4n_0 + 2$ π electrons, where the highest-energy occupied MO of the ground state is characterized, as in Fig. 5.10, by $n = n_0$. The lowest-energy MO transition is produced by excitation of an electron from the MO n_0 to $n_0 + 1$. There are four distinct ways to describe this transition:

$$\varphi_{n_0+1} \leftarrow \varphi_{n_0} \qquad\qquad \varphi_{(n_0+1)+} \leftarrow \varphi_{n_0+}$$

$$\varphi_{n_0+1}{}' \leftarrow \varphi_{n_0} \qquad\qquad \varphi_{(n_0+1)-} \leftarrow \varphi_{n_0+}$$

or, equivalently,

$$\varphi_{n_0+1} \leftarrow \varphi_{n_0}{}' \qquad\qquad \varphi_{(n_0+1)+} \leftarrow \varphi_{n_0-}$$

$$\varphi_{n_0+1}{}' \leftarrow \varphi_{n_0}{}' \qquad\qquad \varphi_{(n_0+1)-} \leftarrow \varphi_{n_0-} \qquad (5.16)$$

In a one-electron approximation, these four possibilities correspond to the

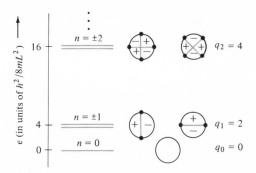

FIG. 5.9. MO energy-level scheme for a few lowest-energy MO's of a closed bond path. Compare with the diagram for an open bond path: In Fig. 5.3, the lowest-energy MO was obtained for $n = 1$, not for $n = 0$. Moreover, in Fig. 5.3, the gaps between any two specific MO's are four times smaller than those shown here.

same transition energy. The simple Hückel LCAO-MO theory also predicts a fourfold degeneracy of the first-excited state. FEMO theory, on the other hand, can easily rationalize the removal of this fourfold degeneracy. The state functions resulting from the MO excitations of Eq. (5.16) are simple products of the specified MO's. If we use the complex MO's of Eq. (5.13b), these products are exponentials—with exponents which equal sums or difference of the component MO exponents. Thus, it is possible to introduce a state nomenclature dependent on a number N with possible values:

$$N = 2n_0 + 1: \qquad (\Phi_{(2n_0+1)+};\ \Phi_{(2n_0+1)-}) \qquad\qquad (5.17a)$$

$$N = 1: \qquad (\Phi_{1+};\ \Phi_{1-}) \qquad\qquad (5.17b)$$

If the perimeter is a circle, these state functions are eigenfunctions of the ring angular momentum operator $\hat{L}_z$ with eigenvalues $\pm N\hbar$. Now, one of Hund's rules specifies that the many-electron state with the largest angular momentum is lowest in energy. Hence, the advantage of using the complex functions of Eq. (5.14b) is that the resulting state functions are diagonal with respect to the electron repulsion operator. The fourfold degenerate set of Eq. (5.16) splits into the two twofold degenerate sets of Eqs. (5.17); $\Phi_{(2n_0+1)\pm}$ corresponds to the lowest-energy excited state.

If the circle is reshaped into the correct polygonal form, the residual twofold degeneracy may be lifted. If the distortion is not very drastic, no significant energy changes should occur and Hund's rules retain considerable validity. Therefore, even in the distorted case, the two $N = 2n_0 + 1$ func-

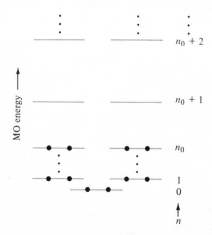

FIG. 5.10.   Electronic configuration of the ground state of a closed bond-path system with $4n_0 + 2\ \pi$ electrons. Values of the quantum number n are on the right-hand side of the figure.

tions of Eq. (5.17), or any linear combination of these, should still correspond to smaller energy than either of the two $N = 1$ functions.

(C) Platt Nomenclature

The discussion of Secs. 2(A) and 2(B) is not of much utility because of the scarcity of stable aromatic polygonal molecules. Nonetheless, Platt developed the ideas of Secs. 2(A) and 2(B) into a viable model for catacondensed hydrocarbons.[7]

Platt ignored the cross links altogether; naphthalene, for instance, is treated as if it were a circle or decagon whose perimeter is of the same length as that of naphthalene. Such an approximation may seem drastic; therefore, we stress that the purpose of the Platt approach is not the provision of a quantitative theory but rather the elaboration of an attitude which allows correlation of the experimental spectra of aromatic ring systems with the predicted spectra of their cyclic homologues (whether circular or regularly polygonal).

Consider an aromatic molecule whose circular equivalent has a ground-state electron configuration identical to that of Fig. 5.10. The quantum number of the highest occupied MO, namely, n_0, is also the number of benzenelike hexagons contained in the aromatic molecule in question. An electron excitation from $n = n_0$ to $n = n_0 + 1$ yields states characterized by $N = 2n_0 + 1$ and $N = 1$; the first of these is termed an L state, the second a B state—regardless of the specific value of n_0. In general, states characterized by $N = 0, 1, 2, \ldots$ are designated $A, B, C, \ldots$; states characterized by $N = 2n_0, 2n_0 + 1, 2n_0 + 2, \ldots$ are designated $K, L, M, \ldots$. Distortion of the circle in order to regenerate the aromatic molecule may lift the double degeneracy of the L state and/or of the B state. If so, the resulting states are distinguished as $L_a, L_b; B_a, B_b;$ etc. The meaning of the subscripts a and b is explained in Sec. 2(D). According to Hund's rules, the L states are always of lower energy than the B states. Furthermore, it has been found experimentally that electronic transitions associated with certain Platt labels (say, $^1L_a \leftarrow {}^1A$, where ^1A denotes the ground state) retain very similar characteristics throughout the whole series of the aromatic hydrocarbons. It is in this facet of the Platt model, namely, its correlative utility, that its importance lies.

(D) Transition Probabilities

If the highest-energy occupied MO in the ground-state electron configuration is characterized by n_0, we must evaluate the one-electron transi-

---

[7] By *catacondensed hydrocarbons* is meant the fused ring systems with general formula $C_{4t+2}H_{2t+4}$, $t = 1, 2, 3, \cdots$. Every carbon atom in these molecules is on the periphery; no carbon atom belongs to more than two rings.

tion probabilities:

$$\varphi_{(n_0+1)+} \longleftarrow \varphi_{n_0+}; \qquad |\,\Delta N\,| = 1$$

$$\varphi_{(n_0+1)-} \longleftarrow \varphi_{n_0+}; \qquad |\,\Delta N\,| = 2n_0 + 1$$

$$\varphi_{(n_0+1)+} \longleftarrow \varphi_{n_0-}; \qquad |\,\Delta N\,| = 2n_0 + 1$$

$$\varphi_{(n_0+1)-} \longleftarrow \varphi_{n_0-}; \qquad |\,\Delta N\,| = 1$$

The first of these transition probabilities is connected to the magnitude of the integral

$$\mathbf{D} = \int_0^L \varphi_{(n_0+1)+}{}^*\hat{\mathbf{r}}\varphi_{n_0+}ds \tag{5.18}$$

where $\hat{\mathbf{r}}$ is the dipole length operator. The x component in the case of a circular bond path is given by

$$D_x = \frac{1}{L}\int_0^L e^{-2\pi i n_0 s/L}x e^{2\pi i(n_0+1)s/L}ds$$

$$= \frac{R}{2L}\int_0^L e^{-2\pi i n_0 s/L}\left(e^{2\pi i s/L} + e^{-2\pi i s/L}\right)e^{2\pi i(n_0+1)s/L}ds = \frac{R}{2} \tag{5.19}$$

where R is the radius of the circular bond path. Similarly, it is found that

$$D_y = \tfrac{1}{2}iR; \qquad D_z = 0 \tag{5.20}$$

Exactly identical results are obtained for the fourth transition, namely, for $\varphi_{(n_0+1)-} \longleftarrow \varphi_{n_0-}$. On the other hand, we find $\mathbf{D} = 0$ for the second and third of the listed transitions. These conclusions illustrate a general selection rule for electric dipole-allowed electronic transitions of a homogeneous circular molecule:

$$\Delta N = \pm 1 \tag{5.21}$$

Therefore, *transitions between the ground state and the B states are allowed, whereas transitions between the ground state and the L states are forbidden.* Equations (5.19) and (5.20) illustrate a general polarization rule: *only light whose electric field vector is parallel to the molecular plane is absorbed.*

The introduction of distortions and cross links vitiates the intensity selection rule to some extent. The higher-energy $B \leftrightarrow A$ transitions, however, probably remain more strongly allowed than the lower-energy $L \leftrightarrow A$ transitions. The polarization selection rule remains valid in its entirety, in that no transition can be other than in-plane polarized. The reshaping of the molecule may remove the double degeneracies. Therefore, a given $B \leftrightarrow A$ or $L \leftrightarrow A$ transition may be either x polarized or y polarized, where we designate x as the *long axis* and y as the *short axis* of the molecule.[8] In the most common alternant hydrocarbons, which usually belong to point groups C_{2v} or D_{2h}, one of the $L \leftrightarrow A$ transitions is x polarized whereas the other is y polarized; the same situation holds for the $B \leftrightarrow A$ transitions. The two states (one L and one B) giving rise to the same polarization are given the same subscript—a or b, according to the following convention: the L_b state is *defined* as having nodes at the atom positions. By means of this definition it is possible to determine the polarization of a specific transition in any given molecule.[6]−[8]

We illustrate using anthracene and phenanthrene. The nodes in the L_b states of these molecules are shown in Fig. 5.11. In anthracene, the L_b state transforms as x because the wave function is symmetric with respect to reflection in the xz plane and antisymmetric with respect to reflection in the yz plane. In phenanthrene, on the other hand, the L_b state is symmetric with respect to reflection in the yz plane. Therefore, the transition $L_b \leftrightarrow A$ (and also $B_b \leftrightarrow A$) is long axis polarized in anthracene and short axis polarized in phenanthrene.

(E) Benzene Example (Comparison with Hückel Theory)

(*i*) *Wave functions:* The results of Hückel theory are shown on the left of Fig. 5.12; the MO energy levels are labeled group theoretically. The FEMO results are shown on the right of Fig. 5.12; the classification here is based on the quantum number n. We now wish to label the FEMO's in a group-theoretic fashion.

The wave functions of Eqs. (5.13) were assumed to describe the situa-

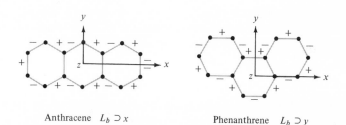

Anthracene $L_b \supset x$ Phenanthrene $L_b \supset y$

FIG. 5.11. Nodal behavior of the L_b states of anthracene and phenanthrene. The nodes at the atom positions are shown as black dots. The sign of the wave function is indicated by plus and minus signs along the perimeter. The z axis is perpendicular to the plane of the paper.

---

[8] In highly symmetrical molecules such as benzene or triphenylene, the x and y directions remain equivalent. In the very unsymmetrical molecules (symmetry group C_s), the transitions possess simultaneous x and y polarizations such that $| M_x |^2 \neq | M_y |^2 \neq 0$.

tion existing along the one-dimensional bond path. However, all of the *free electrons* are π electrons and all orbitals descriptive of their behavior must have complete nodal character (i.e., zero electron density) along the bond path! Thus, we must revise the meaning of the free-electron wave functions. We do this by taking account of the fact that the wave functions must refer to a three-dimensional space. The FEMO wave functions have meaning only in the regions above and below the bond path. The complete wave function has the sign of the FEMO *above* the bond path (positive z), the opposite sign *below* the bond path (negative z), and is zero *in* the bond path (zero z); we symbolize this

$$(\varphi_{n\pm})_{\text{total}} = (\varphi_{n\pm})(u) \qquad (5.22)$$

where the function u is antisymmetric with respect to reflection in the molecular plane. The analytical expression for u is not of interest as long as we consider only symmetry properties: All that matters is that u transforms as the z axis. Obviously, then, the nondegenerate function $\varphi_0 = (1/L)^{1/2}$ transforms like the z axis and is classified as a_{2u} in D_{6h}.

In the general case, the set $\{\varphi_{n+}, \varphi_{n-}\}$ generates a representation in D_{6h}

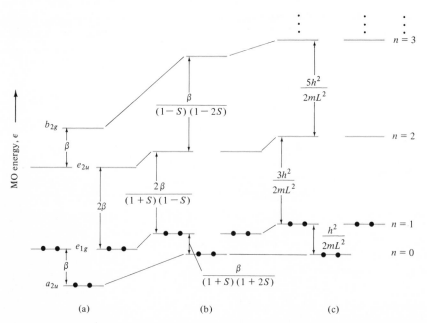

FIG. 5.12.   MO energy-level diagrams for benzene. (a): Hückel LCAO-MO theory neglecting overlap; (b): Hückel LCAO-MO theory including overlap ($\beta \sim 20\,000$ cm^{-1} and $S \sim 0.25$); (c): FEMO theory for a circular bond path.

whose characters are given in Table 5.1. Since the character for the inversion operation is $-2 \cos n\pi$, the representation generated is labeled g when n is odd, and u when n is even. The benzene functions $\varphi_{1\pm}$ and $\varphi_{2\pm}$ form bases for the irreducible representations e_{1g} and e_{2u}, respectively; the functions $\varphi_{3\pm}$ form a basis for a reducible representation which is decomposable to $b_{1g} + b_{2g}$. This order coincides with the predictions of Hückel theory.[9]

(*ii*)   *Orbital Energies*[6]:   The results of three different treatments of benzene are summarized in Fig. 5.12.

Hückel theory, neglecting overlap, predicts MO energy gaps in the ratio $1:2:1$. If overlap is included (i.e., $S = 0.25$), a ratio of approximately $1:4:5$ is obtained for the same MO energy gaps. In both cases, the value of the energy unit β must be determined empirically. According to Eq. (5.12), FEMO theory predicts a ratio $1:3:5$ for the MO energy gaps—in better agreement with that form of the Hückel theory which includes overlap. In addition, FEMO theory provides energy gaps which are independent of any adjustable parameter. In the case of benzene, Eq. (5.12) becomes $\epsilon_n = (17\,000)n^2$, in cm^{-1}.

(*iii*)   *State Energies:*   All theories predict that the first-excited configuration should be $e_{1g}{}^3 e_{2u}$. This configuration gives rise to B_{1u}, B_{2u}, and E_{1u} states. From Platt's point of view, since the transition takes place from an $n = \pm 1$ MO to an $n = \pm 2$ MO, the first-excited states may also be identified with total ring angular momentum quantum numbers $N = \pm 1$ and $N = \pm 3$. In order to make the connection between the N labels and the symmetry designations, we must investigate the transformation behavior of $\Phi_{1\pm}$ and $\Phi_{3\pm}$. The state functions Φ transform as products of MO's. Since the behavior of Φ with respect to reflection in the molecular plane (i.e., σ_h) is determined, *via* Eq. (5.22), by $(u)^6$ instead of (u), it follows that the state functions are all symmetric with respect to σ_h. Table 5.2 gives the character representation generated by the general set of state functions $\{\Phi_{N+}, \Phi_{N-}\}$ under the symmetry operators of D_{6h}. Application of these results to the cases $N = \pm 1, \pm 3$ leads to the conclusion that $\Phi_{1\pm}$ transforms as E_{1u}, and that $\Phi_{3\pm}$ transforms as $B_{1u} + B_{2u}$. Therefore, E_{1u} corresponds to the B state while B_{1u} and B_{2u} correspond to the L_a and L_b states. Because of the arguments presented in Secs. 2(B) and 2(C), the state E_{1u} should be of higher energy than either of the other two. This prediction agrees with experiment. This same conclusion is not easily achievable within the framework of any simple Hückel theory.

---

[9] This type of agreement occurs often but not always. In some aromatic molecules (see Sec. 3), there are definite MO level inversions predicted in passing from one theory to the other (Ref. [5]).

TABLE 5.1. D_{6h} representation generated by set of MO functions $\{\varphi_{n+}, \varphi_{n-}\}$.

D_{6h}	E	$2C_6$	$2C_3$	C_2	$3C_2'$	$3C_2''$	i	$2S_3$	$2S_6$	σ_h	$3\sigma_d$	$3\sigma_v$
	2	$2\cos\frac{1}{3}n\pi$	$2\cos\frac{2}{3}n\pi$	$2\cos n\pi$	0	0	$-2\cos n\pi$	$-2\cos\frac{2}{3}n\pi$	$-2\cos\frac{1}{3}n\pi$	-2	0	0

TABLE 5.2. D_{6h} representation generated by set of functions $\{\Phi_{N+}, \Phi_{N-}\}$.

D_{6h}	E	$2C_6$	$2C_3$	C_2	$3C_2'$	$3C_2''$	i	$2S_3$	$2S_6$	σ_h	$3\sigma_d$	$3\sigma_v$
	2	$2\cos\frac{1}{3}N\pi$	$2\cos\frac{2}{3}N\pi$	$2\cos N\pi$	0	0	$2\cos N\pi$	$2\cos\frac{2}{3}N\pi$	$2\cos\frac{1}{3}N\pi$	2	0	0

3. BRANCHED CHAINS

Insofar as the Platt model represents an acceptable one-electron theory, it is hardly susceptible to fruitful generalization beyond benzene.[10] In order to formulate a more general FEMO theory—one comparable, say, to the generality of the Hückel approach—one must establish a simple means of introducing cross links into the *circular* or *polygonal* wave functions which were considered in Sec. 2. Kuhn[10] and Ruedenberg and Scherr[5] have developed a general one-electron formalism valid for any aromatic hydrocarbon, catacondensed or pericondensed[11]; for any open π system, linear or branched; and for any aromatic molecule consisting of any combination of open and closed π parts. Their theory is an extension of the constant potential case discussed in Secs. 1(A) and 2; it is, therefore, not very good for polyenes [see Sec. 1(D)].

(A) Joint Conditions

A molecular system with branch points and/or cross links contains carbon atoms which are linked to *three other* carbon centers. The first step in the analysis consists of breaking down the molecule into its separate *branches*. Each branch connects two joints, two end points or one joint and one end point. Each joint is the intersection of three branches. Figure 5.13 indicates the decomposition of naphthalene into three branches, denoted 1, 2, and 3. Each of these branches constitutes a one-dimensional space whose variable is the bond-path coordinate s_B ($B = 1, 2, 3, \ldots$). If each s_B runs through its assigned range, the conjugated structure is completely covered.

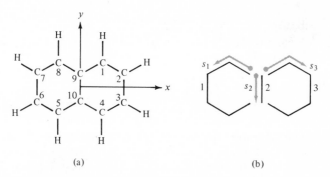

(a)   (b)

FIG. 5.13. Naphthalene. (a) Atom numbering convention. (b) Decomposition of molecular skeleton into three branches.

---

[10] Except, perhaps, for molecules like the porphins (Ref. [9]).
[11] Pericondensed hydrocarbons may contain carbon atoms which belong simultaneously to three different rings (i.e., carbon atoms not lying on the perimeter).

Any molecular orbital φ, therefore, is a function of all the necessary s variables. In other words,

$$\varphi = \varphi(s_1, s_2, s_3, \ldots) \tag{5.23}$$

Thus, in the constant potential case, $U = 0$, the Schrödinger equation has the form

$$-\frac{\hbar^2}{2m}\left[\frac{\partial^2\varphi(s_1, s_2, \ldots)}{\partial s_1^2} + \frac{\partial^2\varphi(s_1, s_2, \ldots)}{\partial s_2^2} + \cdots\right] = \epsilon\varphi(s_1, s_2, \ldots) \tag{5.24}$$

Because of the simple juxtaposition of the different branches, Eq. (5.24) separates into as many equations as there are branches, so that

$$-\frac{\hbar^2}{2m}\frac{d^2\varphi_B(s_B)}{ds_B^2} = \epsilon\varphi_B(s_B); \qquad B = 1, 2, 3, \ldots \tag{5.25}$$

where $\varphi_B(s_B)$ is that part of $\varphi(s_1, s_2, \ldots)$ situated on branch B.

There is no formal difference between Eq. (5.1) and Eqs. (5.25). Therefore, the solutions of each of the Eqs. (5.25) are simple trigonometric functions. However, the boundary conditions are now determined by the manner in which the different branches intersect. These boundary conditions are:

(i)   At a joint (i.e., where three branches meet), the wave function must be single valued. If the three branches are denoted B', B'', and B''', it follows that

$$\varphi_{B'}(s_{B'})_J = \varphi_{B''}(s_{B''})_J = \varphi_{B'''}(s_{B'''})_J \tag{5.26}$$

where the subscript J stands for *joint*.

(ii)   A second condition concerns the derivatives of φ and ensures the continuity of the wave function at a joint; this condition is

$$\left[\sum_B \frac{\partial\varphi_B(s_B)}{ds_B}\right]_J = 0 \tag{5.27}$$

where the summation extends over the three branches all of which meet at or originate at the joint.

(iii)   A third condition applies only to the free end points; this condition is

$$[\varphi_B(s_B)]_E = 0 \tag{5.28}$$

where E stands for *end point*. The end of the free-electron path should be chosen one bond length beyond the terminal atom [see also Sec. 1(A)].

(B) Molecular Orbital Wave Functions

The solutions to Eqs. (5.25) are

$$\varphi_B(s_B) = K_1 \sin \frac{(2m\epsilon)^{1/2}}{\hbar} s_B + K_2 \cos \frac{(2m\epsilon)^{1/2}}{\hbar} s_B \qquad (5.29)$$

where K_1, K_2, and ϵ follow from the boundary conditions of Eqs. (5.26)–(5.28) and from the normalization condition

$$\sum_B \int_0^{L_B} \varphi_B{}^2(s_B)\,ds_B = 1 \qquad (5.30)$$

If we limit ourselves to real wave functions, an alternative form of Eq. (5.29) is

$$\varphi_B(s_B) = a_B \cos\left[\frac{(2m\epsilon)^{1/2}}{\hbar} s_B + \delta_B \right] = a_B \cos(ks_B + \delta_B) \qquad (5.31)$$

where[12]

$$\epsilon = \frac{\hbar^2 k^2}{2m} \qquad (5.32)$$

The quantities a_B and δ_B are determinable by straightforward application of the boundary equations. Solutions are possible only for some specific (eigen-) values of the energy ϵ (i.e., for certain values of k). Therefore, the energy and all the corresponding parameters and functions may be labeled by a distinctive subscript m: ϵ_m, k_m, a_{B_m}, δ_{B_m}, φ_{B_m}, φ_m. Simple addition of all the $\varphi_{B_m}(s_B)$ functions yields $\varphi_m(s_1, s_2, \ldots)$.

We now wish to prove that a solution of the eigenvalue problem is always possible. The proof is based on the fact that there are just as many boundary equations as there are unknown parameters. Let the number of branches be n_B, the number of joints n_J, and the number of free end points n_E. The number of available boundary equations then is $3n_J + n_E + 1$, where the last term (i.e., $+1$) arises from the normalization condition. The number of unknowns is $2n_B + 1$ (a_B and δ_B for each branch, *plus* the energy ϵ). Since $3n_J + n_E = 2n_B$, the number of unknowns equals the number of boundary conditions.

If a_B of Eq. (5.31) is replaced by $-a_B$ and if, at the same time, δ_B is replaced by $\delta_B \pm \pi$, the wave function remains unchanged. In order to avoid

---

[12] In order to obtain proper energy values from Eq. (5.32), k must be expressed in (radians) (cm)$^{-1}$, not in (degrees) (cm)$^{-1}$.

any resultant ambiguity, Ruedenberg and Scherr[5] proposed the phasing convention

$$-\frac{\pi}{2} < \delta_B \le \frac{\pi}{2} \qquad (5.33)$$

Furthermore, simultaneous replacement of k by $-k$ and δ_B by $-\delta_B$, also leaves the wave function unchanged. Therefore, in order to avoid ambiguity, we adopt the convention

$$k \ge 0 \qquad (5.34)$$

It should be noted also that ϵ_m is no longer simply related to an integer quantum number or even to the total number of nodes; m, indeed, is a mere serial index. This can most easily be understood by considering the de Broglie wavelength λ of the free electron. This wavelength λ is that length along the bond path which increases the argument of the cosine function by 2π (i.e., one complete period). Therefore, $k_m(s_B + \lambda_m) + \delta_{B_m} = (k_m s_B + \delta_{B_m}) + 2\pi$ and

$$\lambda_m = \frac{2\pi}{k_m} \qquad (5.35)$$

The boundary conditions for either an unbranched open chain or a simple closed chain require that L be an integer multiple of either $\lambda/2$ or λ, respectively. But there is no reason why either λ or $\lambda/2$ should be contained an integer number of times in L_B which, after all, is an arbitrary part of the total conjugated structure.

(C) Free-Electron Eigenvectors

A general procedure for the evaluation of the FEMO eigenvalues and eigenfunctions is sketched in Sec. 3(B); this procedure requires the intermediacy of the quantities k_m, a_{B_m}, and δ_{B_m}. However, it is more convenient to proceed along different lines. In particular, Ruedenberg and Scherr[5] have developed an alternative formalism which has the following advantages:

(i)   The boundary conditions are incorporated in the formalism.

(ii)   As a result of (i), the computations become routine and are readily adaptable to computer usage.

(iii)   The formalism uses a matrix algebraic approach which makes clear many analogies with the LCAO-MO method.

We now develop the Ruedenberg–Scherr formalism.

Let the distance R between any two neighboring atoms be a constant.

Let the value of the wave function φ_m at the different atom sites be denoted

$$\varphi_m(1), \varphi_m(2), \ldots \varphi_m(Q) \qquad (5.36)$$

where Q is the number of atoms in the conjugated structure. Then, we may define a free-electron eigenvector which corresponds to the FEMO eigenfunction φ_m as

$$[\varphi_m] = \begin{bmatrix} \varphi_m(1) \\ \varphi_m(2) \\ \vdots \\ \varphi_m(Q) \end{bmatrix} \qquad (5.37)$$

For each eigenfunction φ_m we can define a corresponding eigenvector $[\varphi_m]$, so that there exists a one-to-one correspondence between the set of eigenfunctions $\{\varphi_m\}$ and the set of eigenvectors $\{[\varphi_m]\}$.

The constants a_{B_m} and δ_{B_m} associated with any branch B are readily obtained from ϵ_m and any two of the elements of $[\varphi_m]$, say, $\varphi_m(S)$ and $\varphi_m(T)$, which are situated on branch B. More specifically, if the point S is located at $s_B = 0$ and the point T at $s_B = R$, we obtain

$$\varphi_m(S) = a_{B_m} \cos (k_m s_B + \delta_{B_m}) = a_{B_m} \cos \delta_{B_m} \qquad (5.38)$$

and

$$\varphi_m(T) = a_{B_m} \cos (k_m R + \delta_{B_m}) \qquad (5.39)$$

Equations (5.38) and (5.39) determine the two unknowns a_B and δ_B. Therefore, the eigenvectors $[\varphi_m]$ are strictly equivalent to the eigenfunctions φ_m: They contain exactly the same information. Henceforth, we focus attention solely on the eigenvectors.

One disadvantage of the vector $[\varphi_m]$ defined in Eq. (5.37) is that it is not generally normalized, that is,

$$[\varphi_m]'[\varphi_m] = \overline{\varphi_m(1) \varphi_m(2) \ldots \varphi_m(Q)} \begin{bmatrix} \varphi_m(1) \\ \varphi_m(2) \\ \vdots \\ \varphi_m(Q) \end{bmatrix} = \sum_{S=1}^{Q} \varphi_m{}^2(S) \qquad (5.40)$$

does not generally equal unity.[13] The prime denotes the transpose[14] of $[\varphi_m]$. However, it is possible to define a related vector $[\phi_m]$, which is normalized.

[13] Moreover, the product of Eq. (5.40) has the dimensions of a (length)$^{-1}$. For the normalization condition to be satisfied, the product should equal the dimensionless number 1.

[14] The normalization condition as expressed in Eqs. (5.40) and (5.42) is correct for real wave functions only; if the φ_m are complex, the normalization condition becomes $[\varphi_m]^*[\varphi_m] = 1$ where the asterisk denotes the transpose of the conjugate matrix of $[\varphi_m]$.

This vector contains elements $\phi_m(S)$ related to $\varphi_m(S)$ by

$$\phi_m(S) = R^{1/2}\varphi_m(S), \qquad \text{if } S \text{ is not a joint} \qquad (5.41a)$$

$$\phi_m(S) = \left(\frac{3R}{2}\right)^{1/2}\varphi_m(S), \qquad \text{if } S \text{ is a joint} \qquad (5.41b)$$

It is readily shown that $[\phi_m]$ satisfies the condition

$$[\phi_m]'[\phi_m] = \sum_{S=1}^{Q} \phi_m{}^2(S) = 1 \qquad (5.42)$$

which is an alternative form of the normalization condition of Eq. (5.30). The factor $(3/2)^{1/2}$ for an atom at a joint, as compared to $(2/2)^{1/2}$ for a normal atom, can be traced back to the fact that three branches intersect at a joint whereas only two meet at a normal atom.

It is convenient to express the relationship between $[\varphi_m]$ and $[\phi_m]$ in a matrix formulation. First, we define a diagonal matrix $[P]$ whose elements P_{SS} are

$$P_{SS} = 1, \qquad \text{if } S \text{ is not a joint} \qquad (5.43a)$$

$$P_{SS} = (\tfrac{2}{3})^{1/2}, \qquad \text{if } S \text{ is a joint} \qquad (5.43b)$$

Then, from Eqs. (5.41), it follows that

$$[\phi_m] = R^{1/2}[P]^{-1}[\varphi_m] \qquad (5.44)$$

where $[P]^{-1}$ is the inverse matrix of $[P]$.

(D) Free-electron Matrices

The problem of finding eigenfunctions is equivalent to finding the eigenvectors $[\varphi_m]$ or $[\phi_m]$. We now construct a matrix whose eigenvectors are precisely $[\phi_m]$ and whose eigenvalues are simply connected to ϵ_m.

In order to obtain this matrix, we first construct relationships between the elements of the eigenvector $[\varphi_m]$ (i.e., between the values of φ_m at different atoms).

(i) Let the point S on branch B have the coordinate s_B; then its neighbor in the positive direction along s_B has the coordinate $s_B + R$. If we symbolize this second point by S_+, we have

$$\varphi_B(S_+) = \varphi_B(s_B + R)$$

$$= a_B \cos (k s_B + \delta_B) \cos (kR) - a_B \sin (k s_B + \delta_B) \sin (kR)$$

$$= \varphi_B(S) \cos \kappa - a_B \sin \left(\kappa \frac{s_B}{R} + \delta_B \right) \sin \kappa \qquad (5.45)$$

where

$$\kappa = kR \qquad (5.46)$$

A similar relation exists for the point S_- situated one bond length from S in the negative direction along s_B:

$$\varphi_B(S_-) = \varphi_B(S) \cos \kappa + a_B \sin \left(\kappa \frac{s_B}{R} + \delta_B \right) \sin \kappa \qquad (5.47)$$

Addition of Eqs. (5.45) and (5.47) eliminates the term in $\sin (\kappa s_B/R + \delta_B)$ which is not an allowable part of the free-electron eigenvector elements; we obtain

$$\varphi_B(S_-) - (2 \cos \kappa) \varphi_B(S) + \varphi_B(S_+) = 0 \qquad (5.48)$$

Equation (5.48) is valid for any *normal* atom S having two neighbors in the conjugated structure.

    (ii)   If S is an end atom, then S_- is the end of the bond path and $\varphi_B(S_-) = 0$; hence

$$- (2 \cos \kappa) \varphi_B(S) + \varphi_B(S_+) = 0 \qquad (5.49)$$

    (iii)   If S is a joint, it may be shown using the joint conditions of Eqs. (5.26) and (5.27) that

$$\varphi(S_1) + \varphi(S_2) + \varphi(S_3) - (3 \cos \kappa) \varphi(S) = 0 \qquad (5.50)$$

where S_1, S_2, and S_3 are the three points adjacent to the joint S.

    Some one of Eqs. (5.48)–(5.50) pertains to every atom site.

    In summary, there exists a set of Q linear homogeneous equations:

$$\begin{aligned}
c_{11}\varphi(1) + c_{12}\varphi(2) + \cdots + c_{1Q}\varphi(Q) &= 0 \\
c_{21}\varphi(1) + c_{22}\varphi(2) + \cdots + c_{2Q}\varphi(Q) &= 0 \\
\vdots \qquad\qquad \vdots \qquad\qquad\qquad\qquad & \\
c_{Q1}\varphi(1) + c_{Q2}\varphi(2) + \cdots + c_{QQ}\varphi(Q) &= 0
\end{aligned} \qquad (5.51)$$

where the coefficients c are as follows:
The diagonal elements c_{SS} are given by

$$c_{SS} = -2 \cos \kappa \equiv -F, \qquad \text{if } S \text{ is a normal atom}$$
$$\text{or a free end point} \qquad (5.52a)$$
$$c_{SS} = -3 \cos \kappa \equiv -\tfrac{3}{2}F, \qquad \text{if } S \text{ is a joint} \qquad (5.52b)$$

The off-diagonal elements are given by

$$c_{ST} = c_{TS} = 1, \qquad \text{if } T \text{ and } S \text{ are neighbors} \qquad (5.52c)$$
$$c_{ST} = c_{TS} = 0, \qquad \text{if } T \text{ and } S \text{ are not neighbors} \qquad (5.52d)$$

An alternative formulation of the set of equations given by Eq. (5.51) is

$$[C][\varphi] = [0] \qquad (5.53)$$

where $[C]$ is the square matrix containing all the c coefficients and $[0]$ is a null column vector. For this set of equations to have a nontrivial solution[15] it is necessary that the secular equation be satisfied, namely,

$$\det C = 0 \qquad (5.54)$$

where $\det C$ is the determinant of the matrix $[C]$. The elements of $[C]$ are either constants or multiples of $F = 2 \cos \kappa$; therefore, Eq. (5.54) is an equation of Qth degree in $\cos \kappa$ whose solutions yield the eigenvalues κ_m and the energies ϵ_m of the system.

We now reformulate the whole procedure using a matrix eigenvalue equation. We define the matrix $[\hat{F}]$ as

$$[\hat{F}] = [P][C][P] + F[I] \qquad (5.55)$$

where $[I]$ is a unit matrix. It is not difficult to show that the eigenvalue equation

$$[\hat{F}][\phi] = F[\phi] \qquad (5.56)$$

yields exactly the same eigenvalues F_m as Eq. (5.53) and that its eigenvectors are the normalized eigenvectors already defined in Eq. (5.41). From Eq. (5.55), it follows that a matrix element of $[\hat{F}]$, say, $\hat{f}_{TS}$, is given by

$$\hat{f}_{TS} = p_{TT}p_{SS}c_{TS} + F\delta_{T,S} \qquad (5.57a)$$

where $\delta_{T,S}$ is the Kronecker δ. Therefore, the structure of $[\hat{F}]$ is as follows:

—the diagonal elements: $\hat{f}_{SS} = F - F = 0$ $\qquad (5.57b)$

—the off-diagonal elements: $\hat{f}_{ST} = \hat{f}_{TS} = 0$
    if S and T are not neighbors $\qquad (5.57c)$

—if S and T are neighbors and if neither
    of them is a joint: $\hat{f}_{TS} = \hat{f}_{ST} = 1$ $\qquad (5.57d)$

—if S and T are neighbors and one of them
    is a joint: $\hat{f}_{TS} = \hat{f}_{ST} = (\tfrac{2}{3})^{1/2}$ $\qquad (5.57e)$

—if S and T are neighbors and they are
    both joints: $\hat{f}_{TS} = \hat{f}_{ST} = \tfrac{2}{3}$ $\qquad (5.57f)$

Since the $[\hat{F}]$ matrix is symmetric, it follows that its eigenvalues are real and that its eigenvectors are orthonormal. Therefore,

$$[\phi_{m'}]'[\phi_{m''}] = \delta_{m',m''} \qquad (5.58)$$

In summary, $[\hat{F}]$ is a topological matrix which contains all the informa-

---

[15] The so-called trivial solution of Eq. (5.53) is

$$\varphi(1) = \varphi(2) = \cdots = \varphi(Q) = 0$$

tion required for FEMO purposes. For an arbitrary conjugated structure—one which may be very large and very complicated—the matrix $[\hat{F}]$ is deducible by inspection of the molecular skeleton (examples are given in Sec. 4). In order to find the eigenvectors and eigenvalues (i.e., in order to diagonalize the matrix), the size of $[\hat{F}]$ may necessitate use of a computer.

The eigenvalues of $[\hat{F}]$ are cosine functions [see Eqs. (5.52a) and (5.52b)]. Since the energy is related to the argument of the cosine, it follows that if κ_m is a solution, $2\pi\nu \pm \kappa_m$ is also a solution for all integer values of ν. Thus, each $\cos \kappa$ value generates an infinite number of energy eigenvalues. However, for physical reasons, only $\nu = 0$ and values of κ_m such that

$$0 \leq \kappa_m \leq \pi \tag{5.59}$$

are acceptable. If $\nu \geq 1$, one obtains very large energy gaps whose values exceed the ionization potential of the molecule.

If the condition of Eq. (5.59) is fulfilled, the wavelength λ_m given by Eq. (5.35) satisfies

$$2\pi/k_m = \lambda_m \geq 2R \tag{5.60}$$

Therefore, there can be no more than one node between any two neighbors. In order to find the actual position of the nodes, one must calculate, *via* Eqs. (5.38) and (5.39), the complete analytical expression for the wave function φ_m.

4. AN EXAMPLE CALCULATION: NAPHTHALENE

(A) Platt Model

The only parameter is the length $L = 10R$ where R is the average C—C distance shown in Fig. 5.14. Therefore, the solutions are

$$\epsilon_n = \frac{n^2 h^2}{2m(10R)^2} \tag{5.61}$$

$$\varphi_n = \left(\frac{1}{5R}\right)^{1/2} \cos(2\pi ns/10R); \qquad \varphi_n' = \left(\frac{1}{5R}\right)^{1/2} \sin(2\pi ns/10R) \tag{5.62}$$

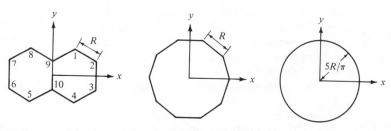

FIG. 5.14. Naphthalene and its perimeter homologues.

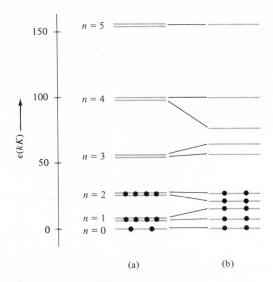

(a)   (b)

FIG. 5.15.   Naphthalene FEMO energy-level scheme. (a) Circular model. (b) Ruedenberg–Scherr model.

By putting $R = 1.39$ Å, the energy becomes

$$\epsilon_n \simeq (6290)\,n^2, \quad \text{in cm}^{-1} \tag{5.63}$$

The resulting FEMO levels and eigenfunctions are shown on the left-hand sides of Figs. 5.15 and 5.16, respectively.

FIG. 5.16.   Nodes (black dots) and the relative signs of the FEMO'S in the Platt circular model (left-hand side) and in the Ruedenberg–Scherr model (right-hand side). The two wave functions corresponding to the same quantum number n $(n > 0)$ are degenerate in the circular model, but not in the more realistic model. The left-most Platt function corresponds to the left-most Ruedenberg–Scherr function; the right-most Platt function corresponds to the right-most Ruedenberg–Scherr function. The number of nodes q and the Platt quantum number n are given in each case. When the amplitude of the wave function in the cross link is zero in the Ruedenberg–Scherr treatment, the cross link is drawn as a heavy line; in these cases, the Ruedenberg–Scherr FEMO is formally identical to the Platt FEMO provided the variable s is adequately defined in both cases. The left-most Ruedenberg–Scherr function corresponding to $n = 5$ (shown in box) is not obtained by diagonalizing the free-electron matrix; it corresponds to the *trivial* solution of Eq. (5.51)—namely, that solution for which all $\varphi(S) = 0$. An explicit discussion of the role of this trivial solution is available (Ref. [5]).

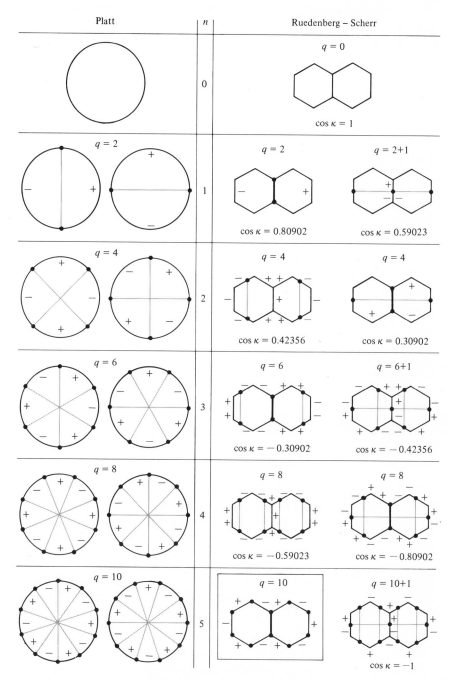

| Platt | n | Ruedenberg – Scherr |

Figure legend is on page 184.

(B) Ruedenberg–Scherr Model

We adopt the atom numbering convention shown in Fig. 5.14. The free-electron matrix $[\hat{F}]$ of Eq. (5.55) is

$$[\hat{F}] = \begin{bmatrix}
0 & 1 & 0 & 0 & 0 & 0 & 0 & 0 & (\tfrac{2}{3})^{1/2} & 0 \\
1 & 0 & 1 & 0 & 0 & 0 & 0 & 0 & 0 & 0 \\
0 & 1 & 0 & 1 & 0 & 0 & 0 & 0 & 0 & 0 \\
0 & 0 & 1 & 0 & 0 & 0 & 0 & 0 & 0 & (\tfrac{2}{3})^{1/2} \\
0 & 0 & 0 & 0 & 0 & 1 & 0 & 0 & 0 & (\tfrac{2}{3})^{1/2} \\
0 & 0 & 0 & 0 & 1 & 0 & 1 & 0 & 0 & 0 \\
0 & 0 & 0 & 0 & 0 & 1 & 0 & 1 & 0 & 0 \\
0 & 0 & 0 & 0 & 0 & 0 & 1 & 0 & (\tfrac{2}{3})^{1/2} & 0 \\
(\tfrac{2}{3})^{1/2} & 0 & 0 & 0 & 0 & 0 & 0 & (\tfrac{2}{3})^{1/2} & 0 & \tfrac{2}{3} \\
0 & 0 & 0 & (\tfrac{2}{3})^{1/2} & (\tfrac{2}{3})^{1/2} & 0 & 0 & 0 & \tfrac{2}{3} & 0
\end{bmatrix} \qquad (5.64)$$

Diagonalization of this (10×10) matrix is readily performed by electronic computer. It is possible, however, to find the solutions by hand by using the symmetry properties of the naphthalene molecule (D_{2h} point group).

(C) Simplification Using Symmetry Considerations[11]

The matrices $[C]$ and $[\hat{F}]$ were obtained from Eqs. (5.48)–(5.50) which interrelate the column elements $\varphi(S)$. The symmetry of the molecule also imposes certain conditions on these elements. Indeed, the symmetry relations are usually simpler than the ones which result from Eqs. (5.48)–(5.50). It is sufficient to consider only the two symmetry planes $\sigma(xz)$ and $\sigma(yz)$. We classify the naphthalene MO's according to their behavior with respect to reflection in these two planes. Table 5.3 indicates how this classification gives rise to four distinct categories of wave function. For the $SS(b_{1u})$ MO's we must have

$$\varphi(1) = \varphi(4) = \varphi(5) = \varphi(8)$$
$$\varphi(2) = \varphi(3) = \varphi(6) = \varphi(7)$$
$$\varphi(9) = \varphi(10) \qquad (5.65)$$

TABLE 5.3. *Transformation behavior of naphthalene MO's with respect to $\sigma(xz)$ and $\sigma(yz)$.* [S means *symmetric* with respect to reflection in a given plane (character: $+1$); A means *antisymmetric* (character: -1). All naphthalene MO's are also antisymmetric with respect to reflection in the $\sigma(xy)$ plane—a characteristic property of π orbitals. From these three characters, all others can be derived; this, the reader should verify using the D_{2h} character table.]

$\sigma(xz)$	$\sigma(yz)$	Representation
S	S	b_{1u}
S	A	b_{2g}
A	S	b_{3g}
A	A	a_u

This provides us with a set of seven equations in ten unknowns. We need three more equations. There are just three nonequivalent atom sites of the naphthalene molecule to which we can apply Eqs. (5.48)–(5.50). If we select atoms 1, 2, and 9, we obtain

$$\varphi(9) - 2(\cos\kappa)\varphi(1) + \varphi(2) = 0$$

$$\varphi(1) - 2(\cos\kappa)\varphi(2) + \varphi(3) = 0$$

$$\varphi(8) + \varphi(10) + \varphi(1) - 3(\cos\kappa)\varphi(9) = 0 \qquad (5.66)$$

Because of Eq. (5.65), these latter are equivalent to

$$-2(\cos\kappa)\varphi(1) + \varphi(2) + \varphi(9) = 0$$

$$\varphi(1) + (1 - 2\cos\kappa)\varphi(2) = 0$$

$$2\varphi(1) + (1 - 3\cos\kappa)\varphi(9) = 0 \qquad (5.67)$$

This leaves us with three homogeneous equations in three unknowns; a solution is possible only when

$$\begin{vmatrix} -2\cos\kappa & 1 & 1 \\ 1 & 1 - 2\cos\kappa & 0 \\ 2 & 0 & 1 - 3\cos\kappa \end{vmatrix} = 0 \qquad (5.68)$$

The solutions are

$$\cos\kappa = 1.0;\ 0.42356;\ -0.59023 \qquad (5.69a)$$

These values of κ correspond to the energy eigenvalues[16]

$$\epsilon = 0;\ 20\ 420;\ 77\ 000\ \text{cm}^{-1} \qquad (5.69\text{b})$$

These solutions, obviously, are found among the ten which result from diagonalization of Eq. (5.64).

For the $SA\,(b_{2g})$ case, symmetry provides eight equations—one more than in the b_{1u} case:

$$\varphi(1) = \varphi(4) = -\varphi(5) = -\varphi(8)$$
$$\varphi(2) = \varphi(3) = -\varphi(6) = -\varphi(7)$$
$$\varphi(9) = \varphi(10) = 0 \qquad (5.70)$$

Therefore, only two relations of the type of Eq. (5.66) are needed. This leads to a 2×2 determinant which yields two eigenvalues.

The $SA\,(b_{3g})$ and $AA\,(a_u)$ cases are equally straightforward; they produce the five remaining eigenvalues.

The collected results are shown in Table 5.4 and on the right-hand side

TABLE 5.4. *FEMO energies for naphthalene.*

	Platt	Ruedenberg–Scherr			
n	ϵ (cm^{-1})	$\cos\kappa$	κ (degrees)	κ (radians)	ϵ^a (cm^{-1})
0	0	1	$0°$	0	0
1	6 290	0.80902	$36°$	0.62832	6 290
	6 290	0.59023	$53°49'37''$	0.93941	14 060
2	25 100	0.42356	$64°56'25''$	1.13356	20 420
	25 100	0.30902	$72°$	1.25664	25 100
3	56 600	-0.30902	$108°$	1.88496	56 600
	56 600	-0.42356	$115°03'35''$	2.00803	64 090
4	100 600	-0.59023	$126°10'23''$	2.20213	77 000
	100 600	-0.80902	$144°$	2.51327	100 600
5	157 000	-1	$180°$	3.14159	157 000
	157 000				

^{a} ϵ was calculated at $R = 1.39$ Å.

[16] Use Eq. (5.32), but note that Eq. (5.32) contains k, *not* $\kappa = kR$. For a given value of κ obtained from Eq. (5.68), the energy ϵ is larger for smaller R.

of Fig. 5.15. The MO degeneracies of the Platt model are removed by the cross links. Furthermore, the splittings produced are important, being of the order of 10 000 cm^{-1}.

Table 5.4 illustrates the pairing theorem for alternant hydrocarbons.[17] This theorem is formulated in FEMO theory as[5]: If cos κ is an eigenvalue, $-\cos \kappa$ is also an eigenvalue; hence, if κ_m corresponds to an allowed energy, $\pi - \kappa_m$ also corresponds to an allowed energy. The levels corresponding to cos κ are occupied in the ground state; those corresponding to $-\cos \kappa$ are empty in the ground state.

The lowest-energy eigenvalue, $\epsilon = 0$ ($F = 2 \cos \kappa = 2$), must occur in all molecules which contain no free end points: It corresponds to a constant value of the wave function. If there is a free end point, on the other hand, $F = 2$ cannot possibly be an eigenvalue because the wave function must possess zero amplitude at the end point; therefore, a constant value for the wave function throughout the molecular extent implies that $\varphi = 0$ everywhere.

(D) Eigenvectors and Eigenfunctions

Solution of Eq. (5.64) yields the ten orthonormal eigenvectors $[\phi]$. Using Eq. (5.41), one calculates the corresponding eigenvectors $[\varphi]$. Finally, using Eqs. (5.38) and (5.39), one obtains the eigenfunctions of Eq. (5.31).

It is more convenient to proceed along the lines of Sec. 4(C). Consider the $b_{1u}(SS)$ MO's whose properties are expressed in Eq. (5.67). Once the eigenvalues of Eq. (5.69) are known, one merely substitutes each of these values of cos κ into Eq. (5.67) and solves for $\varphi(1), \varphi(2)$, and $\varphi(9)$. Solution yields the ratios $\varphi(2)/\varphi(1)$ and $\varphi(9)/\varphi(1)$; however, the absolute values follow from the normalization condition.

Equation (5.42) expresses the normalization condition for $[\phi]$. In order to obtain the equivalent relation for $[\varphi]$, it is convenient[11] to consider R as the unit of length. With this convention, the normalization condition for the b_{1u} naphthalene MO's becomes [see Eq. (5.41)]

$$4\varphi(1)^2 + 4\varphi(2)^2 + 2(\tfrac{3}{2})\varphi(9)^2 = 1 \qquad (5.71)$$

Proceeding in this way, one obtains for, say, cos $\kappa = 0.42356$

$$\varphi(1) = 0.0543; \qquad \varphi(2) = -0.3553; \qquad \varphi(9) = 0.4012 \quad (5.72)$$

Consecutive application of Eqs. (5.38) and (5.39) to each of the three

---

[17] An alternant hydrocarbon is one where the carbon atoms can be divided into two sets, such that any carbon of one set has neighbors belonging only to the other set. Naphthalene is an alternant; azulene is a nonalternant.

branches shown in Fig. 5.13 yields

$$a_1 = a_3 = 0.4211 \qquad a_2 = 0.4757$$

$$\delta_1 = \delta_3 = 17°38'57'' \qquad \delta_2 = -32°28'12'' \qquad (5.73)$$

The resulting wave functions are shown in Fig. 5.17.

The nine other FEMO functions can be derived in a similar way. They are shown on the right-hand side of Fig. 5.16. The FEMO's having zero amplitude on the cross link are completely identical to the Platt *circular* functions. For all FEMO's, the number of nodes *in the perimeter* is identical to the number of nodes predicted by the Platt approach. Therefore, even though the Ruedenberg–Scherr treatment may induce additional nodes in the cross link, the quantum number n of Eq. (5.13) retains a meaning.

The pairing theorem for alternant hydrocarbons has an interesting consequence for the eigenvectors. Consider the FEMO corresponding to $\cos \kappa = -0.42356$ (an AS function); solution yields

$$\varphi(1) = 0.0543; \qquad \varphi(2) = 0.3553; \qquad \varphi(9) = -0.4012 \quad (5.74)$$

Comparison to Eq. (5.72) illustrates a general property of alternants: The two eigenvectors corresponding to $\cos \kappa$ and $-\cos \kappa$ are identical but for

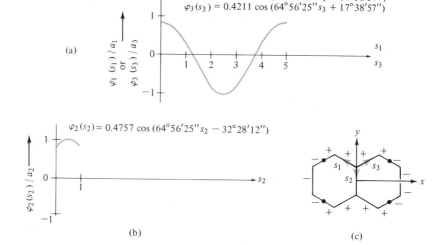

FIG. 5.17. Naphthalene wave function corresponding to $\cos \kappa = 0.42356$ (from Table 5.4, $\kappa = 64°56'25''$). (a) and (b) The wave function on the separate branches; (c) the resulting FEMO. R is defined as the unit of length. The argument of the cosine is given in degrees; for energy calculations, one must convert to radians.

the sign of the amplitude on every other atom. Thus, $\varphi(1)$, $\varphi(3)$, $\varphi(6)$, $\varphi(8)$, and $\varphi(10)$ have the same value in both MO's; $\varphi(2)$, $\varphi(4)$, $\varphi(5)$, $\varphi(7)$, and $\varphi(9)$ are identical also, except for opposite sign, in both MO's.

(E)   Comparison with Hückel Theory

The simple Hückel treatment of naphthalene leads to the equation

$$
|M| = \begin{vmatrix}
x & 1 & 0 & 0 & 0 & 0 & 0 & 0 & 1 & 0 \\
1 & x & 1 & 0 & 0 & 0 & 0 & 0 & 0 & 0 \\
0 & 1 & x & 1 & 0 & 0 & 0 & 0 & 0 & 0 \\
0 & 0 & 1 & x & 1 & 0 & 0 & 0 & 0 & 1 \\
0 & 0 & 0 & 1 & x & 1 & 0 & 0 & 0 & 1 \\
0 & 0 & 0 & 0 & 1 & x & 1 & 0 & 0 & 0 \\
0 & 0 & 0 & 0 & 0 & 1 & x & 1 & 0 & 0 \\
0 & 0 & 0 & 0 & 0 & 0 & 1 & x & 1 & 0 \\
1 & 0 & 0 & 0 & 0 & 0 & 0 & 1 & x & 1 \\
0 & 0 & 0 & 1 & 1 & 0 & 0 & 0 & 1 & x
\end{vmatrix} = 0 \qquad (5.75)
$$

The eigenvalue equation for the $[\hat{F}]$ matrix [Eq. (5.64)] is very similar to Eq. (5.75); the only difference lies in the value of some off-diagonal matrix elements—specifically, those involving joints.[18]

(F)   Comparison with Experiment
(Electron Repulsion Correction)

The four predicted lowest-energy transitions are shown in Table 5.5. The FEMO calculation does not include electron repulsion; thus, it merely provides the barycenter of the singlet and triplet states which arise from the electron configuration in question. The values of these centers of gravity, as well as their assigned Platt labels, are also listed in Table 5.5. The agreement is surprisingly good.

---

[18] As a matter of fact, Ruedenberg has developed (Ref. [12]) an $[\hat{F}]$ matrix variant for Hückel theory and an $[M]$ matrix variant for FEMO theory. Thus, the formal analogy between the two treatments is complete.

It is instructive to scrutinize the results of Table 5.5 more closely. Consider the first-excited state Φ_A. This state results from excitation of an electron from the MO characterized by $\cos \kappa = 0.30902$ to the MO characterized by $\cos \kappa = -0.30902$. The nodal behavior of these FEMO's is shown in Fig. 5.16; both MO's have zero amplitude in the cross link and are formally identical to the real Platt functions of Eq. (5.13a). Thus, if we take R as the unit of length, if we define s as the perimeter bond-path coordinate,[19] and if we consider only two unpaired electrons, the state Φ_A is

$$\Phi_A = \frac{1}{5} \sin \frac{2\pi s(1)}{5} \sin \frac{3\pi s(2)}{5} \tag{5.76}$$

where $s(1)$ and $s(2)$ are the perimeter coordinates of electrons 1 and 2, respectively. Let us now compare Φ_A to the Platt L and B states. The B states correspond[20] to excitation from φ_{2+} to φ_{3+} (or φ_{2-} to φ_{3-}); the L states correspond to excitation from φ_{2+} to φ_{3-} (or φ_{2-} to φ_{3+}). Therefore, using Eq. (5.13b) and the same simplifications that lead to Eq. (5.76), we find

L states:

$$\Phi_{5+} = \tfrac{1}{10} e^{2\pi i s(1)/5} e^{3\pi i s(2)/5} \tag{5.77a}$$

$$\Phi_{5-} = \tfrac{1}{10} e^{-2\pi i s(1)/5} e^{-3\pi i s(2)/5} \tag{5 77b}$$

B states:

$$\Phi_{1+} = \tfrac{1}{10} e^{-2\pi i s(1)/5} e^{3\pi i s(2)/5} \tag{5.77c}$$

$$\Phi_{1-} = \tfrac{1}{10} e^{2\pi i s(1)/5} e^{-3\pi i s(2)/5} \tag{57.7d}$$

Comparison of Eqs. (5.76) and (5.77) yields

$$\Phi_A = \tfrac{1}{2}(\Phi_{1+} + \Phi_{1-}) - \tfrac{1}{2}(\Phi_{5+} + \Phi_{5-}) \tag{5.78}$$

Consequently, the first-excited state of the one-electron Ruedenberg–Scherr theory—which was correlated in Table 5.5 with the Platt L_a state barycenter—is really a 50-50 mixture of the Platt L and B states! Somewhat analogous conclusions hold for the other Ruedenberg–Scherr states of Table 5.5. Thus, the comparison of Table 5.5 is somewhat suspect.

In view of the results of the previous paragraph, it is imperative to attempt separation of the effects of cross-link perturbations from electron repulsion perturbations. Consider, first, *circular* naphthalene (with no cross link) and neglect electron repulsion; in this approximation, the first-excited state is fourfold degenerate and of energy equal to the gap between $\varphi_{2\pm}$ and $\varphi_{3\pm}$ (or between $\{\varphi_2, \varphi_2'\}$ and $\{\varphi_3, \varphi_3'\}$); the results are shown in Fig. 5.18(I). If electron repulsion is now introduced, the fourfold degeneracy

---

[19] The bond-path coordinate along the perimeter, namely, s, replaces both of the branch coordinates s_1 and s_2. Equation (5.76) follows directly from Eq. (5.13a) applied to naphthalene.

[20] See Sec. 2(B).

TABLE 5.5. *Comparison of the results of the Ruedenberg–Scherr model with experiment in the case of naphthalene.*
(Φ_A, Φ_B, etc., are the wavefunctions of the excited states whose energies—with respect to that of the ground state—are listed.)

Transition Energies FEMO Predictions (cm^{-1})	Experimental Energies (Center of Gravity of Singlets and Triplets[a]) (cm^{-1})
31 400 (Φ_A)	28 200 (L_a)
36 100 (Φ_B)	33 200 (L_b)
39 100 (Φ_C)	39 200 (B_b)
43 700 (Φ_D)	54 500 (B_a)

[a] Taken, not without reservation concerning the propriety of assignments, from a compilation by Scherr (Ref. [11]).

is removed [see Sec. 2(B)] and one obtains two twofold degenerate levels L and B, of which the L state corresponds to the smaller energy; the results are shown in Fig. 5.18(II). Finally, upon introduction of distortion and cross linking on top of the interelectronic repulsion effects all degeneracies disappear [13]; the net effect is that the pure L or B character of the circular electronic states is lost; the results are shown in Figs. 5.18(III). On the other hand, if we first shape the molecule properly—following Ruedenberg and Scherr—the fourfold degenerate level splits into four nondegenerate levels, each of which contains L character as well as B character in nearly equal proportions; the results are shown in Fig. 5.18(IV). Finally, under further influence of electron repulsion, [8], [14] the wave functions are scrambled so as to form more nearly pure L or more nearly pure B states; the predominantly L states are stabilized whereas those which are predominantly B are destabilized; the results are shown in Fig. 5.18(IV).

The above considerations emphasize the extent to which an apparently good agreement of experiment and theory can be deceptive—a good correlation of energies does not imply the generation of an acceptable set of wave functions.

(G) Bond Populations and Atom Populations[5], [11], [15], [16]

The atom population $a'(S)$ on atom S is defined as

$$a'(S) \equiv \sum_m g_m \int_{s(M_-)}^{s(M_+)} \varphi_m{}^2(s)\,ds \qquad (5.79)$$

where $s(M_+)$ and $s(M_-)$ are the coordinates of the midpoints of the bonds

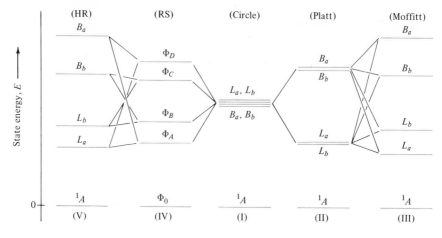

FIG. 5.18. State energy-level schemes for naphthalene in several models of varying complexity. (I) Circular bond path; no electron repulsion. (II) Circular bond path plus electron repulsion {Platt (Ref. [6])}. (III) Introduction of cross links in the Platt state functions {Moffitt (Ref. [13])}. (IV) Correct bond path; no electron repulsion {Rueden-berg and Scherr (Ref. [5])}. (V) Introduction of electron repulsion into the Ruedenberg–Scherr treatment {Ham and Ruedenberg (Refs. [8] and [14]}. The singlet-triplet splits are neglected throughout; only barycenters are shown.

adjacent to atom S, and $g_m = 0, 1, 2$ is the number of electrons in φ_m. If S is a joint, we define

$$a'(S) \equiv \sum_m g_m \sum_{B=1}^{3} \int_{s(S)}^{s_B(M_B)} \varphi_{mB}^2(s_B)\, ds_B \tag{5.80}$$

where M_B is the midpoint of the bond between S and its neighbor on branch B. The bond population between adjacent atoms S and T is defined

$$b'(S, T) = \sum_m g_m \int_{s(S)}^{s(T)} \varphi_m^2(s)\, ds \tag{5.81}$$

The definitions of Eqs. (5.79)–(5.81) involve many integrations and are longsome. With little loss of accuracy, it is possible to introduce the new quantities $a(S)$ and $b(S, T)$ which replace $a'(S)$ and $b'(S, T)$. These quantities are defined as

$$a(S) \equiv \sum_m g_m \phi_m^2(S) \tag{5.82}$$

$$b(S, T) \equiv b(M) \equiv \sum_m g_m \varphi_m^2(M) \tag{5.83}$$

where M is the midpoint between the adjacent centers S and T. The replacement of φ_m by ϕ_m in Eq. (5.82) should be noted; it is this replacement which takes care of the situation which exists at a joint atom. The midpoint amplitude $\varphi(M)$ can be calculated from the eigenvector elements $\varphi(S)$ and $\varphi(T)$ by means of

$$\varphi_m(M) = \frac{\varphi_m(S) + \varphi_m(T)}{2 \cos (\kappa_m/2)} \tag{5.84}$$

Table 5.6 lists the numerical values of the closed-shell MO amplitudes at the atomic centers in both the φ and ϕ representations. The atom

TABLE 5.6. *Atom populations and bond populations in ground state of naphthalene.*

φ, ϕ	1	0.80902	0.59023	0.42356[a]	0.30902	Populations	Bound Valence
$\varphi(1)$	0.301	0.263	0.395	0.054	0.425		
$\varphi(2)$	0.301	0.425	0.181	−0.355	0.263		
$\varphi(3)$	0.301	0.425	−0.181	−0.355	−0.263		
$\varphi(9)$	0.301	0	0.285	0.401	0		
$\varphi(10)$	0.301	0	−0.285	0.401	0		
$\phi(1)$	0.301	0.263	0.395	0.054	0.425		
$\phi(2)$	0.301	0.425	0.181	−0.355	0.263		
$\phi(3)$	0.301	0.425	−0.181	−0.355	−0.263		
$\phi(9)$	0.369	0	0.349	0.491	0		
$\phi(10)$	0.369	0	−0.349	0.491	0		
$\phi^2(1)$	0.091	0.069	0.156	0.003	0.181	$a(1) = 1.000$	
$\phi^2(2)$	0.091	0.181	0.033	0.126	0.069	$a(2) = 1.000$	
$\phi^2(3)$	0.091	0.181	0.033	0.126	0.069	$a(3) = 1.000$	
$\phi^2(9)$	0.137	0	0.122	0.241	0	$a(9) = 1.000$	
$\phi^2(10)$	0.137	0	0.122	0.241	0	$a(10) = 1.000$	
$\varphi(M_{1,2})$	0.301	0.361	0.323	−0.179	0.425		
$\varphi^2(M_{1,2})$	0.091	0.130	0.104	0.033	0.181	$b(1, 2) = 1.078$	
$\varphi(M_{2,3})$	0.301	0.447	0	−0.421	0		
$\varphi^2(M_{2,3})$	0.091	0.200	0	0.177	0	$b(2, 3) = 0.936$	
$\varphi(M_{1,9})$	0.301	0.138	0.381	0.270	0.263		
$\varphi^2(M_{1,9})$	0.091	0.019	0.145	0.073	0.069	$b((1, 9) = 0.794$	
$\varphi(M_{9,10})$	0.301	0	0	0.475	0		
$\varphi^2(M_{9,10})$	0.091	0	0	0.226	0	$b(9, 10) = 0.634$	
							$v(1) = 1.872$
							$v(2) = 2.014$
							$v(9) = 2.222$

[a]. See Eq. 5.72 for calculation of the φ amplitudes.

populations and bond populations are obtained by summation of the appropriate line and multiplication[21] by 2. The electron populations relate to the square of the wave functions; thus, they have the full symmetry of the molecule and it follows that

$$b(1, 2) = b(3, 4) = b(5, 6) = b(7, 8)$$

$$b(1, 9) = b(9, 10) = b(5, 10) = b(8, 9)$$

$$b(2, 3) = b(7, 8) \tag{5.85}$$

It appears from Table 5.6 that all atom populations equal unity. It has been shown[5] that this is generally true for any alternant hydrocarbon. A consequence of this is that alternant hydrocarbons possess a zero π-electron dipole moment. The average bond population, on the other hand, is smaller than unity. This, of course, must be the case because 10 π electrons are divided among 11 bonds—10 peripheral and 1 cross link.

(*i*) *Bond lengths*[15]–[19]: A large bond population means a stronger bond and, therefore, a shorter bond length. Semiempirical relations between bond length and bond population[22] have been established; one of these relations is[11]

$$R(S, T) = 1.665 - 0.1398[1 + b(S, T)] \tag{5.86}$$

where $R(S, T)$ is the distance in Å between atoms S and T. Application of Eq. (5.86) to naphthalene produces the R values given in Table 5.7.

TABLE 5.7.   Calculated and experimental bond lengths in naphthalene.

Atom Pair	R[from Eq. (5.86)] (Å)	R (experimental)[a] (Å)
(1, 2)	1.374	1.365
(2, 3)	1.394	1.404
(1, 9)	1.414	1.425
(9, 10)	1.436	1.393

[a] Data taken from Chap. VIII of Robertson [J. M. Robertson, *Organic Crystals and Molecules* (Cornell University Press, Ithaca, N.Y., 1953)].

[21] For all occupied MO's, g_m equals 2.

[22] All calculations have been performed on the assumption that the bond length R is a constant. Now we ask the eigenvectors obtained in this way to provide us with bond-length information!

The relative magnitude of the interatomic distances is predicted in a surprisingly excellent manner.

(ii)   *Free Valence*[11],[15]−[19]:   The *bound* valence $v(S)$ for atom S is defined as

$$v(S) = \sum_T b(S, T) \qquad (5.87)$$

where the summation runs over all atoms T adjacent to S—2 for a normal atom, 3 for a joint. Table 5.6 lists the different $v(S)$ values. Obviously, symmetry requires

$$v(1) = v(4) = v(5) = v(8); \quad v(2) = v(3) = v(6) = v(7); \quad v(9) = v(10)$$

$$(5.88)$$

The bound valence, as the name implies, constitutes an estimate of the binding power that the atom S has *used* by virtue of its incorporation in the conjugated structure. Therefore, the smaller $v(S)$, the larger is the remaining *free* valence on atom S and the larger is the expected reactivity of atom S. On the basis of Table 5.6 and Eq. (5.88), one would predict positions 1, 4, 5, and 8 to be the most reactive. It is found experimentally that the positions attacked upon oxidation, halogenation, nitration, and sulfonation are precisely those listed. [11]

5.   CHAINS CONTAINING HETEROATOMS

Kuhn[20] has extended the FEMO method to the case of linear dye molecules in which a CH group *internal to the chain is replaced by a nitrogen center*. Two examples of such molecules are given in Fig. 5.19.

The localized orbital description of the four valence electrons of carbon and the five valence electrons of nitrogen is shown in Fig. 5.20. In contrast to the σ bonded electrons of either carbon or nitrogen, the electrons of the nitrogen nonbonding orbital are of comparable energy to the π electrons of the molecular system. Consequently, some caution must be used in making comparisons of theoretical and experimental results. In particular, an experimental classification of the absorption spectrum into MO excitation types—$\pi^* \leftarrow n$ or $\pi^* \leftarrow \pi$—is mandatory. These latter remarks are of especial relevance if the theoretical approach which we now pursue is ever applied to the aromatic azines (i.e., pyridine, pyridazine, etc.).

The nitrogen atom is more electronegative than carbon. Thus, the substituted chain may be represented by the one-dimensional potential well shown in Fig. 5.21. The width of the subwell is set equal to the C—N distance; the depth of the subwell, denoted U_w, may be estimated[20] as $\sim$3–5 eV. The conjugated structure may be divided into three branches,

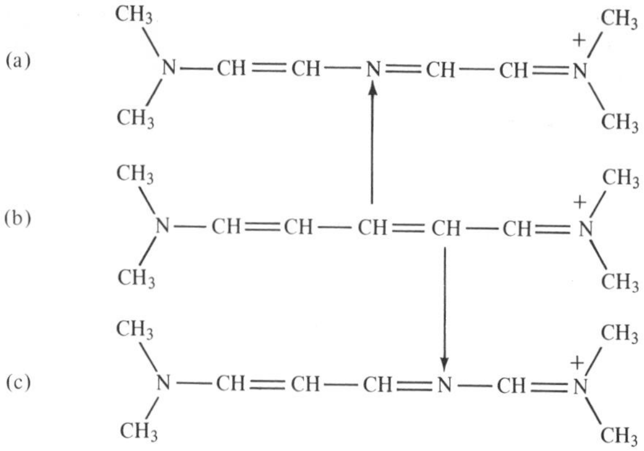

Amidinium dyes

FIG. 5.19. Two N-substituted dye ions, (a) and (c), derive from the unsubstituted dye, (b), by replacement of a CH group by an N atom. Each one of these three molecules resonates between two extreme structures. We show only one of these structures; the other is obtained by interchange of the double and single bonds of the structure shown.

in a way entirely analogous to that of Secs. 3 and 4. Each branch is assigned a separate variable; the Schrödinger equation takes the form

$$\sum_B \frac{d^2\varphi_B}{ds_B{}^2} + \frac{8\pi^2 m}{h^2} [\epsilon - \hat{U}(s)]\varphi = 0 \tag{5.89}$$

with

$$U(s) = 0, \quad \text{for} \quad s(S_1) = 0 \leq s \leq s(S_2) \text{ and } s(S_3) \leq s \leq s(S_4) = L$$

$$U(s) = -U_w, \quad \text{for} \quad s(S_2) \leq s \leq s(S_3)$$

$$U(s) = \infty, \quad \text{for} \quad s > L \text{ and } s < 0 \tag{5.90}$$

The solutions for the three regions are

$$\varphi_1(s_1) = a_1 \cos\left[\frac{(2m\epsilon)^{1/2}}{\hbar} s_1 + \delta_1\right] = a_1 \cos(k_1 s_1 + \delta_1)$$

$$\varphi_2(s_2) = a_2 \cos\left[\frac{[2m(\epsilon + U_w)]^{1/2}}{\hbar} s_2 + \delta_2\right] = a_2 \cos(k_2 s_2 + \delta_2)$$

$$\varphi_3(s_3) = a_3 \cos\left[\frac{(2m\epsilon)^{1/2}}{\hbar} s_3 + \delta_3\right] = a_3 \cos(k_3 s_3 + \delta_3) \tag{5.91}$$

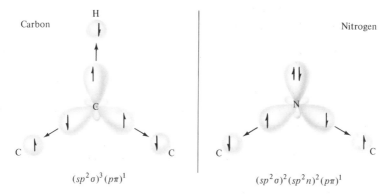

Carbon

H

C

C   C

$(sp^2\sigma)^3(p\pi)^1$

Nitrogen

N

C   C

$(sp^2\sigma)^2(sp^2n)^2(p\pi)^1$

FIG. 5.20. σ and n orbitals of carbon and nitrogen. The π orbitals are perpendicular to the plane of the sp^2-σ hybrids and are not shown. The electron spins are represented by singly barbed arrows.

It should be noted that the wavelength inside of the subwell is different from the wavelength elsewhere (i.e., $k_1 = k_3 \neq k_2$). As in Sec. 3, the boundary conditions of Eqs. (5.26)–(5.28) and the normalization condition of Eq. (5.30) must be satisfied. These impositions introduce a discrete ϵ spectrum; they also ordain certain conditions which the amplitudes and the phases of the different cosine functions must satisfy. A derivation of the eigenvalues and eigenfunctions has been given by Kuhn.[20] Though simple in principle, this derivation involves rather tedious numerical solutions of a number of trigonometric equations. However, a simple perturbation

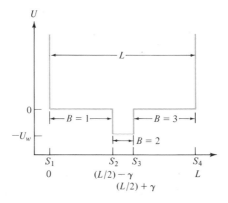

FIG. 5.21. Potential energy U along a one-dimensional bond path representing an N substituted dye cation. The first branch ($B = 1$) is situated between the points S_1 and S_2; the second one ($B = 2$) between S_2 and S_3; and the third one ($B = 3$) between S_3 and S_4.

treatment based on the potential well of Fig. 5.21 often yields results comparable to those of the numeric procedure.

(A) An Example Calculation

The molecule of Fig. 5.19(a) derives from the dye in Fig. 5.19(c) by replacing the middle CH group by an N atom. Dyes of the type illustrated in Fig. 5.19(c) have been discussed in Sec. 1; this dye contains eight π electrons; its four lowest MO's are occupied in the ground state. The first intense absorption band can be assigned to the orbital transition $\varphi_5 \leftarrow \varphi_4$ [see Sec. 1(C)]. The introduction of the nitrogen atom is equivalent to introducing a perturbing potential $U_p = -U_w$ for $(L/2) - \gamma \leq s \leq (L/2) + \gamma$, where 2γ is the width of the subwell.

First-order perturbation theory predicts an energy correction for the highest-energy occupied MO given by

$$\Delta\epsilon_4 = \int_0^L \varphi_4 \hat{U}_p \varphi_4 ds = \int_{(L/2)-\gamma}^{(L/2)+\gamma} - \varphi_4 \hat{U}_w \varphi_4 ds = -\frac{2U_w}{L} \int_{(L/2)-\gamma}^{(L/2)+\gamma} \sin^2 \frac{4\pi s}{L} ds$$

$$= -\frac{4U_w\gamma}{L}\left[\frac{1}{2} - \frac{1}{2}\frac{\sin (8\pi\gamma/L)}{(8\pi\gamma/L)}\right] \tag{5.92}$$

In a similar way, the lowest-energy unoccupied MO shifts by an amount

$$\Delta\epsilon_5 = \int_0^L \varphi_5 \hat{U}_p \varphi_5 ds = -\frac{4U_w\gamma}{L}\left[\frac{1}{2} + \frac{1}{2}\frac{\sin (10\pi\gamma/L)}{(10\pi\gamma/L)}\right] \tag{5.93}$$

Setting the C—N distance equal to the C—C distance (i.e., $\gamma = L/16$) and supposing $U_w \simeq 4$ eV, we find $\Delta\epsilon_4 = -1450$ cm^{-1} and $\Delta\epsilon_5 = -5950$ cm^{-1}. Consequently, the total transition energy is predicted to decrease from 21 800 to about 17 300 cm^{-1} (i.e., by ~4500 cm^{-1}) upon introduction of the central N atom. This prediction agrees with experiment and with the numerical solution [20] of Eq. (5.89).

(B) Comments

It was shown in Sec. 5(A) that the introduction of the nitrogen atom into the dye molecule decreased the energy of the lowest-energy electronic transition. This conclusion should not be accepted as a generalization: In some instances, the nitrogen perturbation causes the lowest-energy electronic transition to *increase* in energy. The effects which occur in any specific example depend not only on the total number of atoms in the conjugated chain but also on the exact position of substitution. [20]

The qualitative direction of the energy shift is readily predictable.

Consider Fig. 5.22 wherein the unperturbed wave functions φ_4 and φ_5 are obtained from Fig. 5.3. The position of N substitution in dyes (a) and (c) of Fig. 5.19 are indicated by points A and B, respectively, in Fig. 5.22. Since φ_4 has zero electron density at A, it is reasonable to assume that the corresponding energy is not affected much by the introduction of the sub-well perturbation at this site. On the other hand, φ_5 has maximal electron density at A; the presence of the subwell perturbation at the site A obviously stabilizes the MO φ_5 and lowers its energy. Thus, this rough and quantitatively incorrect approach[23] leads to the same conclusion for dye (a) of Fig. 5.19 as that obtained in Sec. 5(A). Consider now the molecule depicted in Fig. 5.19(b); this molecule is characterized by N atom substitution at location B of Fig. 5.22. From Fig. 5.22, it is obvious that φ_5 is more or less unaffected and that φ_4 is stabilized; hence, we expect an *increase* in the MO excitation energy $\varphi_5 \leftarrow \varphi_4$ in this instance.

It is now possible to understand why the dyes treated in Sec. 1 could be successfully described by a constant potential assumption. The nitrogen atoms are both situated at ends of the chain, where $\varphi_4{}^2$, *as well as* $\varphi_5{}^2$, is maximal (see Fig. 5.3). Substitution affects the MO energies; however, both effects are in the same direction and tend to cancel.

Continuing this line of reasoning, Kuhn[20] was able to derive a general rule for symmetrical polymethines: *Central substitution of a nitrogen atom*

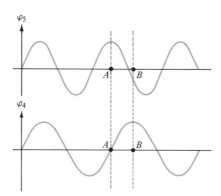

FIG. 5.22. Positions of substitution A and B related to the unperturbed FEMO's φ_4 and φ_5.

[23] This approach is equivalent to letting $\gamma \to 0$ and $U_w \to \infty$ (point perturbation). Since

$$\lim_{\gamma \to 0} \frac{\sin{(n\pi\gamma/L)}}{(n\pi\gamma/L)} = 1,$$

Eqs. (5.92) and (5.93) reduce to: $\Delta\epsilon_4 = 0$ and $\Delta\epsilon_5 = -4U_w\gamma/L$.

leads to a decrease in transition energy when the number of atoms in the conjugated chain is $4\nu - 1$ (where $\nu = 2, 3, 4, \ldots$) and an increase in transition energy when the number of atoms is $4\nu + 1$.

6. CONCLUSION

FEMO theory, restricted as it is to π-electron systems, is more limited with regard to applications than is LCAO-MO theory. However, where applicable, it yields results which are comparable to those of the one-electron LCAO-MO theory *including overlap*.

The FEMO theory possesses an advantage: It contains no parameter which can be varied to suit the investigators pleasure. Moreover, the FEMO approach allows a qualitative account of electron repulsion effects to be made.

EXERCISES

1. Consider the dye cation shown in Fig. 5.23(a). Its total length is $18R$. It exhibits a *cis*-configuration at two locations; it may be considered to consist of three linear portions, each one of length $L/3 = 6R$ [see Fig. 5.23(b)].

(a) Calculate the oscillator strength of the first intense transition.

(b) Compare the results to those of Sec. 1(B). In what way are the polarization and selection rules different from those for the all-*trans*-configuration?

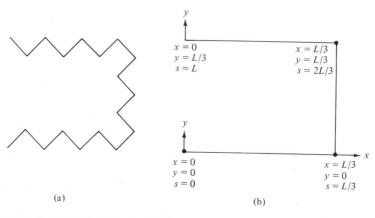

(a)   (b)

FIG. 5.23. Bond paths for a dye cation.

Hint: Since $\Pi = 18$ (see page 164), the lowest-energy MO excitation occurs from $n = 9$ to $n = 10$. The transition moment integrals must be evaluated for both the x and the y components of the electric dipole operator. The relationship between x, y, and s is

$$y = 0; \qquad x = s; \qquad \text{for} \quad 0 \le s \le L/3$$

$$y = s; \qquad x = L/3; \quad \text{for} \quad L/3 \le s \le 2L/3$$

$$y = L/3; \quad x = -s; \quad \text{for} \quad 2L/3 \le s \le L$$

2.   Consider a linear polyene or dye molecule of the type discussed in Sec. 1. Derive Eqs. (5.2) and (5.3) by means of the Ruedenberg–Scherr method.

Hint: The free-electron determinant of Eq. (5.54) has a very simple structure. Because of this simplicity, the determinantal equation can be solved by hand for an arbitrary Nth-order determinant. The solution can be found in Kauzmann.[24]

3.   Derive the benzene eigenfunctions and eigenvalues by means of the Ruedenberg–Scherr approach. Compare the results obtained with those of Eqs. (5.12) and (5.13).

Hints: The secular equation gives rise to a 6×6 cyclic determinant, whose roots are readily obtained by hand (see Hints to Exercise 2).

4.   Derive the complete naphthalene MO energy scheme and check the nodal behavior of the wave functions which are illustrated in Fig. 5.16.

Hint: Follow the procedure outlined in Sec. 4(C).

5.   Calculate the MO energy-level scheme of biphenylene. Use the symmetry properties of the molecule.

Answer: The lowest-energy transitions—in the one-electron approximation—are situated at 17 400, 27 700, 32 300, 32 800, 41 800, and 43 100 cm^{-1}.

6.   Figure 5.15 illustrates an MO energy-level diagram of naphthalene (a) for the circular model and (b) for the true molecular shape. *One* of the two MO's which correspond to the same quantum number n, and which are degenerate in the circular model, retains its energy even after the cross-linking process.

[24] W. Kauzmann, *Quantum Chemistry* (Academic Press Inc., New York, 1959). pp. 49–50.

It is obvious from Fig. 5.16 that the MO whose energy is unaffected has zero electron density in the cross link. The MO whose energy is affected moves *upwards* when n is odd, and *downwards* when n is even. Explain this behavior.

Hint: When n is odd, the one-dimensional Platt function is *ungerade* with respect to inversion in the midpoint; when n is even, the one-dimensional Platt function is *gerade*. Only in the former case, can the MO have a node in the cross link. Determine the influence of this fact on the wavelength λ.

7.   Calculate the bond population in benzene and in hexatriene on the basis of the FEMO wave functions. Compare the results obtained for both molecules.

Hint: Use the approximate expression of Eq. (5.83).

8.   It was pointed out in Sec. 5(B) that the effect of central substitution of an N atom into a conjugated chain was dependent on the number of atoms in the chain. Use perturbation theory to calculate the effect of central substitution on a dye which contains nine atoms in the conjugated chain.

BIBLIOGRAPHY

[1]   H. Kuhn, *Helv. Chim. Acta* **31,** 1441 (1948).
[2]   H. Kuhn, *J. Chem. Phys.* **17,** 1198 (1949).
[3]   H. Kuhn, *Helv. Chim. Acta* **34,** 1308 (1951).
[4]   N. S. Bayliss, *J. Chem. Phys.* **17,** 1853 (1949).
[5]   K. Ruedenberg and C. W. Scherr, *J. Chem. Phys.* **21,** 1565 (1953).
[6]   J. R. Platt, *J. Chem. Phys.* **17,** 484 (1949).
[7]   W. T. Simpson, *J. Chem. Phys.* **16,** 1124 (1948).
[8]   N. S. Ham and K. Ruedenberg, *J. Chem. Phys.* **25,** 13 (1956).
[9]   W. T. Simpson, *J. Chem. Phys.* **17,** 1218 (1949).
[10] H. Kuhn, *Helv. Chim. Acta.* **32,** 2247 (1949).
[11] C. W. Scherr, *J. Chem. Phys.* **21,** 1582 (1953).
[12] K. Ruedenberg, *J. Chem. Phys.* **22,** 1878 (1954).
[13] W. Moffitt, *J. Chem. Phys.* **22,** 320 (1954).
[14] N. S. Ham and K. Ruedenberg, *J. Chem. Phys.* **25,** 1 (1956).
[15] N. S. Ham and K. Ruedenberg, *J. Chem. Phys.* **29,** 1199 (1958).
[16] K. Ruedenberg, *J. Chem. Phys.* **34,** 1884 (1961).
[17] N. S. Ham and K. Ruedenberg, *J. Chem. Phys.* **29,** 1215 (1958).
[18] N. S. Ham, *J. Chem. Phys.* **29,** 1229 (1958).
[19] K. Ruedenberg, *J. Chem. Phys.* **29,** 1232 (1958).
[20] H. Kuhn, *J. Chem. Phys.* **34,** 2371 (1951).

General References

1. J. R. Platt, *J. Chem. Phys.* **17,** 484 (1949).
2. J. R. Platt, in *Encyclopedia of Physics—Handbuch der Physik,* edited by S. Flügge (Springer Verlag, Berlin, Germany, 1961), Vol. 37, Part 2, p. 173.
3. K. Ruedenberg and C. W. Scherr, *J. Chem. Phys.* **21,** 1565 (1953).
4. J. R. Platt *et al., Free Electron Theory of Conjugated Molecules. A Source Book. Papers of the Chicago Group 1949–1961* (John Wiley & Sons, Inc., New York, 1964).
5. H. Martin, H. D. Försterling and H. Kuhn, *Tetrahedron,* **S2,** 243 (1963).
6. N. S. Bayliss, *Quart. Rev.* **6,** 319 (1952).

CHAPTER 6

Composite-Molecule Methods

The Mulliken–Wolfsberg–Helmholz computational scheme of chapter 4 is essentially an *atoms-in-molecules* approach.[1] The spectroscopic information concerning the constituent atoms and their various stages of ionization is utilized—in the guise of valence orbital ionization energies—in an effort to synthesize knowledge of the molecule from the known behavior of its component atoms.

The purpose of this chapter is to discuss *composite-molecule* or, equivalently, *molecules-in-molecules* methods. These methods are based on the assumption that a molecule may be fragmented into lesser *molecular bits*, and that we may utilize information concerning these bits to synthesize a quantum-chemical picture of the *whole molecule*. In its simpler aspects, such a theory would describe biphenyl in terms of our knowledge of benzene, styrene in terms of benzene and ethylene, and trinitrobenzene in terms of C_6H_6 and CH_3NO_2. In a more involved facet of the same approach, we might consider benzene—with all C—C bonds of equal length—to be described in terms of two Kekulé structures—with unequal C—C bond

---

[1] That is, the MWH approach is a *composite-molecule* theory in which the component *bits* are atoms in their various stages of excitation and ionization.

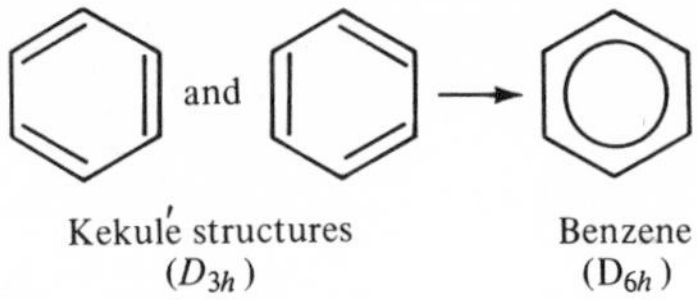

Kekulé structures   Benzene
(D_{3h})   (D_{6h})

FIG. 6.1. Synthesis of benzene from Kekulé structures.

lengths; such a description is schematized in Fig. 6.1. This latter view, of course, leads to the standard resonance formalism of the valence bond theory. However, as far as we are concerned—but with the provision that we can adduce usable information concerning the Kekulé *molecules*—the synthesis of benzenoid information from Kekulé information is merely another type of composite-molecule approach.

Within the outline of the previous paragraph, a great number of variants of the molecules-in-molecules approach is possible. These may differ in the following ways:

(i)   The synthesis of *component* information into *whole-molecule* information may take place at the one-electron molecular-orbital stage or at the many-electron state or configuration stage. Section 1 is concerned with the orbital synthesis; Secs. 2–4 are devoted to configuration mixing.

(ii)   The *lesser component bits* may be supposed to be the standard resonance structures appropriate to the molecule in question. We discuss this topic in Sec. 2.

(iii)   The discussion of Sec. 2 leads to the *molecular exciton theory* of Sec. 3. The primary distinction between Secs. 2 and 3 lies in the types of structure which are supposed to be necessary for a good description of the electronic states of the whole molecule.

(iv)   The approximations used for coupling elements may be quite different. Thus, in Secs. 1 and 4 the coupling elements are identical and quasi-one-electron in nature, whereas in Secs. 2 and 3—particularly, in Sec. 3—the coupling elements are implicitly many electron in nature.

The aim of the composite-molecule approach is to provide a simple description of the ground and excited states of the whole molecule. The value of the approach is determined by its accuracy and the number of basis functions required to produce this accuracy: The fewer this number for a given precision, the better is the approach in question.

1. LINEAR COMBINATION OF MOLECULAR ORBITALS (LCMO) APPROACH

Methods which generate whole-molecule MO's as linear combinations of the MO's of the bits of which the whole molecule is constituted are aptly termed LCMO methods. These methods are usually restricted to the π subset of MO's. The method was evolved by Dewar.[1]

(A) Basis of Method

A given molecule is fragmented into two bits, P and Q, by scission of a formal single σ bond connecting atom p on fragment P with atom q on fragment Q. The MO's of fragment P are denoted φ_m^P and the Hückel Hamiltonian on fragment P is given by

$$\hat{H}^{hP} = \hat{T}^P + \hat{V}^P \tag{6.1}$$

where $\hat{H}^{hP}$ is operative only in the region P of the whole molecule. A collection of self-evident notations is given in Table 6.1. The Hamiltonian for the entire molecule is

$$\hat{H}^h = \hat{T} + \hat{V}^P + \hat{V}^Q + \hat{H}' = \hat{H}^{hP} + \hat{H}^{hQ} + \hat{H}' \tag{6.2}$$

where $\hat{H}'$ represents the interaction of fragments P and Q. Within the spirit of the Hückel theory, we assume that $\hat{H}'$ is localized in the region of the pq single bond and is otherwise inoperative in regions P or Q. Hence, we can define

$$\hat{H}' = \hat{H}_{pq}^h \tag{6.3}$$

The matrix elements of the one-electron Hamiltonian $\hat{H}^h$ are now given by

$$\langle \varphi_m^P \mid \hat{H}^h \mid \varphi_n^P \rangle = \langle \varphi_m^P \mid \hat{H}^{hP} \mid \varphi_n^P \rangle = H_{mn}^P = \epsilon_m^P \delta_{m,n} \tag{6.4}$$

since φ_m^P is, by definition, an eigenvector of $\hat{H}^{hP}$ with eigenvalue ϵ_m^P. Similarly, we find

$$\langle \varphi_m^Q \mid \hat{H}^h \mid \varphi_n^Q \rangle = \langle \varphi_m^Q \mid \hat{H}^{hQ} \mid \varphi_n^Q \rangle = H_{mn}^Q = \epsilon_m^Q \delta_{m,n} \tag{6.5}$$

and

$$\langle \varphi_m^P \mid \hat{H}^h \mid \varphi_n^Q \rangle = \langle \sum_\mu c_{m\mu}^P \chi_\mu^P \mid \hat{H}' \mid \sum_\nu c_{n\nu}^Q \chi_\nu^Q \rangle$$

$$= \langle c_{mp}^P \chi_p^P \mid \hat{H}' \mid c_{nq}^Q \chi_q^Q \rangle = c_{mp}^P c_{nq}^Q \beta_{pq} \tag{6.6}$$

where β_{pq} is a standard Hückel resonance integral.[2]

The whole-molecule molecular-orbital eigenvectors and energies which result from conjugation of the φ_m^P and φ_n^Q fragment MO's across the bond pq are now obtained from the solutions of the secular determinant

$$\begin{vmatrix} \epsilon_m^P - \epsilon & c_{mp}^P c_{nq}^Q k_{pq}\beta \\[2ex] c_{nq}^Q c_{mp}^P k_{pq}\beta & \epsilon_n^Q - \epsilon \end{vmatrix} = 0 \tag{6.7}$$

---

[2] See chapter 3; we can also write $\beta_{pq} = k_{pq}\beta$, where k_{pq} is given in Table 3.3.

TABLE 6.1. *LCMO notation for fragments P and Q and molecule PQ.*
($\hat{H}^{hP}$ acts only in region P; $\hat{H}^{hQ}$ acts only in region Q; $\hat{H}'$ acts only in the pq bond region.)

	Fragment P	Fragment Q	Molecule PQ
MO Wave Function	$\varphi_m^P = \sum\limits_\mu c_{m\mu}^P \chi_\mu^P$	$\varphi_m^Q = \sum\limits_\mu c_{m\mu}^Q \chi_\mu^Q$	$\varphi_m = \sum\limits_\mu c_{m\mu} \chi_\mu$
MO Energy	ϵ_m^P	ϵ_m^Q	ϵ_m
Hamiltonian	$\hat{H}^{hP} = \hat{T}^P + \hat{V}^P$	$\hat{H}^{hQ} = \hat{T}^Q + \hat{V}^Q$	$\hat{H}^h = \hat{T} + \hat{V}^P + \hat{V}^Q + \hat{H}'$ $= \hat{H}^{hP} + \hat{H}^{hQ} + \hat{H}'$
Elements	$H_{mn}^{PP} = H_{mn}^{\ P} = \epsilon_m^P \delta_{m,n}$	$H_{mn}^{QQ} = H_{mn}^{\ Q} = \epsilon_m^Q \delta_{m,n}$	$H_{mn}^{PQ} = (H_{mn}^{\ PQ})'$ $= c_{mp}^P c_{mq}^Q \beta_{pq}$
β_{pq}	$\cdots$	$\cdots$	$\beta_{pq} = \langle \chi_p^P | \hat{H}' | \chi_q^Q \rangle$

The procedure is readily generalized for the case in which all MO's of fragment P are allowed to interact with all MO's of fragment Q.

(B) Butadiene: An Example Calculation

The butadiene molecule may be supposed to consist of two ethylene fragments. The MO wave functions and energies of these fragments are shown in Fig. 6.2. The secular equation is

$$
\begin{array}{c}
\quad\ \ \varphi_{-1} \qquad\quad\ \ \varphi_{1} \qquad\quad\ \ \varphi_{-1'} \qquad\quad\ \varphi_{1'} \\
\begin{array}{c}
\varphi_{-1} \\[6pt]
\varphi_{1} \\[6pt]
\varphi_{-1'} \\[6pt]
\varphi_{1'}
\end{array}
\begin{vmatrix}
\alpha - \beta_{12} - \epsilon & 0 & -\beta_{23}/2 & -\beta_{23}/2 \\[6pt]
0 & \alpha + \beta_{12} - \epsilon & \beta_{23}/2 & \beta_{23}/2 \\[6pt]
-\beta_{23}/2 & \beta_{23}/2 & \alpha - \beta_{34} - \epsilon & 0 \\[6pt]
-\beta_{23}/2 & \beta_{23}/2 & 0 & \alpha + \beta_{34} - \epsilon
\end{vmatrix} = 0 \quad (6.8)
\end{array}
$$

where, for example, the H_{23} matrix element is given by

$$
H_{23} = \langle (\tfrac{1}{2})^{1/2}(\chi_1 + \chi_2) \mid \hat{H}^h \mid (\tfrac{1}{2})^{1/2}(\chi_3 - \chi_4) \rangle
$$

$$
= \tfrac{1}{2}\langle (\chi_1 + \chi_2) \mid \hat{H}' \mid (\chi_3 - \chi_4) \rangle = \tfrac{1}{2}\langle \chi_2 \mid \hat{H}_{23}^{\,h} \mid \chi_3 \rangle = \tfrac{1}{2}\beta_{23} \quad (6.9)
$$

The solutions of the secular determinant of Eq. (6.8) with $\beta_{12} = \beta_{23} = \beta_{34}$ are identical to those of Eq. (3.58a) and Fig. 3.7 when $S = 0$. If we take account of bond alternation, we find $\beta_{12} = \beta_{34}(= 1.1\beta) > \beta_{23}(= 0.87\beta)$; this problem is discussed in Exercise 4, chapter 3.

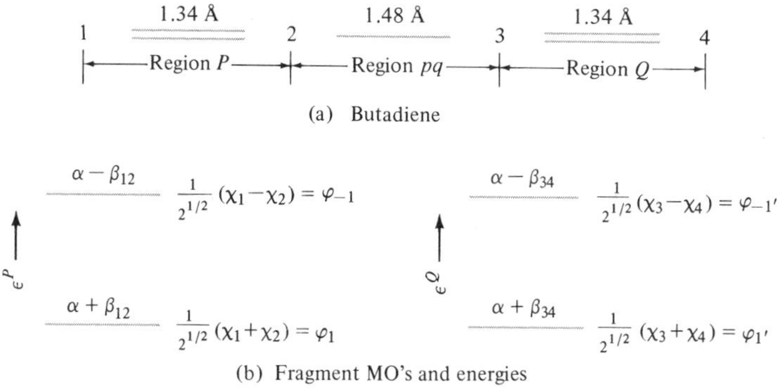

(a) Butadiene

(b) Fragment MO's and energies

FIG. 6.2. Fragmentation of butadiene into two ethylenic bits and the MO eigenvectors and energies of these bits.

(C) Perspective

The simple LCMO method is merely another way of doing simple Hückel theory. It does, however, have a few advantages:

(i)   The fragment MO's need not be simple Hückel MO's; indeed, they might very well be self-consistent field MO's (SCF calculations are feasible for smaller fragments).

(ii)   The fragment MO energies can be normalized to experiment [the diagonal elements of Eq. (6.8) can be replaced by empirical data for the fragments].

(iii)   If interaction is restricted to the highest-energy filled and lowest-energy unfilled MO's of the fragments P and Q, it becomes possible to classify conjugated systems into different types. This topic has been discussed by Suzuki.[2],[3]

(iv)   The LCMO approach provides a neat method of parametrizing the pq bond of a given composite molecule; consequently, extension to other molecules of similar type is facilitated (i.e., an empirical β_{pq} for biphenyl can be used for α- or β-phenylnaphthalene, β-binaphthyl, etc.).

2. METHOD OF RESONANCE

Molecular-orbital theory has superseded valence-bond theory as a semiquantitative tool for the interpretation and systematization of much of chemistry. Molecular-orbital theory is conceptually simpler, it requires less chemical clairvoyance and it is computationally less burdensome—hence, its present dominance.

Nonetheless, the formalism of resonance theory[3] (i.e., the use of resonance structures), still constitutes a widely used language for the communication, if not the formulation, of chemical information. Certain rules for the handling of resonance structures have evolved and the qualitative predictions which result are in reasonable accord with both physical and chemical properties of most ground-state molecules.

The method of resonance hinges on the selection of a *small* set of *chemically reasonable* structures from the full canonical set[4]−[8] of covalent and ionic structures. It is this process of selection which is the difficult element in the method of resonance; it requires usage of the full art of chemistry. We cannot teach this art here; we can merely give examples and refer the reader to any good text on organic chemistry.

---

[3] The method of resonance was originally intended to be a qualitative extrapolation of valence-bond theory [W. T. Simpson, *Theories of Electrons in Molecules* (Prentice-Hall, Inc., Englewood Cliffs, N. J., 1962)].

(A) Basis of Method

It is assumed that a given molecule can be represented by a small number of component resonance structures. These *component resonance structures* are not found by fragmenting the molecule as in Sec. 1; instead, they are obtained by introducing chemical experience into the molecules-in-molecules approach. Thus, benzene may be represented[4] by two Kekulé structures as in

or a simple amide may be supposed to be some hybrid of covalent and ionic structures as in

The resonance structures need not correspond to any *real* molecule. It is only necessary that they obey the usual valence rules and that they be chemically reasonable. In addition, in order for this approach to be useful, it is required that we be able to obtain precise information concerning that interaction of resonance structures which generates a faithful representation of some parent molecule and that we be able to extrapolate this information to considerations of other molecules. Thus, a knowledge of the interaction of the two Kekulé structures—which supposedly leads to a reasonable representation of benzene—may also be used to describe the resonance interactions $\langle$ I | II $\rangle$ and $\langle$ I | III $\rangle$ of the set

which we suppose to be representative of naphthalene: Certainly, it is quite clear that the relationship between the members of either pair of structures is a simple Kekulé-like resonance restricted to the left-most ring in the top pair and to the right-most ring in the bottom pair.

[4] The symbol "⊃" as used here is synonymous with the word *contains*.

We know nothing—at least initially—concerning the wave functions of the resonance structures. We merely know that in principle we can write a wave function the square of which corresponds to the structure. The nature of the approach is such that we need never know this wave function; the only required knowledge is the magnitude of structure wave-function interactions.

(B) Benzene: An Example Calculation

The two Kekulé structures of benzene are:

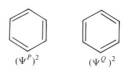

$$(\Psi^P)^2 \qquad\qquad (\Psi^Q)^2$$

These structures correspond to the squares of wave functions Ψ^P and Ψ^Q, respectively. The wave functions are the singlet ground wave functions for each structure. The perturbation energy necessary to convert benzene, where all C—C bonds are equal, to a Kekulé structure, where all C—C bonds are not equal, is denoted δ^P for going to Ψ^P and δ^Q for going to Ψ^Q. Consequently, if $\hat{H}$ be the total Hamiltonian for benzene, that for the Kekulé structure is one of

$$\hat{H}^P = \hat{H} + \delta^P; \qquad \hat{H}^Q = \hat{H} + \delta^Q \qquad (6.10)$$

The energies of both Kekulé structures are obviously identical (i.e., $E^P = E^Q$). Consequently, the interaction of Ψ^P and Ψ^Q may be discussed within the formalism of degenerate perturbation theory. The resulting secular equation is

$$\begin{vmatrix} \langle \Psi^P \mid \hat{H} \mid \Psi^P \rangle - E & \langle \Psi^P \mid \hat{H} \mid \Psi^Q \rangle \\[2mm] \langle \Psi^Q \mid \hat{H} \mid \Psi^P \rangle & \langle \Psi^Q \mid \hat{H} \mid \Psi^Q \rangle - E \end{vmatrix} = 0 \qquad (6.11)$$

where we have assumed orthonormality of the structure wave functions, so that

$$\langle \Psi^P \mid \Psi^Q \rangle = \langle \Psi^Q \mid \Psi^P \rangle = \delta_{P,Q} \qquad (6.12)$$

Now, we simplify the diagonal matrix elements, to find

$$\langle \Psi^P \mid \hat{H} \mid \Psi^P \rangle = \langle \Psi^P \mid \hat{H}^P - \delta^P \mid \Psi^P \rangle = E^P - \langle \Psi^P \mid \delta^P \mid \Psi^P \rangle \quad (6.13)$$

where the last equality results from the fact that Ψ^P is an eigenvector of $\hat{H}^P$ with eigenvalue E^P. Similarly, we find

$$\langle \Psi^Q \mid \hat{H} \mid \Psi^Q \rangle = E^Q - \langle \Psi^Q \mid \delta^Q \mid \Psi^Q \rangle \qquad (6.14)$$

which clearly equals $\langle \Psi^P \mid \hat{H} \mid \Psi^P \rangle$. The off-diagonal elements are

$$\langle \Psi^P \mid \hat{H} \mid \Psi^Q \rangle = \langle \Psi^P \mid \hat{H}^Q \mid \Psi^Q \rangle + \langle \Psi^P \mid -\delta^Q \mid \Psi^Q \rangle$$

$$= \quad\quad 0 \quad\quad + \quad\quad \Delta^{PQ} \quad\quad\quad (6.15)$$

where

$$\Delta^{PQ} \equiv -\langle \Psi^P \mid \delta^Q \mid \Psi^Q \rangle \quad\quad\quad (6.16)$$

The symmetry of the interaction[5] indicates that $\langle \Psi^Q \mid \hat{H} \mid \Psi^P \rangle$ must also equal Δ^{PQ}. Thus, Eq. (6.11) may be rewritten

$$\begin{vmatrix} -E' & \Delta^{PQ} \\ \Delta^{PQ} & -E' \end{vmatrix} = 0 \quad\quad\quad (6.17)$$

where

$$-E' = -E + \langle \Psi^P \mid \hat{H} \mid \Psi^P \rangle = -E + E^P - \langle \Psi^P \mid \delta^P \mid \Psi^P \rangle$$

$$= -E + E^Q - \langle \Psi^Q \mid \delta^Q \mid \Psi^Q \rangle \quad (6.18)$$

The solutions of Eq. (6.17) are

$$E' = \pm\Delta^{PQ}$$

$$E = E^P - \langle \Psi^P \mid \delta^P \mid \Psi^P \rangle \pm \Delta^{PQ} \quad\quad\quad (6.19)$$

The energies and wave functions are given in Fig. 6.3. Thence, it becomes evident that the final states of benzene are obtained by two different energy effects:

(i)   a compression of three bonds and an extension of three bonds. In this manner, the two Kekulé structures of symmetry D_{3h} are converted to two *Kekulé structures* of symmetry D_{6h} and are prepared, as it were, for the final resonance which destroys specific pair bonding. This compression-extension energy equals $-\langle \Psi^P \mid \delta^P \mid \Psi^P \rangle$.

(ii)   a resonance of two Kekulé structures of symmetry D_{6h} to yield two states of benzene. In this formulation, the resonance energy of benzene is simply $\mid \Delta^{PQ} \mid$ and the stability of D_{6h} benzene as opposed to either of the two D_{3h} Kekulé forms is dependent on the inequality

$$\mid \Delta^{PQ} \mid > \mid \langle \Psi^P \mid \delta^P \mid \Psi^P \rangle \mid \quad\quad\quad (6.20)$$

The assumption is now made that the two states which have evolved correspond to the ground and the lowest-energy electronic excited state of

---

[5] More specifically: If Ψ^P and Ψ^Q are real wave functions, then $\Delta^{PQ} = \Delta^{QP}$ because $\hat{H}$ is Hermitean.

benzene. The lowest-energy electronic transition of benzene occurs at 2600 Å and is predicted to equal $-2\Delta^{PQ}$. Thus, the resonance energy of benzene is given by $|\Delta^{PQ}| = 55$ kcal/mole. The experimental value is 36.0 kcal/mole.

(C)   Naphthalene: An Example Calculation

The principal (i.e., Kekulé) resonance forms of naphthalene are

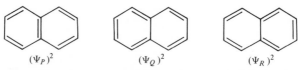

$$(\Psi_P)^2 \qquad\qquad (\Psi_Q)^2 \qquad\qquad (\Psi_R)^2$$

We denote the corresponding wave functions Ψ_P, Ψ_Q, and Ψ_R. The Hamiltonian for D_{2h} naphthalene is denoted $\hat{H}$.

As in Sec. 2(A), the elements $\langle \Psi^P | \hat{H} | \Psi^Q \rangle$ and $\langle \Psi^P | \hat{H} | \Psi^R \rangle$ are benzenoid; thus, they equal -55 kcal/mole (-19 kK).

The element $\langle \Psi_Q | \hat{H} | \Psi_R \rangle$ is, however, of a different nature; it specifies the interaction of two structures which differ by a switch-over of *all* double bonds by one bond unit. To evaluate this element, we note the identity of the interaction required to that which occurs in azulene where the (two)

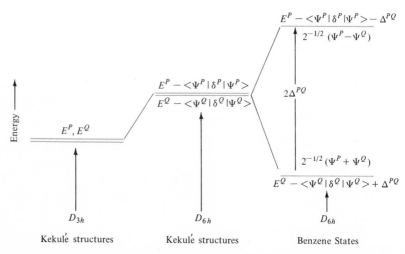

FIG. 6.3.   Evolution of two electronic states of benzene from two Kekulé structures. Note the conventions: $\langle \Psi^P | \delta^P | \Psi^P \rangle < 0$; $\langle \Psi^P | \delta^P | \Psi^Q \rangle > 0$; $\Delta^{PQ} < 0$.

principal resonance structures are

The description of resonance interaction in azulene proceeds along the lines of benzene, where it follows that the lowest-energy electronic transition should occur at $-2\Delta_{azulene}{}^{QR}$. The experimental location of the lowest-energy electronic absorption band of azulene is 7000 Å, whereupon we conclude that

$$\Delta_{azulene}{}^{QR} = -7 \text{ kK} \qquad (6.21)$$

With the assumption that $\Delta_{naphthalene}{}^{QR} = \Delta_{azulene}{}^{QR}$, we now have values available for all the required interaction elements.

The secular determinant for naphthalene is

$$
\begin{array}{ccc}
 & \Psi^P & \Psi^Q & \Psi^R \\
\Psi^P & \begin{vmatrix} -E' & -19 & -19 \\ \\ -19 & -E' & -7 \\ \\ -19 & -7 & -E' \end{vmatrix} & & \\
\Psi^Q & & & = 0 \qquad (6.22) \\
\Psi^R & & &
\end{array}
$$

where all energy units are in kiloKaysers and all three resonance structures are assumed to be equienergetic. The solutions are $E' = -30.5, +7.0$, and 23.5 kK. The lowest-energy electronic transition is predicted to occur at 37.5 kK (i.e., slightly lower in energy than that in benzene). This prediction accords with experiment. The predicted resonance energy is 88.5 kcal/mole; the experimental value is 61.0 kcal/mole.

(D) Purine-Pyrimidine Bases of DNA: Example Calculations

The purine-pyrimidine bases may be supposed[9] to be constructed from a fundamental unit: the amide grouping. This fragmentation is obvious in the complete set of resonance structures for uracil given in Fig. 6.4. The wave functions descriptive of these structures are given by the kets $|1\rangle$, $|2\rangle$, ... Our aim is to construct the secular determinant

$$| \langle i | \hat{H} | j \rangle - E \langle i | j \rangle | = 0 \qquad (6.23)$$

and, by solution, obtain the eigenvalues and eigenvectors appropriate to

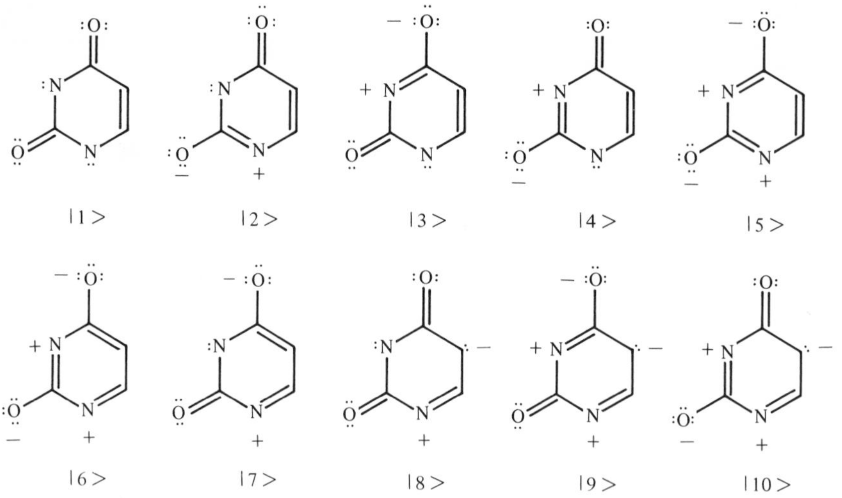

FIG. 6.4.   Resonance structures for the uracil molecule.

uracil. Since, by definition,

$$E\langle i \mid j \rangle = E\delta_{i,j} \tag{6.24}$$

we need concern ourselves only with the elements $\langle i \mid \hat{H} \mid j \rangle$. In order to obtain values for these elements we now consider a series of simple molecules for which the range of resonance structures is correspondingly restricted.

(i)   *A Simple Amide:*   The two commonly accepted resonance structures are

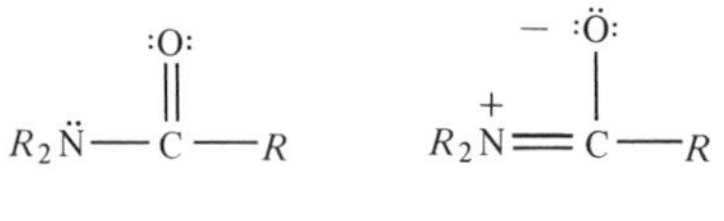

The secular determinant is

$$\begin{vmatrix} \langle 1' \mid \hat{H} \mid 1' \rangle - E & \langle 1' \mid \hat{H} \mid 2' \rangle \\ \langle 2' \mid \hat{H} \mid 1' \rangle & \langle 2' \mid \hat{H} \mid 2' \rangle - E \end{vmatrix} = 0 \tag{6.25a}$$

The zero of energy may be taken to correspond to structure 1'. Thus,

$$\langle 1' \mid \hat{H} \mid 1' \rangle \equiv 0 \tag{6.26}$$

Relative to this zero of energy, we also define

$$\langle 2' \mid \hat{H} \mid 2' \rangle \equiv \alpha \tag{6.27}$$

$$\langle 2' \mid \hat{H} \mid 1' \rangle = \langle 1' \mid \hat{H} \mid 2' \rangle \equiv \beta \tag{6.28}$$

Consequently, Eq. (6.25a) becomes

$$
\begin{array}{cc}
 & |\,1'\rangle \quad |\,2'\rangle \\
\begin{array}{c} \langle 1' | \\[8pt] \langle 2' | \end{array} &
\begin{vmatrix} 0 - E & \beta \\[8pt] \beta & \alpha - E \end{vmatrix} = 0
\end{array}
\tag{6.25b}
$$

The solutions are

$$E = [\alpha \pm (\alpha^2 + 4\beta^2)^{1/2}]/2 \tag{6.29}$$

The first electronic transition is expected to occur at

$$\Delta E = (\alpha^2 + 4\beta^2)^{1/2} \tag{6.30}$$

The experimental value is $\sim$54 kK.

We now note that the elements $\langle 1 \mid \hat{H} \mid 2 \rangle$, $\langle 1 \mid \hat{H} \mid 3 \rangle$, $\langle 1 \mid \hat{H} \mid 4 \rangle$, $\langle 2 \mid \hat{H} \mid 5 \rangle$, and $\langle 3 \mid \hat{H} \mid 5 \rangle$ of uracil (see Fig. 6.4) are simple amidic resonances and that they may be denoted β. The elements $\langle 1 \mid \hat{H} \mid 8 \rangle$, $\langle 3 \mid \hat{H} \mid 9 \rangle$, $\langle 4 \mid \hat{H} \mid 10 \rangle$, etc., of Fig. 6.4 are also similar to the simple amidic resonance; we assume that these elements may also be denoted β. The energy corresponding to the structures $|\,2\rangle$, $|\,3\rangle$, $|\,4\rangle$, and $|\,8\rangle$ is α; that corresponding to the structures $|\,5\rangle$, $|\,6\rangle$, $|\,9\rangle$, $|\,10\rangle$—which contain two ionic amide resonance structures—is taken to be 2α. We approximate the energy of $|\,7\rangle$ by α also; this latter is a poor approximation but, unfortunately, it is the best we can do at this point.

(ii) *Formamidinium ion:* The commonly accepted resonance structures are

$$
\underset{|1''\rangle}{\underset{\displaystyle \underset{H}{|}}{H_2\overset{..}{N} \text{—} C \text{==} \overset{+}{N}H_2, \overset{-}{C}lO_4}}
\qquad\qquad
\underset{|2''\rangle}{\underset{\displaystyle \underset{H}{|}}{H_2\overset{+}{N} \text{==} C \text{—} \overset{..}{N}H_2, \overset{-}{C}lO_4}}
$$

The secular determinant is

$$
\begin{array}{cc}
 & |\,1''\rangle \quad |\,2''\rangle \\
\begin{array}{c} \langle 1'' | \\[8pt] \langle 2'' | \end{array} &
\begin{vmatrix} 0 - E & \gamma \\[8pt] \gamma & 0 - E \end{vmatrix} = 0
\end{array}
\tag{6.31}
$$

where $\langle 1'' \mid \hat{H} \mid 1'' \rangle = \langle 2'' \mid \hat{H} \mid 2'' \rangle$ is taken to be of zero energy. The first electronic excitation should occur at 2γ; the experimental value is ~ 50 kK.

We now note that the element $\langle 2 \mid \hat{H} \mid 4 \rangle$ of Fig. 6.4 is of a simple formamidinium type and may be denoted γ. The element $\langle 7 \mid \hat{H} \mid 9 \rangle$ of Fig. 6.4 is also very similar. It consists of the switching of a lone pair and a double bond about the negatively charged oxygen; the fact that this process generates other charge centers is considered irrelevant. We assume that we may denote $\langle 7 \mid \hat{H} \mid 9 \rangle$ by γ also.

(iii) Succinimide: The commonly accepted resonance structures are

The secular determinant is given, in view of items (i) and (ii), by

$$
\begin{array}{c@{\qquad}ccc}
 & |\,1'''\rangle & |\,2'''\rangle & |\,3'''\rangle \\[4pt]
\langle 1'''| & \begin{vmatrix} 0 - E & \beta & \beta \\[6pt] \beta & \alpha - E & 0 \\[6pt] \beta & 0 & \alpha - E \end{vmatrix} & & = 0 \qquad (6.32) \\[6pt]
\langle 2'''| & & & \\[6pt]
\langle 3'''| & & &
\end{array}
$$

The only questionable entry in Eq. (6.32) is the equivalence $\langle 2''' \mid \hat{H} \mid 3''' \rangle = \langle 3''' \mid \hat{H} \mid 2''' \rangle = 0$. We justify this entry by assuming that $\hat{H}$ is a sum of parts referring to right-hand and left-hand sides of the molecule whereas the wave functions are products of parts referring to the same regions. Thus,

$$
\begin{aligned}
\langle 2''' \mid \hat{H} \mid 3''' \rangle &= \langle \Psi^L \Psi^{R\dagger} \mid \hat{H}^L + \hat{H}^R \mid \Psi^{L\dagger} \Psi^R \rangle \\
&= \langle \Psi^L \mid \hat{H}^L \mid \Psi^{L\dagger} \rangle \langle \Psi^{R\dagger} \mid \Psi^R \rangle + \langle \Psi^{R\dagger} \mid \hat{H}^R \mid \Psi^R \rangle \langle \Psi^L \mid \Psi^{L\dagger} \rangle
\end{aligned}
$$
$$(6.33)$$

where L denotes *left* and R denotes *right*, and where the symbol $\dagger$ denotes the presence of the ionic state. Now Ψ^R and $\Psi^{R\dagger}$ are different eigenfunctions of $\hat{H}^R$, hence $\langle \Psi^R \mid \Psi^{R\dagger} \rangle = 0$. Thus, it follows that $\langle 2''' \mid \hat{H} \mid 3''' \rangle = 0$.

At any rate, the solutions of Eq. (6.32) predict a lowest-energy electronic excitation at

$$
\Delta E = [\alpha + (\alpha^2 + \beta^2)^{1/2}]/2 \qquad (6.34)
$$

The experimental value is ~ 51.5 kK.

(iv) Uracil: The resonance structures are given in Fig. 6.4. The secular determinant is

$$
\begin{array}{c|cccccccccc}
 & |1\rangle & |2\rangle & |3\rangle & |4\rangle & |5\rangle & |6\rangle & |7\rangle & |8\rangle & |9\rangle & |10\rangle \\
\hline
\langle 1| & 0-E & \beta & \beta & \beta & 0 & 0 & 0 & \beta & 0 & 0 \\
\langle 2| & \beta & \alpha-E & 0 & \gamma & \beta & 0 & 0 & 0 & 0 & 0 \\
\langle 3| & \beta & 0 & \alpha-E & 0 & \beta & 0 & 0 & 0 & \beta & 0 \\
\langle 4| & \beta & \gamma & 0 & \alpha-E & 0 & 0 & 0 & 0 & 0 & \beta \\
\langle 5| & 0 & \beta & \beta & 0 & 2\alpha-E & -19 & 0 & 0 & 0 & 0 \\
\langle 6| & 0 & 0 & 0 & 0 & -19 & 2\alpha-E & \beta & 0 & 0 & \beta \\
\langle 7| & 0 & 0 & 0 & 0 & 0 & \beta & \alpha-E & \beta & \gamma & 0 \\
\langle 8| & \beta & 0 & 0 & 0 & 0 & 0 & \beta & \alpha-E & \beta & \beta \\
\langle 9| & 0 & 0 & \beta & 0 & 0 & 0 & \gamma & \beta & 2\alpha-E & 0 \\
\langle 10| & 0 & 0 & 0 & \beta & 0 & \beta & 0 & \beta & 0 & 2\alpha-E \\
\end{array} = 0
$$

$$(6.35)$$

The elements of Eq. (6.35) are obtained as outlined in items (i)–(iii). The elements $\langle 5 \,|\, \hat{H} \,|\, 6 \rangle$ and $\langle 6 \,|\, \hat{H} \,|\, 5 \rangle$ are simple Kekulé-like resonances for which we use the value of -19 kK as in benzene. The values of α, β, and γ are obtained from the information of items (i), (ii), and (iii) relating to electronic excitation energies; they are

$$\alpha = 35.5 \text{ kK}; \qquad \beta = -20.3 \text{ kK}; \qquad \gamma = -25.0 \text{ kK} \qquad (6.36)$$

The two lowest roots of Eq. (6.35) are -37.59 and -1.69 kK. Thus, the lowest-energy electronic excitation of uracil should occur at 35.9 kK. The observed excitation energy is $\sim$38.5 kK (in aqueous solution at a pH of 7.2). A further elaboration of this topic is found in Exercise 1.

3. MOLECULAR EXCITON MODEL

The molecular exciton approach can be considered to be a variant of the resonance structure interaction model. The primary difference between the two theories is that the structures of the latter usually differ from each other in bonding characteristics or in charge distributions associated with different regions of the molecule whereas, in the exciton approach, the structures considered usually differ only in the degree of electronic excitation present in different regions of the molecule. Thus, in butadiene, the resonance structures usually considered are the Kekulé and Thielé forms:

$$CH_2=CH-CH=CH_2 \quad \text{and} \quad \dot{C}H_2-CH=CH-\dot{C}H_2$$
$$\text{(Kekulé)} \qquad\qquad\qquad \text{(Thielé)}$$

The exciton approach, however, considers butadiene to consist of three structures:

$$CH_2=CH-CH=CH_2 \qquad CH_2=CH-\overset{**}{CH=CH_2}$$

$$CH_2=\overset{**}{CH}-CH=CH_2$$

where the asterisks indicate that electronic excitation is present in the specified ethylenic bond. However, even these differences are partially superficial: Certainly, the Thielé structure is *excited* relative to

$$CH_2=CH-CH=CH$$

The advantage of the exciton model is that it allows us to prescribe an exact interaction Hamiltonian[6] whose matrix elements are evaluated from a knowledge of the spectroscopic characteristics of the submolecular units.

The exciton approach is most useful in the following instances:

(i)   Composite molecules which contain identical subunits. Thus, it is useful in butadiene (which contains two ethylenes) or DNA (which contains a structured core of repeating base pairs); it is not obviously useful in either nitrobenzene or uracil.

(ii)   The solid state. It leads to the concept of the quasiparticulate *exciton* which can travel throughout the crystal structure. In this way, the ability to localize excitation energy on any given molecular center is described in proper probabilistic fashion.

(iii)   Certain biological systems. The exciton approach is of crucial importance in considerations of energy transfer in the lamellar phases common in biology.

---

[6] If it is assumed that the regions of submolecular excitation are well separated within the molecule.

(A)   Harmonic Oscillator Analogue of Exciton Interaction

Consider two distinguishable[7] but otherwise identical harmonic oscillators P and Q such as schematized in Fig. 6.5. The eigenfunctions of oscillator P are denoted as follows: P_0, the zero-point vibrational wave function; P_1, containing one unit of vibrational excitation equal to $h\nu$; P_2, containing two units of vibrational excitation each equal to $h\nu$; etc. The vibrational eigenfunctions of Q are denoted $Q_0, Q_1, Q_2, \ldots$. Thus, the composite wave functions for the system of two independent oscillators are

$$P_0Q_0;\; P_1Q_0;\; P_0Q_1; \ldots;\; P_iQ_j; \ldots$$

These eigenfunctions are correct in the limit of no interaction between the two oscillators. The compound state which contains energy $nh\nu$ is n-fold degenerate.

Let the Hamiltonian operator for oscillator P be

$$\hat{H}^P = \hat{T}^P + f(x^P) \tag{6.37}$$

where x^P is defined relative to the left-hand origin of Fig. 6.5. Similarly, for oscillator Q, we have

$$\hat{H}^Q = \hat{T}^Q + f(x^Q) \tag{6.38}$$

The interaction energy of the point charges of Fig. 6.5 is, by inspection

$$\frac{V^{PQ}}{e^2} = \frac{1}{R} - \frac{1}{R - x^P} - \frac{1}{R + x^Q} + \frac{1}{R - x^P + x^Q} \tag{6.39}$$

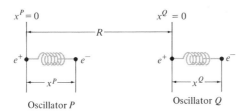

Oscillator P Oscillator Q

FIG. 6.5.   Schematic of two one-dimensional *harmonic oscillators*. Each oscillator consists of an electron of charge $-e$ vibrating with respect to a fixed nucleus of charge $+e$. Two coordinate systems with nucleus-fixed origins are also defined. (The harmonic-oscillator approximation to the Coulomb interaction of an electron and a proton is never strictly valid— but is acceptable when the vibrational quantum number is small (i.e., near the bottom of the potential well).]

---

[7] By virtue of their different locations in space—as in Fig. 6.5.

If it is assumed that R is large relative to either x^P or x^Q, we find

$$\frac{1}{R + x^Q} = \frac{1}{R} \cdot \frac{1}{1 + (x^Q/R)} \simeq \frac{1}{R}\left[1 - \frac{x^Q}{R} + \left(\frac{x^Q}{R}\right)^2\right] \tag{6.40}$$

and

$$\frac{1}{R - x^P} \simeq \frac{1}{R}\left[1 + \frac{x^P}{R} + \left(\frac{x^P}{R}\right)^2\right] \tag{6.41}$$

$$\frac{1}{R - (x^P - x^Q)} \simeq \frac{1}{R}\left[1 + \frac{x^P - x^Q}{R} + \left(\frac{x^P - x^Q}{R}\right)^2\right] \tag{6.42}$$

Insertion of these expressions into Eq. (6.39) yields

$$V^{PQ}/e^2 = -2x^P x^Q/R^3 \tag{6.43}$$

This expression is correct when x^P, $x^Q \ll R$.

If the two interacting one-dimensional oscillators are arbitrarily positioned in three-dimensional space, the expression corresponding to Eq. (6.43) is[10]

$$\begin{aligned}
V^{PQ} &= (-2z^P z^Q + y^P y^Q + x^P x^Q)/R^3 \\
&= (-r^P r^Q/R^3)(2\cos\theta_z^P \cos\theta_z^Q - \cos\theta_y^P \cos\theta_y^Q - \cos\theta_x^P \cos\theta_x^Q)
\end{aligned}$$

$$\tag{6.44}$$

where the notation used has the following meaning: The coordinate system centered on oscillator P is parallel to the coordinate system on Q but displaced from it by a distance R; the z axis is defined along the line of oscillator centers; the parameters $\cos\theta_z^P$, $\cos\theta_y^Q,\ldots$ are direction cosines which describe the orientation of the individual oscillators relative to their respective coordinate systems.

In view of Eqs. (6.43) and (6.44), the total Hamiltonian for the system of two weakly interacting oscillators is

$$\hat{H} = \hat{H}^P + \hat{H}^Q + \hat{H}^{PQ} \tag{6.45}$$

where $\hat{H}^{PQ}$ is the operator equivalent of V^{PQ}.

At this point, we have available a Hamiltonian $\hat{H}$ appropriate to a system of two oscillators which are not too close together and a set of system eigenfunctions which is correct when the two oscillators are infinitely far apart. We now wish to obtain the eigenfunctions of $\hat{H}$ in terms of the available set of zero-order eigenfunctions. In order to do this, we proceed through the usual interaction matrix and we evaluate elements of the type

$\langle P_iQ_j \mid \hat{H} \mid P_kQ_l \rangle$. A few examples of such interaction elements are detailed below:

(i) $\quad \langle P_0Q_0 \mid \hat{H} \mid P_0Q_0 \rangle = \langle P_0Q_0 \mid \hat{H}^P \mid P_0Q_0 \rangle + \langle P_0Q_0 \mid \hat{H}^Q \mid P_0Q_0 \rangle$

$$+ \langle P_0Q_0 \mid \hat{H}^{PQ} \mid P_0Q_0 \rangle \quad (6.46)$$

Since $\hat{H}^Q$ is a function of x^Q only, $\hat{H}^P$ is a function of x^P only, and $\hat{H}^{PQ}$ is given by $2e^2x^Px^Q/R^3$, Eq. (6.46) simplifies to

$$\langle P_0Q_0 \mid \hat{H} \mid P_0Q_0 \rangle = \langle P_0 \mid \hat{H}^P \mid P_0 \rangle \langle Q_0 \mid Q_0 \rangle + \langle P_0 \mid P_0 \rangle \langle Q_0 \mid \hat{H}^Q \mid Q_0 \rangle$$

$$- (2e^2/R^3)\langle P_0 \mid x^P \mid P_0 \rangle \langle Q_0 \mid x^Q \mid Q_0 \rangle \quad (6.47)$$

Evaluation of the three right-hand parts yields

$$= \frac{h\nu}{2} \cdot 1 + 1 \cdot \frac{h\nu}{2} + 0 \quad (6.48)$$

The null value of $\langle P_0 \mid x^P \mid P_0 \rangle$ follows from the fact that the Hermite polynomial representative of P_0 is of even parity about $x^P = 0$, whereas x^P itself is of odd parity about the same origin; therefore, the integrand of $\langle P_0 \mid x^P \mid P_0 \rangle$, as also of $\langle Q_0 \mid x^Q \mid Q_0 \rangle$, is of odd parity and the corresponding integral is zero. Thus, we find

$$\langle P_0Q_0 \mid \hat{H} \mid P_0Q_0 \rangle = h\nu \quad (6.49)$$

(ii) $\quad \langle P_iQ_j \mid \hat{H} \mid P_iQ_j \rangle$: It may be shown by analogous reasoning that

$$\langle P_iQ_j \mid \hat{H} \mid P_iQ_j \rangle = (1 + i + j)h\nu \quad (6.50)$$

where the unit within the last bracket arises from the zero-point energy of a two-oscillator system (i.e., $\frac{1}{2}h\nu + \frac{1}{2}h\nu = 1h\nu$).

(iii) $\quad \langle P_0Q_1 \mid \hat{H} \mid P_1Q_0 \rangle = \langle P_0 \mid \hat{H}^P \mid P_1 \rangle \langle Q_1 \mid Q_0 \rangle + \langle P_0 \mid P_1 \rangle \langle Q_1 \mid \hat{H}^Q \mid Q_0 \rangle$

$$- (2e^2/R^3)\langle P_0 \mid x^P \mid P_1 \rangle \langle Q_1 \mid x^Q \mid Q_0 \rangle \quad (6.51)$$

$$= 0 + 0 - 2 \mid \mathbf{M} \mid^2/R^3 \quad (6.52)$$

where $\mathbf{M}$ is the transition moment[8] of the $1 \leftarrow 0$ excitation of a harmonic oscillator. The transition moment between any two levels of a harmonic oscillator is[11]

$$\langle n' \mid ex \mid m' \rangle = e(n'/2a)^{1/2}\delta_{n',m'+1} \quad (6.53)$$

where n' and m' are the quantum numbers of the combining levels; where $n' > m'$; where $a \equiv 4\pi^2m\nu/h$; where m is the electron mass; and where δ is the Kronecker δ.

---

[8] See chapter 9.

The secular determinant is now written out in full as

$$
\begin{array}{c|cccccc}
 & |P_0Q_0\rangle & |P_0Q_1\rangle & |P_1Q_0\rangle & |P_1Q_1\rangle & |P_2Q_0\rangle & |P_0Q_2\rangle \cdots \\
\hline
\langle P_0Q_0| & h\nu - E & 0 & 0 & -\dfrac{2e^2}{R^3}\cdot\dfrac{1}{2a} & 0 & 0 \cdots \\
\langle P_0Q_1| & 0 & 2h\nu - E & -\dfrac{2e^2}{R^3}\cdot\dfrac{1}{2a} & 0 & 0 & 0 \cdots \\
\langle P_1Q_0| & 0 & -\dfrac{2e^2}{R^3}\cdot\dfrac{1}{2a} & 2h\nu - E & 0 & 0 & 0 \cdots \\
\langle P_1Q_1| & -\dfrac{2e^2}{R^3}\dfrac{1}{2a} & 0 & 0 & 3h\nu - E & -\dfrac{2e^2}{R^3}\dfrac{\sqrt{2}}{2a} & -\dfrac{2e^2}{R^3}\dfrac{\sqrt{2}}{2a} \cdots \\
\langle P_2Q_0| & 0 & 0 & 0 & -\dfrac{2e^2}{R^3}\dfrac{\sqrt{2}}{2a} & 3h\nu - E & 0 \cdots \\
\langle P_0Q_2| & 0 & 0 & 0 & -\dfrac{2e^2}{R^3}\dfrac{\sqrt{2}}{2a} & 0 & 3h\nu - E \cdots \\
 & \vdots & \vdots & \vdots & \vdots & \vdots & \vdots
\end{array} = 0
$$

$$(6.54)$$

In the zero-order calculation, we set $\hat{H}^{PQ} = 0$. Under these conditions the energy matrix is already diagonalized and the energy levels are $h\nu$, $2h\nu$, $3h\nu, \ldots$. The level of energy $nh\nu$ is n-fold degenerate. The eigenfunctions of $\hat{H}$ are simply the $|P_iQ_j\rangle$.

In the first order, where $\hat{H}^{PQ} \neq 0$ but where we only allow interaction of the degenerate levels, the energy matrix decomposes into one 1×1 block, one 2×2 block, one 3×3 block, $\ldots$. Consider the 2×2 block; it is

$$
\begin{array}{c|cc}
 & |P_0Q_1\rangle & |P_1Q_0\rangle \\
\langle P_0Q_1| & 2h\nu - E & -(2/R^3)\,|\mathbf{M}|^2 \\
\langle P_1Q_0| & -(2/R^3)\,|\mathbf{M}|^2 & 2h\nu - E
\end{array} = 0 \qquad (6.55)
$$

The solutions are

$$E_1 = 2h\nu - 2\,|\mathbf{M}|^2/R^3; \qquad \Psi_1 = (\tfrac{1}{2})^{1/2}(P_0Q_1 + P_1Q_0) \qquad (6.56)$$

$$E_2 = 2h\nu + 2\,|\mathbf{M}|^2/R^3; \qquad \Psi_2 = (\tfrac{1}{2})^{1/2}(P_0Q_1 - P_1Q_0) \qquad (6.57)$$

These results are readily visualized using a simple physical model. In Ψ_1, the two oscillatory transition dipoles are exactly in-phase and may be represented as

$$\overrightarrow{\;\;\;\;\;} + \overrightarrow{\;\;\;\;\;} = \overrightarrow{\;\;\;\;\;\;\;\;\;\;\;}$$
$$(P) \qquad\qquad (Q) \qquad\quad (\text{RESULTANT})$$

This situation is attractive and the energy associated with Ψ_1 is lower than that of either P_1Q_0 or P_0Q_1 by an amount equal to the interaction energy of two dipoles phased as shown, namely, $2 \mid \mathbf{M} \mid^2/R^3$. It is also clear that the transition moment associated with the excitation $\Psi_1 \leftarrow P_0Q_0$ is larger than that of $P_1Q_0 \leftarrow P_0Q_0$ by a factor of $(2)^{1/2}$. In Ψ_2, the oscillatory transition dipoles are exactly out-of-phase and may be represented as

$$\overleftarrow{\;\;\;\;\;} + \overrightarrow{\;\;\;\;\;} = \quad 0$$
$$(P) \qquad\qquad (Q) \qquad\quad (\text{RESULTANT})$$

This is a repulsive dipolar orientation. Consequently, $\Psi_2 \leftarrow P_0Q_0$ is forbidden and Ψ_2 is of higher energy than either P_1Q_0 or P_0Q_1 by an amount $2 \mid \mathbf{M} \mid^2/R^3$. The effects described are illustrated in Fig. 6.6. The separation of the two excited levels is termed an exciton splitting; in the case of Fig. 6.6, the exciton splitting is $4M^2/R^3$.

If we consider the interaction of all coupled states (i.e., including initially nondegenerate states), the states P_0Q_0 and P_1Q_1 do interact and a

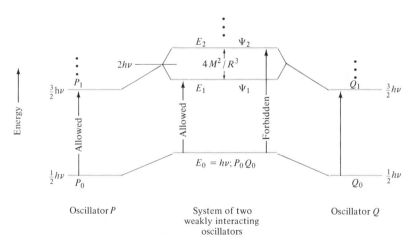

FIG. 6.6.   First-order interaction of two harmonic oscillators P and Q. The two states Ψ_1 and Ψ_2 are given by Eqs. (6.56) and (6.57), respectively. The ground state is P_0Q_0. Interaction is restricted to the two zero-order degenerate wave functions P_0Q_1 and P_1Q_0.

slight energy depression of the ground state results. Indeed, the ground state is now given by

$$\Psi_0 = |\, P_0Q_0 \rangle + (M^2/h\nu R^3)\, |\, P_1Q_1 \rangle$$

$$E_0 = h\nu - 2M^4/h\nu R^6 \tag{6.58}$$

This stabilization of the compound ground state is said to be caused by *dispersion forces*. Thus, dispersion forces are seen to result from a simple exciton effect.[10]

(B) Polyenes: Example Calculations

We suppose that the polyenes consist of ethylenic units and that each of these units may be treated as an electron oscillator.[12] Thus, we identify ethylene with the oscillator P which possesses two energy levels, one at zero energy (i.e., P_0, the ground state) and one at 7.60 eV (i.e., P_1, the first-excited electronic $\pi^* \leftarrow \pi$ singlet state). Since we are interested only in the lower-lying electronic states and since all higher-energy states enter only in a second-order way, we assume that a consideration of only two zero-order states, namely, P_0 and P_1, is adequate.

   (*i*) *trans-Butadiene:*   *trans*-Butadiene consists of two oscillators, P and Q, as shown

The possible compound-oscillator zero-order wave functions which we may write are

$$P_0Q_0,\ P_1Q_0,\ P_0Q_1,\ P_1Q_1$$

The secular equation follows from Eq. (6.54) and is

	$\|P_0Q_0\rangle$	$\|P_1Q_1\rangle$	$\|P_1Q_0\rangle$	$\|P_0Q_1\rangle$
$\langle P_0Q_0\|$	$2W_1 - E$	ρ	0	0
$\langle P_1Q_1\|$	ρ	$2W_2 - E$	0	0
$\langle P_1Q_0\|$	0	0	$W_1 + W_2 - E$	ρ
$\langle P_0Q_1\|$	0	0	ρ	$W_1 + W_2 - E$

$$= 0 \tag{6.59}$$

where ρ is defined as

$$\rho \equiv \langle P_1 Q_0 \mid \hat{H}^{PQ} \mid P_0 Q_1 \rangle = \langle P_0 Q_0 \mid \hat{H}^{PQ} \mid P_1 Q_1 \rangle \qquad (6.60)$$

and where W_1 and W_2 are the energies of the ground and first-excited singlet states, respectively, of ethylene. The energy solutions are

$$E_2 = W_1 + W_2 - \rho; \qquad E_1 = W_1 + W_2 + \rho \qquad (6.61)$$

for the two exciton states containing one unit of excitation, and

$$E_3 = W_1 + W_2 + [(W_1 - W_2)^2 + \rho^2]^{1/2}$$

$$E_0 = W_1 + W_2 - [(W_2 - W_1)^2 + \rho^2]^{1/2} \qquad (6.62)$$

for the ground and doubly excited states. Since the longest-wavelength $\pi^* \leftarrow \pi$ transition of butadiene occurs at 5.93 eV we find[9] $\rho = -1.91$ eV. We may now use this value of ρ to estimate various quantities for butadiene.

The resonance energy (R.E.) may be taken to be the exciton stabilization of the ground state (i.e., the dispersion force interaction); it is given by

$$\text{R.E.} = 2W_1 - \{W_1 + W_2 - [(W_2 - W_1)^2 + \rho^2]^{1/2}\}$$

$$= 0.24 \text{ eV} = 5.5 \text{ kcal/mole} \qquad (6.63)$$

The experimental value is 3.6 kcal/mole.

The transition moment of the lowest-energy electronic excitation of *trans*-butadiene requires a knowledge of Ψ_0 and Ψ_1. These are given by

$$\Psi_0 \simeq \mid P_0 Q_0 \rangle + [\rho/(2W_1 - 2W_2)]\mid P_1 Q_1 \rangle$$

$$\simeq \mid P_0 Q_0 \rangle + 0.125 \mid P_1 Q_1 \rangle \qquad (6.64)$$

$$\Psi_1 = (\tfrac{1}{2})^{1/2}(\mid P_1 Q_0 \rangle + \mid P_0 Q_1 \rangle) \qquad (6.65)$$

The transition moment is given by

$$\langle \Psi_0 \mid e\hat{\mathbf{r}} \mid \Psi_1 \rangle \simeq (\tfrac{1}{2})^{1/2}[\mathbf{M}_P + \mathbf{M}_Q + 0.125(\mathbf{M}_P + \mathbf{M}_Q)]$$

$$= (2)^{1/2}(1.125\mathbf{M}_{\text{ethylene}}) \qquad (6.66)$$

since $\mathbf{M}_P = \mathbf{M}_Q = \mathbf{M}_{\text{ethylene}}$. Thus, the oscillator strength of the lowest-energy transition of butadiene is $2(1.125)^2$ times that of the lowest-energy $\pi^* \leftarrow \pi$ transition in ethylene. The polarization of this transition is parallel to the double bonds since the individual ethylenic oscillators are so polar-

[9] Note that $W_1 + W_2 \equiv 7.60$ eV and $W_2 - W_1 \equiv 7.60$ eV; in other words, the ground-state energy of ethylene is $W_1 \equiv 0$.

ized. This is readily visualized as

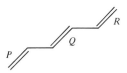

Simple vector addition indicates both x- and z-axis polarization. These predictions accord with experiment. The transition $\Psi_2 \leftarrow \Psi_0$ is, of course, forbidden.

(ii) *cis-Butadiene:*   The electronic transition moments in the two lowest-energy excited electronic states of *cis*-butadiene are shown in Fig. 6.7. Since $\cos 60° = \frac{1}{2}$, the intensity of the lowest-energy $\pi^* \leftarrow \pi$ transition of *cis*-butadiene is predicted to be $(\frac{1}{2})^2 = \frac{1}{4}$ that of *trans*-butadiene when mixing of nondegenerate states is excluded. It is also predicted to be solely x polarized. Furthermore, since the separation of oscillator centroids is smaller in *cis*- than in *trans*-butadiene, it is expected that the *cis*-compound should absorb at longer wavelengths than the *trans*-compound. All of these predictions agree with experiment.

(iii) *trans-Hexatriene:*   In terms of the oscillator identifications

the zero-order wave functions and the secular equation are contained in the matrix of Eq. (6.67) on page 231. The secular determinant of

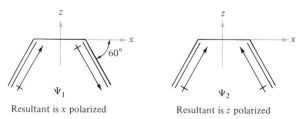

FIG. 6.7.   Polarization of the lowest-energy electronic transitions in *cis*-butadiene.

	$\lvert P_0Q_0R_0\rangle$	$\lvert P_1Q_0R_0\rangle$	$\lvert P_0Q_1R_0\rangle$	$\lvert P_0Q_0R_1\rangle$	$\lvert P_1Q_1R_0\rangle$	$\lvert P_1Q_0R_1\rangle$	$\lvert P_0Q_1R_1\rangle$	$\lvert P_1Q_1R_1\rangle$	
$\langle P_0Q_0R_0\rvert$	$3W_1-E$	0	0	0	ρ	0	ρ	0	
$\langle P_1Q_0R_0\rvert$	0	$2W_1+W_2-E$	ρ	0	0	0	0	ρ	
$\langle P_0Q_1R_0\rvert$	0	ρ	$2W_1+W_2-E$	ρ	0	0	0	0	
$\langle P_0Q_0R_1\rvert$	0	0	ρ	$2W_1+W_2-E$	0	0	0	ρ	$=0$
$\langle P_1Q_1R_0\rvert$	ρ	0	0	0	W_1+2W_2-E	ρ	0	0	
$\langle P_1Q_0R_1\rvert$	0	0	0	0	ρ	W_1+2W_2-E	ρ	0	
$\langle P_0Q_1R_1\rvert$	ρ	0	0	0	0	ρ	W_1+2W_2-E	0	
$\langle P_1Q_1R_1\rvert$	0	ρ	0	ρ	0	0	0	$3W_2-E$	

$$(6.67)$$

Eq. (6.67) is based on the Hamiltonian

$$\hat{H} = \hat{H}^P + \hat{H}^Q + \hat{H}^R + \hat{H}^{PQR} \tag{6.68}$$

where

$$\hat{H}^{PQR} \equiv \hat{H}^{PQ} + \hat{H}^{QR} \tag{6.69}$$

The interaction Hamiltonian $\hat{H}^{PR}$ is neglected because of the $1/R^3$ dependence which indicates that

$$\langle \hat{H}^{PR} \rangle \simeq (1/8)\,(\langle \hat{H}^{PQ} \rangle \text{ or } \langle \hat{H}^{QR} \rangle) \tag{6.70}$$

The eight solutions of the secular determinant are

$$E - 2W_1 - W_2 \text{ (in eV)} = -8.08;\, -2.91;\, 0;\, 2.43;\, 5.17;\, 7.60;\, 10.52;\, 15.69 \tag{6.71}$$

The lowest-energy transition is predicted to occur at 2400 Å; the observed value is 2570 Å. The predicted resonance energy is 11.1 kcal/mole.

(*iv*) *Larger Polyenes:* The secular equation becomes too large and unwieldy for larger polyenes. Recognizing this, and the fact that the interaction between the degenerate singly excited compound-oscillator wave functions is by far the dominant energy effect, we truncate the basis set of functions to those possessing single excitation only. Thus, for decapentaene, the zero-order basis set is

$$P_1Q_0R_0S_0T_0;\ P_0Q_1R_0S_0T_0;\ P_0Q_0R_1S_0T_0;\ P_0Q_0R_0S_1T_0;\ P_0Q_0R_0S_0T_1$$

The secular equation representing the interaction of this fivefold degenerate set is

$$
\begin{array}{cc}
\langle P_1Q_0R_0S_0T_0 | & \\
\langle P_0Q_1R_0S_0T_0 | & \\
\langle P_0Q_0R_1S_0T_0 | & \\
\langle P_0Q_0R_0S_1T_0 | & \\
\langle P_0Q_0R_0S_0T_1 | &
\end{array}
\begin{vmatrix}
-\gamma & 1 & 0 & 0 & 0 \\
1 & -\gamma & 1 & 0 & 0 \\
0 & 1 & -\gamma & 1 & 0 \\
0 & 0 & 1 & -\gamma & 1 \\
0 & 0 & 0 & 1 & -\gamma
\end{vmatrix} = 0 \tag{6.72}
$$

where $\hat{H}^{PQRST} \equiv \hat{H}^{PQ} + \hat{H}^{QR} + \hat{H}^{RS} + \hat{H}^{ST}$ and $(4W_1 + W_2 - E)/\rho \equiv -\gamma$. The solutions to this equation are

$$\gamma = 2\cos(\pi k/6); \quad k = 1, 2, 3, 4, 5 \tag{6.73}$$

In the general case, we find

$$\gamma = 2\cos[\pi k/(n+1)]; \quad k = 1, 2, 3, \ldots, n \tag{6.74}$$

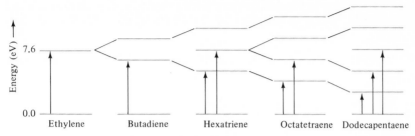

FIG. 6.8. Schematic of the number of electronic transitions with energy equal to or less than $(W_2 - W_1)$ which are predicted for some polyenic molecules.

where n is the number of double bonds in the molecule. The excitation energies are given by

$$\Delta E = W_2 - W_1 + 2\rho \cos\left[\pi k/(n+1)\right]; \qquad k = 1, 2, 3, \ldots, n \quad (6.75)$$

In this manner, the electronic energy levels of any larger polyene are readily predicted.

The prediction of prime importance concerns the *number of transitions of energy less than* $W_2 - W_1$ which is observed in a given polyene. Since the exciton splitting of the lowest-energy excited state is symmetric about the value $W_2 - W_1$ there are $n/2$ transitions with energy less than $W_2 - W_1$ when n is even and $(n-1)/2$ such transitions when n is odd. To the extent that data are available this prediction appears to agree with experiment. The predicted situations are diagrammed in Fig. 6.8.

4. LINEAR COMBINATION OF CONFIGURATIONAL WAVE FUNCTIONS (LCCW) APPROACH

It was assumed in Sec. 1 that the molecular orbitals of a molecule $PQR\ldots$ could be obtained as a linear combination of molecular orbitals localized on the fragments P, Q, $R,\ldots$. This approach was dubbed the LCMO method. There is no reason why we cannot pursue an analogous description at the configuration or state wave-function stage; in other words, we might wish to describe the electronic states of a molecule $PQR\ldots$ in terms of a linear combination of configuration wave functions appropriate to the fragments P, Q, $R,\ldots$. However, we should bear in mind that the discussions of Secs. 2 and 3 demonstrate the necessity of including also non-localized excited structures (or the equivalent electron configurations) of both the charge-transfer and molecular-exciton type.

The purpose of this section is to construct a configuration-interaction description of the ground and lower-energy excited states of the *whole molecule PQR*.... This description should include the following desirable features:

(i)   The basis set of zero-order configuration wave functions must include:

The ground configuration: $P_0Q_0R_0$....

The locally excited configurations: $P_1Q_0R_0$..., $P_0Q_1R_0$..., $P_0Q_0R_1$..., ..., $P_iQ_jR_k$..., etc. If any of the fragments are identical, say, Q and R, then the locally excited states involving excitation of Q or R should be replaced by the corresponding molecular-exciton states—or, equivalently, excitation-resonance states: $P_0(Q_1R_0 \pm Q_0R_1)S_0$..., $P_0(Q_jR_0 \pm Q_0R_j)S_0$..., etc.

The charge-transfer configurations: $P^+Q^-R_0$..., $P^-Q^+R_0$..., $P_0Q^-R^+$..., etc. If any of the fragments are identical, say, Q and R, then the local charge-transfer states should be replaced by charge-resonance states: $P_0(Q^+R^- \pm Q^-R^+)S_0$..., $P_1(Q^+R^- \pm Q^-R^+)S_j$..., etc.

This basis set is defined in more detail in Sec. 4(A) for butadiene.

(ii)   The theory should make maximum use of the available experimental information. In specific, the energies of all configurations specified in item (i) should be available from experiment—and these experimental values should constitute the diagonal elements of the energy matrix based on the set of basis functions of item (i).

(iii)   The interaction elements—the nondiagonal elements of the energy matrix—should be evaluable in the most simple possible way. For this reason, we select a one-electron Hamiltonian of the type given in Eq. (6.2), namely,

$$\hat{H}^h = \hat{H}^{hP} + \hat{H}^{hQ} + \hat{H}^{hR} + (\hat{H}^{PQR\cdots})' \qquad (6.76)$$

In order to clarify the computational procedure usually employed, we now carry through an extended discussion of the butadiene molecule—which we may represent as PQ, P and Q being ethylene fragments.

(A)   Basis Functions for Butadiene[13]

The basis functions are configurational in nature.[10] Those which appear necessary to a proper description of the lower-energy excited states of butadiene are schematized in Fig. 6.9. This set does not include any doubly excited configurations because the energy of such excitation is much too large ($\sim$14 eV). The set does include the ground configuration P_0Q_0, the locally excited configurations P_1Q_0 and P_0Q_1, and the charge-transfer configurations P^+Q^- and P^-Q^+.

---

[10] A reading of Chapters 7 and 8 would be helpful in understanding Secs. 4(A)–4(C).

The locally excited configurations $^1\Psi_{1\to-1}{}^{(1)}$ and $^1\Psi_{1'\to-1'}{}^{(1)}$ are degenerate.[11] Furthermore, our discussion of molecular-exciton theory in Sec. 3 indicates that these configurations do interact to produce excitation-resonance configurations given by

$$^1\Psi_{\mathrm{EXC}}{}^{\pm} = (\tfrac{1}{2})^{1/2}(^1\Psi_{1\to-1}{}^{(1)} \pm {}^1\Psi_{1'\to-1'}{}^{(1)}) \qquad (6.77)$$

with energies denoted

$$E(^1\Psi_{\mathrm{EXC}}{}^{\pm}) = E(^1\Psi_{1\to-1}{}^{(1)}) \pm \rho \qquad (6.78)$$

where ρ is considered negative.

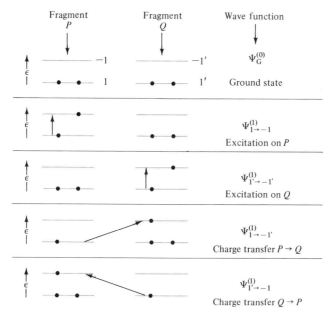

FIG. 6.9. Configuration wave functions (schematic) for butadiene. The mixing of these configurations supposedly provides a good description of the lower-energy electronic states of butadiene. The notation used is described in chapter 8; it has the following meaning: The bracketed right-hand superscript to Ψ denotes the number of electrons, relative to $\Psi_G{}^{(0)}$, which are excited. The wave function $\Psi_G{}^{(0)}$ denotes the ground configuration in which, of course, no electrons are excited. The π MO's of ethylene P are denoted 1 and -1, where the numbers 1 and -1 refer to bonding and antibonding MO's, respectively. The corresponding MO's of ethylene Q are denoted $1'$ and $-1'$. The right subscript to Ψ denotes the MO excitation of $\Psi_G{}^{(0)}$ which produces the configuration in question.

[11] The notation is explained in the caption to Fig. 6.9. The left superscript to Ψ denotes spin multiplicity.

The reason for use of $^1\Psi_{EXC}{}^{\pm}$, as opposed to either $^1\Psi_{1 \to -1}{}^{(1)}$ or $^1\Psi_{1' \to -1}{}^{(1)}$, is that the wave functions $^1\Psi_{EXC}{}^{\pm}$ are symmetry adapted[12] whereas the locally excited wave functions are not. We also note that the wave functions $^1\Psi_{EXC}{}^{\pm}$ are whole molecule rather than fragment in nature—the excitation energy being delocalized over the whole molecule rather than fixated on any one fragment.

Proper symmetry adaptation requires that we also mix the charge-transfer wave functions to obtain

$$^1\Psi_{CR}{}^{\pm} = (\tfrac{1}{2})^{1/2}[^1\Psi_{1 \to -1'}{}^{(1)} \pm {}^1\Psi_{1' \to -1}{}^{(1)}] \tag{6.79}$$

These latter wave functions are termed *charge resonance* or CR wave functions, as opposed to the term *charge transfer* (CT) which we reserve for $^1\Psi_{1 \to -1'}{}^{(1)}$ or $^1\Psi_{1' \to -1}{}^{(1)}$. As is evident from Eq. (6.79), no actual charge transfer from one fragment to another occurs in a CR state, whereas such transfer does occur in a CT state.

(B) Diagonal Elements of Energy Matrix for Butadiene

The energies of the exciton configurations are given by Eq. (6.78).It remains to evaluate ρ. At this point, we note that ρ does not necessarily equal -1.91 eV as found for *trans*-butadiene. This value, after all, is an empirical estimate based on a theory which completely neglects all charge-transfer effects. Indeed, we must return to Eq. (6.44) and use the known transition dipole moment for the $\pi^* \leftarrow \pi$ transition of ethylene in order to estimate ρ. If we use the equation[13]

$$| \mathbf{M} |^2 = 2.126 \times 10^{-30} f \nu^{-1} \text{ (esu)}^2 \text{ cm}^2 \tag{6.80}$$

where f, the oscillator strength of the transition, is approximately 0.3 and $\bar{\nu}$, the frequency of the transition, is taken to be $\sim$56 000 cm^{-1}, we find $\mathbf{M} = 3.4 \times 10^{-18}$ esu cm; the direction associated with $\mathbf{M}$ lies along the C—C axis of ethylene. All information required for the evaluation of ρ, as given by the equation

$$\rho = (M^2/R^3)(-2 \cos^2 \theta + \sin^2 \theta) \tag{6.81}$$

is available in Fig. 6.10. The value obtained[14] is $\rho \simeq -4000$ cm^{-1}.

The energy of a charge-transfer configuration P^+Q^- is given by

$$E(\Psi_{CT}) = I_P - A_Q - C \tag{6.82}$$

---

[12] The topic of symmetry-adapted wave functions is discussed in chapter 7.

[13] See chapter 9.

[14] The two ethylene fragments are too close together to satisfy the requirements of the exciton approach. The value -4000 cm^{-1} = -0.5 eV used here is that obtained by Longuet-Higgins and Murrell (Ref. [13]) by means which they claim are superior to those used in Sec. 3.

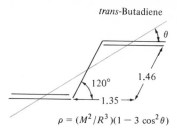

trans-Butadiene

$$\rho = (M^2/R^3)(1 - 3\cos^2\theta)$$

FIG. 6.10. Trigonometric evaluation of the exciton splitting for the lowest-energy $\pi^* \leftarrow \pi$ molecular-exciton configuration of butadiene.

where I_P is the ionization potential of fragment P; A_Q is the electron affinity of fragment Q; and C is the attractive electrostatic interaction of the excess hole on the positively charged fragment P (i.e., P^+) with the excess electron on the negatively charged fragment Q (i.e., Q^-). In the case of butadiene, we find[15]

$$E(^1\Psi_{1\rightarrow-1'}{}^{(1)}) = 10.5 - (-2.8) - C \quad \text{(in eV)} \qquad (6.83)$$

where C is now the Coulomb interaction of the hole in MO 1 with the excess electron in MO $-1'$. In the case of *trans*-butadiene, the charge density map of P^+Q^- is that shown in Fig. 6.11. Consequently, we find

$$C = (e^2/4)[(C_1C_1 \mid C_3C_3) + (C_1C_1 \mid C_4C_4)$$
$$+ (C_2C_2 \mid C_3C_3) + (C_2C_2 \mid C_4C_4)] \qquad (6.84)$$

where C_1 is a $2p\pi$ AO located on carbon number 1, etc. The integral $(C_iC_i \mid C_jC_j)$ is the standard Coulomb repulsion integral in which an electron on carbon i repels an electron on carbon j. If we use a point-charge ap-

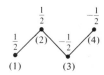

FIG. 6.11. Charge distribution in the $^1\Psi_{1\rightarrow-1'}{}^{(1)}$ configuration of butadiene. Charges are quoted in units of $\mid e \mid$. The atom numbering convention used is given by the bracketed numbers. Thus, C_1 of Eq. (6.84) refers to the $2p\pi$ AO of carbon located on center number 1.

[15] The ionization potential of ethylene is 10.5 eV. The electron affinity of ethylene is calculated as -2.8 eV.

proximation[16] and Coulomb's law, we find $C = 6.4\,\mathrm{eV}$. Hence, $E(^1\Psi_{1\to-1'}{}^{(1)}) = 6.9$ eV. The energy of $E(^1\Psi_{1'\to-1}{}^{(1)})$ is also 6.9 eV. It is shown in Sec. 4(C) that

$$E(^1\Psi_{1\to-1'}{}^{(1)}) = E(^1\Psi_{1'\to-1}{}^{(1)}) = E(^1\Psi_{\mathrm{CR}}{}^+) = E(^1\Psi_{\mathrm{CR}}{}^-) \qquad (6.85)$$

(C)   Off-Diagonal Elements of Butadiene Energy Matrix

Having defined the diagonal matrix elements associated with the charge- and excitation-resonance configurations, we now adopt the Hamiltonian of Eqs. (6.2) and (6.3). This Hamiltonian consists of one-electron parts: $\hat{H}^{hP}$ operative only on fragment P, $\hat{H}^{hQ}$ operative only on fragment Q, and $\hat{H}_{pq}'$ operative only at the junction-bond pq. Consequently, $\hat{H}^h$ cannot mix configurations which differ from each by more than two spin orbitals.[17] Furthermore, since the MO's on P are already orthogonal with respect to $\hat{H}^{hP}$, and those on Q with respect to $\hat{H}^{hQ}$, and since $\hat{H}^{hQ}$ is ineffective in P or $\hat{H}^{hP}$ in Q, we can, henceforth, restrict attention to configuration mixing induced solely by $\hat{H}_{pq}'$.

At this point, it is well to stress that the energies of Eqs. (6.78) and (6.82) for both the charge-transfer and molecular-exciton configurations contain specific input of two-electron information. Hence, neither Eq. (6.78) nor Eq. (6.82) can result from the use of a Hamiltonian such as Eqs. (6.2) or (6.76). Indeed, the formulation of the Hamiltonian of Eqs. (6.2) and (6.76) makes sense only for the off-diagonal elements.

Finally, before departing the subject of matrix elements, it is well to note that, with respect to $\hat{H}_{pq}'$, the two configurations $^1\Psi_{1\to-1'}{}^{(1)}$ and $^1\Psi_{1'\to-1}{}^{(1)}$ do not interact; in other words, we wish to prove the validity of Eq. (6.85). Reference to Fig. 6.9 indicates that these configurations differ in the occupancy of four spin orbitals. Thus, the element $\langle\Psi_{1\to-1'}{}^{(1)} \mid \hat{H}_{pq}' \mid \Psi_{1'\to-1}{}^{(1)}\rangle$ involves the molecular-orbital integral $\langle\varphi_1\varphi_{-1'} \mid \hat{H}_{pq}' \mid \varphi_{-1}\varphi_{1'}\rangle$ which cannot be rendered nonzero unless $\hat{H}_{pq}'$ is, at least, a two-electron operator.

We now illustrate the evaluation of some off-diagonal energy matrix elements.

(i)   *Matrix Element* $\langle^1\Psi_{\mathrm{CR}}{}^+ \mid \hat{H}_{pq}' \mid {}^1\Psi_G{}^{(0)}\rangle$:   This element may be written as

$$\langle^1\Psi_{\mathrm{CR}}{}^+ \mid \hat{H}_{pq}' \mid {}^1\Psi_G{}^{(0)}\rangle = 2^{-1/2}\{\langle^1\Psi_{1\to-1'}{}^{(1)} \mid \hat{H}_{pq}' \mid {}^1\Psi_G{}^{(0)}\rangle$$
$$+ \langle^1\Psi_{1'\to-1}{}^{(1)} \mid \hat{H}_{pq}' \mid {}^1\Psi_G{}^{(0)}\rangle\} \qquad (6.86)$$

---

[16] The electron is supposed to be completely localized at the origin of coordinates of the AO.

[17] See chapter 8.

Reference to chapter 8 enables us to expand the first integral on the right-hand side of Eq. (6.86) as

$$\langle {}^1\Psi_{1\to1'}{}^{(1)} \mid \hat{H}_{pq}' \mid {}^1\Psi_G{}^{(0)} \rangle = 2^{-1/2}\langle \{\mid 1\alpha(1) - 1'\beta(2)\,1'\alpha(3)\,1'\beta(4) \mid$$
$$- \mid 1\beta(1) - 1'\alpha(2)\,1'\alpha(3)\,1'\beta(4) \mid\} \mid \hat{H}_{pq}' \mid$$
$$\times \mid 1\alpha(1)\,1\beta(2)\,1'\alpha(3)\,1'\beta(4) \mid\rangle \qquad (6.87)$$

Straightforward manipulation finally yields

$$\langle {}^1\Psi_{1\to1'}{}^{(1)} \mid \hat{H}_{pq}' \mid {}^1\Psi_G{}^{(0)} \rangle = 2^{-1/2}[\langle -1' \mid \hat{H}_{pq}' \mid 1\rangle + \langle -1' \mid \hat{H}_{pq}' \mid 1\rangle] \quad (6.88)$$

The second integral on the right-hand side of Eq. (6.86), when expanded in determinant form, becomes

$$\langle {}^1\Psi_{1'\to-1}{}^{(1)} \mid \hat{H}_{pq}' \mid {}^1\Psi_G{}^{(0)} \rangle = 2^{-1/2}\langle\{\mid 1\alpha(1)\,1\beta(2)\,1'\alpha(3) - 1\beta(4)\mid$$
$$- \mid 1\alpha(1)\,1\beta(2)\,1'\beta(3) - 1\alpha(4)\mid\} \mid \hat{H}_{pq}' \mid$$
$$\times \mid 1\alpha(1)\,1\beta(2)\,1'\alpha(3)\,1'\beta(4) \mid\rangle \qquad (6.89)$$
$$= 2^{-1/2}[\langle -1 \mid \hat{H}_{pq}' \mid 1'\rangle + \langle -1 \mid \hat{H}_{pq}' \mid 1'\rangle] \qquad (6.90)$$

Collecting all terms into Eq. (6.86), one finally obtains

$$\langle {}^1\Psi_{CR}{}^+ \mid \hat{H}_{pq}' \mid {}^1\Psi_G{}^{(0)} \rangle = \langle -1' \mid \hat{H}_{pq}' \mid 1\rangle + \langle -1 \mid \hat{H}_{pq}' \mid 1'\rangle \quad (6.91)$$

The MO's of the fragments P and Q are given in Fig. 6.2; using these, we expand the MO integrals of Eq. (6.91) into AO integrals, to find[18]

$$\langle {}^1\Psi_{CR}{}^+ \mid \hat{H}_{pq}' \mid {}^1\Psi_G{}^{(0)} \rangle = \tfrac{1}{2}\langle(\chi_3 - \chi_4) \mid \hat{H}_{23}' \mid (\chi_1 + \chi_2)\rangle$$
$$+ \tfrac{1}{2}\langle(\chi_1 - \chi_2) \mid \hat{H}_{23}' \mid (\chi_3 + \chi_4)\rangle$$
$$= \tfrac{1}{2}(\beta_{23} - \beta_{23}) = 0 \qquad (6.92)$$

(ii) Matrix Element $\langle {}^1\Psi_{CR}{}^- \mid \hat{H}_{pq}' \mid {}^1\Psi_G{}^{(0)} \rangle$: The wave function ${}^1\Psi_{CR}{}^-$ differs from ${}^1\Psi_{CR}{}^+$ with respect to but one sign [see Eq. (6.79)]. The effect of this sign difference is such that we find

$$\langle {}^1\Psi_{CR}{}^- \mid \hat{H}_{pq}' \mid {}^1\Psi_G{}^{(0)} \rangle = \langle -1' \mid \hat{H}_{pq}' \mid 1\rangle - \langle -1 \mid \hat{H}_{pq}' \mid 1'\rangle$$
$$= 2\beta_{23}/2 = \beta_{23} \qquad (6.93)$$

---

[18] This same result could have been obtained group theoretically without any necessity of going through the labor of Eqs. (6.86)–(6.92).

(iii)   *Other Matrix Elements:* Straightforward application of the method outlined in items (i) and (ii) yields

$$\langle {}^1\Psi_{CR}{}^- \mid \hat{H}_{pq}{}' \mid {}^1\Psi_{EXC}{}^- \rangle = 0$$

$$\langle {}^1\Psi_{CR}{}^+ \mid \hat{H}_{pq}{}' \mid {}^1\Psi_{EXC}{}^+ \rangle = -\beta_{23}$$

$$\langle {}^1\Psi_{CR}{}^+ \mid \hat{H}_{pq}{}' \mid {}^1\Psi_{EXC}{}^- \rangle = 0$$

$$\langle {}^1\Psi_{CR}{}^- \mid \hat{H}_{pq}{}' \mid {}^1\Psi_{EXC}{}^+ \rangle = 0$$

$$\langle {}^1\Psi_{EXC}{}^+ \mid \hat{H}_{pq}{}' \mid {}^1\Psi_{G}{}^{(0)} \rangle = 0$$

$$\langle {}^1\Psi_{EXC}{}^- \mid \hat{H}_{pq}{}' \mid {}^1\Psi_{G}{}^{(0)} \rangle = 0$$

$$\langle {}^1\Psi_{EXC}{}^+ \mid \hat{H}_{pq}{}' \mid {}^1\Psi_{EXC}{}^- \rangle = 0$$

$$\langle {}^1\Psi_{CR}{}^+ \mid \hat{H}_{pq}{}' \mid {}^1\Psi_{CR}{}^- \rangle = 0 \tag{6.94}$$

(D) Energy Matrix for Butadiene

The energy matrix is

	$\mid {}^1\Psi_{EXC}{}^+ \rangle$	$\mid {}^1\Psi_{CR}{}^+ \rangle$	$\mid {}^1\Psi_{EXC}{}^- \rangle$	$\mid {}^1\Psi_{CR}{}^- \rangle$	$\mid {}^1\Psi_{G}{}^{(0)} \rangle$
$\langle {}^1\Psi_{EXC}{}^+ \mid$	$E({}^1\Psi_{1\to-1}{}^{(1)})+\rho$	$-\beta_{23}$	0	0	0
$\langle {}^1\Psi_{CR}{}^+ \mid$	$-\beta_{23}$	I_P-A_Q-C	0	0	0
$\langle {}^1\Psi_{EXC}{}^- \mid$	0	0	$E({}^1\Psi_{1\to-1}{}^{(1)})-\rho$	0	0
$\langle {}^1\Psi_{CR}{}^- \mid$	0	0	0	I_P-A_Q-C	β_{23}
$\langle {}^1\Psi_{G}{}^{(0)} \mid$	0	0	0	β_{23}	0

$$\tag{6.95}$$

All quantities in this matrix, with the exception of β_{23}, have already been evaluated. We may now pursue two courses: We may solve the secular determinant and obtain the resulting energy levels as functions of β; we may then consider β as a parameter which is to be determined by matching the levels obtained to experiment. A good match with experiment for the lowest-energy electronic transition is obtained when $\beta_{23} = -1.68$ eV. In the second approach, we may simply adopt whatever value or values for β_{23} which have been provided by other authors for the C_2—C_3 bond of butadiene. It is perhaps coincidental that the value of β_{23} used by Pariser and Parr[14] in another connection is identical to the value of -1.68 eV as adduced above.

(E) State Energies and Wave Functions for Butadiene

The singlet state energies and wave functions obtained with $\beta \equiv -1.68$ eV are shown in Fig. 6.12.

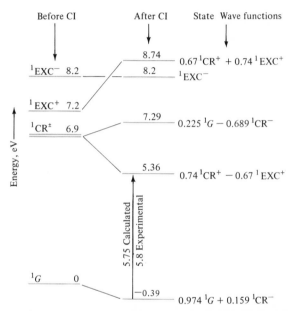

FIG. 6.12. Singlet energies and wave functions for butadiene for the situations prevailing before and after configuration interaction. The notation used is abbreviated, for example, $^1\Psi_{CR}{}^+$ is now denoted CR^+, etc.

EXERCISES

1. The resonance interaction in a simple amide is discussed in Sec. 2(D)(i). Both resonance structures correspond to shifting the loci of the π electrons while keeping all σ electrons fixed. Hence, in this molecule, as well as in all others discussed in Sec. 2, we can evaluate energies corresponding to $\pi^* \leftarrow \pi$ excitations only. Now, it is well known that the lowest-energy excitation of amides is not $\pi^* \leftarrow \pi$, but rather $\pi^* \leftarrow n$ in nature. Given that the $\pi^* \leftarrow n$ excitation of the carbonyl group in acetone is $\sim$37 kK, what predictions can you make concerning the energy

of the $\pi^* \leftarrow n$ transition of a simple amide? (The symbol n denotes a nonbonding atomic orbital containing both s and p character which is localized on the oxygen atom.)

Answer: The $\pi^* \leftarrow n$ excitation of an amide corresponds to

$$\ldots \pi_N{}^2 \pi_{CO}{}^2 n_O{}^2 \rightarrow \ldots \pi_N{}^2 \pi_{CO}{}^2 n_O \pi_{CO}{}^*$$

The bonding in the ground-state electron configuration is identifiable with resonance structure $|\,1'\rangle$; the bonding in the excited configuration is not identifiable with either resonance structure $|\,1'\rangle$ or $|\,2'\rangle$. The situation prior to resonance is depicted on the left of Fig. 6.13. The situation after resonance is depicted on the right of Fig. 6.13. The $\ldots n\pi^*$ configuration is not affected by the resonance process. The predicted and experimental excitation energies are

$\pi^* \leftarrow n \ldots$ predicted $= 46.2$ kK;   observed $\simeq 48$ kK

$\pi^* \leftarrow \pi \ldots$ predicted $= 53.9$ kK;   observed $\simeq 54$ kK   (6.96)

Thus, the blue shift of the $\pi^* \leftarrow n$ transition in amides is associated with a resonance stabilization of the ground states.

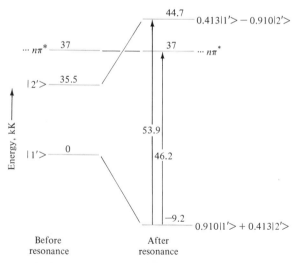

FIG. 6.13.   Resonance interaction in a simple amide. An interesting feature of this diagram, and the discussion pertaining to it, is the manner in which it correlates molecular-orbital and resonance structure viewpoints. Energy units are kiloKaysers (kK).

2. The dipole moments of structures $|1'\rangle$ and $|2'\rangle$ of Sec. 2(D)(i) are readily estimated by methods akin to that of Chap. 3. The estimated values are

$$\mathbf{M}_{1'} = (1 \text{ to } 2) \text{ D}; \quad \mathbf{M}_{2'} \sim 11 \text{ D} \tag{6.97}$$

and both moments are codirectional. Estimate the dipole moment of a simple amide.

Answer: The normalized ground-state wave function is given in Fig. 6.13. The total dipole moment is

$$\langle \hat{\mathbf{M}} \rangle = (0.910\langle 1' | + 0.413\langle 2' |) | \hat{\mathbf{M}} | (0.910 | 1'\rangle + 0.413 | 2'\rangle)$$

$$= 0.83\langle 1' | \hat{\mathbf{M}} | 1'\rangle + 0.17\langle 2' | \hat{\mathbf{M}} | 2'\rangle$$

$$\simeq 0.83 \times 1 + 0.17 \times 11 \simeq 3 \text{ D} \tag{6.98}$$

The observed moment is 3.2 D.

3. (a) Use the resonance interaction approach to write down the secular determinant of urea.
(b) Find the energy of the lowest-energy $\pi^* \leftarrow \pi$ electronic excitation in urea.

Answer: (a) The principal resonance structures are

The secular determinant is

$$
\begin{array}{c}
\quad\quad |1''''\rangle \quad |2''''\rangle \quad |3''''\rangle \\
\begin{array}{c}
\langle 1''''| \\[4pt]
\langle 2''''| \\[4pt]
\langle 3''''|
\end{array}
\begin{vmatrix}
0 - E & \beta & \beta \\
\beta & \alpha - E & \gamma \\
\beta & \gamma & \alpha - E
\end{vmatrix} = 0
\end{array}
\tag{6.99}
$$

(b) The lowest-energy electronic transition should occur at

$$\Delta E = [(\alpha - \gamma)^2 + 8\beta^2]^{1/2} \simeq 58.5 \text{ kK} \tag{6.100}$$

The experimental value is $\sim$58.4 kK.

4. Consider the grouping of three harmonic oscillators shown in Fig. 6.14. The potential energy of interaction of this assemblage is

$$V = \frac{-2e^2x_1x_2}{R^3} - \frac{2e^2x_2x_3}{R^3} - \frac{e^2x_1x_3}{4R^3} \qquad (6.101)$$

(a) Show that the secular equation for the exciton states of the assemblage is

$$
\begin{array}{l}
| \, 1^\dagger 2 \, 3 \rangle \\[2em]
| \, 1 \, 2^\dagger 3 \rangle \\[2em]
| \, 1 \, 2 \, 3^\dagger \rangle
\end{array}
\begin{vmatrix}
\dfrac{5}{2}h\nu - E & -\dfrac{2M^2}{R^3} & -\dfrac{M^2}{4R^3} \\[1.5em]
-\dfrac{2M^2}{R^3} & \dfrac{5}{2}h\nu - E & -\dfrac{2M^2}{R^3} \\[1.5em]
-\dfrac{M^2}{4R^3} & -\dfrac{2M^2}{R^3} & \dfrac{5}{2}h\nu - E
\end{vmatrix} = 0 \qquad (6.102)
$$

where the oscillators are denoted 1, 2, and 3 and where the dagger indicates the presence of one unit of excitation equal to $h\nu$ on the oscillator so superscripted.

(b) What corrections to the *trans*-hexatriene secular equation are suggested?

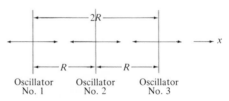

FIG. 6.14. Grouping of oscillators for Exercise 6.4.

5. The absorption spectra of the monomeric and dimeric forms of 1,1'-diethyl-2,2'-pyridocyanine perchlorate are shown in Fig. 6.15. Assume that the pyridocyanine dye may be represented by a large flat platelet as shown and that the monomer transition is in-plane polarized with $M/e = 1.4$ Å. Calculate the separation of molecules in the dimeric form, assuming this form has the structure also shown in Fig. 6.15.

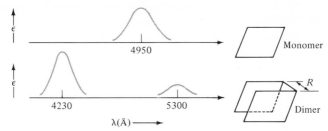

FIG. 6.15. Absorption spectra of monomeric and dimeric forms of 1,1'-diethyl-2,2'-pyridocyanine perchlorate. The monomer- and dimer-structure representations are schematized on the right of the diagram.

Hint: The appropriate coupling potential is

$$V/e^2 = (x_1 x_2 + y_1 y_2)/R^3 \qquad (6.103)$$

This specific problem has been discussed by Levison, Curtis, and Simpson[19]; these authors should be consulted for further detail.

Answer: It is found that R equals 4.6 Å.

6. Evaluate Eq. (6.84) for the specific case of the lowest-energy CT state of *cis*-butadiene. Represent the atomic-orbital charge distribution by a point charge located at the origin of coordinates and use Coulomb's law.

Answer: 6.7 eV.

7. Use the LCCW approach to ascertain energies and wave functions for the lowest-energy triplet state of *trans*-butadiene.

Answer: The relevant secular equation is given by

$$
\begin{array}{c}
{}^3\text{EXC}^+ \\
{}^3\text{CR}^+ \\
{}^3\text{EXC}^- \\
{}^3\text{CR}^-
\end{array}
\begin{vmatrix}
E' - E & -\beta_{23} & 0 & 0 \\
-\beta_{23} & I - A - C - E & 0 & 0 \\
0 & 0 & E' - E & 0 \\
0 & 0 & 0 & I - A - C - E
\end{vmatrix} = 0
$$

$$(6.104)$$

[19] Levinson, Curtis, and Simpson, J. Am. Chem. Soc. **79**, 4314 (1957).

No exciton splitting of the $^3exc^\pm$ configurations is shown. (Why is this so?) If the experimental value of 4.6 eV for the lowest triplet state of ethylene is identified with E', if we follow Sec. 4(D) and set $\beta = -1.68$ eV, and if we take $^3CR^\pm = {}^1CR^\pm = 6.9$ eV, we find the lowest-energy triplet state at 3.7 eV. The experimental value is 3.5 eV. The wave function for the lowest triplet state is $0.88(^3EXC^+) - 0.48(^3CR^+)$.

BIBLIOGRAPHY

[1]   M. J. S. Dewar, *Proc. Cambridge Phil. Soc.* **45,** 639 (1949); *J. Chem. Soc.* 2329 (1950); *J. Am. Chem. Soc.* **74,** 3341 (1952); **74,** 3345 (1952); **74,** 3350 (1952); **74,** 3353 (1952); **74,** 3355 (1952); **74,** 3357 (1952).

[2]   H. Suzuki, *Electronic Absorption Spectra and Geometry of Molecules* (Academic Press Inc., New York, 1967); Chap. 19.

[3]   H. Suzuki, *Bull. Chem. Soc. Japan* **35,** 1853 (1962).

[4]   E. Heilbronner, in *Molecular Orbitals in Chemistry, Physics and Biology* edited by P. O. Löwdin and B. Pullman (Academic Press Inc., New York, 1964), p. 329ff.

[5]   L. Pauling, *J. Chem. Phys.* **1,** 280 (1933).

[6]   G. W. Wheland, *Resonance in Organic Chemistry* (John Wiley & Sons, Inc., New York, 1955).

[7]   O. Klement and O. Mäder, *Helv. Chim. Acta* **46,** 1 (1963).

[8]   M. Simonetta and V. Schomaker *J. Chem. Phys.* **19,** 649 (1951).

[9]   E. J. Rosa and W. T. Simpson in *Physical Processes in Radiation Biology,* edited by L. Augenstein, R. Mason, and B. Rosenberg (Academic Press, New York, 1964), p. 43ff.

[10] W. Kauzmann, *Quantum Chemistry* (Academic Press, Inc., New York, 1957), p. 505ff.

[11] L. Pauling and E. B. Wilson, Jr., *Introduction to Quantum Mechanics* (McGraw-Hill Book Co., Inc., New York, 1935), p. 82ff.

[12] W. T. Simpson, *J. Am. Chem. Soc.* **73,** 5363 (1951); **77,** 6164 (1955).

[13] H. C. Longuet-Higgins and J. N. Murrell, *Proc. Phys. Soc. (London)* **68A,** 601 (1955).

[14] R. Pariser and R. G. Parr, *J. Chem. Phys.* **21,** 466 (1963); **21,** 767 (1963).

General References

1.   H. Suzuki, *Electronic Absorption Spectra and Geometry of Molecules* (Academic Press, Inc., New York, 1967), Chaps. 19–23.

2.   J. N. Murrell, *The Theory of The Electronic Spectra of Organic Molecules* (John Wiley & Sons, Inc., New York, 1963), Chaps. 8–10.

3.   H. C. Longuet-Higgins and J. N. Murrell, *Proc. Phys. Soc. (London)* **A68,** 601 (1955); J. N. Murrell, *J. Chem. Phys.,* **37,** 1162 (1962).

4. S. Nagakura, M. Kojima, and Y. Maruyama, *J. Mol. Spectry.* **13,** 174 (1964).
5. W. T. Simpson, *Theories of Electrons in Molecules* (Prentice-Hall, Inc., Englewood Cliffs, N.J., 1962).
6. A. S. Davydov, *Theory of Molecular Excitons*, translated by M. Oppenheimer and M. Kasha (McGraw-Hill, Inc., New York, 1962).
7. D. P. Craig and S. H. Walmsley, *Excitons in Molecular Crystals* (W. H. Benjamin, Inc., New York, 1969).
8. S. P. McGlynn, A. T. Armstrong, and T. Azumi, *Modern Quantum Chemistry*, edited by O. Sinanoğlu (Academic Press, Inc., New York, 1965), Vol. III, Chap. B6.

CHAPTER 7

Symmetry- and Spin-Adapted Wave Functions

The construction of symmetry- and spin-adapted antisymmetrized wave functions—ones which transform as the Fermi representation and which possess quantum characteristics S, M_S, and Γ—is oftentimes a very important preliminary step in quantum-chemical computations. It is the aim of this chapter to construct a reasonably systematic way for generating such wave functions. Toward this end, we, insofar as is possible, use projection operator[1] techniques. The projection operator concept is based on the fact that one can construct operators $\hat{O}(S_k)$ or $\hat{O}[\Gamma_i]$ which, when allowed to act sequentially on an arbitrary many-electron spin-orbital trial wave function, either project out of it the desired eigenfunction—one with spin angular momentum quantum number S_k and with orbital transformation properties identical to that of the point group representation Γ_i—or completely annihilate it.

The plan of the chapter is as follows:

(i) The construction of symmetry-adapted (one-electron) molecular orbitals is discussed in Sec. 1. These MO's provide the basis set of space

---

[1] The projection operator concept is quite old. It seems to have reached its full maturity in the work of Löwdin on spin-projection operators (Refs. [1] and [2]).

orbitals to be used in generating the final fully adapted many-electron wave functions.

(ii)   Antisymmetrization is discussed in Sec. 2. We do not use the projection operator technique here—the usual exchange antisymmetry requirement (that is, the Pauli principle) leading to the Slater determinant formulation is adequate for our purposes. However, we note in passing that a projection operator capable of generating a wave function which transforms as the Fermi representation of an appropriate permutation group can be constructed and that this operator is useful in certain types of valence-bond calculations.

(iii)   Spin adaptation is broached in Sec. 3.

(iv)   The results of items (i)–(iii) are combined in Sec. 4 in order to generate fully adapted single-configuration wave functions. This exercise is trivial if no orbital degeneracies exist in item (i); if orbital degeneracies do occur, certain difficulties arise which usually require a final orbital-symmetry readaptation. This latter effort can be complicated; as a result, we present an exercise exemplifying degeneracies in Sec. 5.

1.   SYMMETRY-ADAPTED MOLECULAR-ORBITAL WAVE FUNCTIONS

The atomic centers in any molecule can be classified into two types:

(i)   Those atoms which, when subjected to the symmetry operations of the molecular point group, change position and fill molecular sites previously occupied by other atoms of the same kind. An example of such centers is afforded by the two oxygen atoms of $SO_2(C_{2v})$, or the four hydrogen atoms of methane (T_d), or the six carbon atoms of benzene (D_{6h}). A molecule may sometimes contain more than one group of such atoms; a case in point is benzene which consists of two such groups: The six H atoms and the six C atoms. The importance of such collections or groups of atoms lies in the fact that those parts of the molecular orbitals which span the collection are fully determinable from symmetry alone. Thus, in benzene, those parts of the MO's which span the six hydrogen atoms are determinable from symmetry considerations alone, as are also those parts of the MO's which span the six carbon atoms. These *partial* molecular orbitals are termed *group orbitals* (GO). In some instances, the group orbital spans all centers in the molecule. This is the case in $O_2(D_{\infty h})$. When this occurs, the terms *group orbital* and *molecular orbital* become synonymous.

(ii)   Those atoms which, when subjected to the symmetry operations of the molecular point group, do not change position. The carbon atom of methane provides an appropriate example. The orbitals on these centers may also be considered to be group orbitals—ones, however, which span only one center.

(A)   Construction of Group Orbitals

The group orbitals are readily determined using the well-known equation[3]2

$$\varphi(\Gamma_i) = \sum_{\hat{R}} \overline{\chi}_i(\hat{R})\,\hat{R}\chi \tag{7.1}$$

where $\varphi(\Gamma_i)$ denotes an MO which transforms as the *ith* representation Γ_i, $\overline{\chi}_i(\hat{R})$ is the character of the *ith* representation for the operator $\hat{R}$, $\hat{R}$ is any one of the symmetry operators of the group, and χ is any one of the atomic orbitals in the basis set. Equation (7.1) may be written in a projection operator formalism as

$$\varphi(\Gamma_i) = \hat{O}[\Gamma_i]\chi \tag{7.2}$$

where $\hat{O}[\Gamma_i]$ is an operator which projects the $\varphi(\Gamma_i)$ out of the basis function χ. Thus, the projection operator is given by

$$\hat{O}[\Gamma_i] = \sum_{\hat{R}} \overline{\chi}_i(\hat{R})\,\hat{R} \tag{7.3}$$

where the summation is over all operators—not merely classes of operators—of the molecular point group.

Since the generation of group orbitals is a relatively straightforward affair, we do no more than provide a few procedural examples. We treat the water molecule in Sec. 1(B); we have performed MWH calculations on H_2O in chapter 4 and the group-theoretic results to be obtained here have already been referred to there. The σ MO's of H_2 are discussed in Sec. 4(A). The π MO's of benzene, unlike those of H_2 or the GO's of H_2O, contain degeneracies; this produces certain small difficulties which are emphasized in Sec. 5(A) where we generate the benzene π MO's.

(B)   Water Molecule: An Example Calculation

The water molecule is diagrammed in Fig. 7.1. It possesses C_{2v} symmetry. The C_{2v} point group is given in Table 7.1. The basis set of atomic orbitals is

$$1s_{H1},\ 1s_{H2},\ 2s_O,\ 2p_{xO},\ 2p_{yO},\ 2p_{zO}$$

The projection operators obtained from Eq. (7.3) and Table 7.1 are

$$\hat{O}[A_1] = E + C_2 + \sigma_v + \sigma_v'$$

$$\hat{O}[A_2] = E + C_2 - \sigma_v - \sigma_v'$$

$$\hat{O}[B_1] = E - C_2 + \sigma_v - \sigma_v'$$

$$\hat{O}[B_2] = E - C_2 - \sigma_v + \sigma_v' \tag{7.4}$$

---

2 Note the usage of the cap symbol, as in $\overline{\chi}_i$, to distinguish the character of a matrix from an atomic orbital (which we denote χ or χ_μ).

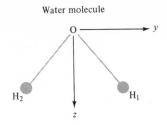

Water molecule

x axis is out of plane

FIG. 7.1. Definition of axes and atom locations for the water molecule.

We now allow these projection operators to act on the basis set of atomic orbitals. We tabulate the results below for the $\hat{O}[A_1]$ projection operator:

(i) $\hat{O}[A_1]1s_{\text{H1}} = E(1s_{\text{H1}}) + C_2(1s_{\text{H1}}) + \sigma_v(1s_{\text{H1}}) + \sigma_v{}'(1s_{\text{H1}})$

$$= 1s_{\text{H1}} + 1s_{\text{H2}} + 1s_{\text{H1}} + 1s_{\text{H2}} \qquad (7.5)$$

Upon normalizing—using the approximation of zero differential overlap—we find

$$\varphi(a_1) = (\tfrac{1}{2})^{1/2}(1s_{\text{H1}} + 1s_{\text{H2}}) \qquad (7.6)$$

(ii)   Operation with $\hat{O}[A_1]$ on $1s_{\text{H2}}$ yields the same group orbital as Eq. (7.6). The $\hat{O}[A_1]1s_{\text{H2}}$ wave function, therefore, is redundant.

(iii) $\hat{O}[A_1]2s_{\text{O}} = 4 \times 2s_{\text{O}} \qquad (7.7a)$

Normalization yields

$$\varphi(a_1) = 2s_{\text{O}} \qquad (7.7b)$$

(iv) $\hat{O}[A_1]2p_{x\text{O}} = 2p_{x\text{O}} - 2p_{x\text{O}} - 2p_{x\text{O}} + 2p_{x\text{O}} = 0 \qquad (7.8)$

TABLE 7.1. C_{2v} *point group.*

C_{2v}	E	C_2	$\sigma_v(xz)$	$\sigma_v{}'(yz)$	
A_1	1	1	1	1	$z,\ x^2,\ y^2,\ z^2$
A_2	1	1	-1	-1	xy
B_1	1	-1	1	-1	$x,\ xz$
B_2	1	-1	-1	1	$y,\ yz$

TABLE 7.2. *Transformation properties of orbital basis set for water.*

Γ_i	Oxygen AO's	Hydrogen GO's
a_1	$2s_O$	$(\frac{1}{2})^{1/2}(2s_{H1} + 2s_{H2})$
	$2p_{zO}$	
a_2	$\cdots$	$\cdots$
b_1	$2p_{xO}$	$\cdots$
b_2	$2p_{yO}$	$(\frac{1}{2})^{1/2}(2s_{H1} - 2s_{H2})$

The $2p_{xO}$ atomic orbital does not contain any A_1 character since it is impossible to project any function of type A_1 out of $2p_{xO}$.

(v) $\hat{O}[A_1]2p_{yO} = 2p_{yO} - 2p_{yO} + 2p_{yO} - 2p_{yO} = 0$ (7.9)

(vi) $\hat{O}[A_1]2p_{zO} = 2p_{zO} + 2p_{zO} + 2p_{zO} + 2p_{zO}$ (7.10a)

Normalization yields

$$\varphi(a_1) = 2p_{zO}$$ (7.10b)

Continuing in like manner with the three remaining projection operators, eliminating redundancies, and normalizing, we find the results of Table 7.2. The results of Table 7.2 are the ultimate insofar as symmetry is concerned. Further progress toward complete MO specification is dependent on the doing of an actual quantum-chemical calculation. In any event,[3] since three linearly independent basis functions transform as a_1, it is clear that three linearly independent MO's of symmetry a_1 result. Two b_2 MO's arise for similar reasons. The $2p_{xO}$ atomic orbital cannot mix with any hydrogen orbitals and it remains nonbonding. The net effect of group theory then, is to reduce the original 6×6 determinant based on Table 4.12 to a set composed of three determinants—one 3×3, one 1×1, and one 2×2. In the case of water, this simplification does not produce any remarkable decrease of labor. However, as the molecular size increases and as the symmetry of the molecule increases, the labor evaded can be significant.

2. DETERMINANTAL WAVE FUNCTIONS

An assignment of electrons to individual molecular orbitals specifies an orbital electron configuration. The assignment in question must conform

---

[3] At this point, we note that $\hat{H}$ transforms as A_1. Consequently, nondiagonal elements in the secular determinant which connect basis functions of different transformation properties are zero. Thus, only basis functions which transform identically can mix.

to the requirement that no more than two electrons may occupy any non-degenerate molecular orbital; furthermore, if two electrons occupy such an MO, these two electrons must have different spin wave functions. A specification of an orbital configuration on which is further superposed a specification of the individual electron spin wave functions, α or β, defines a spin-orbital electron configuration. This last wave function, when properly antisymmetrized, is designated here as a *single-configuration wave function*.

Let us consider, as a first example, a molecule whose MO's are all non-degenerate. These MO's, in order of increasing energy, are designated φ_1, φ_2, $\varphi_3, \ldots \varphi_k, \varphi_l, \ldots, \varphi_n, \ldots, \varphi_s, \ldots$. The symbolization $\varphi_k \alpha$ and $\varphi_k \beta$ denotes spin wave functions $\alpha(\omega)$ or $\beta(\omega)$, respectively, associated with the electron in the orbital φ_k. We assume that the orbital motion of any one electron is independent of the orbital motion of all other electrons and of its own and all other electron spin motions. We assume that the spin of any one electron is independent of the spin of all other electrons and of its own and all other electron orbital motions. These last two assumptions (i.e., an *independent-systems* assumption) imply that the many-electron spin-orbital wave function can be written as a product of one-electron spin-orbital wave functions.

The Pauli principle requires that the many-electron spin-orbital wave function be antisymmetric with respect to the interchange of coordinates of all pairs of electrons. In the specific case of a single-configuration wave function, the Pauli principle dictates that each spin orbital has a maximum occupancy of only one electron. In other words, a given spin orbital is either occupied by one electron or is not occupied at all—a prescription which provides a good initiation point for the generation of a single-configuration wave function. Thus, the lowest-energy many-electron wave function of a two-electron molecule might be written

$$\Psi' = \varphi_1 \alpha(1) \varphi_1 \beta(2) \tag{7.11}$$

This wave function satisfies the Pauli occupation maxim; nonetheless, it is not adequate because it is not antisymmetric with respect to electron interchange. The correct single-configuration wave function, properly normalized, is in fact given by

$$\Psi = (\tfrac{1}{2})^{1/2} [\varphi_1 \alpha(1) \varphi_1 \beta(2) - \varphi_1 \beta(1) \varphi_1 \alpha(2)] \tag{7.12}$$

(A) Slater-Determinant Wave Functions

The many-electron wave function must be antisymmetric with respect to the interchange of coordinates of any two electrons. In other words, it is required that

$$\hat{P}_{ij} \Psi = -\Psi \tag{7.13}$$

where $\hat{P}_{ij}$ is the permutation operator which interchanges the coordinates of any two electrons i and j. For the two-electron case of Eq. (7.11), we find

$$\hat{P}_{12}\Psi' = \varphi_1\alpha(2)\varphi_1\beta(1) = \varphi_1\beta(1)\varphi_1\alpha(2) \tag{7.14}$$

which is by no means identical to Ψ'. However, an appropriate linear combination of these two wave functions, namely, $\Psi = \Psi' - \hat{P}_{12}\Psi'$ leads, when normalized, to Eq. (7.12) and it may be verified that

$$\hat{P}_{12}\Psi = -\Psi \tag{7.15}$$

Thus, the wave function of Eq. (7.12) represents a satisfactory configuration wave function.

The wave function of Eq. (7.12) is naturally expressible in determinant form as

$$\Psi = (2)^{-1/2} \begin{vmatrix} \varphi_1\alpha(1) & \varphi_1\beta(1) \\ \varphi_1\alpha(2) & \varphi_1\beta(2) \end{vmatrix} \tag{7.16}$$

This wave function is known as a Slater-determinant wave function.[4]

In the case of three electrons, the lowest-energy many-electron wave function based on the occupancy criterion is

$$\Psi' = \varphi_1\alpha(1)\varphi_1\beta(2)\varphi_2\alpha(3) \tag{7.17}$$

If this wave function be made antisymmetric with respect to all possible two-electron permutations, namely, $\hat{P}_{12}$, $\hat{P}_{13}$, and $\hat{P}_{23}$, we find

$$\Psi = (6)^{-1/2} \begin{vmatrix} \varphi_1\alpha(1) & \varphi_1\beta(1) & \varphi_2\alpha(1) \\ \varphi_1\alpha(2) & \varphi_1\beta(2) & \varphi_2\alpha(2) \\ \varphi_1\alpha(3) & \varphi_1\beta(3) & \varphi_2\alpha(3) \end{vmatrix} \tag{7.18}$$

In the general case of a $2n$-electron system, the lowest-energy many-electron wave function is readily written as

$$\Psi = [(2n)!]^{-1/2} \begin{vmatrix} \varphi_1\alpha(1) & \varphi_1\beta(1) & \cdots & \varphi_n\beta(1) \\ \varphi_1\alpha(2) & \varphi_1\beta(2) & \cdots & \varphi_n\beta(2) \\ \vdots & & & \vdots \\ \varphi_1\alpha(2n) & \varphi_1\beta(2n) & \cdots & \varphi_n\beta(2n) \end{vmatrix} \tag{7.19a}$$

[4] After the originator: Professor J. C. Slater of MIT-UF. Determinants are naturally antisymmetric: The interchange of any two rows or of any two columns changes the sign of the determinant.

The wave function of Eq. (7.19a) may also be written in a number of other equivalent ways. Some of these are

$$\Psi = | \varphi_1\alpha(1)\varphi_1\beta(2)\ldots\varphi_n\alpha(2n-1)\varphi_n\beta(2n) | = | \Psi_d | \qquad (7.19b)$$

$$\Psi = \hat{A}\Psi_d \qquad (7.19c)$$

$$\Psi = [(2n)!]^{-1/2} \sum_{\hat{P}} (-1)^P \hat{P}\Psi_d \qquad (7.19d)$$

The symbol Ψ_d represents the principal diagonal term of the determinant of Eq. (7.19a) and $\hat{A}$ is a normalized antisymmetrization operator given by

$$\hat{A} = [(2n)!]^{-1/2} \sum_{\hat{P}} (-1)^P \hat{P} \qquad (7.20)$$

where the summation $\sum_{\hat{P}} \hat{P}$ runs over all possible permutations of the electron indices which produces unique terms and P is the number of electron-pair interchanges required to regenerate Ψ_d from the term in question (i.e., from $\hat{P}\Psi_d$).

(B)   Validity of Single-Configuration Approximation

The single-configuration wave function is not usually a very good approximation to the correct state function. The concept of a one-electron orbital is never exactly valid in a many-electron system and any single-configuration wave function based on such spin orbitals must suffer from corresponding inadequacies. A number of methods exist for removing part of these inadequacies; one of these, configuration mixing or configuration interaction (CI), is discussed briefly in chapter 8. It gives rise to multi-configuration wave functions, as opposed to the single-configuration wave functions to be treated in this chapter.

3.   SPIN-ADAPTED ANTISYMMETRIZED WAVE FUNCTIONS

We wish to develop general methods which enable us to obtain:

(i)   The number of independent spin states characterized by the quantum number S which exists for a system of N electrons when the space orbitals occupied by each electron are different. We denote this number: $f(N, S)$.

(ii)   A readily usable form of the operator $\hat{S}^2$. We may expand this operator as

$$\hat{S}^2 = \hat{S}_x^2 + \hat{S}_y^2 + \hat{S}_z^2 \qquad (7.21)$$

If we now define

$$\hat{S}_+ \equiv \hat{S}_x + i\hat{S}_y; \qquad \hat{S}_- \equiv \hat{S}_x - i\hat{S}_y \qquad (7.22)$$

we obtain

$$\begin{aligned}
\hat{S}_-\hat{S}_+ &= \hat{S}_x{}^2 + \hat{S}_y{}^2 + i\hat{S}_x\hat{S}_y - i\hat{S}_y\hat{S}_x \\
&= \hat{S}_x{}^2 + \hat{S}_y{}^2 + i[\hat{S}_x, \hat{S}_y] \\
&= \hat{S}_x{}^2 + \hat{S}_y{}^2 - \hbar\hat{S}_z
\end{aligned} \qquad (7.23)$$

where we have used the commutator relationship $[\hat{S}_x, \hat{S}_y] = i\hbar\hat{S}_z$. Insertion of Eq. (7.23) into Eq. (7.21) yields

$$\hat{S}^2 = \hat{S}_z{}^2 + \hbar\hat{S}_z + \hat{S}_-\hat{S}_+ \qquad (7.24a)$$

An equivalent form, readily verifiable, is

$$\hat{S}^2 = \hat{S}_z{}^2 - \hbar\hat{S}_z + \hat{S}_+\hat{S}_- \qquad (7.24b)$$

At this point, we also note that

$$\hat{S}_+ \equiv \sum_j \hat{s}_+(j); \qquad \hat{S}_- \equiv \sum_j \hat{s}_-(j); \qquad \hat{s}_+(j) \equiv \hat{s}_x(j) + i\hat{s}_y(j);$$

$$\hat{s}_-(j) \equiv \hat{s}_x(j) - i\hat{s}_y(j) \qquad (7.25)$$

where j is an electron numbering index and $\hat{s}$ is a one-electron operator. These spin operators (i.e., $\hat{S}_+$, $\hat{s}_+$, $\hat{S}_-$, $\hat{s}_-$) are known as *ladder operators* or *step-up–step-down operators* because their effects on the spin functions are

$$\hat{s}_-(j)\alpha(j) = \hbar\beta(j); \qquad \hat{s}_-(j)\beta(j) = 0$$

$$\hat{s}_+(j)\alpha(j) = 0; \qquad \hat{s}_+(j)\beta(j) = \hbar\alpha(j) \qquad (7.26)$$

The operator $\hat{s}_-$ steps down α to β but annihilates β whereas $\hat{s}_+$ annihilates α but steps up β to α. If the operands of $\hat{S}^2$ are limited to Slater determinants or to simple products of spin orbitals, the Eqs. (7.24) can be simplified further.

(iii)   A general method for construction of wave functions with given S and M_S characteristics. For this purpose, we again use projection operator techniques.

(A)   Number of Independent Spin Wave Functions, $f(N, S)$

Consider a many-electron spin function which is a simple product of N one-electron spin wave functions. Since each one-electron spin wave function is restricted to either α or β type, it follows that 2^N different many-electron spin wave functions may be generated. Each of these, of necessity, is an eigenfunction of $\hat{S}_z$.

For a three-electron system, the possible values of S are $\frac{3}{2}$ and $\frac{1}{2}$. The values of $f(N, S)$ for this system are readily seen to be

$$f(3, \tfrac{3}{2}) = \frac{3!}{3!0!} = 1 \tag{7.27}$$

$$f(3, \tfrac{1}{2}) = \frac{3!}{2!1!} - 1 = 2 \tag{7.28}$$

In Eq. (7.27), we have in essence calculated the number of possible spin orientations which yield $M_S = \frac{3}{2}$; the state $S = \frac{3}{2}$ also contains three other spin components with $M_S = \frac{1}{2}$, $-\frac{1}{2}$, and $-\frac{3}{2}$. Thus, in Eq. (7.28) where we calculate the number of possible spin orientations with $M_S = \frac{1}{2}$, before we associate the number 3 with the number of independent spin states $S = \frac{1}{2}$ we must subtract out that one spin orientation with $M_S = \frac{1}{2}$ belonging to $S = \frac{3}{2}$—hence, Eq. (7.28). Thus, the three-electron system possesses one quartet spin wave function and two doublet spin wave functions. The total number of wave functions is $8 = 2^3 = 1 \times 4 + 2 \times 2$.

For a four-electron system, the values of S are 2, 1, and 0. The values of $f(N, S)$ are

$$f(4, 2) = \frac{4!}{4!0!} = 1 \qquad \text{(1 quintet)} \tag{7.29}$$

$$f(4, 1) = \frac{4!}{3!1!} - 1 = 3 \qquad \text{(3 triplets)} \tag{7.30}$$

$$f(4, 0) = \frac{4!}{2!2!} - 3 - 1 = 2 \qquad \text{(2 singlets)} \tag{7.31}$$

The total number of wave functions is $16 = 2^4 = 1 \times 5 + 3 \times 3 + 2 \times 1$.

The results for the three- and four-electron systems are now generalized in analogical fashion to yield

$$f(N, S) = \frac{N!}{(N/2 - S)!(N/2 + S)!} - \frac{N!}{(N/2 - S - 1)!(N/2 + S + 1)!} \tag{7.32}$$

Thus, by insertion of the appropriate values of N and S into $f(N, S)$ we can evaluate the number of independent spin states of a given multiplicity available for an N-electron system. The results of such an evaluation are oftentimes presented in graphic form, in a construct such as Fig. 7.2 termed a *branching diagram*.

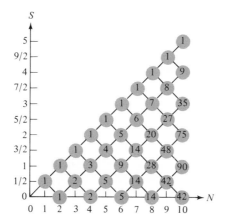

FIG. 7.2. Branching diagram for systems of N nonequivalent electrons. The number of electrons N is plotted along one axis, the resultant spin quantum number S is plotted along the other. Each possible spin state is represented by a circle with coordinates N and S. The value of $f(N, S)$ or, equivalently, the number of independent spin states for the given N and S is inscribed within the circle.

It is important to emphasize that the only restriction placed on the spin wave functions is that they be either α or β. In view of this, it follows that the space functions associated with each electron must be different. In other words, our conclusions here and in Fig. 7.2 apply only to a system of *nonequivalent* electrons.

(B) Analytic Form of $\hat{S}^2$

Insofar as we are concerned, the operands of $\hat{S}^2$ are always Slater determinants or linear combinations of such determinants. The analytic form of $\hat{S}^2$ given in Eqs. (7.24) can be simplified when the operand class is restricted as specified.

Consider the determinant

$$\Psi = |\ \varphi_1\alpha(1)\varphi_2\beta(2)\varphi_3\alpha(3)\ldots\varphi_n\alpha(N)\ | \qquad (7.33a)$$

in which each electron is associated with a different space function. Since the spin operators do not affect the orbital parts and since no restriction need be placed on the spin function associated with a given space function (i.e., the spin function may be either α or β) it follows that the determinantal wave function of Eq. (7.33a) may be abbreviated to

$$\Psi = |\ \alpha(1)\beta(2)\alpha(3)\ldots\alpha(N)\ | \qquad (7.33b)$$

Indeed, with no loss of information, we can use the condensations

$$\Psi = \alpha(1)\beta(2)\alpha(3) \tag{7.33c}$$

$$\Psi = \alpha\beta\alpha \tag{7.33d}$$

as long as we remember that all of these constitute a *short hand* for Eq. (7.33a).

The results of operating on Ψ of Eq. (7.33) with $\hat{S}_z$ and $\hat{S}_z{}^2$ are

$$\hat{S}_z\Psi = \tfrac{1}{2}(n_\alpha - n_\beta)\hbar\Psi \tag{7.34}$$

$$\hat{S}_z{}^2\Psi = \tfrac{1}{4}(n_\alpha - n_\beta)^2\hbar^2\Psi \tag{7.35}$$

We may rewrite $\hat{S}_-\hat{S}_+$ in the form

$$\hat{S}_-\hat{S}_+ = \sum_i^N \hat{s}_-(i)\hat{s}_+(i) + \sum_j^N \sum_{i\neq j}^N \hat{s}_-(i)\hat{s}_+(j) \tag{7.36}$$

and consider each of the parts on the right-hand side of Eq. (7.36) with respect to their effects on Ψ. Consider first

$$\sum_i^N \hat{s}_-(i)\hat{s}_+(i)\Psi = \sum_i^N \hat{s}_-(i)\hat{s}_+(i)\,|\,\alpha(1)\beta(2)\ldots\alpha(N)\,| \tag{7.37}$$

By virtue of the one-electron nature of the $\hat{s}_-(i)\hat{s}_+(i)$ operator, Eq. (7.37) reduces to

$$\sum_i^N \hat{s}_-(i)\hat{s}_+(i)\Psi = n_\beta\hbar^2\Psi \tag{7.38}$$

where the last equality follows from the noncommutativity of $\hat{s}_-$ and $\hat{s}_+$ and from

$$\hat{s}_-(i)\hat{s}_+(i)\alpha(i) = 0; \qquad \hat{s}_-(i)\hat{s}_+(i)\beta(i) = \hbar^2\beta(i) \tag{7.39}$$

Consider next the two-electron operator part $\sum_j \sum_{i\neq j} \hat{s}_-(i)\hat{s}_+(j)$ and note that the result

$$\hat{s}_-(i)\hat{s}_+(j)\alpha(i)\beta(j) = \hbar^2\beta(i)\alpha(j) \tag{7.40}$$

implies a formal identity of the operators

$$\sum_j \sum_{i\neq j} \hat{s}_-(i)\hat{s}_+(j) = \hbar^2 \sum_{\hat{P}} \hat{P}^{\alpha\beta} \tag{7.41}$$

where $\hat{P}^{\alpha\beta}$ is that subclass of all electron permutation operators which leads to an interchange of α and β spins. Thus, we find

$$\hat{S}_-\hat{S}_+\Psi = (\sum_{\hat{P}} \hat{P}^{\alpha\beta} + n_\beta)\hbar^2\Psi \tag{7.42}$$

If we now collect all the terms of Eqs. (7.34), (7.35), and (7.42) and insert them into Eqs. (7.24), we obtain

$$\hat{S}^2\Psi = \{\sum_{\hat{P}} \hat{P}^{\alpha\beta} + \tfrac{1}{4}[(n_\alpha - n_\beta)^2 + 2(n_\alpha + n_\beta)]\}\hbar^2\Psi$$

$$= \{\sum_P \hat{P}^{\alpha\beta} + \tfrac{1}{4}[(n_\alpha - n_\beta)^2 + 2N]\}\hbar^2\Psi \tag{7.43}$$

Thus, given Ψ as an eigenfunction of $\hat{S}^2$, it is a relatively simple matter to determine S.

(C) Spin-Projection Operators

We now turn our attention to the generation of eigenfunctions of $\hat{S}^2$. Suppose we start with a trial wave function Ψ_t which satisfies one condition—namely, that it be resolvable into a linear combination of eigenfunctions of $\hat{S}^2$

$$\Psi_t = \sum_k c_{tk}\Psi(S_k) \tag{7.44}$$

where the summation runs over all possible values of S_k consistent with the specific number of electrons under consideration and where

$$\hat{S}^2\Psi(S_k) = S_k(S_k + 1)\hbar^2\Psi(S_k) \tag{7.45}$$

The coefficients c_{tk} need not be known; all that is required is that Ψ_t be a function in the domain spanned by the basis set $\Psi(S_k)$. It then follows that

$$[\hat{S}^2 - S_l(S_l + 1)\hbar^2]\Psi_t = \sum_k{}' d_k\Psi(S_k) \tag{7.46}$$

where the primed summation sign denotes that the summation runs over all spin eigenvalues except S_l. We may consider $[\hat{S}^2 - S_l(S_l + 1)]$ to be an operator which eliminates any admixture of an S_l wave function from Ψ_t. Successive application of such *annihilators* finally yields one pure spin state of specified multiplicity—if such a spin state exists in Ψ_k. The required operator is

$$\hat{O}(S_k) = \prod_{l \neq k} [\hat{S}^2 - S_l(S_l + 1)\hbar^2] \tag{7.47}$$

This operator projects a wave function of the required spin quantum number (i.e., S_k) out of the general spin space; $\hat{O}(S_k)$ is a spin-projection operator.

(D) Spin Eigenfunctions of a Three-Electron System:
An Example Calculation

The eigenfunctions of $\hat{S}_z$ may be written immediately. In a simple short-hand form, they are

$$\alpha\alpha\alpha \qquad M_S = \tfrac{3}{2}$$

$$\alpha\alpha\beta \qquad M_S = \tfrac{1}{2}$$

$$\alpha\beta\alpha \qquad M_S = \tfrac{1}{2}$$

$$\beta\alpha\alpha \qquad M_S = \tfrac{1}{2}$$

$$\alpha\beta\beta \qquad M_S = -\tfrac{1}{2}$$

$$\beta\alpha\beta \qquad M_S = -\tfrac{1}{2}$$

$$\beta\beta\alpha \qquad M_S = -\tfrac{1}{2}$$

$$\beta\beta\beta \qquad M_S = -\tfrac{3}{2} \qquad (7.48)$$

where, for example, $\alpha\beta\alpha$ denotes $\mid \varphi_1\alpha(1)\varphi_2\beta(2)\varphi_3\alpha(3) \mid$ with $\varphi_1 \neq \varphi_2 \neq \varphi_3$. Inspection of the branching diagram of Fig. 7.2 indicates that we should obtain one quartet spin state $(S = \tfrac{3}{2})$ and two doublet spin states $(S = \tfrac{1}{2})$. The projection operators are

$$\hat{O}[\tfrac{3}{2}] = \hat{S}^2 - 3/4$$

$$\hat{O}[\tfrac{1}{2}] = \hat{S}^2 - 15/4 \qquad (7.49)$$

where, for brevity, we have eliminated the factor $\hbar^2$ from the operator $\hat{S}^2$. Since we have made no attempt to normalize the spin-projection operators, we fully expect that the spin-projected wave functions require normalization.

(*i*) *Spin Eigenfunctions with* $S = \tfrac{3}{2}, M_S = \tfrac{3}{2}, \tfrac{1}{2}, -\tfrac{1}{2}, -\tfrac{3}{2}$: We first let $\hat{O}[\tfrac{3}{2}]$ operate on $\alpha\alpha\alpha$. The result is

$$\hat{O}[\tfrac{3}{2}]\alpha\alpha\alpha = [0 + \tfrac{1}{4}(3^2 + 2 \times 3)]\alpha\alpha\alpha - \tfrac{3}{4}\alpha\alpha\alpha = 3\alpha\alpha\alpha \qquad (7.50)$$

Thus, with normalization, it is clear that $\alpha\alpha\alpha$ is a satisfactory eigenfunction[5] with $S = \tfrac{3}{2}, M_S = \tfrac{3}{2}$.

---

[5] In order to exemplify the manner in which unwanted functions are eliminated, consider the result of $\hat{O}[\tfrac{1}{2}]$ operating on $\alpha\alpha\alpha$. The result is

$$\hat{O}[\tfrac{1}{2}]\alpha\alpha\alpha = [0 + \tfrac{1}{4}(3^2 + 2 \times 3)]\alpha\alpha\alpha - \tfrac{15}{4}\alpha\alpha\alpha = 0$$

Thus, as evidenced by our inability to project a doublet spin eigenfunction out of $\alpha\alpha\alpha$, there is no doublet character in $\alpha\alpha\alpha$. This, of course, had to be the case—$\alpha\alpha\alpha$ is, after all, a quartet eigenfunction.

We next construct the spin wave function with $S = \frac{3}{2}$, $M_S = \frac{1}{2}$. Clearly, this eigenfunction may be a linear combination of $\alpha\alpha\beta$, $\alpha\beta\alpha$, and $\beta\alpha\alpha$ only. We start with $\alpha\alpha\beta$ to find

$$\hat{O}[\tfrac{3}{2}]\alpha\alpha\beta = (\sum_{\hat{P}} \hat{P}_{\alpha\beta} + \tfrac{1}{4}[(n_\alpha - n_\beta)^2 + 2n_\alpha + 2n_\beta])\alpha\alpha\beta - \tfrac{3}{4}\alpha\alpha\beta$$

$$= \alpha\beta\alpha + \beta\alpha\alpha + \tfrac{1}{4}[1^2 + 2 \times 2 + 2 \times 1]\alpha\alpha\beta - \tfrac{3}{4}\alpha\alpha\beta$$

$$= \alpha\beta\alpha + \beta\alpha\alpha + \alpha\alpha\beta \tag{7.51a}$$

Normalization yields

$$\Psi(\tfrac{3}{2}, \tfrac{1}{2}) = (\tfrac{1}{3})^{1/2}(\alpha\beta\alpha + \beta\alpha\alpha + \alpha\alpha\beta) \tag{7.51b}$$

The spin wave function with $S = \frac{3}{2}$, $M_S = -\frac{1}{2}$ is constructed in like manner from $\alpha\beta\beta$, $\beta\alpha\beta$, and $\beta\beta\alpha$. It is found to be

$$\Psi(\tfrac{3}{2}, -\tfrac{1}{2}) = (\tfrac{1}{3})^{1/2}(\alpha\beta\beta + \beta\alpha\beta + \beta\beta\alpha) \tag{7.52}$$

The last of these four spin wave functions is

$$\Psi(\tfrac{3}{2}, -\tfrac{3}{2}) = \beta\beta\beta \tag{7.53}$$

(ii) **Spin Eigenfunctions with** $S = \frac{1}{2}$, $M_S = \pm\frac{1}{2}$: There are two doublet spin wave functions. Those with $M_S = \frac{1}{2}$ are constructed from $\alpha\alpha\beta$, $\alpha\beta\alpha$, and $\beta\alpha\alpha$. Letting $\hat{O}[\frac{1}{2}]$ operate on each of these in turn, we find

$$\hat{O}[\tfrac{1}{2}]\alpha\alpha\beta = (1/6)^{1/2}(\alpha\beta\alpha + \beta\alpha\alpha - 2\alpha\alpha\beta)$$

$$\hat{O}[\tfrac{1}{2}]\alpha\beta\alpha = (1/6)^{1/2}(\alpha\alpha\beta + \beta\alpha\alpha - 2\alpha\beta\alpha)$$

$$\hat{O}[\tfrac{1}{2}]\beta\alpha\alpha = (1/6)^{1/2}(\alpha\beta\alpha + \alpha\alpha\beta - 2\beta\alpha\alpha) \tag{7.54}$$

Thus, three wave functions with $S = \frac{1}{2}$, $M_S = \frac{1}{2}$ have been constructed. However, one of them is linearly dependent on the other two and they are not orthogonal. Redundancy can be eliminated and orthogonality produced if we note the cyclic symmetry exhibited by the three wave functions of Eqs. (7.54) and make maximum use of it. It is clear that these three wave functions can be represented as three in-plane vectors separated by 120°. Thus, the difference of any two functions is orthogonal to the remaining one. A possible set of orthonormal wave functions, then, is

$$\Psi(\tfrac{1}{2}, \tfrac{1}{2}) = (1/6)^{1/2}(\alpha\beta\alpha + \beta\alpha\alpha - 2\alpha\alpha\beta)$$

$$\Psi(\tfrac{1}{2}, \tfrac{1}{2}) = (1/2)^{1/2}(\beta\alpha\alpha - \alpha\beta\alpha) \tag{7.55}$$

The generation of the two wave functions $\Psi(\frac{1}{2}, -\frac{1}{2})$ is left as an exercise.

4. CONFIGURATION WAVE FUNCTIONS

A number of methods may be used to generate symmetry- and spin-adapted single-configuration wave functions. No procedure is particularly difficult, but many are quite laborious. Hence, we choose that procedure which yields the desired results in the shortest possible time. The initial step consists of writing Slater determinants which are eigenfunctions of $\hat{S}_z$ and which contain MO parts which are symmetry adapted; one then projects eigenfunctions of $\hat{S}^2$ out of these; finally, one subjects these eigenfunctions of $\hat{S}_z$ and $\hat{S}^2$ to the space-group projection operators in order to validate symmetry adaptation or, if necessary, to determine the proper orbital symmetry-adapted wave functions. We now illustrate this procedure in a few instances.

(A) Hydrogen Molecule, H_2: An Example Calculation

The basis set of atomic orbitals which we use for H_2 consists of a $1s$ AO on $H(1)$, denoted $1s_1$ and a $1s$ AO on H_2, denoted $1s_2$. The relevant symmetry of H_2, insofar as the specified AO basis set is concerned, may be reduced to C_i. The C_i point group representation table is given in Table 7.3.

The projection operators for C_i symmetry-adapted wave functions are

$$\hat{O}[\Sigma_g] = E + i$$

$$\hat{O}[\Sigma_u] = E - i \tag{7.56}$$

Using these, the MO's are found to be

$$\varphi_g = (\tfrac{1}{2})^{1/2}(1s_1 + 1s_2)$$

$$\varphi_u = (\tfrac{1}{2})^{1/2}(1s_1 - 1s_2) \tag{7.57}$$

We now write all properly antisymmetrized eigenfunctions of $\hat{S}^2$ for this two-electron system. They are

$$M_S$$

$$\Psi_1 = \mid \varphi_g\alpha(1)\varphi_g\beta(2) \mid \qquad 0$$

$$\Psi_2 = \mid \varphi_g\alpha(1)\varphi_u\alpha(2) \mid \qquad 1$$

$$\Psi_3 = \mid \varphi_g\alpha(1)\varphi_u\beta(2) \mid \qquad 0$$

$$\Psi_4 = \mid \varphi_g\beta(1)\varphi_u\alpha(2) \mid \qquad 0$$

$$\Psi_5 = \mid \varphi_g\beta(1)\varphi_u\beta(2) \mid \qquad -1$$

$$\Psi_6 = \mid \varphi_u\alpha(1)\varphi_u\beta(2) \mid \qquad 0 \tag{7.58}$$

The eigenfunctions of $\hat{S}^2$ which are required are singlets and triplets as is indicated in Fig. 7.2. The projection operators are

$$\hat{O}(0) = \hat{S}^2 - 2$$

$$\hat{O}(1) = \hat{S}^2 \qquad (7.59)$$

The results of the spin-projection process are as follows:

$$\hat{S}^2\Psi_1 = (\sum_P \hat{P}^{\alpha\beta} + \tfrac{1}{4}[(n_\alpha - n_\beta)^2 + 2n_\alpha + 2n_\beta]) \mid \varphi_g\alpha(1)\varphi_g\beta(2) \mid$$

$$= \mid \varphi_g\beta(1)\varphi_g\alpha(2) \mid + \tfrac{1}{4}[0^2 + 2 \times 1 + 2 \times 1] \mid \varphi_g\alpha(1)\varphi_g\beta(2) \mid$$

$$= \mid \varphi_g\beta(1)\varphi_g\alpha(2) \mid + \mid \varphi_g\alpha(1)\varphi_g\beta(2) \mid$$

$$= 0 \qquad (7.60)$$

$$(\hat{S}^2 - 2)\Psi_1 = -2 \mid \varphi_g\alpha(1)\varphi_g\beta(2) \mid = -2\Psi_1 \qquad (7.61)$$

$$\hat{S}^2\Psi_2 = 2\Psi_2 \qquad (7.62)$$

$$(\hat{S}^2 - 2)\Psi_2 = 0 \qquad (7.63)$$

$$\hat{S}^2\Psi_3 = \hat{S}^2 \mid \varphi_g\alpha(1)\varphi_u\beta(2) \mid = \mid \varphi_g\beta(1)\varphi_u\alpha(2) \mid + \mid \varphi_g\alpha(1)\varphi_u\beta(2) \mid$$

$$= \Psi_4 + \Psi_3 \qquad (7.64)$$

$$(\hat{S}^2 - 2)\Psi_3 = \Psi_4 - \Psi_3 \qquad (7.65)$$

$$\hat{S}^2\Psi_5 = 2\Psi_5 \qquad (7.66)$$

$$(\hat{S}^2 - 2)\Psi_5 = 0 \qquad (7.67)$$

$$\hat{S}^2\Psi_6 = 0 \qquad (7.68)$$

$$(\hat{S}^2 - 2)\Psi_6 = -2\Psi_6 \qquad (7.69)$$

The normalized spin-projected eigenfunctions are collected in Table 7.4.

The final task is to ensure that the orbital parts of the eigenfunctions of Table 7.4 transform properly in the C_i point group. The symmetry projection operators are given in Eq. (7.56); operation with them affects

TABLE 7.3. *C_i point group.*

C_i	E	i
Σ_g	1	1
Σ_u	1	-1

TABLE 7.4. *Spin and symmetry-adapted eigenfunctions for H_2.*

Eigenfunction	M_S	S	Symmetry Classification (Γ)	State Classification $(^{2S+1}\Gamma)$
Ψ_1	0	0	Σ_g	$^1\Sigma_g$
Ψ_2	1	1	Σ_u	$^3\Sigma_u$
$(\frac{1}{2})^2(\Psi_3 + \Psi_4)$	0	1	Σ_u	$^3\Sigma_u$
$(\frac{1}{2})^{1/2}(\Psi_3 - \Psi_4)$	0	0	Σ_u	$^1\Sigma_u$
Ψ_5	-1	1	Σ_u	$^3\Sigma_u$
Ψ_6	0	0	Σ_g	$^1\Sigma_g$

the orbital parts of the wave function only. Consider, as an example,

$$\hat{O}[\Sigma_g](\Psi_3 + \Psi_4) = (E + i)\Psi_3 + (E + i)\Psi_4$$

$$= (E + i) \mid \varphi_g\alpha(1)\varphi_u\beta(2) \mid + (E + i) \mid \varphi_g\beta(1)\varphi_u\alpha(2) \mid$$

$$= \mid \{E\varphi_g\alpha(1)\}\{E\varphi_u\beta(2)\} \mid + \mid \{i\varphi_g\alpha(1)\}\{i\varphi_u\beta(2)\} \mid$$

$$\quad + \mid \{E\varphi_g\beta(1)\}\{E\varphi_u\alpha(2)\} \mid + \mid \{i\varphi_g\beta(1)\}\{i\varphi_u\alpha(2)\} \mid$$

$$= \mid \varphi_g\alpha(1)\varphi_u\beta(2) \mid - \mid \varphi_g\alpha(1)\varphi_u\beta(2) \mid$$

$$\quad + \mid \varphi_g\beta(1)\varphi_u\alpha(2) \mid - \mid \varphi_g\beta(1)\varphi_u\alpha(2) \mid$$

$$= 0 \tag{7.70}$$

$$\hat{O}[\Sigma_u](\Psi_3 + \Psi_4) = (E - i)\Psi_3 + (E - i)\Psi_4$$

$$= 2(\Psi_3 + \Psi_4) \tag{7.71}$$

Thus, $\Psi_3 + \Psi_4$ transforms as Σ_u. The other wave functions of Table 7.4 may be shown, in like manner, to be already symmetry adapted. Their transformation properties are also listed in Table 7.4.

5. ELECTRON CONFIGURATIONS CONTAINING PARTIALLY FILLED DEGENERATE MOLECULAR ORBITALS

No difficulty is encountered in forming symmetry- and spin-adapted configuration wave functions if:

(i)   all populated molecular orbitals are completely filled; or

(ii)   none of the partially-filled molecular orbitals are degenerate; or
(iii)   only one degenerate molecular orbital is partially filled and it contains either one electron or $2g - 1$ electrons where g is the molecular-orbital degeneracy.

In these instances, any configuration wave function based on symmetry-adapted molecular orbitals is already fully symmetry adapted in an orbital sense. Consequently, one needs only ensure that the wave function is an eigenfunction of $\hat{S}^2$. This was the situation for H_2 in Sec. 4(A).

The conclusions of the previous paragraph are proven as follows: Consider the determinantal wave function descriptive of the one-electron MO excitation $s \leftarrow k$ and given by

$$\Psi_{k \to s}^{(1)} = |\, \varphi_1 \alpha(1) \varphi_1 \beta(2) \ldots \varphi_k \alpha(2k-1) \varphi_s \beta(2k) \ldots \varphi_n \beta(2n) \,| \quad (7.72)$$

This wave function possesses orbital transformation properties given by the direct product

$$\Gamma_\Psi = \Gamma_{\varphi_1} \times \Gamma_{\varphi_1} \times \ldots \times \Gamma_{\varphi_k} \times \Gamma_{\varphi_s} \times \ldots \times \Gamma_{\varphi_n} \quad (7.73)$$

If all of the MO's are symmetry adapted and nondegenerate, it follows that $\Gamma_\Psi = \Gamma_{\varphi_k} \times \Gamma_{\varphi_s}$ and that Γ_Ψ is a nondegenerate representation of the same molecular point group. Thus the wave function in question, $\Psi_{k \to s}^{(1)}$, forms a basis for the irreducible representation Γ_Ψ and for no other. In other words, Γ_Ψ is a pure nondegenerate representation of the molecular point group; $\Psi_{k \to s}^{(1)}$ is unadulterated by any parts which transform as any representation other than Γ_Ψ.

If, on the other hand, Γ_{φ_k} and Γ_{φ_s} are different degenerate representations of the molecular point group, and if all MO's φ_1 to φ_n (other than φ_k and φ_s) are filled, we again find $\Gamma_\Psi = \Gamma_{\varphi_k} \times \Gamma_{\varphi_s}$—but with the difference that now Γ_Ψ is a reducible representation. For example, in the lowest-energy excited state of benzene which is of MO excitation type $e_{2u} \leftarrow e_{1g}$ and where

$$\Psi = |\ldots a_{2u}\alpha(1) a_{2u}\beta(2) e_{1g}\alpha(3) e_{1g}\beta(4) e_{1g}(\alpha \text{ or } \beta)(5) e_{2u}(\alpha \text{ or } \beta)(6) \,| \quad (7.74)$$

we find, using the *hole-electron* formalism,[3] that

$$\Gamma_\Psi = e_{1g} \times e_{2u} = B_{1u} + B_{2u} + E_{1u} \quad (7.75)$$

Thus, under these conditions, it is seen that Ψ may form a basis for more than one irreducible representation of the point group. Consequently, except for the situations specified in items (i)–(iii) above, a determinantal wave function based on symmetry adapted MO's is not necessarily a properly symmetry-adapted many-electron wave function.

The procedure to be followed in the latter instance is straightforward: One must reuse the orbital-projection operators—this time on the deter-

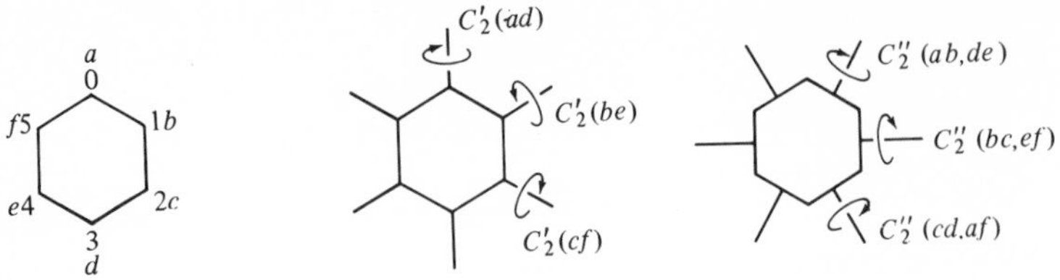

FIG. 7.3. Numbering conventions for the carbon atoms of benzene. The operations of classes C_2' and C_2'' are also defined. The numbering convention can serve for carbon atom site and/or π-AO identification purposes.

minantal eigenfunctions of $\hat{S}_z$ and $\hat{S}^2$. In order to demonstrate the procedure involved in the case of an electron configuration containing partially filled degenerate molecular orbitals, we now discuss the ground-state and the lower-energy $\pi^* \leftarrow \pi$ excited states of benzene. A further example relating to the $e^* \leftarrow e$ transition of a C_{4v} molecule is found in Sec. 5 of chapter 9.

(A) Benzene: An Example Calculation

The lowest-energy electronic transition of benzene is of $\pi^* \leftarrow \pi(e_{2u} \leftarrow e_{1g})$ molecular-orbital origin. The lowest-energy excited configurations are given by $\ldots a_{2u}^2 e_{1g}^3 e_{2u}^1$; they involve two unfilled degenerate MO's.

The purpose of this section is to elicit symmetry-adapted single-configuration wave functions for the $^{1,3}B_{1u}$, $^{1,3}B_{2u}$, and $^{1,3}E_{1u}$ states which derive from $\ldots a_{2u}^2 e_{1g}^3 e_{2u}^1$. Since we are concerned solely with the six π electrons, we can limit discussion to the D_6 point group shown in Table 7.5.

We first obtain the symmetry-adapted π MO's of benzene. In order to conserve effort we proceed as follows: We first determine those representations of the D_6 point group for which the six π atomic orbitals form a

TABLE 7.5. *D_6 point group.*

(Operations of classes C_2' and C_2'' are defined in Fig. 7.3.)

D_6	E	C_2	$2C_3$	$2C_6$	$3C_2'$	$3C_2''$
A_1	1	1	1	1	1	1
$A_2\ \hat{R}_z, z$	1	1	1	1	-1	-1
B_1	1	-1	1	-1	1	-1
B_2	1	-1	1	-1	-1	1
$E_1\ (x, y), (\hat{R}_x, \hat{R}_y)$	2	-2	-1	1	0	0
E_2	2	2	-1	-1	0	0

basis; we then construct the corresponding projection operators and, using these, evolve the π MO's.

(i) **Symmetry-Adapted π MO's of Benzene:** The effect of the symmetry operations of the D_6 point group on the π AO's of benzene is shown in Table 7.6. The π MO on carbon j is denoted χ_j. The character of the 6×6 matrix descriptive of the transformation

$$\hat{R} \begin{bmatrix} \chi_0 \\ \chi_1 \\ \chi_2 \\ \chi_3 \\ \chi_4 \\ \chi_5 \end{bmatrix} = [6 \times 6] \begin{bmatrix} \chi_0 \\ \chi_1 \\ \chi_2 \\ \chi_3 \\ \chi_4 \\ \chi_5 \end{bmatrix} \tag{7.76}$$

is denoted $\overline{\chi}(\hat{R})$ and is displayed on the extreme right of Table 7.6. The representation Γ which consists of the elements $\overline{\chi}(\hat{R})$ is, in general, reducible.

TABLE 7.6. *Effect of D_6 symmetry operations on the six π AO's of benzene.*

(The notations *ccw* and *cw* denote counterclockwise and clockwise, respectively; the axes *ad*, *be*, etc., are defined in Fig. 7.3.)

$\hat{R}$	χ_0	χ_1	χ_2	χ_3	χ_4	χ_5	$\overline{\chi}(\hat{R})$
E	χ_0	χ_1	χ_2	χ_3	χ_4	χ_5	6
C_2	χ_3	χ_4	χ_5	χ_0	χ_1	χ_2	0
C_3; *ccw*	χ_4	χ_5	χ_0	χ_1	χ_2	χ_3	0
cw	χ_2	χ_3	χ_4	χ_5	χ_0	χ_1	0
C_6; *ccw*	χ_5	χ_0	χ_1	χ_2	χ_3	χ_4	0
cw	χ_1	χ_2	χ_3	χ_4	χ_5	χ_0	0
C_2'; *ad*	$-\chi_0$	$-\chi_5$	$-\chi_4$	$-\chi_3$	$-\chi_2$	$-\chi_1$	-2
be	$-\chi_2$	$-\chi_1$	$-\chi_0$	$-\chi_5$	$-\chi_4$	$-\chi_3$	-2
cf	$-\chi_4$	$-\chi_3$	$-\chi_2$	$-\chi_1$	$-\chi_0$	$-\chi_5$	-2
C''; *ab, de*	$-\chi_1$	$-\chi_0$	$-\chi_5$	$-\chi_4$	$-\chi_3$	$-\chi_2$	0
bc, ef	$-\chi_3$	$-\chi_2$	$-\chi_1$	$-\chi_0$	$-\chi_5$	$-\chi_4$	0
cd, af	$-\chi_5$	$-\chi_4$	$-\chi_3$	$-\chi_2$	$-\chi_1$	$-\chi_0$	0

The reduction of Γ is achieved using[3]

$$n(\Gamma_i) = \frac{1}{h} \sum_{\hat{R}} \bar{\chi}_i(\hat{R})\bar{\chi}(\hat{R}) \tag{7.77}$$

where $n(\Gamma_i)$ is the number of times the irreducible representation Γ_i occurs in Γ; h is the dimension of the group; and the summation is over all operators of the group. Such analysis yields

$$\Gamma = A_2 + B_2 + E_1 + E_2 \tag{7.78}$$

The projection operators required[6] are

$$\hat{O}[A_2] = E + C_2 + [2C_3] + [2C_6] - [3C_2'] - [3C_2'']$$

$$\hat{O}[B_2] = E - C_2 + [2C_3] - [2C_6] - [3C_2'] + [3C_2'']$$

$$\hat{O}[E_1] = 2E - 2C_2 - [2C_3] + [2C_6]$$

$$\hat{O}[E_2] = 2E + 2C_2 - [2C_3] - [2C_6] \tag{7.79}$$

Let us consider the effects of $\hat{O}[A_2]$ on the π AO situated on carbon number zero:

$$\hat{O}[A_2]\chi_0 = E\chi_0 + C_2\chi_0 + [2C_3]\chi_0 + [2C_6]\chi_0 - [3C_2']\chi_0 - [3C_2'']\chi_0$$

$$= \chi_0 + \chi_3 + (\chi_4 + \chi_2) + (\chi_5 + \chi_1) + (\chi_0 + \chi_2 + \chi_4)$$

$$+ (\chi_1 + \chi_3 + \chi_5)$$

$$= 2(\chi_0 + \chi_1 + \chi_2 + \chi_3 + \chi_4 + \chi_5) \tag{7.80a}$$

Normalization yields

$$\varphi_0 \equiv a_2 \text{ MO} = (1/6)^{1/2}(\chi_0 + \chi_1 + \chi_2 + \chi_3 + \chi_4 + \chi_5) \tag{7.80b}$$

The b_2 MO is obtained using $\hat{O}[B_2]$ and yields, when normalized

$$\varphi_3 \equiv b_2 \text{ MO} = (1/6)^{1/2}(\chi_0 - \chi_1 + \chi_2 - \chi_3 + \chi_4 - \chi_5) \tag{7.81}$$

Some ambiguity occurs with the e MO's. Consider $\hat{O}[E_1]$ operating on the AO χ_0

$$\hat{O}[E_1]\chi_0 = 2E\chi_0 - 2C_2\chi_0 - [2C_3]\chi_0 + [2C_6]\chi_0$$

Normalization of this function yields the projected MO $\varphi_p(e_1)$

$$\varphi_p(e_1) = (1/12)^{1/2}[2\chi_0 - 2\chi_3 - (\chi_4 + \chi_2) + (\chi_5 + \chi_1)] \tag{7.82a}$$

Consider $\hat{O}[E_1]$ operating on the AO χ_1

$$\hat{O}[E_1]\chi_1 = 2\chi_1 - 2\chi_4 - (\chi_5 + \chi_3) + (\chi_0 + \chi_2)$$

---

[6] We use the square-bracket notation, for example, $[3C_2']$, to signify the totality of operations of the given class. In the above example, we signify

$$[3C_2'] = C_2'; ad + C_2'; be + C_2'; cf$$

Again, normalization yields

$$\varphi_q(e_1) = (1/12)^{1/2}[2\chi_1 - 2\chi_4 - (\chi_5 + \chi_3) + (\chi_0 + \chi_2)] \quad (7.82a')$$

These MO's are not orthogonal; in fact, they can be represented by two vectors angled at 60° in the same plane. In view of this, it follows that the two vectors obtained by addition and by subtraction of these are necessarily orthogonal. These are, when normalized

$$\varphi_a(e_1) = (1/3)^{1/2}[\varphi_p(e_1) + \varphi_q(e_1)] = (1/2)(\chi_0 + \chi_1 - \chi_3 - \chi_4)$$

$$\varphi_b(e_1) = \varphi_p(e_1) - \varphi_q(e_1)$$

$$= (1/12)^{1/2}(\chi_0 - \chi_1 - 2\chi_2 - \chi_3 + \chi_4 + 2\chi_5) \quad (7.82b)$$

One finds, similarly,

$$\varphi_a(e_2) = (1/2)(\chi_0 - \chi_1 + \chi_3 - \chi_4)$$

$$\varphi_b(e_2) = (1/12)^{1/2}(\chi_0 + \chi_1 - 2\chi_2 + \chi_3 + \chi_4 - 2\chi_5) \quad (7.83)$$

The MO's of Eqs. (7.82b) and (7.83) may be used to construct Slater determinants and symmetry-adapted state functions. However, consider the effect of the operator C_6 on $\varphi_a(e_1)$; we find

$$C_6\varphi_a(e_1) = C_6[\tfrac{1}{2}(\chi_0 + \chi_1 - \chi_3 - \chi_4)] = \tfrac{1}{2}(\chi_5 + \chi_0 - \chi_2 - \chi_3) \quad (7.84)$$

which is a linear combination of both $\varphi_a(e_1)$ and $\varphi_b(e_1)$. When several MO's in a determinant are transformed into such linear combinations by a symmetry operation, the total effect of the operation on the determinant is obviously rather complicated. It turns out in this case that it is easier to work with certain specific linear combinations of φ_a and φ_b rather than with φ_a and φ_b of Eqs. (7.82b) and (7.83) themselves. Now, consider the MO's

$$\varphi_1 = N[\varphi_a(e_1) - i\varphi_b(e_1)] = (1/6)^{1/2} \sum_{\mu=0}^{5} e^{2\pi i\mu/6}\chi_\mu \quad (7.85)$$

$$\varphi_{-1} = N'[\varphi_a(e_1) + i\varphi_b(e_1)] = (1/6)^{1/2} \sum_{\mu=0}^{5} e^{-2\pi i\mu/6}\chi_\mu \quad (7.86)$$

where N and N' are normalization factors.[7] The right-most expressions of $\varphi_{\pm 1}$ are analytically convenient and similar to the free-electron molecular orbitals of chapter 5. As a matter of fact, all MO's of Eqs. (7.80)–(7.83) are given by

$$\varphi_l = (1/6)^{1/2} \sum_{\mu} e^{2\pi i\mu l/6}\chi_\mu = (1/6)^{1/2} \sum_{\mu} \omega^{\mu l}\chi_\mu \quad (7.87)$$

where

$$\omega = e^{2\pi i/6}; \quad \mu = 0, 1, 2, 3, 4, 5; \quad l = 0, \pm 1, \pm 2, 3 \quad (7.88)$$

---

[7] We find $N = (1/8)^{1/2}[i + (3)^{1/2}]$ and $N' = (1/8)^{1/2}[-i + (3)^{1/2}]$.

As is obvious from the previous discussion, φ_1 and φ_{-1} [obtained by letting $l = \pm 1$ in Eq. (7.87)] form basis for the e_1 irreducible representation. Similarly, it can be shown that $\varphi_{\pm 2}(l = \pm 2)$ transform as e_2, while $\varphi_0(l = 0)$ and $\varphi_3(l = 3)$ form basis for the a_2 and b_2 representations, respectively. The advantage of using the φ_l functions rather than the real MO's [Eq. (7.82a), for example] is that the effect of all symmetry operations can be written in a very simple analytic way: The results of Table 7.7 indicate that $\hat{R}\varphi_l$ always equals some multiple of φ_l or of φ_{-l} and that one never obtains a linear combination of different φ_l functions. As an example of the way Table 7.7 is constructed, consider the result of the C_6 operation on φ_l. The C_6 operation changes the coefficient of each AO into the coefficient of its left (or right) neighbor. Hence, $\omega^{\mu l}$ becomes $\omega^{(\mu+1)l}$ or $\omega^{(\mu-1)l}$; therefore

$$C_6\varphi_l = \omega^{\pm l}\varphi_l \qquad (7.89)$$

where the $\pm$ sign is determined by the clockwise or counterclockwise nature of the C_6 rotation.

(ii) **Eigenfunctions of $\hat{S}_z$:** The eigenfunctions of $\hat{S}_z$, appropriately antisymmetrized, are given by single 6×6 Slater determinants. For the ground state, only one such determinant exists; it is given by

$$\Psi_0 = |\ \varphi_0\alpha(1)\varphi_0\beta(2)\varphi_1\alpha(3)\varphi_1\beta(4)\varphi_{-1}\alpha(5)\varphi_{-1}\beta(6)\ |;\qquad M_S = 0 \quad (7.90)$$

TABLE 7.7.   *Result of the $\hat{R}$ operation on φ_l.*

(The notations *cw* and *ccw* denote clockwise and counterclockwise, respectively; the axes *ad*, *be*, ... are defined in Fig. 7.3.)

Operator $(\hat{R})$	Result $(\hat{R}\varphi_l)$
E	φ_l
C_2	$\omega^{3l}\varphi_l$
C_3; *ccw*	$\omega^{2l}\varphi_l$
cw	$(\omega^{-2l}\ or\ \omega^{4l})\varphi_l$
C_6; *ccw*	$\omega^{l}\varphi_l$
cw	$(\omega^{-l}\ or\ \omega^{5l})\varphi_l$
C_2'; *ad*	$-\varphi_{-l}$
be	$-\omega^{2l}\varphi_{-l}$
cf	$-\omega^{4l}\varphi_{-l}$
C_2''; *ab, de*	$-\omega^{l}\varphi_{-l}$
bc, ef	$-\omega^{3l}\varphi_{-l}$
cd, af	$-\omega^{5l}\varphi_{-l}$

For the excited states it is quite another matter altogether. A total of 16 possibilities exist; they are given in Table 7.8.

(*iii*) **Eigenfunctions of $\hat{S}^2$:** The eigenvectors of $\hat{S}^2$ are obtained by repeated usage of the spin-projection operators $\hat{O}[0]$ and $\hat{O}[1]$. The results are

$$\Psi_1, \Psi_5, \Psi_9, \Psi_{13}; \qquad\qquad S = 1; \quad M_S = 1$$

$$\Psi_2 + \Psi_3, \Psi_6 + \Psi_7, \Psi_{10} + \Psi_{11}, \Psi_{14} + \Psi_{15}; \quad S = 1; \quad M_S = 0$$

$$\Psi_4, \Psi_8, \Psi_{12}, \Psi_{16}; \qquad\qquad S = 1; \quad M_S = -1$$

$$\Psi_2 - \Psi_3, \Psi_6 - \Psi_7, \Psi_{10} - \Psi_{12}, \Psi_{14} - \Psi_{15}; \quad S = 0; \quad M_S = 0 \qquad (7.91)$$

(*iv*) **Symmetry Adaptation of the Eigenfunctions of $\hat{S}_z$ and $\hat{S}_z^2$:** The determinantal wave functions of Eq. (7.91) are not fully symmetry adapted. This situation arises because the two partially empty molecular orbitals are individually degenerate. In Ψ_9, for example, the MO

TABLE 7.8. *Eigenfunctions of $\hat{S}_z$ which derive from the molecular excitation $e_2 \leftarrow e_1$.*

[This table is to be read in the manner exemplified, for instance, by
$$\Psi_9 = |\ \varphi_0\alpha(1)\varphi_0\beta(2)\varphi_{-1}\alpha(3)\varphi_{-1}\beta(4)\varphi_1\alpha(5)\varphi_1\beta(6)|.]$$

Ψ_i	Electron 1	Electron 2	Electron 3	Electron 4	Electron 5	Electron 6	M_S
Ψ_1	$\varphi_0\alpha$	$\varphi_0\beta$	$\varphi_1\alpha$	$\varphi_1\beta$	$\varphi_{-1}\alpha$	$\varphi_2\alpha$	1
Ψ_2					$\varphi_{-1}\alpha$	$\varphi_2\beta$	0
Ψ_3					$\varphi_{-1}\beta$	$\varphi_2\alpha$	0
Ψ_4					$\varphi_{-1}\beta$	$\varphi_2\beta$	-1
Ψ_5					$\varphi_{-1}\alpha$	$\varphi_{-2}\alpha$	1
Ψ_6					$\varphi_{-1}\alpha$	$\varphi_{-2}\beta$	0
Ψ_7					$\varphi_{-1}\beta$	$\varphi_{-2}\alpha$	0
Ψ_8					$\varphi_{-1}\beta$	$\varphi_{-2}\beta$	-1
Ψ_9			$\varphi_{-1}\alpha$	$\varphi_{-1}\beta$	$\varphi_1\alpha$	$\varphi_2\alpha$	1
Ψ_{10}					$\varphi_1\alpha$	$\varphi_2\beta$	0
Ψ_{11}					$\varphi_1\beta$	$\varphi_2\alpha$	0
Ψ_{12}					$\varphi_1\beta$	$\varphi_2\beta$	-1
Ψ_{13}			$\varphi_{-1}\alpha$	$\varphi_{-1}\beta$	$\varphi_1\alpha$	$\varphi_{-2}\alpha$	1
Ψ_{14}					$\varphi_1\alpha$	$\varphi_{-2}\beta$	0
Ψ_{15}					$\varphi_1\beta$	$\varphi_{-2}\alpha$	0
Ψ_{16}					$\varphi_1\beta$	$\varphi_{-2}\beta$	-1

φ_2 occurs; now φ_2 taken alone does not form a basis for the representation e_2—only φ_2 and φ_{-2} taken together constitute a proper basis. Thus, we might not be surprised if symmetry adaptation produced considerable further mixing.

The electron distribution $\ldots a_2{}^2 e_1{}^3 e_2{}^1$ transforms as $E_1 \times E_2$; this yields orbital representations $B_1 + B_2 + E_1$ and states $^{1,3}B_1$, $^{1,3}B_2$, and $^{1,3}E_1$. The corresponding projection operators are

$$\hat{O}[B_1] = E - C_2 + [2C_3] - [2C_6] + [3C_2{}'] - [3C_2{}'']$$

$$\hat{O}[B_2] = E - C_2 + [2C_3] - [2C_6] - [3C_2{}'] + [3C_2{}'']$$

$$\hat{O}[E_1] = 2E - 2C_2 - [2C_3] + [2C_6] \tag{7.92}$$

However, in order to use these for the generation of symmetry-adapted configuration wave functions, we must know the manner in which the determinantal wave functions transform under the various symmetry operations of D_6.

We now outline the effects of the various symmetry operators $\hat{R}$ on Ψ_1. We tabulate the results below:

$$E\Psi_1 = \Psi_1 \tag{7.93}$$

$$C_2\Psi_1 = \mid C_2\varphi_0\alpha \times C_2\varphi_0\beta \times C_2\varphi_1\alpha \times C_2\varphi_1\beta \times C_2\varphi_{-1}\alpha \times C_2\varphi_2\alpha \mid$$

$$= \mid \omega^0\varphi_0\alpha \times \omega^0\varphi_0\beta \times \omega^3\varphi_1\alpha \times \omega^3\varphi_1\beta \times \omega^{-3}\varphi_{-1}\alpha \times \omega^6\varphi_2\alpha \mid$$

$$= (\omega^0\omega^0\omega^3\omega^3\omega^{-3}\omega^6)\Psi_1$$

$$= \omega^9\Psi_1 = \omega^3\Psi_1 \tag{7.94}$$

$$\text{ccw}; \ C_3\Psi_1 = \omega^{(0+0+2+2-2+4)}\Psi_1 = \Psi_1 \tag{7.95}$$

$$\text{cw}; \ C_3\Psi_1 = \Psi_1 \tag{7.96}$$

$$\text{ccw}; \ C_6\Psi_1 = \omega^3\Psi_1 \tag{7.97}$$

$$\text{cw}; \ C_6\Psi_1 = \omega^{-3}\Psi_1 = \omega^3\Psi_1 \tag{7.98}$$

$$ad; \ C_2{}'\Psi_1 = \varphi_0\alpha\varphi_0\beta\varphi_{-1}\alpha\varphi_{-1}\beta\varphi_1\alpha\varphi_{-2}\alpha = \Psi_{13} \ (\text{see Table 7.8}) \tag{7.99}$$

$$be; \ C_2{}'\Psi_1 = \Psi_{13} \tag{7.100}$$

$$cf; \ C_2{}'\Psi_1 = \Psi_{13} \tag{7.101}$$

$$ab, \ de; \ C_2{}'\Psi_1 = \omega^3\Psi_{13} \tag{7.102}$$

$$bc, \ ef; \ C_2{}''\Psi_1 = \omega^9\Psi_{13} = \omega^3\Psi_{13} \tag{7.103}$$

$$cd, \ af; \ C_2{}''\Psi_1 = \omega^{15}\Psi_{13} = \omega^3\Psi_{13} \tag{7.104}$$

Straightforward application of $\hat{O}[B_1]$ to Ψ_1 yields

$$\hat{O}[B_1]\Psi_1 = E\Psi_1 - C_2\Psi_1 + [2C_3]\Psi_1 - [2C_6]\Psi_1 + [3C_2']\Psi_1 - [3C_2'']\Psi_1$$

$$= \Psi_1 - \omega^3\Psi_1 + 2\Psi_1 - 2\omega^3\Psi_1 + 3\Psi_{13} - 3\omega^3\Psi_{13}$$

$$= 3(1 - \omega^3)(\Psi_1 + \Psi_{13}) \qquad (7.105)$$

Normalization yields $(\frac{1}{2})^{1/2}(\Psi_1 + \Psi_{13})$ which transforms as B_1 and has $S = 1$, $M_S = 1$. Similarly, it is found that

$$\hat{O}[B_2]\Psi_1 = 3(1 - \omega^3)(\Psi_1 - \Psi_{13}) \qquad (7.106)$$

so that $(\frac{1}{2})^{1/2}(\Psi_1 - \Psi_{13})$ transforms as B_2 and has $S = 1$, $M_S = 1$. Finally, and as might have been expected, one finds

$$\hat{O}[E_1]\Psi_1 = 0 \qquad (7.107)$$

so that Ψ_1 is shown to contain no part which transforms as E_1. One proceeds in this way through all the eigenfunctions of $\hat{S}^2$ tabulated in Eq. (7.91). The results are given in Table 7.9.

The process of complete adaptation can be quite lengthy; it cannot, however, be considered complicated. Finally, we wish to stress that the use of analytic MO's, as in Sec. 5(A)(iv), is not necessary. An example of

TABLE 7.9. *Symmetry-adapted eigenfunctions of $\hat{S}_z$ and $\hat{S}^2$ for benzene.*

$^{(2S+1)}\Gamma$	S	M_S	Wave function
$^3B_1$	1	1	$(\frac{1}{2})^{1/2}(\Psi_1 + \Psi_{13})$
		0	$\frac{1}{2}(\Psi_2 + \Psi_3 + \Psi_{14} + \Psi_{15})$
		-1	$(\frac{1}{2})^{1/2}(\Psi_4 + \Psi_{16})$
$^1B_1$	0	0	$\frac{1}{2}(\Psi_2 - \Psi_3 + \Psi_{14} - \Psi_{15})$
$^3B_2$	1	1	$(\frac{1}{2})^{1/2}(\Psi_1 - \Psi_{13})$
		0	$\frac{1}{2}(\Psi_2 + \Psi_3 - \Psi_{14} - \Psi_{15})$
		-1	$(\frac{1}{2})^{1/2}(\Psi_4 - \Psi_{16})$
$^1B_2$	0	0	$\frac{1}{2}(\Psi_2 - \Psi_3 - \Psi_{14} + \Psi_{15})$
$^3E_1$	1	1	Ψ_5, Ψ_9
		0	$(\frac{1}{2})^{1/2}(\Psi_6 + \Psi_7),\ (\frac{1}{2})^{1/2}(\Psi_{10} + \Psi_{11})$
		-1	Ψ_8, Ψ_{12}
$^1E_1$	0	0	$(\frac{1}{2})^{1/2}(\Psi_6 - \Psi_7),\ (\frac{1}{2})^{1/2}(\Psi_{10} - \Psi_{11})$

complete adaptation which avoids use of functional forms for the MO's is given in Sec. 5 of chapter 9.

EXERCISES

1. The molecule SO_2 is of symmetry C_{2v}. A reasonably complete basis set of atomic orbitals for MO calculations might be

$$2s_{O1}, 2s_{O2}, 2p_{xO1}, 2p_{xO2}, 2p_{yO1}, 2p_{yO2}, 2p_{zO1}, 2p_{zO2}$$

$$3s_S, 3p_{xS}, 3p_{yS}, 3p_{zS}, 4s_S$$

Construct a table of symmetry-adapted group orbitals similar to that of Table 7.2.

Answer: If we drop the identifying subscripts O and S, the result is given in Table 7.10.

TABLE 7.10.   *Transformation properties of symmetry-adapted GO's of SO_2.*

Γ_i	Sulfur AO's	Oxygen GO's
a_1	$3s$	$(\frac{1}{2})^{1/2}(2s_1 + 2s_2)$
	$3p_z$	$(\frac{1}{2})^{1/2}(2p_{z1} + 2p_{z2})$
	$4s$	$(\frac{1}{2})^{1/2}(2p_{y1} + 2p_{y2})$
a_2		$(\frac{1}{2})^{1/2}(2p_{x1} - 2p_{x2})$
b_1	$3p_x$	$(\frac{1}{2})^{1/2}(2p_{x1} + 2p_{x2})$
b_2	$3p_y$	$(\frac{1}{2})^{1/2}(2s_1 - 2s_2)$
		$(\frac{1}{2})^{1/2}(2p_{z1} - 2p_{z2})$
		$(\frac{1}{2})^{1/2}(2p_{y1} - 2p_{y2})$

2. The molecule bicyclohexatriene is shown in Fig. 7.4. Construct the representation for which the set of π AO's $\{\chi_a, \chi_b, \chi_c, \chi_d, \chi_e, \chi_f\}$ forms a basis. Reduce this representation. Construct the two a_1 and the one b_2 symmetry-adapted molecular orbitals.

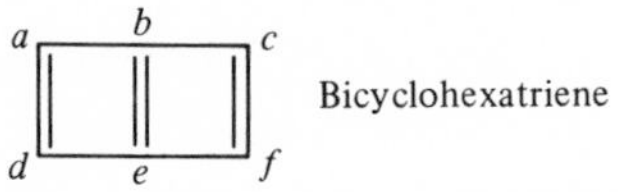

FIG. 7.4.   Bicyclohexatriene molecule. Atom sites are designated $a, b, \ldots$.

Answers: $\Gamma_{\text{red}} = 2A_1 + B_2 + A_2 + 2B_1$ (7.108)

$\varphi(a_1) = \frac{1}{2}(\chi_a + \chi_c + \chi_d + \chi_f)$

$\varphi(a_1) = (\frac{1}{2})^{1/2}(\chi_b + \chi_e)$

$\varphi(b_2) = \frac{1}{2}(\chi_a - \chi_c + \chi_d - \chi_f)$ (7.109)

3. (a) Write down the M_S eigenfunctions of a two-electron system in which each electron occupies a different MO.
(b) Construct the spin-projection operators.
(c) Generate the eigenfunctions of $\hat{S}^2$.

Answer: (a) $\alpha(1)\alpha(2), M_S = 1;$ $\alpha(1)\beta(2), M_S = 0$

$\beta(1)\alpha(2), M_S = 0;$ $\beta(1)\beta(2), M_S = 1$ (7.110)

(b) $\hat{O}[0] = \hat{S}^2 - 2$

$\hat{O}[1] = \hat{S}^2$ (7.111)

(c)
	$M_S = 1$	$S = 1$
$\alpha(1)\alpha(2);$		
$\alpha(1)\beta(2) + \alpha(2)\beta(1);$	0	1
$\beta(1)\beta(2);$	-1	1
$\alpha(1)\beta(2) - \alpha(2)\beta(1);$	0	0

(7.112)

where, for example, $\alpha(1)\beta(2)$ is *short hand* for $| \varphi_1\alpha(1)\varphi_2\beta(2) |$.

4. (a) Derive the three eigenfunctions of $\hat{S}^2$ with $S = \frac{1}{2}$ resulting from the $\alpha\beta\beta$, $\beta\alpha\beta$, and $\beta\beta\alpha$ eigenfunctions of $\hat{S}_z$.
(b) Find the two eigenfunctions (of the above three) which are linearly independent and orthonormal.

Answer: (a) $(1/6)^{1/2}(\beta\alpha\beta + \beta\beta\alpha - 2\alpha\beta\beta)$

$(1/6)^{1/2}(\alpha\beta\beta + \beta\beta\alpha - 2\beta\alpha\beta)$

$(1/6)^{1/2}(\alpha\beta\beta + \beta\alpha\beta - 2\beta\beta\alpha)$ (7.113)

(b) $(1/6)^{1/2}(\beta\alpha\beta + \beta\beta\alpha - 2\alpha\beta\beta)$

$(1/2)^{1/2}(\beta\beta\alpha - \beta\alpha\beta)$ (7.114)

where, for example, $\beta\beta\alpha$ is *short hand* for $| \varphi_1\beta(1)\varphi_2\beta(2)\varphi_3\alpha(3) |$.

5. (a) Use Eq. (7.32) to ascertain the number of independent eigenfunctions of $\hat{S}^2$ for a four-electron system. Verify using the branching diagram of Fig. 7.2.
(b) Write down the possible spin-projection operators.

Answer: $\hat{O}[0] = (\hat{S}_2 - 2)(\hat{S}_2 - 6)$

$\hat{O}[1] = \hat{S}^2(\hat{S}^2 - 6)$

$\hat{O}[2] = \hat{S}^2(\hat{S}^2 - 2)$ (7.115)

6. The $e^* \leftarrow e$ transition of a C_{4v} molecule is discussed in chapter 9, Sec. 5. Determine the symmetry-adapted configurational eigenfunctions of $\hat{S}_z$ and $\hat{S}^2$ using the projection operator techniques of the present chapter.

Answer: The answers are given in chapter 9, Sec. 5.

BIBLIOGRAPHY

[1]   P. O. Löwdin, *Rev. Mod. Phys.* **36,** 966 (1964); *Phys. Rev.* **97,** 1509 (1956).
[2]   R. Pauncz, *Alternant Molecular Orbital Method* (W. B. Saunders Co., Philadelphia, Pa., 1967).
[3]   F. A. Cotton, *Chemical Applications of Group Theory* (John Wiley & Sons, Inc., New York, 1963).

General References

1.   F. A. Cotton, *Chemical Applications of Group Theory* (John Wiley & Sons, Inc., New York, 1963).
2.   R. Pauncz, *Alternant Molecular Orbital Method* (W. B. Saunders Co., Philadelphia, Pa., 1967).

CHAPTER 8

Energy of Many-Electron Systems

The purpose of this chapter is to discuss the energy associated with a many-electron wave function—a wave function which is assumed to be symmetry adapted, spin adapted, and antisymmetrized. We do not concern ourselves with the nuclear motion problem (that is, rotations and vibrations): nor do we question the separability of electronic and nuclear motions (that is, the Born–Oppenheimer approximation). We simply assume the existence of a purely electronic energy and we seek to evaluate it.

Our discussion is elementary. Our purpose, in providing it, is associated with a desire for some degree of textual completeness, a wish to render certain parts of chapters 1, 5, and 6 more readable, and a compulsion to provide an access to more detailed writings on many-electron energies.

The plan of the chapter is as follows:

(i) The types of fully adapted many-electron single-configuration wave functions with which we have most concern are specified in Sec. 1.

(ii) Since the wave functions of Sec. 1 consist of a linear combination of Slater-determinant wave functions—a fact which also obtains for the multiconfiguration wave functions which result after configuration interaction—we detail the matrix representation of the energy and overlap operators in a Slater-determinant basis in Sec. 2.

(iii)   Thereafter, in Sec. 3, we present an example calculation of the energy associated with the ground configuration of butadiene. We do this without recourse to the rules evolved in Sec. 2—by so doing, we hope simultaneously to clarify these rules and to introduce certain pertinent notations.

(iv)   Section 4 provides a listing of energies associated with the wave functions of Sec. 1.

(v)   The topic of configuration interaction (CI) is broached in Sec. 5.

(vi)   The chapter concludes in Sec. 6 with a discussion of the reduction of MO integrals to AO integrals; this topic is pursued further in Appendix E.

(vii)   Exercises 3–7, in which the reader is asked to solve the ethylene π-electron problem up to and past the CI stage, is considered to be an integral part of this chapter.

1. WAVE FUNCTIONS

This topic has been discussed at some length in chapter 7. We assume that all MO's are nondegenerate and symmetry adapted.

(A) Ground Configuration

The ground configuration is that distribution of electrons among molecular orbitals which possesses the lowest energy. If the molecule contains an even number of electrons, $2n$, the corresponding single-configuration wave function is

$$^1\Psi_G{}^{(0)} = \hat{A}[\varphi_1\alpha(1)\varphi_1\beta(2)\ldots\varphi_n\alpha(2n-1)\varphi_n\beta(2n)] \qquad (8.1)$$

where the left superscript denotes the spin multiplicity $2S + 1$ and the right superscript denotes the degree of MO excitation relative to that of the ground configuration. This wave function is characterized by $S = 0$, $M_S = 0$.

If the molecule contains an odd number of electrons, $2n + 1$, the corresponding single-configuration wave functions are

$$^2\Psi_G{}^{(0)} = \hat{A}[\varphi_1\alpha(1)\varphi_1\beta(2)\ldots\varphi_n\alpha(2n-1)\varphi_n\beta(2n)\varphi_p\alpha(2n+1)] \quad (8.2a)$$

with $S = \frac{1}{2}$, $M_S = \frac{1}{2}$; and

$$^2\Psi_G{}^{(0)} = \hat{A}[\varphi_1\alpha(1)\varphi_1\beta(2)\ldots\varphi_n\alpha(2n-1)\varphi_n\beta(2n)\varphi_p\beta(2n+1)] \quad (8.2b)$$

with $S = \frac{1}{2}$, $M_S = -\frac{1}{2}$. These are the two components of a doublet state.

(B) Singly Excited Configurations

Consider a singly excited configuration in which an electron has been promoted from the filled MO φ_k to the unfilled[1] MO φ_s. If the number of electrons is $2n$, the corresponding singlet and triplet single-configuration wave functions are

$${}^1\Psi_{k \to s}{}^{(1)}$$

$$= (\tfrac{1}{2})^{1/2} \{ \hat{A}[\varphi_1\alpha(1)\varphi_1\beta(2)\ldots\varphi_k\alpha(2k-1)\varphi_s\beta(2k)\ldots\varphi_n\alpha(2n-1)\varphi_n\beta(2n)]$$

$$- \hat{A}[\varphi_1\alpha(1)\varphi_1\beta(2)\ldots\varphi_k\beta(2k-1)\varphi_s\alpha(2k)\ldots\varphi_n\alpha(2n-1)\varphi_n\beta(2n)]\}$$

$$(8.3)$$

with $S = 0$, $M_S = 0$; and

$${}^3\Psi_{k \to s}{}^{(1)}$$

$$= (\tfrac{1}{2})^{1/2} \{ \hat{A}[\varphi_1\alpha(1)\varphi_1\beta(2)\ldots\varphi_k\alpha(2k-1)\varphi_s\beta(2k)\ldots\varphi_n\alpha(2n-1)\varphi_n\beta(2n)]$$

$$+ \hat{A}[\varphi_1\alpha(1)\varphi_1\beta(2)\ldots\varphi_k\beta(2k-1)\varphi_s\alpha(2k)\ldots\varphi_n\alpha(2n-1)\varphi_n\beta(2n)]\}$$

$$(8.4)$$

with $S = 1$, $M_S = 0$. The generation of the other two components of the triplet state, those with $M_S = \pm 1$, constitutes Exercise 1.

(C) Doubly Excited Configuration

The simplest type of doubly excited configuration is obtained by promoting two electrons from one filled MO of ${}^1\Psi_G{}^{(0)}$ to one unfilled MO. The corresponding single-configuration wave function is

$${}^1\Psi_{k \to s}{}^{(2)} = \hat{A}[\varphi_1\alpha(1)\varphi_1\beta(2)\ldots\varphi_s\alpha(2k-1)\varphi_s\beta(2k)\ldots\varphi_n\alpha(2n-1)\varphi_n\beta(2n)]$$

$$(8.5)$$

with $S = 0$, $M_S = 0$.

2. MATRIX ELEMENTS IN A BASIS OF SLATER DETERMINANTS

Configuration wave functions consist of a linear combination of Slater determinants. The single-configuration wave functions of this chapter consisted of one [see Eq. (8.1)] or two [see Eqs. (8.3) and (8.4)] Slater determinants; the single-configuration wave functions for benzene consisted

---

[1] The terms *filled* and *unfilled* have meaning only with reference to the ground configuration.

of one, two, or four Slater determinants (see Table 7.9) ; the multiconfiguration wave functions which result from configuration interaction are also linear combinations of many Slater determinants. Thus, the evaluation of matrix elements in a single- or multi-configuration wave-function basis eventually devolves on the evaluation of the same elements in a basis consisting of single Slater determinants.

Slater determinants are eigenfunctions of $\hat{S}_z$. Since neither the overlap operator (which we may take to be unity or a unit matrix) nor the Hamiltonian operator [see Eq. (8.16)] contain any spin-dependent parts, it follows that

$$\langle \Psi_M(M_S) \mid \Psi_N(M_S') \rangle = \langle \Psi_M(M_S) \mid \hat{H} \mid \Psi_N(M_S') \rangle = 0 \qquad (8.6)$$

if $M_S \neq M_S'$. Similarly, since $\hat{H}$ transforms as the totally symmetric representation of the molecular point group, it follows that

$$\langle \Psi_M(\Gamma) \mid \Psi_N(\Gamma') \rangle = \langle \Psi_M(\Gamma) \mid \hat{H} \mid \Psi_N(\Gamma') \rangle = 0 \qquad (8.7)$$

unless Γ, the representation for which Ψ_M serves as basis, is identical to Γ'.

(A) Overlap Matrix

Slater determinants are usually constructed from an orthonormal set of MO's. Under such conditions, it may be shown that the Slater determinants constitute an orthonormal set. If the diagonal term of the Slater determinant Ψ_M be denoted $\Psi_{d,M}$, the proof of orthonormality proceeds as follows[2]:

$$\Delta_{MN} = \langle \Psi_M \mid \Psi_N \rangle = \langle (\hat{A}\Psi_{d,M}) \mid (\hat{A}\Psi_{d,N}) \rangle$$
$$= \langle \Psi_{d,M} \mid \hat{A}^\dagger \hat{A}\Psi_{d,N} \rangle = \langle \Psi_{d,M} \mid \hat{A}^2\Psi_{d,N} \rangle \qquad (8.8)$$

We may write $\hat{A}^2$ as

$$\hat{A}^2\Psi_d = (2n!)^{-1/2} \sum_{\hat{P}} (-1)^P \hat{P}\hat{A}\Psi_d \qquad (8.9)$$

Since $\hat{A}\Psi_d$ is an antisymmetrized wave function, it follows that $\hat{P}_{ij}(\hat{A}\Psi_d)$ equals $\hat{A}\Psi_d$ multiplied by -1 (that is, $-\hat{A}\Psi_d$); by the same token, multiple permutations merely produce $(-1)^P \hat{A}\Psi_d$. Since a total of $(2n!)$ permutations are possible, it follows that

$$\hat{A}^2\Psi_d = (2n!)^{1/2} \sum_{\hat{P}} (-1)^{2P} \hat{A}\Psi_d = \sum_{\hat{P}} (-1)^P \hat{P}\Psi_d \qquad (8.10)$$

where we have made the substitution $(-1)^P = (-1)^{3P}$. Consequently,

---

[2] Since $\hat{A}$ is a real Hermitian operator, $\hat{A}^\dagger = \hat{A}^* = \hat{A}$.

Eq. (8.8) may be rewritten as

$$\Delta_{MN} = \sum_{\hat{P}} (-1)^P \langle \Psi_{d,M} \mid \hat{P} \Psi_{d,N} \rangle \qquad (8.11)$$

If $\Psi_{d,M} = \Psi_{d,N}$ and if, for the moment, we restrict ourselves to the identity permutation (that is, $\hat{P}_0 \equiv 1$), we may expand the integral $\langle \Psi_{d,M} \mid \Psi_{d,M} \rangle$ as

$$\prod_{k=1}^{2n} \langle \varphi_k(\alpha \text{ or } \beta) \mid \varphi_k(\alpha \text{ or } \beta) \rangle$$

This latter quantity, by definition, equals unity. Any of the other permutations yield terms in the expansion of $\langle \Psi_{d,M} \mid \Psi_{d,M} \rangle$ which contains either spin orthogonalities such as $\langle \alpha \mid \beta \rangle$ or MO orthogonalities such as $\langle \varphi_k \mid \varphi_l \rangle$; all of these, by definition, equal zero. Therefore, $\Delta_{MM} = \Delta_{NN} = 1$. When $\Psi_{d,M} \neq \Psi_{d,N}$, all terms in the expansion of Eq. (8.11) contain at least one multiplicative factor of the form $\langle \varphi_k \mid \varphi_l \rangle$ or $\langle \alpha \mid \beta \rangle$, where $l \neq k$, and equal zero. In general then, it follows that

$$\Delta_{MN} = \delta_{M,N} \qquad (8.12)$$

where $\delta_{M,N}$ is the Kronecker δ.

Therefore, in a basis of Slater determinants constructed from an orthonormal set of MO's, the overlap matrix is a unit matrix.

(B) Energy Matrix

The matrix elements of $\hat{H}$ are given by

$$H_{MN} = \langle \Psi_M \mid \hat{H} \mid \Psi_N \rangle \qquad (8.13)$$

$$= \langle \Psi_{d,M} \mid \hat{H} \mid \hat{A}^2 \Psi_{d,N} \rangle \qquad (8.14)$$

$$= \sum_{\hat{P}} (-1)^P \langle \Psi_{d,M} \mid \hat{H} \mid \hat{P} \Psi_{d,N} \rangle \qquad (8.15)$$

Equation (8.13) implies normality of the Slater determinant basis; Eq. (8.14) implies commutativity of $\hat{H}$ and $\hat{A}$—which follows from the fact that $\hat{H}$ is independent of electron numbering whereas $\hat{A}$ is merely an electron-renumbering operator; Eq. (8.15) then follows from the same arguments as used in deriving Eq. (8.11). Further reduction beyond Eq. (8.15) depends on a more precise specification of the Hamiltonian $\hat{H}$.

The molecular Hamiltonian is written in the form

$$\hat{H} = \sum_{i=1}^{2n} \hat{H}_i^{\text{core}} + \sum_{i=1}^{2n} \sum_{j>i}^{2n} e^2/r_{ij} \qquad (8.16)$$

where the core represents the nuclear framework of the molecule plus all inner-shell (that is, non-valence-shell) electrons. The term $\hat{H}_i^{core}$ represents the kinetic energy of electron i and the potential energy of attraction of this electron to the core. The term e^2/r_{ij} is the potential energy of the repulsion between electron i and j. Thus, the Hamiltonian contains two types of terms: one-electron terms $\hat{H}_i^{core}$ and two-electron terms e^2/r_{ij}.

(C)   One-Electron Operators

It is easier to specify the rules for the one- and two-electron parts of $\hat{H}$ if we rewrite Ψ_d in terms of one-electron molecular spin orbitals ψ_r rather than in terms of the $\varphi_r\alpha$ or $\varphi_r\beta$ functions. Hence, Eq. (8.15) becomes

$$H_{MN} = \sum_{\hat{P}} (-1)^P$$

$$\times \langle \psi_1(1)\psi_2(2)\ldots\psi_r(i)\ldots\psi_{2n}(2n) \mid \hat{H} \mid \hat{P}\psi_1'(1)\psi_2'(2)\ldots\psi_r'(i)\ldots\psi_{2n}'(2n) \rangle$$

$$(8.17)$$

where the primes distinguish the molecular spin orbitals (MSO's) of $\Psi_{d,N}$ from those of $\Psi_{d,M}$. Now, the one-electron operator $\hat{H}_i^{core}$ depends only on the coordinates of electron i. Hence, $\langle \Psi_M \mid \hat{H}_i^{core} \mid \Psi_N \rangle$ is zero unless some permutation $\hat{P}'$ exists which renders

$$\hat{P}'[\psi_1'(1)\psi_2'(2)\ldots\psi_r'(i)\ldots\psi_{2n}'(2n)] = \psi_1(1)\psi_2(2)\ldots\psi_r'(i)\ldots\psi_{2n}(2n)$$

$$(8.18)$$

In other words, the set of primed functions must contain at least $2n - 1$ members of the unprimed set—say, $\psi_1, \psi_2, \ldots, \psi_{r-1}, \psi_{r+1}, \ldots \psi_{2n}$; equivalently, Ψ_M may differ from Ψ_N by no more than one MSO, ψ_r. In this case, we find

$$\langle \Psi_M \mid \hat{H}_i^{core} \mid \Psi_N \rangle = \pm \langle \psi_r(i) \mid \hat{H}_i^{core} \mid \psi_r'(i) \rangle \qquad (8.19)$$

where the sign is determined by the evenness or oddness of the permutation $\hat{P}'$. Thus, a number of cases may now be distinguished.

(*i*) $\Psi_M = \Psi_N$: Each term $\hat{H}_i^{core}$ of $\hat{H}$ yields a contribution of the form $\langle \psi_r(i) \mid \hat{H}_i^{core} \mid \psi_r(i) \rangle$. Hence

$$\langle \Psi_M \mid \sum_i \hat{H}_i^{core} \mid \Psi_M \rangle = \sum_{r=1}^{2n} \langle \psi_r(i) \mid \hat{H}_i^{core} \mid \psi_r(i) \rangle \qquad (8.20)$$

(*ii*) $\Psi_M \neq \Psi_N$; Ψ_N *differs from* Ψ_M *by no more than one*

MSO: From Eq. (8.19), the result is given by

$$\langle \Psi_M \mid \sum_i \hat{H}_i^{\text{core}} \mid \Psi_N \rangle = \pm \langle \psi_r(i) \mid \hat{H}_i^{\text{core}} \mid \psi_r{}'(i) \rangle \tag{8.21}$$

(iii) $\Psi_M \neq \Psi_N$; Ψ_N ***differs from*** Ψ_M ***by more than one***
MSO: In this case, we find

$$\langle \Psi_M \mid \sum_i \hat{H}_i^{\text{core}} \mid \Psi_N \rangle = 0 \tag{8.22}$$

(D) Two-Electron Operators

The two-electron operators are of the form $e^2/\mid \mathbf{r}_i - \mathbf{r}_j \mid$ and depend on the coordinates of both electrons i and j. Arguments analogous to those for the one-electron $\hat{H}_i^{\text{core}}$ case lead to the following rules:

(i) $\Psi_M \neq \Psi_N$; Ψ_N ***differs from*** Ψ_M ***by more than two***
MSO's: In this case, $\psi_r \neq \psi_r{}'$, $\psi_s \neq \psi_s{}'$ and $\psi_t \neq \psi_t{}'$

$$\langle \Psi_M \mid \sum_i \sum_{j>i} e^2/r_{ij} \mid \Psi_N \rangle = 0 \tag{8.23}$$

(ii) $\Psi_M \neq \Psi_N$; Ψ_N ***differs from*** Ψ_M ***by no more than two***
MSO's: In this case, $\psi_r \neq \psi_r{}'$ and $\psi_t \neq \psi_t{}'$ and there exists a permutation $\hat{P}''$ which yields

$$\hat{P}''[\psi_1{}'(1)\psi_2{}'(2)\ldots\psi_r{}'(i)\ldots\psi_t{}'(j)\ldots\psi_{2n}{}'(2n)]$$

$$= \psi_1(1)\psi_2(2)\ldots\psi_r{}'(i)\ldots\psi_t{}'(j)\ldots\psi_{2n}(2n) \tag{8.24}$$

Hence, we find a contribution to the matrix element given by

$$\langle \psi_r(i)\psi_t(j) \mid e^2/r_{ij} \mid \psi_r{}'(i)\psi_t{}'(j) \rangle$$

and which is prefixed by a $\pm$ sign determined by the parity of $\hat{P}''$. However, we also note that $\hat{P}_{ij}(\hat{P}''\psi_M)$ also yields a possible nonvanishing term $\langle \psi_r(i)\psi_t(j) \mid e^2/r_{ij} \mid \psi_r{}'(j)\psi_t{}'(i) \rangle$ and that $\hat{P}_{ij}$ is of odd parity. Therefore, we find a total of

$$\langle \Psi_M \mid \sum_i \sum_{j>i} e^2/r_{ij} \mid \Psi_N \rangle = \pm(1 - \hat{P}_{ij}) \langle \psi_r(i)\psi_t(j) \mid e^2/r_{ij} \mid \psi_r{}'(i)\psi_t{}'(j) \rangle$$

$$\tag{8.25}$$

where $\hat{P}_{ij}$ is restricted to operation on the ket.

(iii) $\Psi_N \neq \Psi_M$; Ψ_M ***differs from*** Ψ_N ***by only one MSO:***
In this case, $\psi_t \neq \psi_t{}'$; consequently, the choice of the index r is not impor-

tant—except insofar as $r \neq t$. We find

$$\langle \Psi_M \mid \sum_i \sum_{j>i} e^2/r_{ij} \mid \Psi_N \rangle = \pm \sum_{r \neq t}^{2n} (1 - \hat{P}_{ij}) \langle \psi_r(i)\psi_t(j) \mid e^2/r_{ij} \mid \psi_r(i)\psi_t{}'(j) \rangle$$

$$(8.26)$$

   (*iv*) $\Psi_N = \Psi_M$: In this instance, we need not impose any restrictions on the index. Hence, we find

$$\langle \Psi_M \mid \sum_i \sum_{j>i} e^2/r_{ij} \mid \Psi_N \rangle = \sum_t \sum_{r \neq t} (1 - \hat{P}_{ij}) \langle \psi_r(i)\psi_t(j) \mid e^2/r_{ij} \mid \psi_r(i)\psi_t(j) \rangle$$

$$(8.27)$$

It is clear that the H matrix in a Slater-determinant basis is not diagonal. The rules we have derived aid in determining those matrix elements which are zero. If, based on the quoted rules, there is a possibility that a given element be nonzero, the value of this element must be determined by actual calculation.

3. GROUND CONFIGURATION OF BUTADIENE: AN EXAMPLE CALCULATION

We now present an extended development of the energy of a four-electron system characterized by a wave function $^1\Psi_G{}^{(0)}$. The system involved need not be any specified molecule—He_2, Li_2^{++}, or HLi are suitable candidates. However, insofar as it aids visualization, we choose to let the problem represent the π-electron system of butadiene for which

$$^1\Psi_G{}^{(0)} = \hat{A}[\varphi_1\alpha(1)\varphi_1\beta(2)\varphi_2\alpha(3)\varphi_2\beta(4)]$$

$$= \hat{A}[\psi_1(1)\psi_2(2)\psi_3(3)\psi_4(4)] \tag{8.28}$$

The rules of Sec. 2 may be applied to the wave function of Eq. (8.28) and the element $\langle ^1\Psi_G{}^{(0)} \mid \hat{H} \mid ^1\Psi_G{}^{(0)} \rangle$ evaluated. We expect the reader to do this. However, we use the middle wave function of Eq. (8.28); we do not use any rules. We believe that such a procedure—extended development in the MO formulation and use of the rules of Sec. 2 in the MSO formulation—leads to a deeper understanding of both procedures.

  Since

$$\hat{H} = \sum_i \hat{H}_i{}^{\text{core}} + \sum_i \sum_{j>i} e^2/r_{ij}$$

it is convenient to break the evaluation of $\langle \hat{H} \rangle$ into three parts: We evaluate $\langle \sum_i \hat{H}_i{}^{\text{core}} \rangle$ in Sec. 3(A), $\langle \sum_i \sum_{i<j} e^2/r_{ij} \rangle$ in Sec. 3(B), and we add both of these expectation values to obtain $\langle \hat{H} \rangle$ in Sec. 3(C).

(A) Evaluation of $\langle {}^1\Psi_G{}^{(0)} \mid \sum_i \hat{H}_i{}^{\text{core}} \mid {}^1\Psi_G{}^{(0)} \rangle$

Consider the expectation value of $\hat{H}_i{}^{\text{core}}$ in the identity permutation, $\hat{P}_0 \equiv 1$. The appropriate part of Eq. (8.15), in this instance, is

$$\langle \Psi_{d,G} \mid \sum_{i=1}^{4} \hat{H}_i{}^{\text{core}} \mid \Psi_{d,G} \rangle$$

$$= \langle \varphi_1\alpha(1)\varphi_1\beta(2)\varphi_2\alpha(3)\varphi_2\beta(4) \mid \hat{H}_1{}^{\text{core}} + \hat{H}_2{}^{\text{core}} + \hat{H}_3{}^{\text{core}} + \hat{H}_4{}^{\text{core}}$$

$$\times \mid \varphi_1\alpha(1)\varphi_1\beta(2)\varphi_2\alpha(3)\varphi_2\beta(4) \rangle$$

$$= \langle \varphi_1\alpha(1) \mid \hat{H}_1{}^{\text{core}} \mid \varphi_1\alpha(1) \rangle\langle \varphi_1\beta(2) \mid \varphi_1\beta(2) \rangle\langle \varphi_2\alpha(3) \mid \varphi_2\alpha(3) \rangle$$

$$\times \langle \varphi_2\beta(4) \mid \varphi_2\beta(4) \rangle + \langle \varphi_1\alpha(1) \mid \varphi_1\alpha(1) \rangle\langle \varphi_1\beta(2) \mid \hat{H}_2{}^{\text{core}} \mid \varphi_1\beta(2) \rangle$$

$$\times \langle \varphi_2\alpha(3) \mid \varphi_2\alpha(3) \rangle\langle \varphi_2\beta(4) \mid \varphi_2\beta(4) \rangle + \langle \varphi_1\alpha(1) \mid \varphi_1\alpha(1) \rangle$$

$$\times \langle \varphi_1\beta(2) \mid \varphi_1\beta(2) \rangle\langle \varphi_2\alpha(3) \mid \hat{H}_3{}^{\text{core}} \mid \varphi_2\alpha(3) \rangle\langle \varphi_2\beta(4) \mid \varphi_2\beta(4) \rangle$$

$$+ \langle \varphi_1\alpha(1) \mid \varphi_1\alpha(1) \rangle\langle \varphi_1\beta(2) \mid \varphi_1\beta(2) \rangle\langle \varphi_2\alpha(3) \mid \varphi_2\alpha(3) \rangle$$

$$\times \langle \varphi_2\beta(4) \mid \hat{H}_4{}^{\text{core}} \mid \varphi_2\beta(4) \rangle \tag{8.29}$$

Since all overlap integrals $\langle \varphi_r(\alpha \text{ or } \beta) \mid \varphi_s(\alpha \text{ or } \beta) \rangle = \delta_{rs}\delta_{\alpha,\beta}$, it follows that

$$\langle \Psi_{d,G} \mid \sum_{i=1}^{4} \hat{H}_i{}^{\text{core}} \mid \Psi_{d,G} \rangle = 2h_1 + 2h_2 \tag{8.30}$$

where

$$h_r \equiv \langle \varphi_r(i) \mid \hat{H}_i{}^{\text{core}} \mid \varphi_r(i) \rangle \tag{8.31}$$

Consider the expectation value of $\sum \hat{H}_i{}^{\text{core}}$ in all the remaining permutations. The permutation operators to be considered are $\hat{P}_{12}$, $\hat{P}_{13}$, $\hat{P}_{14}$, $\hat{P}_{23}$, $\hat{P}_{24}$, and $\hat{P}_{34}$. Of these, $\hat{P}_{12}$, $\hat{P}_{23}$, $\hat{P}_{34}$, and $\hat{P}_{14}$ introduce spin orthogonalities which cannot be removed by an operator which contains no spin-dependent parts. Thus, we find

$$\langle {}^1\Psi_G{}^{(0)} \mid \sum_i \hat{H}_i{}^{\text{core}} \mid {}^1\Psi_G{}^{(0)} \rangle = - \langle \Psi_{d,G} \mid \sum_{i=1}^{4} \hat{H}_i{}^{\text{core}} \mid (-\hat{P}_0 + \hat{P}_{13} + \hat{P}_{24})\Psi_{d,G} \rangle$$

$$\tag{8.32}$$

The $\hat{P}_0$ permutation yields Eq. (8.30). The $\hat{P}_{13}$ permutation yields

$$\langle \Psi_{d,G} \mid \sum_{i=1}^{4} \hat{H}_i{}^{\text{core}} \mid \hat{P}_{13}\Psi_{d,G} \rangle$$

$$= - \langle \varphi_1\alpha(1)\varphi_1\beta(2)\varphi_2\alpha(3)\varphi_2\beta(4) \mid \hat{H}_1{}^{\text{core}} + \hat{H}_2{}^{\text{core}} + \hat{H}_3{}^{\text{core}} + \hat{H}_4{}^{\text{core}} \mid$$

$$\times \mid \varphi_1\alpha(3)\varphi_1\beta(2)\varphi_2\alpha(1)\varphi_2\beta(4) \rangle \tag{8.33}$$

The operator $\hat{H}_1^{\text{core}}$ can remove the orbital orthogonality of $\varphi_1\alpha(1)$ and $\varphi_2\alpha(1)$ *via* the integral $\langle \varphi_1\alpha(1) \mid \hat{H}_1^{\text{core}} \mid \varphi_2\alpha(1) \rangle$; however, the total integral on the right-hand side of Eq. (8.33) also contains the multiplicative factor $\langle \varphi_2\alpha(3) \mid \varphi_1\alpha(3) \rangle = 0$. In this manner, it is readily shown that the total of Eq. (8.33) is zero. It follows, similarly, that

$$\langle \Psi_{d,G} \mid \sum_{i=1}^{4} \hat{H}_i^{\text{core}} \mid \hat{P}_{24}\Psi_{d,G} \rangle = 0.$$

Adding all terms, we find

$$\langle {}^1\Psi_G{}^{(0)} \mid \sum_{i=1}^{4} \hat{H}_i^{\text{core}} \mid {}^1\Psi_G{}^{(0)} \rangle = 2h_1 + 2h_2 \tag{8.34}$$

The only contributions to $\langle {}^1\Psi_G{}^{(0)} \mid \sum_i \hat{H}_i^{\text{core}} \mid {}^1\Psi_G{}^{(0)} \rangle$ arise from the identity permutation of $\Psi_{d,G}$. This, of course, is also the conclusion of Eq. (8.20).

(B)   Evaluation of $\langle {}^1\Psi_G{}^{(0)} \mid \sum_{j>i} \sum_i e^2/r_{ij} \mid {}^1\Psi_G{}^{(0)} \rangle$

Consider the expectation value of $\sum_i \sum_{j>i} e^2/r_{ij}$ in the identity permutation. The integral to be evaluated is

$$\langle \Psi_{d,G} \mid \sum_i^4 \sum_{j>i} e^2/r_{ij} \mid \hat{P}_0\Psi_{d,G} \rangle = \langle \varphi_1\alpha(1)\,\varphi_1\beta(2)\,\varphi_2\alpha(3)\,\varphi_2\beta(4) \mid e^2/r_{12}$$

$$+ e^2/r_{13} + e^2/r_{14} + e^2/r_{23} + e^2/r_{24}$$

$$+ e^2/r_{34} \mid \varphi_1\alpha(1)\,\varphi_1\beta(2)\,\varphi_2\alpha(3)\,\varphi_2\beta(4) \rangle \quad (8.35)$$

This integral decomposes into six nonvanishing terms:

$$\langle \varphi_1\alpha(1)\,\varphi_1\beta(2) \mid e^2/r_{12} \mid \varphi_1\alpha(1)\,\varphi_1\beta(2) \rangle \langle \varphi_2\alpha(3) \mid \varphi_2\alpha(3) \rangle \langle \varphi_2\beta(4) \mid \varphi_2\beta(4) \rangle$$

$$\equiv \langle 11 \mid 11 \rangle \equiv (11 \mid 11)$$

$$\langle \varphi_2\alpha(3)\,\varphi_2\beta(4) \mid e^2/r_{34} \mid \varphi_2\alpha(3)\,\varphi_2\beta(4) \rangle \langle \varphi_1\alpha(1) \mid \varphi_1\alpha(1) \rangle \langle \varphi_1\beta(2) \mid \varphi_1\beta(2) \rangle$$

$$\equiv \langle 22 \mid 22 \rangle \equiv (22 \mid 22) \quad (8.36)$$

which are termed *self-interaction integrals*, and

$$\langle \varphi_1\alpha(1)\,\varphi_2\alpha(3) \mid e^2/r_{13} \mid \varphi_1\alpha(1)\,\varphi_2\alpha(3) \rangle \cdot S_{11} \cdot S_{22} = \langle 12 \mid 12 \rangle \equiv (11 \mid 22)$$

$$\langle \varphi_1\alpha(1)\,\varphi_2\beta(4) \mid e^2/r_{14} \mid \varphi_1\alpha(1)\,\varphi_2\beta(4) \rangle \cdot S_{11} \cdot S_{22} = \langle 12 \mid 12 \rangle \equiv (11 \mid 22)$$

$$\langle \varphi_1\beta(2)\,\varphi_2\alpha(3) \mid e^2/r_{23} \mid \varphi_1\beta(2)\,\varphi_2\alpha(3) \rangle \cdot S_{11} \cdot S_{22} = \langle 12 \mid 12 \rangle \equiv (11 \mid 22)$$

$$\langle \varphi_1\beta(2)\,\varphi_2\beta(4) \mid e^2/r_{24} \mid \varphi_1\beta(2)\,\varphi_2\beta(4) \rangle \cdot S_{11} \cdot S_{22} = \langle 12 \mid 12 \rangle \equiv (11 \mid 22)$$

$$(8.37)$$

which are termed *Coulomb integrals*. The abbreviation of Eq. (8.36) to $(kk \mid kk)$ utilizes the orthonormality of the MO's, namely, $\langle \varphi_r \mid \varphi_s \rangle = \delta_{r,s}$, and the definition

$$\langle \varphi_r(i)\varphi_s(j) \mid e^2/r_{ij} \mid \varphi_t(i)\varphi_u(j) \rangle = \langle rs \mid tu \rangle \equiv (rt \mid su) \qquad (8.38)$$

The derivation of Eq. (8.37) utilizes all the above and, in addition, abbreviates $\langle \varphi_r \mid \varphi_r \rangle$ to S_{rr}.

Consider now the expectation value of $\sum_i \sum_{j>i} e^2/r_{ij}$ in all remaining permutations. As in Sec. 3(A), because this operator contains no spin-dependent parts, the only nonvanishing terms derive from the permutation operators $\hat{P}_{13}$ and $\hat{P}_{24}$ applied to $\Psi_{d,G}$. The resulting terms are

$$- \langle \varphi_1\alpha(1)\varphi_2\alpha(3) \mid e^2/r_{13} \mid \varphi_2\alpha(1)\varphi_1\alpha(3) \rangle = - \langle 12 \mid 21 \rangle \equiv - (12 \mid 21)$$

$$- \langle \varphi_1\beta(2)\varphi_2\beta(4) \mid e^2/r_{24} \mid \varphi_2\beta(2)\varphi_1\beta(4) \rangle = - \langle 12 \mid 21 \rangle \equiv - (12 \mid 21)$$

$$\qquad (8.39)$$

which are termed *exchange integrals*.

(C) Evaluation of $\langle {}^1\Psi_G{}^{(0)} \mid \hat{H} \mid {}^1\Psi_G{}^{(0)} \rangle$

Addition of Eqs. (8.34), (8.36), (8.37), and (8.39) yields the energy associated with ${}^1\Psi_G{}^{(0)}$ as

$${}^1E_G{}^{(0)} = 2h_1 + 2h_2 + (11 \mid 11) + (22 \mid 22) + 4(11 \mid 22) - 2(12 \mid 21) \qquad (8.40)$$

(D) Notation

A number of notations have been introduced in Secs. 3(B) and 3(C). It now seems proper to generalize this notation, collect it for further use, and correlate it with similar notations used in chapter 1. In this way, the similarity of the many-electron problem in both orbital basis sets, MO and AO, is made obvious.

In the AO basis set of chapter 1, the Coulomb integral was defined as

$$J_{\mu\nu} = \langle \chi_\mu(i)\chi_\nu(j) \mid e^2/r_{ij} \mid \chi_\mu(i)\chi_\nu(j) \rangle = \langle \chi_\mu\chi_\nu \mid \chi_\mu\chi_\nu \rangle = \langle \mu\nu \mid \mu\nu \rangle \qquad (1.33)$$

In the MO basis set of this chapter, we define J_{rs} as

$$J_{rs} = \langle \varphi_r(i)\varphi_s(j) \mid e^2/r_{ij} \mid \varphi_r(i)\varphi_s(j) \rangle = \langle \varphi_r\varphi_s \mid \varphi_r\varphi_s \rangle$$

$$= \langle rs \mid rs \rangle = (rr \mid ss) \qquad (8.37')$$

The Coulomb integral may be considered to represent the repulsion of two electrons, one in the MO φ_r and the other in the MO φ_s.

In the AO basis set of chapter 1, the exchange integral was defined as

$$K_{\mu\nu} = \langle \chi_\mu(i)\chi_\nu(j) \mid e^2/r_{ij} \mid \chi_\nu(i)\chi_\mu(j) \rangle = \langle \mu\nu \mid \nu\mu \rangle \qquad (1.34)$$

In the MO basis set, the corresponding definition is

$$K_{rs} = \langle \varphi_r(i)\varphi_s(j) \mid e^2/r_{ij} \mid \varphi_s(i)\varphi_r(j) \rangle = \langle rs \mid sr \rangle = (rs \mid sr) \qquad (8.39')$$

The exchange integral has no classical analogue; it may be considered to represent the repulsion of two electrons each of which, about a given point, possesses a density equal to the overlap $\varphi_r^*\varphi_s d\tau$ or $\chi_\mu^*\chi_\nu^* d\tau$ about that same point.

The *general repulsion integral* is given, in both instances, by

$$\langle \chi_\mu(i)\chi_\nu(j) \mid e^2/r_{ij} \mid \chi_\zeta(i)\chi_\eta(j) \rangle = \langle \mu\nu \mid \zeta\eta \rangle$$

$$\langle \varphi_r(i)\varphi_s(j) \mid e^2/r_{ij} \mid \varphi_t(i)\varphi_u(j) \rangle = \langle rs \mid tu \rangle = (rt \mid su) \qquad (8.38)$$

The *core integrals* are defined as

$$I_{\mu\nu} \equiv \langle \chi_\mu(i) \mid \hat{H}_{(i)}{}^c \mid \chi_\nu(i) \rangle \qquad (1.32)$$

$$h_{rs} = \langle \varphi_r(i) \mid \hat{H}_i{}^{\text{core}} \mid \varphi_s(i) \rangle \qquad (8.31')$$

4.   GENERAL ENERGY EXPRESSIONS

The energy expressions quoted in this section refer to the single-configuration wave functions of Sec. 1. For a molecule containing $2n$ electrons, we may rewrite Eq. (8.40) as

$$^1E_G{}^{(0)} = 2 \sum_{k=1}^{n} h_k + \sum_{k=1}^{n} (kk \mid kk) + 4 \sum_{k<l}^{n} \sum_{l}^{n} (kk \mid ll) - 2 \sum_{k<l}^{n} \sum_{l}^{n} (kl \mid lk)$$

$$= 2 \sum_{k=1}^{n} h_k + \sum_{k=1}^{n} J_{kk} + 4 \sum_{k<l}^{n} \sum_{l}^{n} J_{kl} - 2 \sum_{k<l}^{n} \sum_{l}^{n} K_{kl}$$

$$= 2 \sum_{k=1}^{n} h_k + (2 \sum_{k}^{n} \sum_{l}^{n} J_{kl} - \sum_{k=1}^{n} J_{kk}) - 2 \sum_{k<l}^{n} \sum_{l}^{n} K_{kl} \qquad (8.41)$$

where

$$J_{kl} \equiv (kk \mid ll); \qquad K_{kl} \equiv (kl \mid lk) \qquad (8.42)$$

From the definition of Eq. (8.42); it follows that $J_{kl} = J_{lk}$, $K_{kl} = K_{lk}$, and $J_{kk} = K_{kk}$. Thus, since $J_{kk} = K_{kk}$, we may further abbreviate Eq. (8.41) to

$$^1E_G{}^{(0)} = 2 \sum_{k=1}^{n} h_k + \sum_{k}^{n} \sum_{l}^{n} (2J_{kl} - K_{kl}) \qquad (8.43)$$

where the summation runs over all fully occupied MO's 1 through n. If the molecule contains an odd number of electrons, $2n + 1$, and the extra electron is in the pth MO, we find

$$^2E_G^{(0)} = 2 \sum_{k=1}^{n} h_k + h_p + \sum_{k}^{n} \sum_{l}^{n} (2J_{kl} - K_{kl}) + \sum_{k=1}^{n} (2J_{kp} - K_{kp}) \quad (8.44)$$

The energy of the singly excited singlet state is

$$^1E_{k \to s}^{(1)} = 2 \sum_{m=1}^{n} h_m + (h_s - h_k) + \sum_{m}^{n} \sum_{l}^{n} (2J_{ml} - K_{ml})$$

$$- \sum_{l=1}^{n} [(2J_{kl} - K_{kl}) - (2J_{ls} - K_{ls})] - J_{ks} + 2K_{ks} \quad (8.45)$$

and the energy difference from the ground state is

$$^1E_{k \to s}^{(1)} - {}^1E_G^{(0)} = h_s - h_k$$

$$- \sum_{l=1}^{n} [(2J_{kl} - K_{kl}) - (2J_{ls} - K_{ls})] - J_{ks} + 2K_{ks} \quad (8.46)$$

The energy of the singly excited triplet state is

$$^3E_{k \to s}^{(1)} = {}^1E_{k \to s}^{(1)} - 2K_{ks} \quad (8.47)$$

and the energy difference from the ground state is

$$^3E_{k \to s}^{(1)} - {}^1E_G^{(0)} = h_s - h_k - \sum_{l=1}^{n} [(2J_{kl} - K_{kl}) - (2J_{ls} - K_{ls})] - J_{ks}$$

$$(8.48)$$

The energy of the most simple type of doubly excited configuration is

$$^1E_{k \to s}^{(2)} = 2 \sum_{l=1}^{n} h_l + 2(h_s - h_k) + \sum_{m}^{n} \sum_{l}^{n} (2J_{ml} - K_{ml})$$

$$- \sum_{l=1}^{n} [(2J_{kl} - K_{kl}) - (2J_{ls} - K_{ls})] + J_{ss} - 2J_{ks} + K_{ks} \quad (8.49)$$

and the energy difference from the ground state is

$$^1E_{k \to s}^{(2)} - {}^1E_G^{(0)} = 2(h_s - h_k) - \sum_{l=1}^{n} [(2J_{kl} - K_{kl}) - (2J_{ls} - K_{ls})]$$

$$+ J_{ss} - 2J_{ks} + K_{ks} \quad (8.50)$$

5. CONFIGURATION INTERACTION

The single-configuration wave functions constitute an orthonormal set. In other words,

$$S_{M,N}{}^{SC} = \langle \Psi_M(S, M_S, \Gamma) \mid \Psi_N(S', M_{S'}, \Gamma') \rangle = \delta_{M,N} \qquad (8.51)$$

where the superscript SC on S denotes *single configuration* and where $\delta_{M,N}$ contains the subspecification $\delta_{S,S'}\delta_{M_S,M_{S'}}\delta_{\Gamma,\Gamma'}$. This conclusion follows from the discussion of Sec. 2 and the fact that the specifications $S = S'$, $M_S = M_{S'}$, and $\Gamma = \Gamma'$ require each member of the linearly independent set of $\Psi_M(S, M_S, \Gamma)$ wave functions to contain at least one MO which is different from those contained in any other member of the same set of $\Psi(S, M_S, \Gamma)$ wave functions.

The energy matrix, however, is not diagonal. Indeed, following Sec. 2, we find

$$\langle \Psi_M(S, M_S, \Gamma) \mid \hat{H} \mid \Psi_N(S', M_{S'}, \Gamma') \rangle = H_{MN}{}^{SC}\delta_{S,S'}\delta_{M_S,M_{S'}}\delta_{\Gamma,\Gamma'} \qquad (8.52)$$

where $H_{MN}{}^{SC}$ can be evaluated, per Sec. 2, by decomposing it into its component Slater-determinant matrix elements H_{MN}.

Diagonalization of the matrix

$$[H_{MN}{}^{SC} - E\delta_{M,N}] = [\langle \Psi_M(S, M_S, \Gamma) \mid \hat{H} \mid \Psi_N(S', M_{S'}, \Gamma') \rangle] - E[I]$$

$$(8.53)$$

where the unit matrix $[I]$ is also the overlap matrix, yields a new set of energies $E_J{}^{CI}$ and wave functions

$$\Psi_J{}^{CI} = \sum_M C_{JM}\Psi_M(S, M_S, \Gamma) \qquad (8.54)$$

which are multiconfigurational in nature. The process leading to the $\Psi_J{}^{CI}$ is known as configuration interaction and the wave functions are often said to be *CI corrected*. The Variation theorem dictates that the lowest-energy multiconfiguration wave function of given S, M_S, and Γ is of lower energy than the lowest-energy single-configuration wave function of the same S, M_S, and Γ. It is in this sense that the multiconfiguration wave functions are considered *better than*, or *corrected relative to*, the single-configuration wave functions.

(A) Off-Diagonal Elements $H_{MN}{}^{SC}$

Consider the interaction of the ground single-configuration wave function and a singly excited single-configuration wave function. The interaction element is

$$H_{G,k \to s}{}^{SC} = \langle {}^1\Psi_G{}^{(0)} \mid \hat{H} \mid {}^1\Psi_{k \to s}{}^{(1)} \rangle \qquad (8.55)$$

The two single configurations contain, together, no more than two differently populated MO's φ_k and φ_s. Therefore, both $\langle \sum_i \hat{H}_i^{\text{core}} \rangle$ and $\langle \sum_i \sum_{j>i} e^2/r_{ij} \rangle$ are nonzero. Rewriting Eq. (8.55) in a form which involves the diagonal parts of the Slater determinants, and utilizing only the identity permutation, we find that $H_{G,k \to s}{}^{\text{SC}}$ contains the contribution

$$H_{G,k \to s}{}^{\text{SC}}(\hat{P}_0) = 2^{-1/2} \langle \varphi_1 \alpha(1) \ldots \varphi_k \alpha(2k-1) \varphi_k \beta(2k) \ldots \varphi_n \beta(2n) \mid$$

$$\times \mid \sum_i^{2n} \hat{H}_i^{\text{core}} + \sum_i^{2n} \sum_{j>i}^{2n} e^2/r_{ij} \mid$$

$$\times \mid [\varphi_1 \alpha(1) \ldots \varphi_k \alpha(2k-1) \varphi_s \beta(2k) \ldots \varphi_n \beta(2n)$$

$$- \varphi_1 \alpha(1) \ldots \varphi_k \beta(2k-1) \varphi_s \alpha(2k) \ldots \varphi_n \beta(2n)] \rangle \quad (8.56)$$

The integral involving the first term of the ket on the right-hand side of Eq. (8.56) is evaluated as

$$2^{-1/2} [h_{ks} + \sum_{m \neq k} 2(mm \mid ks) + (kk \mid ks)] \quad (8.57)$$

The integral involving the second term of the ket of Eq. (8.59) is zero—because $\hat{H}$ cannot remove the spin orthogonality $\langle \alpha \mid \beta \rangle = 0$ associated with the electrons numbered $2k$ or $2k-1$.

We now consider the permutation operator $(-1)\hat{P}_{2k-1,2k}$ acting on the second term of the ket in Eq. (8.56). The result is formally identical to the first term in the same ket. Consequently, we find another set of integrals identical to that of Eq. (8.57). Furthermore, the operator $\hat{P}_{2k-1,2k}$—which belongs to the subclass of permutation operators, $\hat{P}^{\alpha\beta}$, which interchange electrons with different spin components—is seen to reduce Eq. (8.55) to the form

$$H_{G,k \to s}{}^{\text{SC}} = \sum_{\hat{P}} 2^{1/2}(-1)^P$$

$$\times \langle \varphi_1 \alpha(1) \ldots \varphi_k \alpha(2k-1) \varphi_k \beta(2k) \ldots \varphi_n \beta(2n) \mid \hat{H}\hat{P} \mid$$

$$\times \mid \varphi_1 \alpha(1) \ldots \varphi_k \alpha(2k-1) \varphi_s \beta(2k) \ldots \varphi_n \beta(2n) \rangle \quad (8.58)$$

In the identity permutation, $\hat{P}_0 \equiv 1$, the evaluation of this integral yields a result which is clearly twice that given in Eq. (8.57).

Any e^2/r_{ij} operator for which $\langle e^2/r_{ij} \rangle$ is nonzero must remove the orthogonality of the φ_k and φ_s molecular orbitals. Consequently, the only permutations which yield nonzero results in Eq. (8.58) are those restricted to the subclass $\sum_{i \neq 2k} \hat{P}_{i,2k}{}^{\beta\beta}$ of permutation operators. The evaluation of

Eq. (8.58), with

$$\hat{P} \equiv \sum_{i \neq 2k}^{2n} \hat{P}_{i,2k}{}^{\beta\beta}$$

yields

$$-2^{1/2}\Big[\sum_m (ms \mid km) + (ks \mid kk)\Big] \tag{8.59}$$

Adding twice the results of Eq. (8.57) to those of Eq. (8.59), we find

$$H_{G,k\rightarrow s}{}^{\mathrm{SC}} = 2^{1/2}\{h_{ks} + \sum_{m=1}^{n} [2(mm \mid ks) - (mk \mid sm)]\} \tag{8.60}$$

As a second example of configuration mixing, let us consider the interaction of the two singly excited single-configuration wave functions ${}^{1}\Psi_{k\rightarrow s}{}^{(1)}$ and ${}^{1}\Psi_{l\rightarrow r}{}^{(1)}$, where $k \neq l$ and $r \neq s$. These two configurations contain a total of four differently populated MO's: φ_k, φ_l, φ_r, and φ_s. Therefore, only the expectation value of $\sum_i \sum_{j>i} e^2/r_{ij}$ can be nonzero. In fact, it is found that

$$\langle {}^{1}\Psi_{k\rightarrow s}{}^{(1)} \mid \hat{H} \mid {}^{1}\Psi_{l\rightarrow r}{}^{(1)} \rangle = 2(lr \mid ks) - (lk \mid sr) \tag{8.61}$$

General formulas for a number of other types of off-diagonal terms are

$$\langle {}^{1}\Psi_{k\rightarrow s}{}^{(1)} \mid \hat{H} \mid {}^{1}\Psi_{k\rightarrow r}{}^{(1)} \rangle = h_{rs} + \sum_{m=1}^{n} [2(mm \mid rs) - (mr \mid sm)]$$

$$+ 2(kr \mid sk) - (kk \mid rs) \tag{8.62}$$

$$\langle {}^{1}\Psi_{k\rightarrow s}{}^{(1)} \mid \hat{H} \mid {}^{1}\Psi_{l\rightarrow s}{}^{(1)} \rangle = -h_{lk} - \sum_{m=1}^{n} [2(mm \mid lk) - (ml \mid km)]$$

$$+ 2(ls \mid ks) - (lk \mid ss) \tag{8.63}$$

$$\langle {}^{3}\Psi_{k\rightarrow s}{}^{(1)} \mid \hat{H} \mid {}^{3}\Psi_{l\rightarrow r}{}^{(1)} \rangle = \delta_{l,k}\{h_{rs} + \sum_{m=1}^{n} [2(mm \mid rs) - (mr \mid sm)]\}$$

$$- \delta_{r,s}\{h_{lk} + \sum_{m=1}^{n} [2(mm \mid lk) - (ml \mid km)]\}$$

$$- 2(lr \mid ks) - (lk \mid sr) \tag{8.64}$$

(B)   CI Secular Determinant

The secular determinant is

$$\mid H_{MN}{}^{\mathrm{SC}} - E\delta_{M,N} \mid = 0 \tag{8.65}$$

The elements of this determinant are obtained from Sec. 5(A). The determinant is solved by standard means for $E_J{}^{\text{CI}}$; the unnormalized wave functions $\Psi_J{}^{\text{CI}}$ are then obtained by insertion of the individual $E_J{}^{\text{CI}}$ into the secular equations.

The CI method provides one means of introducing electron-correlation effects (see chapter 1) into the multielectron wave function. In fact, of the commonly used methods for introducing correlation, CI probably provides the best results for a reasonable amount of input effort. In the case of large molecules, the CI approach to the correlation problem, is probably the only one which is feasible. However, it does suffer from one major defect: The dimension of the CI secular equation in a reasonably large molecule becomes prohibitively big and the basis set of single-configuration wave functions must be restricted. The usual curtailment is subjective and may consist, for example, of the arbitrary exclusion of all doubly excited configurations.

6. EVALUATION OF MO INTEGRALS

The molecular-orbital integrals must be expanded over AO's in order to facilitate their computation. Since the MO φ_m is given by

$$\varphi_m = \sum_\mu c_{m\mu}\chi_\mu \tag{8.66}$$

where χ_μ is an atomic orbital, it follows that

$$h_{kl} = \left\langle \sum_\mu c_{k\mu}\chi_\mu \,\middle|\, \sum_i \hat{H}_i{}^{\text{core}} \,\middle|\, \sum_\nu c_{l\nu}\chi_\nu \right\rangle$$

$$= \sum_\mu \sum_\nu c_{k\mu}{}^* c_{l\nu} \left\langle \chi_\mu(1) \,\middle|\, -\frac{\hbar^2}{2m}\nabla^2(1) - \sum_A Z_A e^2/r_{1A} \,\middle|\, \chi_\nu(1) \right\rangle$$

$$= \sum_\mu \sum_\nu c_{k\mu}{}^* c_{l\nu} h_{\mu\nu} \tag{8.67}$$

where $-(\hbar^2/2m)\nabla^2(1)$ is the kinetic-energy operator of electron 1; where $-Z_A e^2/r_{1A}$ represents the potential energy between nucleus A plus its attendant nonvalence electrons and electron 1; and where $h_{\mu\nu}$ is $I_{\mu\nu}$ of Eq. (1.32).

The standard electron repulsion integral is given by

$$(kl \mid rs) = \sum_\mu \sum_\nu \sum_\zeta \sum_\eta c_{k\mu}{}^* c_{r\zeta} c_{l\nu}{}^* c_{s\eta} \langle \chi_\mu(1)\chi_\nu(2) \mid e^2/r_{12} \mid \chi_\zeta(1)\chi_\eta(2) \rangle$$

$$= \sum_\mu \sum_\nu \sum_\zeta \sum_\eta c_{k\mu}{}^* c_{r\zeta} c_{l\nu}{}^* c_{s\eta} (\mu\zeta \mid \nu\eta) \tag{8.68}$$

The evaluation of these integrals requires a knowledge of the analytical atomic functions χ_μ and the coefficients $c_{k\mu}$, or available atomic spectroscopic data from which values for $(\mu\zeta \mid \nu\eta)$ can be inferred, or some special insight and clairvoyance—which some persons apparently possess—which enables the investigator not only to guess at reasonable values for $(\mu\zeta \mid \nu\eta)$ but even for $(kl \mid rs)$.

It is at this point that we depart the subject of many-electron wave functions. Since it was our intention merely to familiarize the reader with the multielectron problem, with determinantal wave function, and with expectation values of spin-less one- and two-electron operators, we feel our requirements satisfied. However, in order to provide further access to the literature relating to atomic integrals of types $I_{\mu\nu}$ and $(\mu\zeta \mid \nu\eta)$, we have listed a number of different approaches, and references pertaining to these, in Appendix E.

EXERCISES

1. Write the two configuration wave functions for ${}^3\Psi_{k\to s}{}^{(1)}$ characterized by $M_S = 1$ and $M_S = -1$.

Answer: ${}^3\Psi_{k\to s}{}^{(1)}(M_S = 1)$

$$= \hat{A}[\varphi_1\alpha(1)\ldots\varphi_k\alpha(2k-1)\varphi_s\alpha(2k)\ldots\varphi_n\beta(2n)]$$

${}^3\Psi_{k\to s}{}^{(1)}(M_S = -1)$

$$= \hat{A}[\varphi_1\alpha(1)\ldots\varphi_k\beta(2k-1)\varphi_s\beta(2k)\ldots\varphi_n\beta(2n)] \quad (8.69)$$

2. Expand Eq. (8.32) in full and verify that

$$\langle\Psi_{d,G} \mid \sum_i^{2n} \hat{H}_i{}^{\text{core}}(\hat{P}_{13} + \hat{P}_{24}) \mid \Psi_{d,G}\rangle = 0 \quad (8.70)$$

3. Write the configurational wave functions for all π-electron states of ethylene. Ethylene possesses two π-electrons; there are two π-MO's which we denote φ_1 and φ_2 in order of increasing energy.

Answer: ${}^1\Psi_G{}^{(0)} = \mid \varphi_1\alpha(1)\varphi_1\beta(2) \mid$

${}^1\Psi_{1\to2}{}^{(1)} = (2)^{-1/2}\{\mid \varphi_1\alpha(1)\varphi_2\beta(2) \mid - \mid \varphi_1\beta(1)\varphi_2\alpha(2) \mid\}$

${}^3\Psi_{1\to2}{}^{(1)}(M_S = 1) = \mid \varphi_1\alpha(1)\varphi_2\alpha(2) \mid$

${}^3\Psi_{1\to2}{}^{(1)}(M_S = 0) = (2)^{-1/2}\{\mid \varphi_1\alpha(1)\varphi_2\beta(2) \mid + \mid \varphi_1\beta(1)\varphi_2\alpha(2) \mid\}$

${}^3\Psi_{1\to2}{}^{(1)}(M_S = -1) = \mid \varphi_1\beta(1)\varphi_2\beta(2) \mid$

${}^1\Psi_{1\to2}{}^{(2)} = \mid \varphi_2\alpha(1)\varphi_2\beta(2) \mid \quad (8.71)$

4. Evaluate the energies of the wave functions of Exercise 3.

Answer: $E(^1\Psi_G^{(0)}) = 2h_1 + J_{11}$

$$E(^1\Psi_{1\to2}^{(1)}) = h_1 + h_2 + J_{12} + K_{12}$$

$$E(^3\Psi_{1\to2}^{(1)}) = h_1 + h_2 + J_{12} - K_{12}$$

$$E(^1\Psi_{1\to2}^{(2)}) = 2h_2 + J_{22} \tag{8.72}$$

5. Write down the complete CI secular equation for the four π-electron single-configuration wave functions of ethylene.

Hint: The $^3\Psi_{1\to2}^{(1)}$ single-configuration wave function is spin orthogonal to all others; hence, $^3\Psi_{1\to2}^{(1)}$ is unaffected by CI. The ground and the doubly excited configurations are both of symmetry $^1A_{1g}$; the two corresponding single-configuration wave functions may interact; however, since they differ in the orbital occupancy of four MSO's $\varphi_1\alpha$, $\varphi_1\beta$, $\varphi_2\alpha$, and $\varphi_2\beta$, no terms of type h appear in the interaction element. The $^1\Psi_{1\to2}^{(1)}$ wave function transforms as $^1B_{3u}$; hence, it is not involved in the CI process.

Answer:

	$^1\Psi_G^{(0)}$	$^1\Psi_{1\to2}^{(1)}$	$^3\Psi_{1\to2}^{(1)}$	$^1\Psi_{1\to2}^{(2)}$	
$^1\Psi_G^{(0)}$	$(2h_1+J_{11}-E)$	0	0	K_{12}	
$^1\Psi_{1\to2}^{(1)}$	0	$(h_1+h_2+J_{12}+K_{12}-E)$	0	0	
$^3\Psi_{1\to2}^{(1)}$	0	0	$(h_1+h_2+J_{12}-K_{12}-E)$	0	$=0$
$^1\Psi_{1\to2}^{(2)}$	K_{12}	0	0	$(2h_2+J_{22}-E)$	

$$\tag{8.73}$$

6. Adopt the following values for the MO integrals:

$$J_{11} = 13.076 \text{ eV} \qquad h_1 = 0$$

$$J_{12} = 13.011 \text{ eV} \qquad h_2 = 6.064 \text{ eV}$$

$$J_{22} = 13.439 \text{ eV} \qquad K_{12} = 4.157 \text{ eV} \tag{8.74}$$

Thence, construct an appropriate energy diagram.

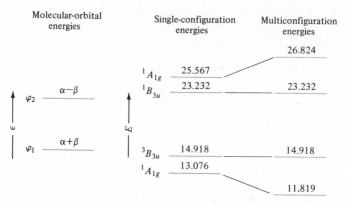

FIG. 8.1. MO, single-configuration, and multiconfiguration energies (in eV) for the ethylene π-electron system.

Hint: The only part of the CI secular determinant we need concern ourselves with is

$$
\begin{array}{cc}
 & {}^{1}\Psi_{G}{}^{(0)} \qquad\qquad {}^{1}\Psi_{1\to2}{}^{(2)}
\end{array}
$$

$$
\begin{array}{c}
{}^{1}\Psi_{G}{}^{(0)} \\[2mm]
{}^{1}\Psi_{1\to2}{}^{(2)}
\end{array}
\begin{vmatrix}
2h_1 + J_{11} - E & K_{12} \\[2mm]
K_{12} & 2h_2 + J_{22} - E
\end{vmatrix} = 0 \qquad (8.75)
$$

Answer: The answer is given in Fig. 8.1.

    7.   Evaluate the CI-corrected wave function for the ground and doubly excited π-electron states of ethylene.

Answer: The normalized wave function for the ground state is

$$(0.958)\,{}^{1}\Psi_{G}{}^{(0)} - (0.288)\,{}^{1}\Psi_{1\to2}{}^{(2)}$$

The normalized wave function for the excited state is

$$(0.288)\,{}^{1}\Psi_{G}{}^{(0)} + (0.958)\,{}^{1}\Psi_{1\to2}{}^{(2)}$$

General References

1. C. Sandorfy, *Electronic Spectra and Quantum Chemistry* (Prentice-Hall, Inc., Englewood Cliffs, N.J., 1964), p. 272–281.
2. H. Suzuki, *Electronic Absorption Spectra and Geometry of Molecules* (Academic Press Inc., New York, 1967).
3. R. G. Parr, *Quantum Theory of Molecular Electronic Structure* (W. A. Benjamin, Inc., New York, 1963).
4. F. L. Pilar, *Elementary Quantum Chemistry* (McGraw-Hill Book Co., Inc., New York, 1968).
5. M. J. S. Dewar, *The Molecular Orbital Theory of Organic Chemistry* (McGraw-Hill Book Co., Inc., New York, 1969).

CHAPTER 9

Electric-Dipole Transition Probabilities

The probability of a transition between two electronic states is connected with integrals of the form

$$\mathbf{M}_{hl} = \langle \Psi_h \mid \hat{\mathbf{M}} \mid \Psi_l \rangle \tag{9.1}$$

where Ψ_h and Ψ_l are electronic wave functions for the higher- and lower-energy states, respectively, and $\hat{\mathbf{M}}$ is a vector operator depending on the transition mechanism under consideration. $\mathbf{M}_{hl}$ is known as the *transition moment* connecting the states h and l.

The purpose of this chapter is the evaluation of electronic transition probabilities when

  (i)   the operator of Eq. (9.1) refers to an electric-dipole transition mechanism; and
  (ii) the electronic wave functions are expressed in the molecular-orbital formalism of Chaps. 7 and 8.

1. TRANSITIONS BETWEEN NONDEGENERATE STATES

When both states are *nondegenerate*, the probability of the spontaneous emission $h \rightarrow l$ is

$$1/\tau^0 = \frac{64\pi^4 \bar{\nu}^3}{3h} \left\{ |\langle \Psi_h | er | \Psi_l \rangle|^2 + \left| \left\langle \Psi_h \left| \frac{e}{2mc} \mathbf{r} \times \mathbf{p} \right| \Psi_l \right\rangle \right|^2 \right.$$

$$\left. + \frac{3\pi^2}{10} \bar{\nu}^2 |\langle \Psi_h | er \cdot \mathbf{r} | \Psi_l \rangle|^2 \right\} \quad (9.2)$$

where τ^0 is the intrinsic emissive lifetime; $\bar{\nu}$ is the wavenumber (cm^{-1}) of the $h \rightarrow l$ luminescence; and the operators $\mathbf{r}$ and $\mathbf{p}$ are electron position and electron momentum operators,[1],[2] respectively. The three terms on the right-hand side of Eq. (9.2) correspond to electric-dipole, magnetic-dipole, and electric-quadrupole transition mechanisms, respectively. Since the first term is usually of the order of 10^5 times larger than either of the other two, it is a good approximation to assume that the transition operator is adequately given by the electric-dipole part er. To this approximation, the transition moment may be correlated, in a classical way, with the idea of an oscillating electric dipole which radiates electromagnetic energy.

A.   Experimental Description of Transition Probability

The *oscillator strength f* of the transition $h \leftarrow l$ is

$$f = \frac{8\pi^2 \bar{\nu} cm}{3he^2} |\mathbf{M}_{hl}|^2 \quad (9.3)$$

If we define a *transition-dipole length D* as $|D|^2 = |\mathbf{M}|^2/e^2$, we find

$$f = 1.085 \times 10^{-5} \bar{\nu} |D|^2 \quad (9.4)$$

where D is in Angstroms. Absorption bands in solution are generally broad, often structureless, and near Gaussian in shape; consequently, f is often approximated as

$$f = 4.32 \times 10^{-9} \int \epsilon(\bar{\nu}) d\bar{\nu}$$

$$\simeq 4.32 \times 10^{-9} \epsilon_{max} \bar{\nu}_{1/2} \quad (9.5)$$

where ϵ_{max} is the molar decadic extinction coefficient of the $h \leftrightarrow l$ band maximum, $\bar{\nu}_{1/2}$ is the half-width of the absorption band (i.e., the width at $\epsilon = \frac{1}{2}\epsilon_{max}$), and $\int \epsilon(\bar{\nu}) d\bar{\nu}$ is the *integrated extinction*. Equations (9.2) and (9.3) indicate that f (obtained from absorption process $h \leftarrow l$) is related to τ^0 (obtained from emission $h \to l$) by

$$\tau^0 = \frac{mc}{8\pi^2 e^2} \cdot \frac{1}{f\bar{v}^2} \tag{9.6}$$

In connecting the theoretical formulas to the two experimental quantities τ_0 and f, two points should be kept in mind:

(i)   The measured luminescence lifetime τ is equal to τ^0 only when the quantum yield[1] of the emissive process, $\Phi(h \to l)$ is unity. Since quantum yields are usually not unity, it becomes necessary to measure both τ and $\Phi(h \to l)$, and to obtain τ^0 from

$$\tau^0 = \tau/\Phi(h \to l) \tag{9.7}$$

(ii)   In the expressions for $\mathbf{M}_{hl}$, it was assumed that the states h and l were completely characterized by the *electronic* wave functions Ψ_h and Ψ_l. However, in order to describe the states adequately, the wave function should also contain a vibrational part. In the Born–Oppenheimer approximation, an appropriate wave function for the state h is

$$\Psi_h' = \Psi_h \Xi_h{}^k \tag{9.8}$$

where $\Xi_h{}^k$ is the vibrational wave function of the kth normal mode in state h. If the state l is supposed to have zero-point energy (i.e., vibrational quantum number zero in all normal modes), the transition from l to h (summed over all vibrational levels of h), is

$$\mathbf{M}_{hl}' \simeq \sum_k \langle \Xi_h{}^k \mid \Xi_l{}^0 \rangle \langle \Psi_h \mid \sum_i e\mathbf{r}_i \mid \Psi_l \rangle = \mathbf{M}_{hl} \sum_k \langle \Xi_h{}^k \mid \Xi_l{}^0 \rangle \tag{9.9}$$

The integrals $\langle \Xi_h{}^k \mid \Xi_l{}^0 \rangle$ are vibrational overlap integrals; they are also known as *Franck–Condon* integrals. Only when $\Xi_l{}^0$ happens to be identical with $\Xi_h{}^0$, do we have a sharp quasiatomic electronic transition. This is normally not the case. We limit ourselves solely to the calculation of the *total* transition probability between the two *pure* electronic states h and l (i.e., to $\mathbf{M}_{hl}$). However, it is well to emphasize that $\mid \mathbf{M}_{hl} \mid^2$ also determines the total intensity of the electronic-vibrational band $\Psi_h' \leftarrow \Psi_l'$, as defined in Eq. (9.9), and that the Franck–Condon integrals merely modulate this intensity. In other words, the factor $\mid \langle \Xi_h{}^k \mid \Xi_l{}^0 \rangle \mid^2$ determines how much of the purely electronic intensity of the $h \leftarrow l$ transition is apportioned out to the concurrent vibrational excitation $k \leftarrow 0$. Thus, Franck–Condon integrals determine band shape and band structure. We are not concerned with this topic here.

---

[1] The term quantum yield is defined as the number of output (that is, emitted) $h \to l$ photons divided by the number of input (that is, absorbed) $h \leftarrow l$ photons. This type of quantum yield is often referred to as a *two-state quantum yield*.

2. SOME SIMPLE THEOREMS

Theorem 1: The transition probability between states of different spin multiplicity is zero: States of different multiplicity are eigenfunctions of $\hat{S}^2$ belonging to different eigenvalues; they are therefore orthogonal. Since the operator $e\mathbf{r}$ is not a function of the spin coordinates, it cannot remove this spin orthogonality. Therefore singlet $\leftrightarrow$ triplet, doublet $\leftrightarrow$ quartet, etc., transitions are strictly forbidden.[2]

Theorem 2: The transition probability between states reduces to an integral over those MO's which are differently populated in the two states: The operator $\sum_i e\mathbf{r}_i$ is a sum of one-electron operators. The wave functions Ψ_h and Ψ_l, whether single or multiconfiguration in nature, are sums of Slater determinants. Consequently, the rules of Sec. 2, chapter 8 apply. If d_1 and d_2 are two Slater determinants constructed from molecular spin orbitals (MSO's) $\psi_1, \psi_2, \ldots \psi_k, \psi_l, \ldots$ these rules are:

(i)   If d_1 and d_2 differ by more than one MSO, we find

$$\langle d_1 \mid \sum_i \mathbf{r}_i \mid d_2 \rangle = 0 \qquad (9.10)$$

(ii)   If d_1 and d_2 differ by only one MSO, say, ψ_r in d_1 against ψ_t in d_2, the determinants may be rearranged so that as many of the same MSO's as is possible occur at identical locations in both d_1 and d_2. Then

$$\langle d_1 \mid \sum_i \mathbf{r}_i \mid d_2 \rangle = \pm \langle \psi_r(i) \mid \mathbf{r}_i \mid \psi_t(i) \rangle \qquad (9.11)$$

where the plus or minus sign is determined by the evenness or oddness of the number of permutations necessary to obtain the required MSO ordering.

Let Ψ_l correspond to ${}^1\Psi_G{}^{(0)}$; let Ψ_h correspond to ${}^1\Psi_{r \to t}{}^{(1)}$; and let all MO's be nondegenerate. In the single-configuration wave-function approximation, we have

$${}^1\Psi_G{}^{(0)} = \mid \varphi_1\alpha(1)\varphi_1\beta(2)\ldots\varphi_r\alpha(2s-3)\varphi_r\beta(2s-2)\varphi_s\alpha(2s-1)\varphi_s\beta(2s) \mid$$
$$(9.12)$$

$${}^1\Psi_{r \to t}{}^{(1)} = (\tfrac{1}{2})^{1/2}\{\mid \varphi_1\alpha(1)\ldots\varphi_r\alpha(2s-3)\varphi_t\beta(2s-2)\varphi_s\alpha(2s-1)\varphi_s\beta(2s)\mid$$
$$- \mid \varphi_1\alpha(1)\ldots\varphi_r\beta(2s-3)\varphi_t\alpha(2s-2)\varphi_s\alpha(2s-1)\varphi_s\beta(2s) \mid\}$$
$$(9.13)$$

---

[2] The experimental observation of *spin-forbidden* transitions is caused by admixture of ${}^{2S+1}\Psi$ into ${}^{2S'+1}\Psi$, where $2S + 1 = 2S' + 1 \pm 2$, *via* spin-orbit coupling. These transitions are usually of very low probability (i.e., small f or large τ^0); they are discussed in chapter 11.

Thus, knowing the transition energy, the calculation of transition probability devolves on evaluation of $\langle \Psi_h \mid \sum_i e\mathbf{r}_i \mid \Psi_l \rangle$. Using Eq. (9.11), this integral, in an obvious shorthand notation, is

$$\langle \Psi_h \mid \sum_i e\mathbf{r}_i \mid \Psi_l \rangle = \langle \Psi_{r \to t}{}^{(1)} \mid \sum_{i=1}^{2s} e\mathbf{r}_i \mid {}^1\Psi_G{}^{(0)} \rangle$$

$$= (\tfrac{1}{2})^{1/2} \langle \mid 1\alpha1\beta \ldots r\alpha t\beta s\alpha s\beta \mid$$
$$- \mid 1\alpha1\beta \ldots r\beta t\alpha s\alpha s\beta \mid \mid \sum_i e\mathbf{r}_i \mid \mid 1\alpha1\beta \ldots r\alpha r\beta s\alpha s\beta \mid \rangle$$

$$= (\tfrac{1}{2})^{1/2} \langle \mid 1\alpha1\beta \ldots r\alpha t\beta s\alpha s\beta \mid$$
$$+ \mid 1\alpha1\beta \ldots t\alpha r\beta s\alpha s\beta \mid \mid \sum_i e\mathbf{r}_i \mid \mid 1\alpha1\beta \ldots r\alpha r\beta s\alpha s\beta \mid \rangle$$

$$= (\tfrac{1}{2})^{1/2} \{ \langle t\beta \mid e\mathbf{r} \mid r\beta \rangle + \langle t\alpha \mid e\mathbf{r} \mid r\alpha \rangle \} = (2)^{1/2} \langle t \mid e\mathbf{r} \mid r \rangle$$

$$(9.14)$$

The oscillator strength for the absorption $\Psi_h \leftarrow \Psi_l$ is proportional to $\mid \mathbf{M} \mid^2$; it is, therefore, twice the oscillator strength one would obtain from consideration of a single MO jump $\varphi_t \leftarrow \varphi_r$. The factor of 2 relates to the presence in the φ_r orbital of two electrons, either of which can make the quantum jump in question.[3],[4]

Theorem 3:   The value of $\mathbf{M}$ *is independent of the origin of the molecular coordinate system:* The transition probability is a fundamental property of the two combining states; therefore, it is independent of any particular coordinate origin: The adoption of one origin defines a set of electron position vectors $\mathbf{r}$, the adoption of another origin defines the new set $\mathbf{r}'$. The two sets are related by the transformation $\mathbf{r} = \mathbf{r}' + \mathbf{r}_0$, where $\mathbf{r}_0$ is a simple displacement vector connecting the origins of the two vector sets; r_0, therefore, is a constant. It follows that

$$\mathbf{M} = \langle \Psi_h \mid \sum_i e\mathbf{r}_i \mid \Psi_l \rangle = \langle \Psi_h \mid \sum_i e(\mathbf{r}_i' + \mathbf{r}_0) \mid \Psi_l \rangle$$

$$= \langle \Psi_h \mid \sum_i e\mathbf{r}_i' \mid \Psi_l \rangle + e\mathbf{r}_0 \langle \Psi_h \mid \Psi_l \rangle \qquad (9.15)$$

Since Ψ_h and Ψ_l are eigenfunctions of the same Hamiltonian belonging to different eigenvalues, the second term is zero; consequently,

$$\mathbf{M} = \langle \Psi_h \mid \sum_i e\mathbf{r}_i' \mid \Psi_l \rangle \qquad (9.16)$$

3. SYMMETRY CONSIDERATIONS

For a transition to be allowed, it is necessary that the direct product $\Gamma(\Psi_h) \times \Gamma(\mathbf{r}) \times \Gamma(\Psi_l)$ contains the totally symmetric representation Γ_1.

In other words, we require

$$\Gamma(\Psi_h) \times \Gamma(\mathbf{r}) \times \Gamma(\Psi_l) \supset \Gamma_1 \tag{9.17}$$

If the integral $\langle \Psi_h \,|\, \mathbf{r} \,|\, \Psi_l \rangle$, or some additive part of it, does not transform as Γ_1, its value and that of all its additive parts are changed by at least one symmetry operation—unless the value of the integral happens to be zero. Since a symmetry operation must not change the value of a physical property, any transition moment integral not containing an additive part which transforms as Γ_1 is zero. By virtue of Theorem 2, it is possible to reduce Eq. (9.17) to

$$\Gamma(\varphi_l) \times \Gamma(\mathbf{r}) \times \Gamma(\varphi_r) \supset \Gamma_1 \tag{9.18}$$

In many instances, the direct product of Eqs. (9.17) or (9.18) is totally symmetric only for one of the components of $\mathbf{r}$, say, x. In that case, we could describe the system classically as an oscillation solely in the x direction. Such an oscillation results in a polarized wave whose electric field vector is parallel to the x axis.

4. F_2CO: AN EXAMPLE CALCULATION

The coordinate system and structural parameters of F_2CO are given in Fig. 9.1. This molecule has C_{2v} symmetry; hence, all molecular orbitals are nondegenerate. A computed[5] set of MO's, as well as the lowest-energy electronic configuration, is shown in Fig. 9.2.

The atomic-orbital basis set is of the single-ζ type described in chapter 1. The orbital exponents are given by Slater's rules as outlined in Appendix B. The analytic form of the $2p_y$ AO, for example, is

$$2p_y = N_{2p}ye^{-Zr/2}; \qquad N_{2p} = (Z^5/32\pi)^{1/2} \tag{9.19}$$

where N_{2p} is a normalization factor and x and r are in atomic units of length. The effective nuclear charges are

$2p$ AO on carbon: $Z_C = 6 - (2 \times 0.85) - (3 \times 0.35) = 3.25$

$2p$ AO on oxygen: $Z_O = 8 - (2 \times 0.85) - (5 \times 0.35) = 4.55$

$2p$ AO on fluorine: $Z_F = 9 - (2 \times 0.85) - (6 \times 0.35) = 5.20$

$2s$ AO on fluorine: $Z_F = 5.20$ $\hfill (9.20)$

x axis is out of plane
$R_{C-O} = 1.32\text{Å} = 2.215$ a.u.
$R_{C-F} = 1.17\text{Å} = 2.499$ a.u.
$\alpha = \angle F_2CF_1 = 112.5°$

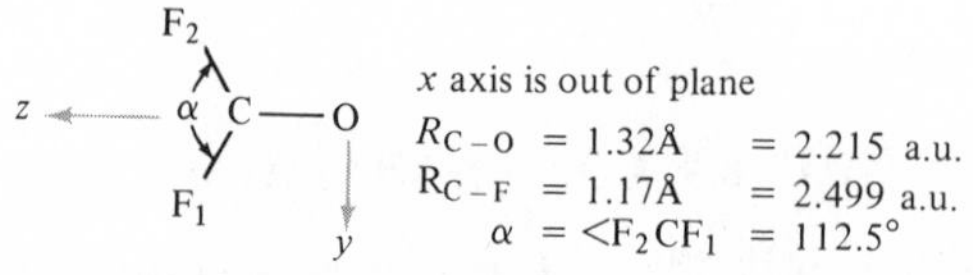

FIG. 9.1. Coordinate system and structural parameters for F_2CO. The origin of coordinates is on the carbon atom.

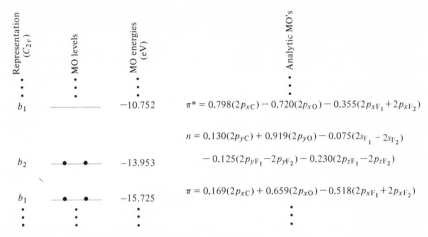

FIG. 9.2. Partial energy-level scheme (Ref. [5]) for F_2CO.

(A) Application of Group Theory

The C_{2v} point group and the transformation properties of all relevant atomic orbitals, group orbitals, and molecular orbitals are shown in Table 9.1. The results of symmetry considerations are:

(*i*) $\pi^* \leftarrow n$ *excitation:* The configurational excitation $\ldots \pi^2 n \pi^* \leftarrow \ldots \pi^2 n^2$ is variously designated: $\pi^* \leftarrow n$; $^1\Gamma_{n\pi^*} \leftarrow {}^1\Gamma_1$; etc. The representation generated by $^1\Gamma_{n\pi^*}$ is

$$\Gamma(n) \times \Gamma(\pi^*) = b_2 \times b_1 = A_2 \qquad (9.21)$$

TABLE 9.1. *Character table for C_{2v} and the transformation characteristics of r, $\hat{R}$ (the rotations), AO's, GO's, and MO's appropriate to F_2CO.*

(axes are defined in Fig. 9.1).

C_{2v}	E	$C_2(z)$	$\sigma_v(xz)$	$\sigma_v'(yz)$	r	$\hat{R}$	AO's	GO's	MO's
a_1	1	1	1	1	z		$2p_{zC}, 2p_{zO}, 2s_C, 2s_O$		
a_2	1	1	-1	-1		$\hat{R}_z$			
b_1	1	-1	1	-1	x	$\hat{R}_y$	$2p_{xC}, 2p_{xO}$	$2p_{xF_1} + 2p_{xF_2}$	π, π^*
b_2	1	-1	-1	1	y	$\hat{R}_x$	$2p_{yC}, 2p_{yO}$	$\begin{cases} 2s_{F_1} - 2s_{F_2} \\ 2p_{yF_1} - 2p_{yF_2} \\ 2p_{zF_1} - 2p_{zF_2} \end{cases}$	n

An alternative designation of this transition is, therefore, $^1A_2 \leftarrow {}^1A_1$. Applying Theorem 2, we find

$$\mathbf{M} = \langle {}^1A_2 \mid \sum_i e\mathbf{r}_i \mid {}^1A_1 \rangle = (2)^{1/2} \langle n \mid e\mathbf{r} \mid \pi^* \rangle \qquad (9.22)$$

Thus, the totality of the group-theoretical information is contained in

$$\begin{aligned}
\Gamma(\mathbf{M}) &= A_2 \times \Gamma(x, y, \text{or } z) \times A_1 \\
&= [A_2 \times \Gamma(x)] + [A_2 \times \Gamma(y)] + [A_2 \times \Gamma(z)] \\
&= B_2 + B_1 + A_2 \not\supset A_1 \qquad (9.23)
\end{aligned}$$

Therefore, this transition is forbidden. Computation must yield

$$(2)^{1/2} \langle n \mid e\mathbf{r} \mid \pi^* \rangle = 0$$

(ii) $\pi^* \leftarrow \pi$ **transition** corresponding to $\ldots \pi\pi^* n^2 \leftarrow \ldots \pi^2 n^2$ or $\Gamma_{\pi\pi^*} \leftarrow \Gamma_1$ can be treated similarly. Indeed

$$\mathbf{M} = \langle \Gamma_{\pi\pi^*} \mid \sum_i e\mathbf{r}_i \mid \Gamma_1 \rangle = (2)^{1/2} \langle \pi^* \mid e\mathbf{r} \mid \pi \rangle \qquad (9.24)$$

Since $\Gamma(\pi) \times \Gamma(\pi^*) = b_1 \times b_1 = A_1$ and since z transforms as a_1, we conclude that the $\pi^* \leftarrow \pi$ transition is allowed and polarized along the molecular axis (that is, along the C—O bond). We now evaluate this transition moment.

(B)   Reduction to AO Matrix Elements

Inserting the AO expansions of π and π^* given in Fig. 9.2 into Eq. (9.24), and using $\mathbf{D} = \mathbf{M}/e$, we find

$$\begin{aligned}
D_z = (2)^{1/2}\{ &(0.169)(0.798) \langle 2p_{xC} \mid z \mid 2p_{xC} \rangle - (0.659)(0.720) \langle 2p_{xO} \mid z \mid 2p_{xO} \rangle \\
&+ 2(0.518)(0.355) \langle 2p_{xF_1} \mid z \mid 2p_{xF_1} \rangle - (0.169)(0.720) \langle 2p_{xC} \mid z \mid 2p_{xO} \rangle \\
&- 2(0.169)(0.355) \langle 2p_{xC} \mid z \mid 2p_{xF_1} \rangle + (0.659)(0.798) \langle 2p_{xO} \mid z \mid 2p_{xC} \rangle \\
&- 2(0.659)(0.355) \langle 2p_{xO} \mid z \mid 2p_{xF_1} \rangle - 2(0.518)(0.798) \langle 2p_{xF_1} \mid z \mid 2p_{xC} \rangle \\
&+ 2(0.518)(0.720) \langle 2p_{xF_1} \mid z \mid 2p_{xO} \rangle + 2(0.518)(0.355) \langle 2p_{xF_1} \mid z \mid 2p_{xF_2} \rangle \}
\end{aligned}$$
$$(9.25)$$

The vector component length z is referred to an origin on the carbon nucleus; therefore, z might be symbolized z_C. On the other hand, the AO origins are fixed on the individual nuclear centers to which the AO's pertain. For example, the AO matrix element $\langle 2p_{xO} \mid z \mid 2p_{xF_1} \rangle$ contains three factors which, reading from the left, are defined with respect to origins on oxygen, carbon, and fluorine, respectively. We also note that the order of the factors in the element $\langle AO_1 \mid z \mid AO_2 \rangle$ is irrelevant; the operator z is a simple multiplier, and may be made to operate on either AO_1 or AO_2 as one sees fit.

(C) Diagonal AO Matrix Elements

There are three different elements of this sort in Eq. (9.25); they are $\langle 2p_{xC} \mid z_C \mid 2p_{xC} \rangle$, $\langle 2p_{xO} \mid z_C \mid 2p_{xO} \rangle$, and $\langle 2p_{xF_1} \mid z_C \mid 2p_{xF_1} \rangle$. We consider each of these in turn:

(i) The element $\langle 2p_{xC} \mid z_C \mid 2p_{xC} \rangle$: For every volume element where $z_C = a$, there exists another volume element with $z_C = -a$, where the integral has exactly the opposite value. Therefore, integration over all space must yield zero.

(ii) The element $\langle 2p_{xO} \mid z_C \mid 2p_{xO} \rangle$: Two different origins are implied in the corresponding integral. Therefore, we transform to a common origin. If z_O denotes the z coordinate referred to the oxygen nucleus, we may write

$$\langle 2p_{xO} \mid z_C \mid 2p_{xO} \rangle = \langle 2p_{xO} \mid z_O - R_{C-O} \mid 2p_{xO} \rangle$$

$$= \langle 2p_{xO} \mid z_O \mid 2p_{xO} \rangle - R_{C-O} \langle 2p_{xO} \mid 2p_{xO} \rangle$$

$$= 0 - 2.215 \langle 2p_{xO} \mid 2p_{xO} \rangle$$

$$= -2.215 \text{ a.u.} \tag{9.26}$$

(iii) The element $\langle 2p_{xF_1} \mid z_C \mid 2p_{xF_1} \rangle$ which, by symmetry, equals $\langle 2p_{xF_2} \mid z_C \mid 2p_{xF_2} \rangle$: A transformation, per item (ii) above, yields

$$\langle 2p_{xF_1} \mid z_C \mid 2p_{xF_1} \rangle = \langle 2p_{xF_1} \mid z_{F_1} + R_{C-F_1} \cos (\alpha/2) \mid 2p_{xF_1} \rangle$$

$$= \langle 2p_{xF_1} \mid z_{F_1} \mid 2p_{xF_1} \rangle + (2.499)(0.556) \langle 2p_{xF_1} \mid 2p_{xF_1} \rangle$$

$$= (2.499)(0.556) \text{ a.u.} \tag{9.27}$$

(D) Off-Diagonal AO Matrix Elements

There are twelve such elements [see Eq. (9.25)]; however, not all of these integrals are different: for example, the symmetry of the molecule implies that $\langle 2p_{xC} \mid z \mid 2p_{xF_1} \rangle = \langle 2p_{xC} \mid z \mid 2p_{xF_2} \rangle$, etc.

(i) The element $\langle 2p_{xC} \mid z_C \mid 2p_{xO} \rangle$: The analytic form is

$$\langle 2p_{xC} \mid z_C \mid 2p_{xO} \rangle = \langle N_{2p_C} x_C e^{-3.25 r_C/2} \mid z_C \mid N_{2p_O} x_O e^{-4.55 r_O/2} \rangle \tag{9.28}$$

where z_C and $2p_{xC}$ are defined with respect to the same origin. Therefore, we let z_C operate to the left to yield

$$\langle 2p_{xC} \mid z_C \mid 2p_{xO} \rangle = \langle N_{2p_C} (xz)_C e^{-3.25 r_C/2} \mid N_{2p_O} x_O e^{-4.55 r_O/2} \rangle \tag{9.29}$$

We now note that the functional form of the $3d_{xz}$ AO is

$$3d_{xz} = N_{3d}(2xz)e^{-Zr/3}; \qquad N_{3d} = [Z^7/2\pi(3)^8]^{1/2} \qquad (9.30)$$

Inspection indicates that the bra on the right-hand side of Eq. (9.29) is of the same functional form. Thus, if we equate $\mathbf{Z}'/3 = 3.25/2$, we obtain the modified $3d$ orbital

$$3d_{xz}' = N_{3d}'2(xz)_C e^{-(3.25)(3)rC/(2)(3)} \qquad (9.31)$$

where

$$N_{3d}' = [(\mathbf{Z}')^7/2\pi(3)^8]^{1/2} = [(3)^7(3.25)^7/\pi(6)^8]^{1/2} \qquad (9.32)$$

Therefore

$$\langle 2p_{xC} \mid z_C \mid 2p_{xO} \rangle = \langle 3d_{xzC}' \mid 2p_{xO} \rangle (N_{2pO}/2N_{3dC}')$$

$$= [(3.25)^5(2)^6(3)^8\pi/(32\pi(3)^7(3.25)^7)]^{1/2}S(2p\pi, 3d\pi)$$

$$= -\frac{(6)^{1/2}}{3.25} S(2p\pi, 3d\pi) = -\frac{(6)^{1/2}}{3.25} (0.415)$$

$$= -0.313 \text{ a.u.} \qquad (9.33)$$

where $S(2p\pi, 3d\pi)$ is the overlap integral evaluated at a C—O separation of 2.215 a.u., as found in overlap tables.[6]−[10]

The problem of evaluating this off-diagonal element is seen to reduce to the evaluation of an overlap integral coupled with a renormalization process. The argument leading to Eq. (9.33) can be shown to be valid for any $\mathbf{Z}$; therefore, analytical expressions can be derived to facilitate the renormalization process. Such expressions are listed in Appendix F.

It is well to emphasize that Eq. (9.33) contains a minus sign. The reasons for this are schematized in Fig. 9.3. It is obvious from the figure that $\langle 2p_{xC} \mid z_C \mid 2p_{xO} \rangle$ must be negative. However, the values of $S(2p\pi, 3d\pi)$ found in the overlap tables are positive: Overlap tables are calculated using the convention that the overlap refers to a *bonding* situation (see chapter 2).

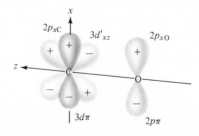

FIG. 9.3. The $2p_{xC}$, $2p_{xO}$ and $3d_{xz}'$ atomic orbitals. It is clear that the overlap $\langle 3d\pi \mid 2p\pi \rangle$ must be negative.

(ii)   The element $\langle 2p_{x\mathrm{O}} \mid z_{\mathrm{C}} \mid 2p_{x\mathrm{C}} \rangle$: Because of commutativity, this element is identical to $\langle 2p_{x\mathrm{C}} \mid z_{\mathrm{C}} \mid 2p_{x\mathrm{O}} \rangle$ of (i) above.

(iii)   The element $\langle 2p_{x\mathrm{C}} \mid z_{\mathrm{C}} \mid 2p_{x\mathrm{F}_1} \rangle$: Evaluation of this element is equivalent, in its essentials, to the evaluation of an overlap $\langle 3d_{xz\mathrm{C}}' \mid 2p_{x\mathrm{F}_1} \rangle$. The orbitals $3d_{xz\mathrm{C}}'$ and $2p_{x\mathrm{F}_1}$ are illustrated in Fig. 9.4. The xz plane is a symmetry plane for the modified d orbital which is generated by the z operator. However, with respect to this same plane, reflection sends $2p_{x\mathrm{F}_1}$ into $2p_{x\mathrm{F}_2}$; therefore, the xz plane is not a symmetry plane of the $2p_{x\mathrm{F}_1}$ AO. Now, the tables of overlap integrals are constructed on the assumption that the xz plane is a common symmetry plane (complete π bonding). In chapter 2, we have shown how to make the required angular corrections. The result, as is also evident from inspection of Fig. 9.4, is

$$S(2p_x, 3d_{zz}') = S(2p\pi, 3d\pi) \cos (\alpha/2) \qquad (9.34)$$

Therefore, as the reader may verify

$$\langle 2p_{x\mathrm{C}} \mid z_{\mathrm{C}} \mid 2p_{x\mathrm{F}_1} \rangle = (6^{1/2}/3.25)\, S(2p\pi, 3d\pi) \cos (\alpha/2) \qquad (9.35)$$

(iv)   The element $\langle 2p_{x\mathrm{O}} \mid z \mid 2p_{x\mathrm{F}_1} \rangle$: Following principles already outlined one finds

$$\langle 2p_{x\mathrm{O}} \mid z \mid 2p_{x\mathrm{F}_1} \rangle = \langle 2p_{x\mathrm{O}} \mid z_{\mathrm{O}} - 2.215 \mid 2p_{x\mathrm{F}_1} \rangle$$

$$= \langle 2p_{x\mathrm{O}} \mid z_{\mathrm{O}} \mid 2p_{x\mathrm{F}_1} \rangle - 2.215 \langle 2p_{x\mathrm{O}} \mid 2p_{x\mathrm{F}_1} \rangle$$

$$= (6^{1/2}/4.55)\, S(2p\pi, 3d\pi) \cos (\beta/2) - 2.215 S(2p\pi, 2p\pi)$$

$$(9.36)$$

where β is the angle $\mathrm{F}_1\mathrm{OF}_2$. Note also that the symbol $S(2p\pi, 3d\pi)$, appearing in both Eqs. (9.35) and (9.36), does not have the same meaning in the two cases: The distances and the orbital exponents to be considered are different in each instance.

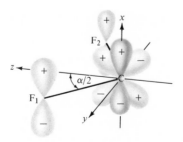

FIG. 9.4.   The $2p_{x\mathrm{F}}$ and $2p_{x\mathrm{C}}$ atomic orbitals. The atomic orbital $3d_{zz\mathrm{C}}'$ is obtained from $\langle 2p_{x\mathrm{C}} \mid z$, as in item (i) above.

(v)   The element $\langle 2p_{x\mathrm{F}_1} \mid z_{\mathrm{C}} \mid 2p_{x\mathrm{F}_2} \rangle$: Straightforward manipulation yields

$$\langle 2p_{x\mathrm{F}_1} \mid z_{\mathrm{C}} \mid 2p_{x\mathrm{F}_2} \rangle = \langle 2p_{x\mathrm{F}_1} \mid z_{\mathrm{F}} + (2.499)(0.556) \mid 2p_{x\mathrm{F}_2} \rangle$$

$$= (1.389)\,S(2p\pi, 2p\pi) \qquad (9.37)$$

(vi)   Other elements of Eq. (9.25): These are equal to some one of the elements (i)–(v).

(E) $\pi^* \leftarrow \pi$ Transition Moment

The results of Secs. 4(B)–4(D) are summarized in Table 9.2. The different AO matrix elements are multiplied by their coefficients from Eq. (9.25) and summed. The final result is $D_z = (2^{1/2})(1.312)$ a.u. The energy of the $^1\Gamma_{\pi\pi^*} \leftarrow {}^1\Gamma_1$ transition, from Fig. 9.3, is 4.973 eV, or $\bar{\nu} = 40\ 122$ cm^{-1}. Therefore, $f = (2)(1.085)(40\ 122)(1.312)^2(0.5292)^2 10^{-5} \sim 0.42$; the transition is predicted to be z polarized.

The experimental value for the oscillator strength of the $\pi^* \leftarrow \pi$ transition is not known, but the corresponding transition in H_2CO—whose electronic structure is not grossly different—has an f value of about 0.3; it is also z polarized. Considering that the calculation is based on an approximate set of wave functions, the agreement must be considered good.

(F)   Comments

One should expect little more from this type of calculation than an order-of-magnitude agreement with experiment.[3] However, this is a significant achievement because the observed values of f vary commonly between 1 and 10^{-6}.

In the f value calculated for the $\pi^* \leftarrow \pi$ transition of F_2CO, the off-diagonal matrix elements contribute only about 15% of the total value. Considering the inherent inaccuracies of the calculation, it would appear reasonable to limit oneself to evaluation of diagonal elements. However, there are a number of objections to such a procedure:

(i)   A transition moment calculation which is limited to diagonal terms yields results which are dependent on the chosen origin of coordinates. In other words, Theorem 3 is valid only when all component AO integrals in **M** are evaluated exactly.

(ii)   In some instances, the off-diagonal matrix elements can be quite large. This is particularly true if the off-diagonal element is of one-center

---

[3] Some authors (Refs. [11]–[13]) multiply the results obtained by empirical factors which vary between 0.25 and 0.40; experience indicates that the oscillator strengths calculated from LCAO MO's tend to be somewhat too large.

nature. No such elements occur in the F$_2$CO computation. However, the one-center integral $\langle 2s \mid x \mid 2p_x \rangle$ occurs in the $\sigma^* \leftarrow \sigma$ transition moment of second-row diatomics. Indeed, such one-center off-diagonal elements are probably more common than not. They may be evaluated using Appendix G.

(iii) There are instances where *only* the off-diagonal matrix elements of two-center nature make any contribution to D.

From items (i)–(iii), we can conclude that it is dangerous to neglect the off-diagonal elements of **M**. Any neglect of these elements should be preceded by a qualitative examination of *all* elements involved in **M**.

TABLE 9.2. *Summation of results of Secs.* $4(B)$, $4(C)$, *and* $4(E)$.

Matrix Element	Coefficient from (Eq. (9.25)	Contribution to dipole length D_z (except for the factor $2^{1/2}$)
Diagonal Elements (in a.u.)		
$\langle 2p_{xC} \mid z \mid 2p_{xC} \rangle =\ \ \ 0.0$	$+0.134$	0.0
$\langle 2p_{xO} \mid z \mid 2p_{xO} \rangle = -2.215$	-0.474	$+1.049$
$\langle 2p_{xF_1} \mid z \mid 2p_{xF_1} \rangle = +1.389$	$+0.183$	$+0.254$
$\langle 2p_{xF_2} \mid z \mid 2p_{xF_2} \rangle = +1.389$	$+0.183$	$+0.254$
	Sub-total	$= 1.557$ a.u.
Off-diagonal Elements (in a.u.)		
$\langle 2p_{xC} \mid z \mid 2p_{xO} \rangle = -0.313$	-0.121	$+0.037$
$\langle 2p_{xC} \mid z \mid 2p_{xF_1} \rangle = +0.126$	-0.059	-0.007
$\langle 2p_{xC} \mid z \mid 2p_{xF_2} \rangle = +0.126$	-0.059	-0.007
$\langle 2p_{xO} \mid z \mid 2p_{xC} \rangle = -0.313$	$+0.525$	-0.164
$\langle 2p_{xO} \mid z \mid 2p_{xF_1} \rangle = -0.001$	-0.233	0.000
$\langle 2p_{xO} \mid z \mid 2p_{xF_2} \rangle = -0.001$	-0.233	0.000
$\langle 2p_{xF_1} \mid z \mid 2p_{xC} \rangle = +0.126$	-0.413	-0.052
$\langle 2p_{xF_1} \mid z \mid 2p_{xO} \rangle = -0.001$	$+0.372$	0.000
$\langle 2p_{xF_1} \mid z \mid 2p_{xF_2} \rangle = -0.004$	$+0.183$	0.000
$\langle 2p_{xF_2} \mid z \mid 2p_{xC} \rangle = +0.126$	-0.413	-0.052
$\langle 2p_{xF_2} \mid z \mid 2p_{xO} \rangle = -0.001$	$+0.372$	0.000
$\langle 2p_{xF_2} \mid z \mid 2p_{xF_1} \rangle = +0.004$	$+0.183$	0.000
	Sub-total	$= -0.245$ a.u.
Total: $D_z = (2^{1/2})(1.312)$ a.u.		

5. TRANSITIONS BETWEEN NONDEGENERATE STATES ARISING FROM AN EXCITATION INVOLVING DEGENERATE MOLECULAR ORBITALS

Sections 1–4 were concerned with electronic excitations involving two nondegenerate MO's. We now show that the calculation of a transition probability between two nondegenerate states, which arise from an excitation involving degenerate MO's, requires application of the same formulas and procedures as were used in Secs. 1–4. We choose the $e^* \leftarrow e\,(^1A_1 \leftarrow {}^1A_1)$ transition of a C_{4v} molecule. A schematic MO diagram for this molecule is shown in Fig. 9.5. The only difficulty in the calculation is associated with the generation of properly adapted many-electron wave functions.

(A) Wave Functions

The (x, y) pair of axes of Fig. 9.5 transforms as e. Consequently, the two component wave functions of each degenerate MO may be chosen to transform as x and y, respectively. We denote these components: e_x, e_y and e_x^*, e_y^*.

The ground wave function is

$$\Psi_G^{(0)} = |\, e_x\alpha(1)e_x\beta(2)e_y\alpha(3)e_y\beta(4)\,|\,; {}^1A_1 \tag{9.38}$$

The presence of the inner core orbitals is implicit in the determinant of Eq. (9.38), but since they are not necessary to further discussion they are not written out.

The excited wave functions resulting from the configuration e^3e^* transform as $e \times e = A_1 + A_2 + B_1 + B_2$. Thus, the four singlet excited wave functions transform as ${}^1A_1$, ${}^1A_2$, ${}^1B_1$, and ${}^1B_2$. From the character table of

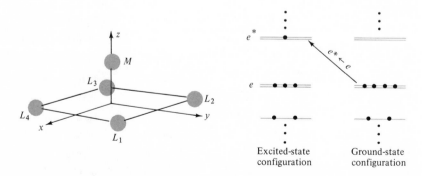

FIG. 9.5. Possible MO energy diagram for a C_{4v} molecule. Axes are also defined. The central M atom does not lie in the plane defined by the four ligand atoms.

C_{4v}, it is readily verified that the only allowed transition is $^1A_1 \leftrightarrow {}^1A_1$ (z polarization). The possible single-determinant wave functions, with $M_S = 0$, corresponding to the $e^* \leftarrow e$ excitation are

$$d_1 = |\, e_x\alpha(1)e_x\beta(2)e_y\alpha(3)e_y{}^*\beta(4)\,|$$

$$d_2 = |\, e_x\alpha(1)e_x\beta(2)e_y{}^*\alpha(3)e_y\beta(4)\,|$$

$$d_3 = |\, e_x\alpha(1)e_x\beta(2)e_y\alpha(3)e_x{}^*\beta(4)\,|$$

$$d_4 = |\, e_x\alpha(1)e_x\beta(2)e_x{}^*\alpha(3)e_y\beta(4)\,|$$

$$d_5 = |\, e_x\alpha(1)e_x{}^*\beta(2)e_y\alpha(3)e_y\beta(4)\,|$$

$$d_6 = |\, e_x{}^*\alpha(1)e_x\beta(2)e_y\alpha(3)e_y\beta(4)\,|$$

$$d_7 = |\, e_x\alpha(1)e_y{}^*\beta(2)e_y\alpha(3)e_y\beta(4)\,|$$

$$d_8 = |\, e_y{}^*\alpha(1)e_x\beta(2)e_y\alpha(3)e_y\beta(4)\,| \qquad (9.39)$$

These determinants, however, are not eigenfunctions of $\hat{S}^2$. The four singlet eigenfunctions of $\hat{S}^2$, obtained by methods outlined in Chap. 7, are

$$\Psi_1 = (\tfrac{1}{2})^{1/2}(d_1 + d_2)$$

$$\Psi_2 = (\tfrac{1}{2})^{1/2}(d_3 + d_4)$$

$$\Psi_3 = (\tfrac{1}{2})^{1/2}(d_5 + d_6)$$

$$\Psi_4 = (\tfrac{1}{2})^{1/2}(d_7 + d_8) \qquad (9.40)$$

These four wave functions, however, are not symmetry adapted; they do not transform properly in the C_{4v} point group. Since these four wave functions span the complete singlet space, linear combinations which do transform properly may be found. In order to find the singlet having A_1 symmetry, we simply check the transformation properties of the four Ψ wave functions under the symmetry operations of C_{4v}. Consider, for instance, the result of the two C_4 operations[4] on $\Psi_1 = (\tfrac{1}{2})^{1/2}(d_1 + d_2)$. The effect of C_4 is $x \to \pm y;\ y \to \mp x$, and consequently $e_x \to \pm e_y;\ e_y \to \mp e_x$. Therefore

$$C_4 d_1 = C_4\, |\, e_x\alpha(1)e_x\beta(2)e_y\alpha(3)e_y{}^*\beta(4)\,|$$

$$= |\, e_y\alpha(1)e_y\beta(2)e_x\alpha(3)e_x{}^*\beta(4)\,|$$

$$= -|\, e_x\alpha(1)e_y\beta(2)e_y\alpha(3)e_x{}^*\beta(4)\,|$$

$$= |\, e_x\alpha(1)e_x{}^*\beta(2)e_y\alpha(3)e_y\beta(4)\,|$$

$$= d_5 \qquad (9.41)$$

---

[4] One C_4 rotation is clockwise, the other is counterclockwise. The topmost sign of the combination $\pm$ or $\mp$ represents the effects of the clockwise rotation.

TABLE 9.3.   *Transformation properties of the singlet wave functions of Eq. (9.40).*

			OPERATOR		
WF	E	$2C_4$	C_2	$2\sigma_v$	$2\sigma_d$
Ψ_1	Ψ_1	Ψ_3	Ψ_1	Ψ_1	Ψ_3
Ψ_2	Ψ_2	$-\Psi_4$	Ψ_2	$-\Psi_2$	Ψ_4
Ψ_3	Ψ_3	Ψ_1	Ψ_3	Ψ_3	Ψ_1
Ψ_4	Ψ_4	$-\Psi_2$	Ψ_4	$-\Psi_4$	Ψ_2

It is found similarly that $C_4 d_2 = d_6$. Consequently, $C_4 \Psi_1 = \Psi_3$. The results of all the symmetry operation of C_{4v} on the four functions of Eq. (9.40) are indicated in Table 9.3. Inspection of this table indicates that the desired singlet wave function of symmetry A_1 is

$$\Psi(^1A_1) = (\tfrac{1}{2})^{1/2}(\Psi_1 + \Psi_3) \tag{9.42}$$

(B)   Transition Probability

The $^1A_1 \leftarrow {}^1A_1$ transition moment is given by

$$M_z = \langle \Psi_G \mid \sum_i ez_i \mid (\tfrac{1}{2})^{1/2}(\Psi_1 + \Psi_3) \rangle \tag{9.43}$$

Upon substitution of the expressions for the Ψ functions and application of Eqs. (9.10) and (9.11), one finds

$$M_z = \tfrac{1}{2}\langle d_1 + d_2 + d_5 + d_6 \mid \sum_i ez_i \mid \Psi_G \rangle$$

$$= \langle e_y{}^* \mid ez \mid e_y \rangle + \langle e_x{}^* \mid ez \mid e_x \rangle \tag{9.44}$$

The MO matrix elements of Eq. (9.44) are readily evaluated by procedures of Sec. 4.

6.   TRANSITIONS BETWEEN DEGENERATE STATES

If the excitation between degenerate molecular orbitals gives rise to degenerate states, the basic formulas of Eqs. (9.2)–(9.6) must be modified. The intrinsic lifetime of an excited state, given by Eq. (9.2) previously, now becomes

$$1/\tau^0(h \rightarrow l) = \frac{64\pi^4 \bar{\nu}^3}{3h} G_l \mid \langle \Psi_h \mid \sum_i e\mathbf{r}_i \mid \Psi_l \rangle \mid^2 \tag{9.45}$$

where G_l is the number of *suitable* final wave functions, Ψ_l, which pertain to the lowest energy level and with which *any one* suitable function Ψ_h of the initial higher energy level can combine.[3] The term *suitable wave functions* implies that all wave functions are chosen in the simplest possible way with respect to the axes x, y, z, and that they are orthonormal. In other words, a degenerate E_{1u} level should be described by wave functions Ψ_x and Ψ_y, transforming as x and y—and not by any other orthogonal linear combination such as $c_1\Psi_x + c_2\Psi_y$, $c_2\Psi_x - c_1\Psi_y$.

In a similar way, the oscillator strength of an absorption is no longer given by Eq. (9.3), but by

$$f_{h\leftarrow l} = \frac{8\pi^2\bar{\nu}cm}{3h}\, G_h \,|\, \langle\Psi_h\,|\sum_i e\mathbf{r}_i\,|\,\Psi_l\rangle\,|^2 \tag{9.46}$$

where G_h is now the number of *suitable* final wave functions, Ψ_h, with which any one initial Ψ_l can combine.

The relation between the intrinsic lifetime of an emission and the oscillator strength of the corresponding absorption now becomes

$$\tau_{h\to l}{}^0 = \frac{mc}{8\pi^2 e^2}\frac{G_h}{G_l}\frac{1}{\bar{\nu}^2 f_{h\leftarrow l}} \tag{9.47}$$

In the simple case of nondegeneracy, this equation reduces to Eq. (9.6).

In evaluating the integrals, one follows the same principles as outlined in the nondegenerate case. The theorems of Sec. 2, as well as the group-theoretical results of Sec. 3, remain valid.

Consider the example of two degenerate wave functions $^1\Psi_x$ and $^1\Psi_y$, which pertain to an upper state of symmetry E_{1u}, and a nondegenerate wave function $^1\Psi_G$, which relates to a totally symmetry ground state $^1A_{1g}$. Benzene, with D_{6h} symmetry, is a case in point.

(*i*) *Emission:* Both Ψ_x and Ψ_y can emit; however, since they can connect with only one final state Ψ_G, we must set $G_l = 1$ in Eq. (9.45). Whether we evaluate $\langle\Psi_x\,|\sum_i e\mathbf{r}_i\,|\,\Psi_G\rangle$ or $\langle\Psi_y\,|\sum_i e\mathbf{r}_i\,|\,\Psi_G\rangle$, we find the same numerical result for $|\,\mathbf{M}\,|$. However, the first integral yields a vector along the x axis, $\mathbf{M} = \mathbf{i}M_x$, whereas the second integral yields a vector along the y axis, $\mathbf{M} = \mathbf{j}M_y$. The transition, in fact, is (x, y) allowed; therefore, the emitted light is polarized perpendicular to the z axis. One might think of the excited electron as *spending* one-half of its time in Ψ_x and one-half in Ψ_y.

(*ii*) *Absorption:* The absorption process initiates in Ψ_G solely, but may connect with both Ψ_x and Ψ_y. To use vernacular: The electron may *jump* to either one of Ψ_x or Ψ_y; therefore $G_h = 2$. The total absorption probability is doubled with respect to the emissive probability.

It is necessary to stress that in the expression $\langle \Psi_h \mid \sum_i e\mathbf{r}_i \mid \Psi_l \rangle$ of Eqs. (9.45) and (9.46), Ψ_h and Ψ_l now mean *any one* of the set of appropriate degenerate functions. One may take, for instance, $\langle \Psi_x \mid \sum_i e\mathbf{r}_i \mid \Psi_G \rangle$ or $\langle \Psi_y \mid \sum_i e\mathbf{r}_i \mid \Psi_G \rangle$, *but not both*. If the functions satisfy our imposed requirement of suitability, these integrals are equal, and the degeneracy is properly accounted for via the use of G factors.

If the restriction of suitability is removed, the appropriate equations become less simple, though at the same time more general. Indeed the various matrix elements $\langle \Psi_h \mid \sum_i e\mathbf{r}_i \mid \Psi_l \rangle$ are no longer equal and it becomes impossible to use a degeneracy factor G. One has to sum over all final-state wave functions and average over all initial-state wave functions. The summation corresponds to the *multiplicity of choice* of the electron with regard to where it may go, whereas the averaging process takes account of the fact that the electron is only partly described by any one of the initial wave functions, yet by each one to the same extent. For example, Eq. (9.46) becomes

$$f_{h \leftarrow l} = 1.085 \times 10^{-5} \bar{\nu} \left[Av(l) \sum_h^{G_h} \mid \langle \Psi_h \mid \sum_i \mathbf{r}_i \mid \Psi_l \rangle \mid^2 \right] \qquad (9.48)$$

where the averaging operator is defined as

$$Av(l) \equiv 1/G_l \sum_l^{G_l}$$

7. TWO-CENTER INTEGRALS

Two-center integrals of type $\langle \chi_A \mid \mathbf{r} \mid \chi_B \rangle$ where χ_A and χ_B are Slater-type functions on A and B, respectively, have been discussed in Sec. 4(D). Their evaluation presents no difficulty. A compendium of transformed Slater STF's $\langle \chi \mid \mathbf{r}$, is provided in Appendix F.

However, various approximation procedures for evaluation of these integrals are in common use. Therefore, while noting possible redundancy, we now discuss approximate methods for two-center integrals.

Consider the special case of two *identical* atomic orbitals on two *different* centers. By identical is meant that they have the same n and l quantum numbers, as well as the same orientation with respect to the molecular coordinate system. Then,

$$\mathbf{M}_{AB} = \langle \chi_A \mid e\mathbf{r} \mid \chi_B \rangle = \langle \chi_A \mid e\mathbf{r}' + e \left[\tfrac{1}{2}(\mathbf{R}_A + \mathbf{R}_B) \right] \mid \chi_B \rangle \qquad (9.49)$$

where $\mathbf{M}_{AB}$ is the contribution of this particular two-center term to the transition moment; where $\mathbf{R}_A$ and $\mathbf{R}_B$ are the position vectors of the centers

A and B with respect to the origin; where, consequently, $\frac{1}{2}(\mathbf{R}_A + \mathbf{R}_B)$ is the position vector of the midpoint of the bond; and where $\mathbf{r}'$ represents the dipole operator referred to this same midpoint. Expanding, we find

$$\mathbf{M}_{AB} = \langle \chi_A \mid e\mathbf{r}' \mid \chi_B \rangle + \tfrac{1}{2}(\mathbf{R}_A + \mathbf{R}_B)e\langle \chi_A \mid \chi_B \rangle \qquad (9.50)$$

The first integral is zero for symmetry reasons: The product $\chi_A\chi_B$ is invariant with respect to inversion in the bond midpoint, whereas $\mathbf{r}'$ transforms into $-\mathbf{r}'$ under the same operation. Therefore,

$$\mathbf{M}_{AB} = \tfrac{1}{2}(\mathbf{R}_A + \mathbf{R}_B)eS_{AB} \qquad (9.51)$$

Although Eq. (9.51) is valid only under very specific conditions, it is often used as a first approximation for *any* two-center integral. If, for instance, the two atomic orbitals are both $2p_z$ and if their orbital exponents are only slightly different, $\mathbf{M}_{AB}$, as given by Eq. (9.51), does not deviate much from the correct value. The more the two atomic orbitals differ, the worse the approximation becomes.

EXERCISES

1. Investigate the manner in which the transition moment of a doublet $\leftrightarrow$ doublet transition is connected to the transition moment integral between molecular orbitals.

Answer: $\mathbf{M}(^2\Psi_1 \leftarrow {}^2\Psi_0) = \langle \varphi_1 \mid e\mathbf{r} \mid \varphi_0 \rangle$ \qquad (9.52)

Comments: The electron in a doublet state can be considered to have 50% α spin and 50% β spin; however, this does not entail that the total transition probability should be twice that given by Eq. (9.52). Indeed, if the electron has α spin it can *only* connect with another α state and *not* with a β state. The detailed argument is similar to that leading to the transition probability between degenerate states in Sec. 6.

2. In breaking down a transition moment integral into atomic matrix elements one finds elements of the type $\langle \chi_A \mid \mathbf{r} \mid \chi_B \rangle$ where χ_A and χ_B are atomic orbitals on the two centers A and B. Show that the example $\langle 2p_x{}_{\mathrm{C}} \mid z_{\mathrm{C}} \mid 2p_x{}_{\mathrm{O}} \rangle$, which was evaluated in Sec. 4(D)(i) by using a *modified* $3d$ orbital on the carbon nucleus, can be evaluated just as well by using a modified $3d$ orbital on the oxygen nucleus.

Hint: $\langle 2p_x{}_{\mathrm{C}} \mid z_{\mathrm{C}} \mid 2p_x{}_{\mathrm{O}} \rangle = \langle 2p_x{}_{\mathrm{C}} \mid z_{\mathrm{O}} - R_{\mathrm{C-O}} \mid 2p_x{}_{\mathrm{O}} \rangle$ \qquad (9.53)

3. Find the numerical value of the one-center integral $\langle 2s \mid z \mid 2p_z \rangle$ for carbon using Slater orbitals for $2s$ and $2p_z$.

Answer: 0.890 a.u.

Hint: Use the Slater rules of Appendix B and follow the procedures outlined in Sec. 4(D). Check the intermediary analytical results with Appendixes F and G.

4. Suppose that ${}^1\Psi_x$ and ${}^1\Psi_y$ are two degenerate excited singlet wave functions which transform as x and y. They may, for convenience, be considered as bases for an E_{1u} representation in D_{6h}. Show that the absorption probability between a nondegenerate ground state ${}^1\Psi_0$ and the excited ${}^1E_{1u}$ state is given by Eq. (9.46). What is the precise nature of G_h? Show that the same result is obtained if one describes the ${}^1E_{1u}$ state by the equivalent set of two orthogonal wave functions: ${}^1\Psi_1 = c_1 {}^1\Psi_x + c_2 {}^1\Psi_y$ and ${}^1\Psi_2 = c_2 {}^1\Psi_x - c_1 {}^1\Psi_y$, where $c_1^2 + c_2^2 = 1$.

Hint: In this instance, one must use the more general expression of Eq. (9.48). Indeed, one may show that the oscillator strength $f({}^1\Psi_1 \leftarrow {}^1\Psi_0)$ is *not generally equal* to $f({}^1\Psi_2 \leftarrow {}^1\Psi_0)$ provided that $f({}^1\Psi_x \leftarrow {}^1\Psi_0) = f({}^1\Psi_y \leftarrow {}^1\Psi_0)$.

5. Consider the transition from one doubly degenerate state (Ψ_x, Ψ_y) to another doubly degenerate state $(\Psi_{x'}, \Psi_{y'})$. In a point group without inversion center (for example, C_{4v}) where (x, y) is classified E, such a transition is allowed and polarized along z. Calculate the transition probability.

Hint: The four transition moments involved are *not* all equal. One does not have to bother about possible inequalities if one uses Eq. (9.48); however, if one uses Eq. (9.46), one has to take $G_h = 1$. Indeed, in this latter instance, only *one* final wave function Ψ_h can combine with *any one* initial Ψ_l: $\langle \Psi_{x'} \mid \sum_i ez_i \mid \Psi_y \rangle = \langle \Psi_{y'} \mid \sum_i ez_i \mid \Psi_x \rangle = 0$.

BIBLIOGRAPHY

[1] G. Herzberg, *Atomic Spectra and Atomic Structure* (Dover Press, New York, 1944), p. 51ff.

[2] W. Kauzmann, *Quantum Chemistry* (Academic Press Inc., New York, 1957), p. 615ff.

[3] R. S. Mulliken, *J. Chem. Phys.* **7,** 14 (1939).

[4] H. Kuhn, *J. Chem. Phys.* **29,** 958 (1958).

[5] D. G. Carroll, L. G. Vanquickenborne, and S. P. McGlynn, *J. Chem. Phys.* **45,** 2777 (1966).

[6] R. S. Mulliken, C. A. Rieke, D. Orloff, and H. Orloff, *J. Chem. Phys.* **17,** 1248 (1949).

[7] H. H. Jaffé, *J. Chem. Phys.* **21,** 258 (1953).

[8] J. L. Roberts, and H. H. Jaffé, *J. Chem. Phys.* **27,** 883 (1957).

[9] D. P. Craig, A. Maccoll, R. S. Nyholm, L. E. Orgel, and L. E. Sutton, *J. Chem. Soc.* 359 (1955).

[10] A. C. Wahl, P. E. Cade, and C. C. J. Roothaan, *J. Chem. Phys.* **41,** 2578 (1964).

[11] R. S. Mulliken, *J. Chem. Phys.* **7,** 20 (1939).

[12] W. Moffitt, *J. Chem. Phys.* **22,** 320 (1954).

[13] C. J. Hoijtink, *Molecular Orbitals in Chemistry, Physics and Biology,* edited by P.-O. Löwdin and B. Pullman (Academic Press Inc., New York, 1964), p. 471.

General References

1. R. S. Mulliken, *J. Chem. Phys.* **7,** 14 (1939); **7,** 20 (1939).

2. J. C. Slater, *Quantum Theory of Atomic Structure* (McGraw-Hill Book Co., Inc., New York, 1960), Vols. I and II.

3. R. S. Mulliken and C. A. Rieke, *Rept. Progr. Phys.* **8,** 231 (1941).

CHAPTER 10

Static Electric-Dipole Moments

This chapter is concerned with evaluation of electric-dipole moments in the molecular-orbital formalism. The formally correct (i.e., exact) MO procedure is very lengthy; because of this, the exact formalism is usually applied to the valence electrons only—the nonvalence electrons being assumed to collapse into their respective nuclei and to be part of the nuclear-dipole problem. Even with this *collapsed-core* approximation, the electronic-dipole calculation remains formidable and a point-charge approximation based on a Mulliken population analysis[1] is often used. We discuss all of these methods (that is, exact, collapsed-core, and population-based point charge). We conclude that no approximation procedure is altogether satisfactory and that even the exact method may not be of much value unless the basis wave functions employed are highly refined.

In view of the above, it is not surprising that semiempirical methods are in wide use: We outline only the most common of these, namely, the Bond-Moment approximation.

1. Fundamental Concepts

A molecule has zero electric-dipole moment when the barycenter of the nuclear charge is coincident with the barycenter of the electronic charge.

When the two barycenters are not coincident, the molecule is dipolar. The barycenter of the nuclear charge is

$$\mathbf{D}_{\text{nuclear}} = \sum_K \mathbf{R}_K Q_K / \sum_K Q_K \qquad (10.1)$$

where $\mathbf{R}_K$ and Q_K are the position vector and the charge of nucleus K, respectively. The charge barycenter of the electronic *cloud* of a molecule, when in its ground electronic state, is

$$\mathbf{D}_{\text{electronic}} = \langle \Psi_G \mid \sum_{i=1}^{N} e\mathbf{r}_i \mid \Psi_G \rangle / eN \qquad (10.2)$$

where N is the number of electrons and $\mathbf{r}_i$ is the displacement vector for the *ith* electron. If the molecule is neutral, it follows that

$$\sum_K Q_K = -eN \qquad (10.3)$$

We adopt the following conventions:

(i)   The vector representative of the dipole points toward the negatively charged end of the dipole as in

$$\overset{\boldsymbol{\mu}}{\oplus \longrightarrow \ominus}$$

Under these conditions, the electric-dipole moment of a neutral molecule is

$$\boldsymbol{\mu} = eN(\mathbf{D}_{\text{nuclear}} - \mathbf{D}_{\text{electronic}}) \qquad (10.4)$$

(ii)   Since the vector lengths are usually of the order of 1 Å and the charges usually of the order of 10^{-10} esu ($e = -4.80 \times 10^{-10}$ esu), it follows that a Debye unit, where

$$1 \text{ Debye} \equiv 10^{-18} \text{ esu cm} \qquad (10.5)$$

provides a convenient unit for electric-dipole-moment magnitudes.

The primary attention of this chapter focuses on the evaluation of $\mathbf{D}_{\text{electronic}}$. Thus, it is well to note that the integral of Eq. (10.2) is closely related to the transition moment integrals evaluated in chapter 9. Indeed, the numerator of Eq. (10.2) is merely a *diagonal matrix element* of the operator $\sum_i e\mathbf{r}_i$ in a basis set of state wave functions; in the calculation of transition-dipole moments, on the other hand, one is concerned with *off-diagonal matrix elements* of the same operator in the same basis set.

(A)   Group-Theoretic Considerations

If a polar molecule possesses an n-fold symmetry axis, the dipole moment vector must be aligned along this axis. If the moment were not so directed, a rotation by $2\pi/n$ about this axis would not be a symmetry operation; indeed, the situations before and after rotation would be distinguishable.

If a molecule possesses two or more noncoincident symmetry axes, that molecule is nonpolar. The conclusions of the prior paragraph require that μ be directed along all symmetry axes. Such alignment is impossible when there are two or more symmetry axes—except when μ is zero.

Similar argument leads to the conclusion that any molecule possessing a symmetry plane perpendicular to a C_n axis, or any molecule in which there is an improper symmetry axis S_n (where, in both cases, $n = 2, 3, 4, \ldots$), has a null dipole moment.

Since a large number of molecules belong to point groups in which one or more of these symmetry requirements are fulfilled, the situation $\mu = 0$ occurs quite frequently. Indeed, the only molecules which can exhibit a nonzero dipole moment are those which belong to point groups C_n or C_{nv} ($n = 1, 2, 3, \ldots \infty$).

2. MOLECULAR-ORBITAL THEORY

(A) Exact Procedure

The wave function Ψ_G can usually be written as a single Slater determinant.[1] The expansion of the integral $\langle \Psi_G \mid \sum_{i=1} e\mathbf{r}_i \mid \Psi_G \rangle$ into component MO integrals, proper note being taken of the one-electron nature of the operator $\mathbf{r}_i$, leads to

$$\mathbf{D}_{\text{electronic}} = \frac{1}{eN} \sum_{t=1}^{N} \langle \psi_t \mid e\mathbf{r} \mid \psi_t \rangle \tag{10.6}$$

where the ψ_t are the N orthogonal spin orbitals used in the construction of the determinant. In the most common situation, namely, that in which the ground state consists of $N/2$ doubly occupied nondegenerate molecular orbitals, we have $\psi_1 = \varphi_1\alpha$, $\psi_2 = \varphi_1\beta, \ldots, \psi_{N-1} = \varphi_{N/2}\alpha$, $\psi_N = \varphi_{N/2}\beta$, and

$$\mathbf{D}_{\text{electronic}} = \frac{2}{N} \sum_{t=1}^{N/2} \langle \varphi_t \mid \mathbf{r} \mid \varphi_t \rangle \tag{10.7}$$

These integrals can be evaluated in the same way as the transition moment integrals of chapter 9. The right-hand side of Eq. (10.7) consists of a sum of matrix elements of the operator $\mathbf{r}$ which are diagonal in a molecular-orbital basis. If the MO-LCAO approximation is used, the expansion of $\langle \varphi_t \mid e\mathbf{r} \mid \varphi_t \rangle$ yields AO matrix elements of types $\langle \chi_\mu \mid \mathbf{r} \mid \chi_\nu \rangle$ and $\langle \chi_\mu \mid \mathbf{r} \mid \chi_\mu \rangle$.

---

[1] The restriction to a single Slater determinant can be removed; such removal introduces no new complications beyond those already present—it merely increases the labor.

Inspection of Eq. (10.4) indicates that μ is related to the difference of two terms, one of which is a sum of the contributions made by all populated molecular orbitals. Since the number of populated MO's can be quite large, the evaluation of a dipole moment can be a very tedious affair—even for small molecules. In view of this, many simplifications have been proposed. We now discuss some of the more important of these simplifications.

(B)   Collapsed-Core Approximation

The core electrons are relatively unaffected by chemical bonding. It is usual, therefore, to assume that the total system of core electrons *plus* nucleus can be treated as a formal point charge

$$Q' = Q + eN_c \qquad (10.8)$$

where Q is the nuclear charge and N_c is the number of core electrons.

The collapsed-core assumption has been quite successful in some categories of molecule—the hydrogen halides,[2] for example. Similar attitudes have been used successfully in the treatment[3] of aromatic molecules,[2] where one calculates the dipole moment due to the π electrons *only*—the so-called *mesomeric moment*; the result is then compared to the difference of dipole moments of the aromatic and the corresponding saturated compound.

On the negative side, we note that Ellison and Shull[4],[5] have performed two different self-consistent-field calculations on H_2O. In one calculation, they took all ten electrons into account; in the second, the $1s$ electrons of oxygen were *collapsed* into the oxygen nucleus. Dipole moment calculations in the former case yielded $1.52D$, whereas the collapsed-core approximation yielded $1.21D$. It appears that the most *acceptable* sort of simplification can have a marked influence on the computed value of a dipole moment.

(C)   Point-Charge Model[3]

If a doubly occupied molecular orbital of a diatomic molecule A—B is given by

$$\varphi_t = c_{t\mu}\chi_\mu + c_{t\nu}\chi_\nu \qquad (10.9)$$

and if χ_μ and χ_ν are atomic orbitals on atoms A and B, respectively, one finds that these two centers possess electronic charges[1]

$$q_{At} = e(2c_{t\mu}{}^2 + 2c_{t\mu}c_{t\nu}S_{\mu\nu}) \qquad (10.10)$$

$$q_{Bt} = e(2c_{t\nu}{}^2 + 2c_{t\mu}c_{t\nu}S_{\mu\nu}) \qquad (10.11)$$

---

[2] See also chapter 3, Sec. 3(G).
[3] See Sec. 8, chapter 1 and Secs. 5(D) and 6(H) of chapter 4.

The total electronic charge on each of the various centers is

$$q_A = \sum_t q_{At} \tag{10.12}$$

$$q_B = \sum_t q_{Bt} \tag{10.13}$$

where each of the populated MO's φ_t is assumed to contain two electrons. It is now possible to assign *formal charges* Q_f to the various atoms:

$$Q_{fA} = Q_A + q_A \tag{10.14}$$

$$Q_{fB} = Q_B + q_B \tag{10.15}$$

It appears reasonable to treat the molecule as a system of point charges Q_{fK} situated at $\mathbf{R}_K$. Such a supposition enables us to write the dipole moment as

$$\boldsymbol{\mu} = -\sum_K \mathbf{R}_K Q_{fK} \tag{10.16}$$

The approximation of Eq. (10.16) is very simple and very easy to use. There are, however, two significant objections[2],[6] to it: It ignores the existence of electronic dipoles caused by overlap asymmetry and it disregards hybridization dipoles completely.

(D) Asymmetry Dipole

The Mulliken population analysis makes equal division of the overlap charge density (which amounts to $4c_{t\mu}c_{t\nu}\langle \chi_\mu \mid \chi_\nu \rangle$ for a doubly occupied φ_t MO) between the two atomic centers of the molecule A—B. Now, the interatomic overlap density is not centrally located in the A—B bond except when $A = B$. Consequently, the population analysis neglects the dipole associated with the asymmetric distribution of overlap charge in the A—B bond when $A \neq B$. This neglected dipole can amount to $1D$; we designate it the *asymmetry dipole*. We now detail the origins of this asymmetry dipole.

Consider the HeH+ molecule of Fig. 10.1. This molecule contains two electrons; in the point-charge approximation, q_{He} and q_H are given by Eqs.

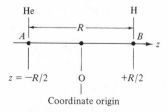

FIG. 10.1.   Coordinates and geometry for AB and HeH+ molecules.

(10.12) and (10.13); the formal charges are $Q_{f(\text{He})} = 2\,|\,e\,| + q_{\text{He}}$ and $Q_{f\text{H}} = 1\,|\,e\,| + q_{\text{H}}$. Using Eq. (10.16), we find (after simplification) that

$$\mathbf{\mu}_{\text{point charge}} = \mathbf{k}\,|\,e\,|\,R(\tfrac{1}{2} - c_{t\mu}{}^2 + c_{t\nu}{}^2) \qquad (10.17)$$

where $\mathbf{k}$ is the unit vector in the z direction of Fig. 10.1.

Let us now evaluate the same dipole moment exactly. Using Eq. (10.7) and noting that $N = 2$, we find a total electronic-dipole length

$$2\mathbf{D}_{\text{electronic}} = 2\mathbf{k}\langle\varphi_t\,|\,z\,|\,\varphi_t\rangle$$

$$= \mathbf{k}(2c_{t\mu}{}^2\langle s_{\text{He}}\,|\,z\,|\,s_{\text{He}}\rangle + 2c_{t\nu}{}^2\langle s_{\text{H}}\,|\,z\,|\,s_{\text{H}}\rangle + 4c_{t\mu}c_{t\nu}\langle s_{\text{He}}\,|\,z\,|\,s_{\text{H}}\rangle)$$

$$(10.18)$$

where s_{He} is a $1s$ AO on helium; s_{H} is a $1s$ AO on hydrogen; and we have used the correspondences $\chi_\mu \equiv s_{\text{He}}$ and $\chi_\nu \equiv s_{\text{H}}$. Since the element $\langle s_{\text{He}}\,|\,z\,|\,s_{\text{He}}\rangle$ is the mean position of the charge cloud of the $1s$ AO on helium (that is, $-R/2$), we rewrite Eq. (10.18) as

$$2\mathbf{D}_{\text{electronic}} = \mathbf{k}(-c_{t\mu}{}^2R + c_{t\nu}{}^2R + 4c_{t\mu}c_{t\nu}\langle s_{\text{He}}\,|\,z\,|\,s_{\text{H}}\rangle) \qquad (10.19)$$

Noting that $\sum_K Q_K$ of Eq. (10.1) equals 3 and using Eqs. (10.1), (10.2), (10.4), and (10.19), we find

$$\mathbf{\mu}_{\text{exact}} = \mathbf{k}\,|\,e\,|\,R\left[\tfrac{1}{2} - c_{t\mu}{}^2 + c_{t\nu}{}^2 + \left(\frac{4c_{t\mu}c_{t\nu}}{R}\right)\langle s_{\text{He}}\,|\,z\,|\,s_{\text{H}}\rangle\right] \qquad (10.20)$$

or

$$\mathbf{\mu}_{\text{exact}} - \mathbf{\mu}_{\text{point charge}} = 4\mathbf{k}\,|\,e\,|\,c_{t\mu}c_{t\nu}\langle s_{\text{He}}\,|\,z\,|\,s_{\text{H}}\rangle \qquad (10.21)$$

It is this last quantity to which the term *asymmetry dipole* makes reference. Now, the element $\langle s_A\,|\,z\,|\,s_B\rangle$ is zero only when $Z_A = Z_B$; under these conditions, any positive contribution to the integral, obtained at $z = z_1$, is exactly cancelled by an equal contribution of opposite sign arising at $z = -z_1$. If $Z_A > Z_B$, s_A assumes any arbitrary fixed value at a relatively smaller distance from the A center than s_B from the B center. Since z is negative on the A side of the origin, it follows that the negative contributions to $\langle s_A\,|\,z\,|\,s_B\rangle$ predominate over the positive contributions and that the barycenter of the φ_t *charge cloud* is closer to the *smaller* orbital on A (that is, He). Hence, we find an overlap asymmetry dipole of direction $A \leftarrow B$ or He $\leftarrow$ H. The neglect of the electronic asymmetry dipole is not justifiable.

(E) Hybridization Dipole

When a molecular orbital is formed from two s electrons, as seen above, a dipole moment contribution can arise from the *asymmetrical distribution of the overlap population* in the bond region. In contrast to this, if the molecular orbital contains *hybrid* atomic orbitals, a dipole moment contribution can arise from the *asymmetric distribution of the atomic populations themselves.* Indeed, as can be seen from Fig. 10.2, a hybrid orbital is not centrally symmetric and the barycenter of an electron in such an orbital is not coincident with the nucleus to which the atomic orbital refers.

Consider an sp-hybrid AO on atom A, of the type

$$\chi_h = c_1(2s) + c_2(2p_z); \qquad c_1^2 + c_2^2 = 1 \tag{10.22}$$

Let this hybrid be one of the constituent parts of a filled MO

$$\varphi = c_A(\chi_h)_A + c_B\chi_B \tag{10.23}$$

If the chosen coordinate system is that shown in Fig. 10.1, we find

$$\langle \chi_h \mid z \mid \chi_h \rangle = c_1^2(-\tfrac{1}{2}R) + c_2^2(-\tfrac{1}{2}R) + 2c_1c_2\langle 2s \mid z \mid 2p_z \rangle$$
$$= -\tfrac{1}{2}R + 2c_1c_2\langle 2s \mid z \mid 2p_z \rangle \tag{10.24}$$

The last term of Eq. (10.24) is zero only when one of the coefficients c_1 or c_2 equals zero (that is, for either a pure s orbital or a pure p orbital). When $c_1c_2 > 0$, the barycenter of the electronic charge cloud is displaced towards the neighboring B atom. Using the results of Appendix G and chapter 9, we find

$$\langle 2s \mid z \mid 2p_z \rangle = \left(\frac{Z_{2s}{}^5 Z_{2p}{}^5}{3}\right)^{1/2} \frac{5(2^6)}{(Z_{2s} + Z_{2p})^6} \text{ bohrs} \tag{10.25}$$

Following Slater, we set $Z_{2s} = Z_{2p} = Z$, and obtain

$$2c_1c_2\langle 2s \mid z \mid 2p_z \rangle = \frac{10c_1c_2}{3^{1/2}Z} \text{ bohrs} \tag{10.26}$$

The one-center term of Eq. (10.26) has a maximum value when $c_1 = c_2$ (that is, digonal hybridization). The resulting dipole moment contribution is readily evaluated using Eqs. (10.1), (10.2), and (10.4);

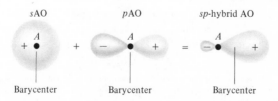

FIG. 10.2. Influence of hybridization on the barycenter of the electronic charge cloud.

the hybridization dipole per electron in φ of Eq. (10.23) is given by $\mathbf{k} \mid e \mid (4.8)c_A{}^2(10c_1c_2)(1/3^{1/2}Z)(0.529)D$. This dipole is an *atomic* dipole; in the case of fluorine it may amount to $1.1D$ per *sp*-hybridized bonding electron. In a calculation based on a population analysis, this atomic dipole would have been completely ignored: The total *intra-atomic overlap population* $2c_A{}^2c_1c_2\langle 2s \mid 2p_z \rangle$ would have been assigned to atom A (at $z = -\frac{1}{2}R$).

Hybridization effects are particularly important if one of the hybrids is nonbonding (that is, $c_A = 1$) and occupied by two electrons. Consider H_2O or NH_3. The lone-pair electrons on either oxygen or nitrogen occupy a hybridized orbital aligned along the principal symmetry axis and directed away from the hydrogen atoms. The dipole moment associated with the noncentrality of such lone pairs is large. Clearly, any analysis which neglects these effects may yield[6] very poor results!

3. SEMIEMPIRICAL BOND MOMENTS

Most of the argument of Sec. 2 concerns diatomic molecules. Similar considerations of polyatomic molecules become complicated. As a result, various semiempirical approaches have evolved for the computation of dipole moments of such molecules. The most common of these is the Bond-Moment approximation.

It is assumed that a complicated molecule can be decomposed into smaller diatomic parts—one part for every bonded pair of atomic centers in the polyatomic molecule. For example, chloroform may be represented by

$$CHCl_3 \rightarrow 1(C\!-\!H) + 3(C\!-\!Cl)$$

If we associate a dipole moment with each of the diatomic parts (that is, a bond moment), it is clear that vector addition of these moments—in a manner consistent with the geometry of $CHCl_3$—yields a dipole moment of the chloroform molecule. The crux of this procedure lies in the assumption that we can obtain values for bond moments, and that these moments are then transferable from molecule to molecule.

Bond moments for the individual bonds of a polyatomic molecule are simply not measurable—it is only the total molecular dipole is measurable. Bond moments are deduced from the known dipole moments of certain small molecules[4] and are then assumed to apply to the corresponding bond in the polyatomic. Such assumptions have had some success[10]—usually when

---

[4] Several tables of bond moments, adduced as described, do exist (Refs. [7]–[9]). The work of Orgel (Ref. [10]) on the σ moments of heterocyclic molecules provides a good example of the use of such bond moment tables. Unknown bond moments can be estimated using parallelisms (Ref. [3]) between bond moments and the electronegativity differences of the bonded centers.

applied to a series of homologous or related molecules. In general, the procedure should be used with caution because:

(i)   The transferability of a bond moment from one molecule to another has little justification. The different bonds of a given molecule are not independent: The charge distribution in one bond assuredly affects that in another. It has been argued,[2],[6] for instance, that the C—H moment in CH_4 has polarity C^+—H^- whereas in aromatics such as ethylene and acetylene the polarity is probably C^-—H^+!

(ii)   The additivity rule can not be expected to apply if lone-pair electrons are present in the molecule and occupy hybridized orbitals. In water, for instance, the most important contribution to the dipole moment probably arises from the in-plane lone pair of electrons on the oxygen atom. To effectively incorporate this effect into two O—H bonds (that is, to obtain an O—H bond moment from the known dipole moment and geometry of water) hardly engenders confidence in the transferability of such a bond moment to other molecules containing the O—H grouping.

4.   CARBON MONOXIDE: AN EXAMPLE CALCULATION

We have chosen the carbon monoxide molecule because it is small and because a set of simple LCAO-MO-SCF wave functions for the molecule has been generated.[11] The coordinate system is shown in Fig. 10.3.

Sahni[11] used an atomic-orbital basis set which consisted of a single Slater representative for each AO. By virtue of the nonorthogonality of Slater functions with the same l and m_l, he performed a Schmidt orthogonalization of the $1s$ and $2s$ atomic orbitals on each center. This orthogonal basis set of atomic orbitals is given in Table 10.1.

The $1s_O$ and $1s_C$ atomic orbitals are assumed to remain nonbonding throughout the energy calculations. Thus, mixing is restricted to the $2s_O'$, $2s_C'$, $2p_{qO}$, and $2p_{qC}$ (where $q = x$, y, or z) set of atomic orbitals. The

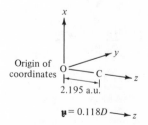

FIG. 10.3.   Coordinate system, interatomic distance, and dipole moment of carbon monoxide. Some experimental evidence in favor of the polarity C^-—O^+ does exist (Ref. [12]).

resultant set of MO's is shown in Table 10.2. The molecular orbitals, arranged in order of increasing energy, are

$$t\sigma < s\sigma < (\pi_x, \pi_y) < u\sigma$$

The ground configuration of carbon monoxide, which contains 14 electrons, is

$$(1s_O)^2(1s_C)^2(t\sigma)^2(s\sigma)^2(\pi_{x,y})^4(u\sigma)^2$$

TABLE 10.1. *Atomic-orbital basis set*

(This basis set is taken from Sahni.[11] The nonprimed orbitals are the usual Slater functions. The primed orbitals are orthogonalized with respect to the corresponding $1s$ functions. The symbol q is a general representative for x, y, or z. The AO's of carbon and oxygen are distinguished by suffixes C and O, respectively.)

Carbon

$$1s_C = \left[\frac{(5.7)^3}{\pi}\right]^{1/2} e^{-5.7r}$$

$$2s_C' = \frac{2s_C - [S(1s_C, 2s_C)]1s_C}{[1 - S^2(1s_C, 2s_C)]^{1/2}}$$

$$= 1.025\left[\frac{(3.25)^5}{96\pi}\right]^{1/2} re^{-1.625r} - 0.226\left[\frac{(5.7)^3}{\pi}\right]^{1/2} e^{-5.7r}$$

$$2p_{qC} = \left[\frac{(3.25)^5}{32\pi}\right]^{1/2} qe^{-1.625r}$$

Oxygen

$$1s_O = \left[\frac{(7.7)^3}{\pi}\right]^{1/2} e^{-7.7r}$$

$$2s_O' = \frac{2s_O - [S(1s_O, 2s_O)]1s_O}{[1 - S^2(1s_O, 2s_O)]^{1/2}}$$

$$= 1.028\left[\frac{(4.55)^5}{96\pi}\right]^{1/2} re^{-2.275r} - 0.240\left[\frac{(7.7)^3}{\pi}\right]^{1/2} e^{-7.7r}$$

$$2p_{qO} = \left[\frac{(4.55)^5}{32\pi}\right] qe^{-2.275r}$$

We are now in a position to evaluate the dipole moment of carbon monoxide. In doing so, we *collapse* the 1s electrons into the carbon and oxygen nuclei.

(A) Point-Charge Model

The relevant overlap integrals are listed in Tables 10.3–10.5. In order to evaluate overlap populations we must generalize Eqs. (10.10)–(10.13) for the situation where the MO consists of more than two atomic orbitals. In specific, Eq. (10.10) becomes

$$q_{At} = 2e \sum_{\mu}^{A} \sum_{\nu} c_{t\mu} c_{t\nu} S_{\mu\nu} \tag{10.27}$$

where the summation symbol $\sum^{A}$ implies consideration of only those terms in which the AO χ_μ is located on center A and where it is assumed that the MO φ_t contains two electrons. The use of Tables 10.1–10.5 and Eq. (10.27)

TABLE 10.2. *Molecular orbitals of carbon monoxide.*[11]

(Only those MO's which are populated in the lowest-energy configuration of CO are shown.)

AO				Molecular Orbitals			
	$1s_O$	$1s_C$	$u\sigma$	$s\sigma$	$t\sigma$	π_x	π_y
$1s_O$	1	0	0	0	0	0	0
$1s_C$	0	1	0	0	0	0	0
$2s_O'$	0	0	0.187	0.7176	0.675	0	0
$2s_C'$	0	0	0.6145	−0.4926	0.270	0	0
$2p_{zO}$	0	0	−0.189	−0.6065	0.231	0	0
$2p_{zC}$	0	0	0.7626	0.168	−0.227	0	0
$2p_{xO}$	0	0	0	0	0	0.8145	0
$2p_{xC}$	0	0	0	0	0	0.4162	0
$2p_{yO}$	0	0	0	0	0	0	0.8145
$2p_{yC}$	0	0	0	0	0	0	0.4162

TABLE 10.3. σ *Overlaps of Slater functions.*

	$1s_O$	$1s_C$	$2s_O$	$2s_C$	$2p_{zO}$	$2p_{zC}$
$1s_O$	1	0.0001	0.2335	0.0452	0	-0.0771
$1s_C$	0.0001	1	0.0425	0.2205	0.0713	0
$2s_O$	0.2335	0.0425	1	0.4052	0	-0.4854
$2s_C$	0.0452	0.2205	0.4052	1	0.3226	0
$2p_{zO}$	0	0.0713	0	0.3226	1	-0.3017
$2p_{zC}$	-0.0071	0	-0.4854	0	-0.3017	1

TABLE 10.4. σ *Overlaps of orthogonalized Slater functions of Table* 10.1.

(The asterisk symbol denotes matrix elements which are not
needed for the overlap population calculations.)

	$1s_O$	$1s_C$	$2s_O'$	$2s_C'$	$2p_{zO}$	$2p_{zC}$
$1s_O$	1	0.0001	0	*	0	*
$1s_C$	0.0001	1	*	0	*	0
$2s_O'$	0	*	1	0.4060	0	-0.4805
$2s_C'$	*	0	0.4060	1	0.3146	0
$2p_{zO}$	0	*	0	0.3146	1	-0.3017
$2p_{zC}$	*	0	-0.4805	0	-0.3017	1

TABLE 10.5. π *Overlaps of Slater functions of Table* 10.1.

	$2p_{xO}$	$2p_{xC}$	$2p_{yO}$	$2p_{yC}$
$2p_{xO}$	1	0.2409	0	0
$2p_{xC}$	0.2409	1	0	0
$2p_{yO}$	0	0	1	0.2409
$2p_{yC}$	0	0	0.2409	1

leads to the required population analysis.[5] The results of this analysis are given in Table 10.6.

The net charge on the oxygen center is $0.088e$ and is negative. The net charge on the carbon center is $0.088e$ and is positive. Therefore, the dipole moment is predicted to be 0.088×2.195 a.u. of dipole moment[6] or $0.490D$. There is discrepancy, not only in magnitude but also in direction, with respect to the best available knowledge—as schematized in Fig. 10.3.

(B) Exact Procedure

The expanded forms of the MO's—for instance, $u\sigma$ of Eq. (10.29)— must be used in order to evaluate all the necessary integrals of the type

TABLE 10.6. *Mulliken population analysis for carbon monoxide ground state.*
[The charges are in atomic units $(e = 1$ a.u.$)$.]

φ_t MO	q_{Ot}	q_{Ct}	MO Population
$1s_O$	2.000	0.000	2.000
$1s_C$	0.000	2.000	2.000
$u\sigma$	0.112	1.888	2.000
$s\sigma$	1.612	0.388	2.000
$t\sigma$	1.384	0.616	2.000
π_x	1.490	0.510	2.000
π_y	1.490	0.510	2.000
	$q_O = 8.088$	$q_C = 5.912$	$N = 14.000$

[5] In this regard, it is well to note that the MO which we have denoted $u\sigma$ is given by Table 10.2 as

$$u\sigma = (0.187)2s_{O}' + (0.6145)2s_{C}' - (0.189)2p_{zO} + (0.7626)2p_{zC} \qquad (10.28)$$

or, using both Tables 10.1 and 10.2, as

$$u\sigma = (0.192)2s_O - (0.0449)1s_O + (0.6299)2s_C - (0.1389)1s_C$$

$$- (0.189)2p_{zO} + (0.7626)2p_{zC} \qquad (10.29)$$

Thus, in the case of $q_{A(u\sigma)}$, the population analysis would utilize either Eq. (10.28) in conjunction with Table 10.4 or Eq. (10.29) in conjunction with Table 10.5. The former combination—namely, Eq. (10.28) *plus* Table (10.4)—provides the most direct computational route.

[6] Note that 1 a.u. of dipole moment = 1 electronic charge $\times$ 0.5292 Å = $4.80 \times 0.5292 \times 10^{-18}$ esu cm = $2.54D$.

$\langle \chi_\mu \mid z \mid \chi_\nu \rangle$. In the case of $u\sigma$, one obtains

$$\langle u\sigma \mid z \mid u\sigma \rangle$$

$$= \langle u\sigma \mid z_0 \mid u\sigma \rangle = (0.192)^2 \langle 2s_0 \mid z_0 \mid 2s_0 \rangle$$

$$+ (0.0449)^2 \langle 1s_0 \mid z_0 \mid 1s_0 \rangle + (0.6299)^2 \langle 2s_C \mid z_0 \mid 2s_C \rangle$$

$$+ (0.1389)^2 \langle 1s_C \mid z_0 \mid 1s_C \rangle + (0.189)^2 \langle 2p_{z0} \mid z_0 \mid 2p_{z0} \rangle$$

$$+ (0.7626)^2 \langle 2p_{zC} \mid z_0 \mid 2p_{zC} \rangle - 2(0.192)(0.0449) \langle 2s_0 \mid z_0 \mid 1s_0 \rangle$$

$$+ 2(0.192)(0.6299) \langle 2s_0 \mid z_0 \mid 2s_C \rangle - 2(0.192)(0.1389) \langle 2s_0 \mid z_0 \mid 1s_C \rangle$$

$$- 2(0.192)(0.189) \langle 2s_0 \mid z_0 \mid 2p_{z0} \rangle + 2(0.192)(0.7626) \langle 2s_0 \mid z_0 \mid 2p_{zC} \rangle$$

$$- 2(0.0449)(0.6299) \langle 1s_0 \mid z_0 \mid 2s_C \rangle + 2(0.0449)(0.1389) \langle 1s_0 \mid z_0 \mid 1s_C \rangle$$

$$+ 2(0.0449)(0.189) \langle 1s_0 \mid z_0 \mid 2p_{z0} \rangle - 2(0.0449)(0.7626) \langle 1s_0 \mid z_0 \mid 2p_{zC} \rangle$$

$$- 2(0.6299)(0.1389) \langle 2s_C \mid z_0 \mid 1s_C \rangle - 2(0.6299)(0.189) \langle 2s_C \mid z_0 \mid 2p_{z0} \rangle$$

$$+ 2(0.6299)(0.7626) \langle 2s_C \mid z_0 \mid 2p_{zC} \rangle + 2(0.1389)(0.189) \langle 1s_C \mid z_0 \mid 2p_{z0} \rangle$$

$$- 2(0.1389)(0.7626) \langle 1s_C \mid z_0 \mid 2p_{zC} \rangle - 2(0.189)(0.7626) \langle 2p_{z0} \mid z_0 \mid 2p_{zC} \rangle$$

$$(10.30)$$

where z_0 means that z is referred to an origin located on the oxygen nucleus. The individual integrals of Eq. (10.30) are evaluated in the manner of chapter 9.

Consider, as an example, the integral $\langle 2s_0 \mid z_0 \mid 2s_C \rangle$. We have

$$\langle 2s_0 \mid z_0 \mid 2s_C \rangle = \frac{(10)^{1/2}}{4.55} \langle 3p_{z0}' \mid 2s_C \rangle = \frac{(10)^{1/2}}{4.55} S(2s_C, 3p_{z0}')$$

$$= \frac{(10)^{1/2}}{4.55} (0.4652) = 0.323 \text{ a.u.} \qquad (10.31)$$

However, since $z_0 = z_C + 2.195$ (see Fig. 10.3), we could equally well proceed as follows:

$$\langle 2s_0 \mid z_0 \mid 2s_C \rangle = \langle 2s_0 \mid z_C \mid 2s_C \rangle + 2.195 \langle 2s_0 \mid 2s_C \rangle$$

$$= -\frac{(10)^{1/2}}{3.25} S(2s_0, 3p_{zC}') + (2.195)(0.4052)$$

$$= -\frac{(10)^{1/2}}{3.25} (0.5819) + (2.195)(0.4052) = 0.323 \text{ a.u.}$$

$$(10.32)$$

The results are identical in both instances—as, of course, they must.

The numerical values of all the $\langle \chi_\mu \mid z \mid \chi_\nu \rangle$ integrals which are needed in the calculation of $\boldsymbol{\mu}$ are listed in Table 10.7. Multiplication of these values by the appropriate coefficients—as in Eq. (10.30), for example—yields the results of Table 10.8. Since each of the listed MO's is occupied by two electrons, the total electronic contribution is 13.564 a.u. and the electronic barycenter is located at (13.564/14) a.u. The nuclear barycenter is located

TABLE 10.7.   *Values of the integrals* $\langle \chi_\mu \mid z \mid \chi_\nu \rangle$.

(The origin of coordinates is located on the oxygen nucleus.
All numbers are in atomic units of length.)

$$\langle 2s_O \mid z \mid 2s_O \rangle = \langle 1s_O \mid z \mid 1s_O \rangle = \langle 2p_{zO} \mid z \mid 2p_{zO} \rangle = 0$$

$$\langle 2s_C \mid z \mid 2s_C \rangle = \langle 1s_C \mid z \mid 1s_C \rangle = \langle 2p_{zC} \mid z \mid 2p_{zC} \rangle = 2.195$$

$$\langle 2s_O \mid z_O \mid 1s_O \rangle = 0$$

$$\langle 2s_O \mid z \mid 2s_C \rangle = 0.323$$

$$\langle 2s_O \mid z \mid 1s_C \rangle = 0.084$$

$$\langle 2s_O \mid z \mid 2p_{zO} \rangle = 0.634$$

$$\langle 2s_O \mid z \mid 2p_{zC} \rangle = -0.214$$

$$\langle 1s_O \mid z \mid 2s_C \rangle = 0.0034$$

$$\langle 1s_O \mid z \mid 1s_C \rangle = 0.0001$$

$$\langle 1s_O \mid z \mid 2p_{zO} \rangle = 0.054$$

$$\langle 1s_O \mid z \mid 2p_{zC} \rangle = -0.0057$$

$$\langle 2s_C \mid z \mid 1s_C \rangle = 0.484$$

$$\langle 2s_C \mid z \mid 2p_{zO} \rangle = 0.5182$$

$$\langle 2s_C \mid z \mid 2p_{zC} \rangle = 0.888$$

$$\langle 1s_C \mid z \mid 2p_{zO} \rangle = 0.142$$

$$\langle 1s_C \mid z \mid 2p_{zC} \rangle = 0.0695$$

$$\langle 2p_{zO} \mid z \mid 2p_{zC} \rangle = -0.395$$

$$\langle 2p_{xO} \mid z \mid 2p_{xO} \rangle = \langle 2p_{yO} \mid z \mid 2p_{yO} \rangle = 0$$

$$\langle 2p_{xC} \mid z \mid 2p_{xC} \rangle = \langle 2p_{yC} \mid z \mid 2p_{yC} \rangle = 2.195$$

$$\langle 2p_{xO} \mid z \mid 2p_{xC} \rangle = \langle 2p_{yO} \mid z \mid 2p_{yC} \rangle = 0.2063$$

at $(6 \times 2.195)/14$ a.u. $= (13.170/14)$ a.u. and thus

$$\mu = 0.394 \text{ a.u. of dipole moment} = 1.001D \qquad (10.33)$$

It is especially noteworthy that the negative end of the dipole is predicted to be on the carbon atom. The population analysis predicted a quite opposite polarity!

(C) Evaluation of Results

The simplest sorts of electronegativity considerations indicate that the polarity should be C^+—O^-. The experimental evidence[12] suggests that the polarity actually is C^-—O^+. The results of the exact calculation agree with this latter polarity but the expected magnitude $1.001D$ is considerably greater than that observed, namely, $0.118D$. On the other hand, the population-analysis approach yields a wrong polarity of about the right magnitude, namely, $0.490D$.

Our computations indicate that the Sahni calculation,[11] although carried out very carefully, does not provide wave functions of sufficient precision for dipole moment evaluations. We surmise that the inherent imprecision of these wave functions arises from the neglect of inner-shell–outer-shell mixing, from the small size of the atomic-orbital basis set—no $3s$ or $3p$ atomic orbitals—which he used, and the single Slater representation which purportedly describes each atomic orbital in this basis set.

TABLE 10.8.   *Values of the integrals* $\langle \varphi_t \,|\, z \,|\, \varphi_t \rangle$.

(The values of the integrals are quoted in atomic units of length.)

Integral	Value
$\langle 1s_O \,|\, z_O \,|\, 1s_O \rangle$	0
$\langle 1s_C \,|\, z_O \,|\, 1s_C \rangle$	2.195
$\langle u\sigma \,|\, z_O \,|\, u\sigma \rangle$	2.907
$\langle s\sigma \,|\, z_O \,|\, s\sigma \rangle$	-0.010
$\langle t\sigma \,|\, z_O \,|\, t\sigma \rangle$	0.650
$\langle \pi_x \,|\, z_O \,|\, \pi_x \rangle$	0.520
$\langle \pi_y \,|\, z_O \,|\, \pi_y \rangle$	0.520
	Total $= 6.782$

EXERCISES

1. Calculate the value of the atomic-dipole moment associated with a tetrahedrally hybridized carbon orbital. Use Slater orbitals.

Hint: The simplest expression for a tetrahedral sp^3 orbital on carbon is $\chi = \frac{1}{2}[2s + (3)^{1/2}2p]$.

Answer: $\mu = 1.95D$; the dipole is directed away from the carbon.

2. Derive Table 10.4 from Table 10.3.

3. One of the integrals of Table 10.7 is $\langle 1s_0 \mid z_0 \mid 2p_{z0}\rangle$. Find three different ways to obtain the numerical value of this one-center integral, without using any interpolation procedure.

Answer: (a) $\langle 1s \mid z \mid 2p_z\rangle = \left(\dfrac{Z_{1s}^{3}Z_{2p}^{5}}{2}\right)^{1/2}\dfrac{2^8}{(2Z_{1s}+Z_{2p})^5}$

(chapter 9, Appendix F).

(b) $\langle 1s_0 \mid z_0 \mid 2p_{z0}\rangle = \dfrac{1}{7.7}\langle 2p_{z0}' \mid 2p_{z0}\rangle = \dfrac{1}{7.7}(1-t^2)^{5/2}$

(see Ref. [13]).

(c) $\langle 1s_0 \mid z_0 \mid 2p_{z0}\rangle = \dfrac{(10)^{1/2}}{4.55}\langle 1s_0 \mid 3s_0'\rangle$

$= \dfrac{(10)^{1/2}}{4.55}[(0.4)(1+t)^3(1-t)^7]^{1/2}$

(see Ref. [13]).

BIBLIOGRAPHY

[1] R. S. Mulliken, *J. Chem. Phys.* **23**, 1833 (1955).
[2] C. A. Coulson, *Valence* (Oxford University Press, Fair Lawn, N.J., 1961).
[3] K. Higasi, H. Baba, and A. Rembaum, *Quantum Organic Chemistry* (Interscience Publishers, New York, 1965).
[4] F. O. Ellison and H. Shull, *J. Chem. Phys.* **21**, 1420 (1953).
[5] F. O. Ellison and H. Shull, *J. Chem. Phys.* **23**, 2348 (1955).
[6] C. A. Coulson, *Trans. Faraday Soc.* **38**, 433 (1942).
[7] C. P. Smyth, *Dielectric Behavior and Structure* (McGraw-Hill Book Co., Inc., New York, 1955).

[8]   J. W. Smith, *Electric Dipole Moments* (Butterworths, London, England, 1955).
[9]   L. E. Sutton, in *Determination of Organic Structures by Physical Methods*, edited by E. A. Braude and F. C. Nachod (Academic Press, New York, 1955).
[10]   L. Orgel, T. Cottrell, W. Dick, and L. E. Sutton, *Trans. Faraday Soc.* **47,** 113 (1951).
[11]   R. C. Sahni, *Trans. Faraday Soc.* **49,** 1246 (1953).
[12]   J. C. Slater, *Quantum Theory of Molecules and Solids* (McGraw-Hill Book Co., Inc., New York, 1963), Vol. 1.
[13]   R. S. Mulliken, C. A. Rieke, D. Orloff, and H. Orloff, *J. Chem. Phys.* **17,** 1248 (1949).

General References

1.   C. A. Coulson, *Valence* (Oxford University Press, Fair Lawn, N.J., 1961).
2.   J. C. Slater, *Quantum Theory of Molecules and Solids* (McGraw-Hill Book Co., Inc., New York, 1963), Vol. 1.
3.   A. L. McClellan, *Tables of Experimental Dipole Moments* (W. H. Freeman and Co., Inc., San Francisco, Calif., 1963).
4.   V. I. Minkin, O. A. Osipov, and Yu. A. Zhdanov, *Dipole Moments in Organic Chemistry* (Plenum Press, New York, 1970).

CHAPTER 11

Spin-Orbit Coupling

In almost all calculations of chapters 1–10, we have either neglected electron-spin considerations or we have assumed that each state wave function was an eigenfunction of $\hat{S}^2$ (that is, was pure singlet, pure doublet, and so forth). These assertions force exact energy degeneracy of all M_S components of a given spin multiplet.

The approximations cited are valid when the Hamiltonian contains no parts which are dependent on the spin coordinates. Such a Hamiltonian is reasonably accurate in certain instances; however, it is never valid in any exact sense and, in many cases, it is grossly wrong. This chapter and the next are concerned with instances where proper description of the system requires a Hamiltonian which is spin dependent. In specific, this chapter is concerned with the magnetic interaction of the spins of the electrons with their own orbital motion whereas the next is concerned with the magnetic interactions of the spin systems among themselves.

The purpose of the present chapter is to introduce the reader to the fundamentals of spin-orbit coupling theory. To that end, we consider the manner in which molecular energy levels split under the influence of spin-orbit coupling; the manner in which spin-orbit coupling invalidates the concept of a spin quantum number; the manner in which spin-orbit coupling complicates the nature of the Landé g factor; and the way in which the results of such calculations can be related to experimental data. The discussion is restricted to one- and two-electron systems; any more replete treatment of spin-orbit coupling is outside the scope of this text.

1. SPIN-ORBIT COUPLING OPERATOR

(A) One-Electron Atom

For a single electron in a central field with potential energy $U(r)$, the term descriptive of the magnetic interaction between the spin and orbital motion of that electron is[1]

$$\hat{H}_{so} = \frac{1}{2m^2c^2}\left(\frac{1}{r}\frac{\partial U(r)}{\partial r}\right)\hat{\mathbf{l}}\cdot\hat{\mathbf{s}} = \xi(r)\hat{\mathbf{l}}\cdot\hat{\mathbf{s}} \tag{11.1}$$

where $\hat{\mathbf{l}}$ and $\hat{\mathbf{s}}$ are the orbital and spin angular momentum operators, respectively.

For the hydrogen atom, the total Hamiltonian becomes

$$\hat{H} = -\frac{\hbar^2}{2m}\nabla^2 - \frac{e^2}{r} + \xi(r)\hat{\mathbf{l}}\cdot\hat{\mathbf{s}} = \hat{H}_0 + \hat{H}_{so} \tag{11.2}$$

where $\hat{H}_0$ is the spin-free Hamiltonian and $\hat{H}_{so}$ is the spin-orbit Hamiltonian. The eigenfunctions of the spin-free Hamiltonian, $\hat{H}_0$, do not satisfy the eigenvalue problem defined by $\hat{H}$. However, the eigenfunctions of $\hat{H}_0$ can be taken as the zero-order functions appropriate to a perturbation procedure based on $\hat{H}_{so}$.

In calculating the atomic matrix elements of $\hat{H}_{so}$, namely, $\langle\chi_\mu\,|\,\hat{H}_{so}\,|\,\chi_\nu\rangle$, it is convenient to separate out the radial part and to adopt the notation

$$\zeta_{nl} \equiv \hbar^2 \int_0^\infty R^2(n,\,l)\xi(r)\,dr \tag{11.3}$$

where $R(n,\,l)$ is the radial part of χ_μ and χ_ν, both of which are characterized by the same n and same l quantum numbers.[1] The quantity ζ_{nl} is the spin-orbit coupling constant for these particular orbitals; it has dimensions of energy; its size determines the magnitude of the first-order energetic effects caused by spin-orbit coupling. In the case of the Coulomb potential of a bare nucleus with charge $+Ze$, it can be shown that

$$\zeta_{nl} = \frac{e^2\hbar^2}{2m^2c^2a_0^3}\cdot\frac{Z^4}{n^3l(l+\frac{1}{2})(l+1)} \tag{11.4}$$

---

[1] If χ_μ and χ_ν have different n or l, the matrix elements of $\hat{H}_{so}$ are very nearly zero: See Ref. [1] of text, pp. 121, 122.

where a_0 is the Bohr radius. In the general central-field case, ζ_{nl} may be treated as an empirical parameter whose magnitude is determinable from experiment.

(B) Many-Electron Atom

For a many-electron case, each electron being subject to a central-field potential, we find[2]

$$\hat{H}_{so} = \sum_i \xi(r_i)\hat{\mathbf{l}}_i \cdot \hat{\mathbf{s}}_i \tag{11.5}$$

where the index i runs over all electrons. Thus, in enumerating the matrix elements of Eq. (11.5), we obtain a sum of terms in each of which the radial integral can be represented by the relevant constant ζ_{nl}.

It may be shown that the matrix elements of $\sum_i \xi(r_i)\hat{\mathbf{l}}_i \cdot \hat{\mathbf{s}}_i$ are proportional to the matrix elements of $\hat{\mathbf{L}} \cdot \hat{\mathbf{S}}$, where $\hat{\mathbf{L}} = \sum_i \hat{\mathbf{l}}_i$ and $\hat{\mathbf{S}} = \sum_i \hat{\mathbf{s}}_i$. The proportionality factor between the two types of element is denoted λ, where λ is a spin-orbit coupling constant characteristic of the atom state characterized by the quantum numbers L and S. It is important to note that, whereas ζ is inherently positive, λ may be either positive or negative. The quantity λ is usually treated as an empirical parameter; however, theoretical expressions which specify λ as a function of the one-electron ζ's do exist.[1]

Expansion of the operator for the one-electron one-center case yields

$$\zeta_{nl}\hat{\mathbf{l}} \cdot \hat{\mathbf{s}} = \zeta_{nl}(\hat{l}_x\hat{s}_x + \hat{l}_y\hat{s}_y + \hat{l}_z\hat{s}_z) \tag{11.6a}$$

where the $\hat{\mathbf{l}}$ components operate only on the orbital parts of the wave function and the $\hat{\mathbf{s}}$ components operate only on the spin parts. It is sometimes convenient to use the alternative expansion:

$$\zeta_{nl}\hat{\mathbf{l}} \cdot \hat{\mathbf{s}} = \zeta_{nl}(\tfrac{1}{2}\hat{l}_+\hat{s}_- + \tfrac{1}{2}\hat{l}_-\hat{s}_+ + \hat{l}_z\hat{s}_z) \tag{11.6b}$$

where, for instance, $\hat{l}_+ \equiv \hat{l}_x + i\hat{l}_y$, $\hat{s}_- \equiv \hat{s}_x - i\hat{s}_y$, etc.

(C) Many-Electron Molecule

For a many-electron many-atom system, one may write the approximate expression

$$\hat{H}_{so} = \sum_i \sum_K \xi_K(r_i)\hat{\mathbf{l}}_{iK} \cdot \hat{\mathbf{s}}_i \tag{11.7}$$

where K denotes an atomic center and i denotes an electron. Thus, in calculating one-center matrix elements, we may use the same spin-orbit coupling constants appropriate to the case of atoms; these may be symbolized by $\zeta_{nl,K}$ for atom K, etc.

---

[2] Note that Eq. (11.5) neglects interactions of the type *spin other orbit* which depend on terms such as $\hat{\mathbf{l}}_i \cdot \hat{\mathbf{s}}_j$.

The formalism of Eqs. (11.3) and (11.4) is obviously not valid for two-center terms. However, because of the strong inverse dependence[3] of ξ on r, the two-center terms are generally assumed to be small; they are neglected henceforth. Deviations from the central-field approximation of Eq. (11.7)[4] may be incorporated, at least in part, into the empirical parameters ζ.

In any event, the many-electron many-center problem has been dissected so that the resulting matrix elements can be treated as simple sums of one-electron one-center terms.

(D)   One-Electron Spin Operators

The operators $\hat{s}_x$, $\hat{s}_y$, and $\hat{s}_z$ are defined

$$\hat{s}_x \begin{pmatrix} \alpha \\ \beta \end{pmatrix} = \frac{\hbar}{2} \begin{pmatrix} 0 & 1 \\ 1 & 0 \end{pmatrix} \begin{pmatrix} \alpha \\ \beta \end{pmatrix}$$

$$\hat{s}_y \begin{pmatrix} \alpha \\ \beta \end{pmatrix} = \frac{\hbar}{2} \begin{pmatrix} 0 & i \\ -i & 0 \end{pmatrix} \begin{pmatrix} \alpha \\ \beta \end{pmatrix}$$

$$\hat{s}_z \begin{pmatrix} \alpha \\ \beta \end{pmatrix} = \frac{\hbar}{2} \begin{pmatrix} 1 & 0 \\ 0 & -1 \end{pmatrix} \begin{pmatrix} \alpha \\ \beta \end{pmatrix} \tag{11.8}$$

Therefore, the matrix elements between the two orthonormal spin functions α and β (in units of $\hbar$) have the values listed in Table 11.1.

(E)   Many-Electron Spin Operators

Since $\hat{H}_{so}$ is a sum of one-electron operators, extension to a many-electron problem presents no difficulty. Consider the case of two electrons. The spin wave functions are

$$\Theta_{00} = (\tfrac{1}{2})^{1/2}(\alpha_1\beta_2 - \alpha_2\beta_1) \qquad S = 0 \quad M_S = 0$$

$$\Theta_{1-1} = \beta_1\beta_2 \qquad S = 1 \quad M_S = -1$$

$$\Theta_{10} = (\tfrac{1}{2})^{1/2}(\alpha_1\beta_2 + \alpha_2\beta_1) \qquad S = 1 \quad M_S = 0$$

$$\Theta_{11} = \alpha_1\alpha_2 \qquad S = 1 \quad M_S = 1 \tag{11.9}$$

---

[3] For a Coulomb potential, ξ is proportional to

$$\frac{1}{r}\frac{\partial U(r)}{\partial r} = \frac{1}{r}\frac{\partial(-Ze^2/r)}{\partial r} = \frac{Ze^2}{r^3};$$

hence, $\xi \sim r^{-3}$.

[4] This point is discussed by Marshall and Stuart (Ref. [2]).

TABLE 11.1. *Matrix elements of the spin operators (in units of $\hbar$).*

	$\hat{s}_x\,\lvert\alpha\rangle$	$\hat{s}_y\,\lvert\alpha\rangle$	$\hat{s}_z\,\lvert\alpha\rangle$	$\hat{s}_x\,\lvert\beta\rangle$	$\hat{s}_y\,\lvert\beta\rangle$	$\hat{s}_z\,\lvert\beta\rangle$
$\langle\alpha\rvert$	0	0	$\dfrac{1}{2}$	$\dfrac{1}{2}$	$-\dfrac{i}{2}$	0
$\langle\beta\rvert$	$\dfrac{1}{2}$	$\dfrac{i}{2}$	0	0	0	$-\dfrac{1}{2}$

	$\hat{s}_+\,\lvert\alpha\rangle$	$\hat{s}_-\,\lvert\alpha\rangle$	$\hat{s}_z\,\lvert\alpha\rangle$	$\hat{s}_+\,\lvert\beta\rangle$	$\hat{s}_-\,\lvert\beta\rangle$	$\hat{s}_z\,\lvert\beta\rangle$
$\langle\alpha\rvert$	0	0	$\dfrac{1}{2}$	1	0	0
$\langle\beta\rvert$	0	1	0	0	0	$-\dfrac{1}{2}$

The spin operators are $\hat{S}_x = \hat{s}_{x_1} + \hat{s}_{x_2}$, and so forth, where $\hat{s}_{x_1}$ operates only on the spin coordinates of electron 1, and $\hat{s}_{x_2}$ operates only on the spin coordinates of electron 2. In this instance, we find

$$\hat{S}_x\Theta_{11} = (\hat{s}_{x_1} + \hat{s}_{x_2})\alpha_1\alpha_2 = \tfrac{1}{2}\hbar(\beta_1\alpha_2 + \alpha_1\beta_2) = (\tfrac{1}{2})^{1/2}\hbar\Theta_{10} \quad (11.10)$$

As in the one-electron case, the many-electron spin operators may transform wave functions with different M_S quantum numbers into each other. As is shown later, the effect of the operator $\hat{\mathbf{L}}\cdot\hat{\mathbf{S}}$ on the total wave function (spin *plus* orbital), may be to transform singlet wave functions into triplet wave functions and *vice versa*. This, of course, is one cause of spin multiplicity mixing.

(F) Symmetry Properties of Orbital Operators

In a right-handed coordinate system, the components of $\hat{\mathbf{l}}$ for the one-electron one-center case are

$$\hat{l}_x = -i\hbar\left(y\frac{\partial}{\partial z} - z\frac{\partial}{\partial y}\right) = -i\hbar\left(-\sin\phi\frac{\partial}{\partial\theta} - \cot\theta\cos\phi\frac{\partial}{\partial\phi}\right)$$

$$\hat{l}_y = -i\hbar\left(z\frac{\partial}{\partial x} - x\frac{\partial}{\partial z}\right) = -i\hbar\left(\cos\phi\frac{\partial}{\partial\theta} - \cot\theta\sin\phi\frac{\partial}{\partial\phi}\right)$$

$$\hat{l}_z = -i\hbar\left(x\frac{\partial}{\partial y} - y\frac{\partial}{\partial x}\right) = -i\hbar\frac{\partial}{\partial\phi} \quad (11.11)$$

Consider a single atom in a C_{4v} environment. The manner in which the components of $\hat{\mathbf{l}}$ transform in C_{4v} is given in Table 11.2; inspection indicates the same transformation properties as those of rotations around the corresponding Cartesian axes. This property, although illustrated here only for the specific case of a C_{4v} atom, is quite general. Consequently, the symmetry classification of the angular momentum operators for an atom in any environment can be found by inspection of the relevant character table: One merely looks up the representations generated by the rotation operators $\hat{R}_x$, $\hat{R}_y$, and $\hat{R}_z$.

Consider spin-orbit coupling effects in a many-atom system. The appropriate one-electron operator is $\sum_K \hat{\mathbf{l}}_K \cdot \hat{\mathbf{s}}$. We now show that the orbital part of this operator, $\sum_K \hat{\mathbf{l}}_K$, possesses the same symmetry properties as the term $\hat{\mathbf{l}}$ in the single-atom case: The three Cartesian components of $\sum_K \hat{\mathbf{l}}_K$ behave in the same way as do the rotations around the corresponding axes of the molecular coordinate system. Each individual $\mathbf{l}_K$ operator, taken separately, need not form a basis for any irreducible representation of the group. For instance, in a C_{2v} molecule of the type shown in Fig. 11.1, we find

$$\hat{l}_{zA} = -i\hbar\left(x_A\frac{\partial}{\partial y_A} - y_A\frac{\partial}{\partial x_A}\right) = -i\hbar\left(x\frac{\partial}{\partial y} - y\frac{\partial}{\partial x}\right) = \hat{l}_z$$

$$\hat{l}_{zB} = -i\hbar\left(x_B\frac{\partial}{\partial y_B} - y_B\frac{\partial}{\partial x_B}\right) = -i\hbar\left[(x+R)\frac{\partial}{\partial y} - y\frac{\partial}{\partial x}\right] = \hat{l}_z - i\hbar R\frac{\partial}{\partial y}$$

$$\hat{l}_{zC} = -i\hbar\left(x_C\frac{\partial}{\partial y_C} - y_C\frac{\partial}{\partial x_C}\right) = -i\hbar\left[(x-R)\frac{\partial}{\partial y} - y\frac{\partial}{\partial x}\right] = \hat{l}_z + i\hbar R\frac{\partial}{\partial y}$$

$$(11.12)$$

where x_A is the x coordinate of a given point referred to atom A as origin. If we also take A as the origin of the molecular coordinate system, we find $x_A = x$, $x_B = x + R$, $x_C = x - R$, and $y_A = y_B = y_C = y$. Then, since C_2

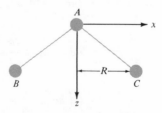

FIG. 11.1.   *ABC* molecule with C_{2v} symmetry. The atoms B and C are identical but are represented by different symbols in order to facilitate the discussion initiating in Eq. (11.12).

TABLE 11.2. *Transformation properties of Cartesian components of orbital angular momentum operator in C_{4v}.*

C_{4v}	E	$2C_4 \begin{pmatrix} x \to \pm y \\ y \to \mp x \end{pmatrix}$	$C_2 \begin{pmatrix} x \to -x \\ y \to -y \end{pmatrix}$	$2\sigma_v \begin{pmatrix} x \to \pm x \\ y \to \mp y \end{pmatrix}$	$2\sigma_d \begin{pmatrix} x \to \pm y \\ y \to \pm x \end{pmatrix}$	
$\hat{l}_x$	$\hat{l}_x$	$\hat{l}_y$	$-\hat{l}_x$	$-\hat{l}_x$	$-\hat{l}_y$	
$\hat{l}_y$	$\hat{l}_y$	$-\hat{l}_x$	$-\hat{l}_y$	$\hat{l}_y$	$-\hat{l}_x$	
$\hat{l}_z$	$\hat{l}_z$	$\hat{l}_z$	$\hat{l}_z$	$-\hat{l}_z$	$-\hat{l}_z$	
$(\hat{l}_x, \hat{l}_y)$	2	0	-2	0	0	$E(\hat{R}_x, \hat{R}_y)$
$\hat{l}_z$	1	1	1	-1	-1	$A_2(\hat{R}_z)$

entails $x \to -x$ and $y \to -y$, we have $C_2 \hat{l}_{zA} = \hat{l}_{zA}$, $C_2 \hat{l}_{zB} = \hat{l}_z + i\hbar R \, \partial/\partial y = \hat{l}_{zB}$, and $C_2 \hat{l}_{zC} = \hat{l}_z - i\hbar R \, \partial/\partial y = \hat{l}_{zB}$. Therefore, $C_2 \sum_K \hat{l}_{zK} = \sum_K \hat{l}_{zK}$; similarly, $\sigma_{xz} \sum_K \hat{l}_{zK} = -\sum_K \hat{l}_{zK}$ and $\sigma_{yz} \sum_K \hat{l}_{zK} = -\sum_K \hat{l}_{zK}$. Thus, the operator sum $\sum_K \hat{l}_{zK}$ transforms as $\hat{R}_z(A_2)$, while neither $\hat{l}_{zB}$, nor $\hat{l}_{zC}$ separately possess the proper transformation coefficients.

(G)   Matrix Elements of Orbital Operators

If χ_μ and χ_ν are atomic orbitals on a given center, the element $\langle \chi_\mu \,|\, \hat{l}_p \,|\, \chi_\nu \rangle$—where $p = x, y$, or z—can be nonzero only if the direct product $\Gamma(\chi_\mu) \times \Gamma(\hat{l}_p) \times \Gamma(\chi_\nu)$ contains the totally symmetric representation. If φ_r and φ_s are molecular orbitals, the element $\langle \varphi_r \,|\, \sum_K \hat{l}_{pK} \,|\, \varphi_s \rangle$ is nonzero only if the direct product $\Gamma(\varphi_r) \times \Gamma(\sum_K \hat{l}_{pK}) \times \Gamma(\varphi_s)$ contains the totally symmetric representation.

The $\hat{l}_p$ operators are purely imaginary [see Eq. (11.11)] and Hermitean. The Hermitean character is defined by

$$\langle \Phi_R \,|\, \sum_i \hat{l}_{ip} \,|\, \Phi_S \rangle = \langle \Phi_S^* \,|\, \sum_i \hat{l}_{ip}^* \,|\, \Phi_R^* \rangle \tag{11.13}$$

In the case of real wave functions, Eq. (11.13) yields

$$\langle \Phi_R \,|\, \sum_i \hat{l}_{ip} \,|\, \Phi_S \rangle = -\langle \Phi_S \,|\, \sum_i \hat{l}_{ip} \,|\, \Phi_R \rangle \tag{11.14}$$

Hence, the *diagonal matrix elements* of an $\hat{l}_p$ operator are zero for any real wave function.

(H)   Effects of Orbital Operators on Atomic Wave Functions

A molecular-orbital matrix element reduces to

$$\langle \varphi_r \,|\, \sum_K \hat{l}_{pK} \,|\, \varphi_s \rangle = \sum_K \sum_\mu \sum_\nu c_{r\mu}^* c_{s\nu} \langle \chi_\mu \,|\, \hat{l}_{pK} \,|\, \chi_\nu \rangle \tag{11.15}$$

In the one-center approximation, each matrix element where either μ or ν belongs to a center different from K is zero. Thus, it is useful to have expressions available for the atomic matrix elements. It follows from Eq. (11.3) that the radial part of the matrix elements is incorporated into the spin-orbit coupling constant ζ_{nl}. Consequently, it is only necessary to consider the matrix elements of the angular parts of the wave functions. Furthermore, we can limit ourselves to the interaction between wave functions within the same (n, l) shell[1] (i.e., WF's which differ only in the quantum number m_l).

(i)   *s orbitals:*   It follows from the definition of an s AO that

$$\hat{l}_x \,|\, s \rangle = \hat{l}_y \,|\, s \rangle = \hat{l}_z \,|\, s \rangle = 0 \tag{11.16}$$

where $|\, s \rangle$ represents the angular part of an s orbital.

TABLE 11.3. *Effects of orbital angular momentum operators on real p orbitals in a right-handed coordinate system (units of $\hbar$).*

	$\mid p_x \rangle$	$\mid p_y \rangle$	$\mid p_z \rangle$
$\hat{l}_x$	0	$i \mid p_z \rangle$	$-i \mid p_y \rangle$
$\hat{l}_y$	$-i \mid p_z \rangle$	0	$i \mid p_x \rangle$
$\hat{l}_z$	$i \mid p_y \rangle$	$-i \mid p_x \rangle$	0

(ii) p orbitals: The effect of the $\hat{l}$ operators on the real wave functions $\mid p_x \rangle$, $\mid p_y \rangle$, and $\mid p_z \rangle$ is shown in Table 11.3. In a right-handed coordinate system, it is seen that the effect of a given angular momentum operator is simply to multiply the p orbital upon which it operates by $i\hbar$, and to rotate it counterclockwise by 90° about the axis specified by the operator subscript.

In the case of complex wave functions $\mid p_1 \rangle$, $\mid p_0 \rangle$, and $\mid p_{-1} \rangle$, it is more convenient to use the set of operators: $(\hat{l}_+ = \hat{l}_x + i\hat{l}_y; \hat{l}_- = \hat{l}_x - i\hat{l}_y; \hat{l}_z)$. The results are given in Table 11.4. Tables 11.3 and 11.4 are readily connected using

$$\mid p_0 \rangle = \mid p_z \rangle$$
$$\mid p_1 \rangle = -(\tfrac{1}{2})^{1/2}(\mid p_x \rangle + i \mid p_y \rangle)$$
$$\mid p_{-1} \rangle = (\tfrac{1}{2})^{1/2}(\mid p_x \rangle - i \mid p_y \rangle) \tag{11.17}$$

It is readily seen that $\hat{l}_+$ increases m_l by 1, while $\hat{l}_-$ decreases m_l by 1; these operators are named *step-up* and *step-down* operators. The complex p orbitals of Eq. (11.17) satisfy the Condon–Shortley sign convention which facilitates expression of the effects of the step-up and step-down operators

TABLE 11.4. *Effects of $\hat{l}_+$, $\hat{l}_-$, and $\hat{l}_z$ operators on complex p orbitals in a right-handed coordinate system (units of $\hbar$).*

	$\mid p_1 \rangle$	$\mid p_0 \rangle$	$\mid p_{-1} \rangle$
$\hat{l}_+$	0	$2^{1/2} \mid p_1 \rangle$	$2^{1/2} \mid p_0 \rangle$
$\hat{l}_-$	$2^{1/2} \mid p_0 \rangle$	$2^{1/2} \mid p_{-1} \rangle$	0
$\hat{l}_z$	$\mid p_1 \rangle$	0	$- \mid p_{-1} \rangle$

in a simple general form

$$\hat{l}_{+}\chi(l, m_l) = \hbar[(l + m_l + 1)(l - m_l)]^{1/2}\chi(l, m_l + 1)$$

$$\hat{l}_{-}\chi(l, m_l) = \hbar[(l - m_l + 1)(l + m_l)]^{1/2}\chi(l, m_l - 1) \quad (11.18)$$

where $\chi(l, m_l)$ is the atomic orbital characterized by the quantum numbers l and m_l. Equation (11.18) is also valid for many-electron atoms if one replaces l by L, m_l by M_L, and $\hat{l}_{\pm}$ by $\hat{L}_{\pm}$.

(*iii*) *d and f orbitals:* Tables similar to Table 11.3 can be constructed for d and f orbitals and are given in Appendix H.

(I) Matrix Elements of Spin-Orbit Coupling Operator

The evaluation of the matrix elements of $\hat{H}_{so}$ is quite straightforward. We illustrate, using the element $\langle \Psi_S | \hat{H}_{so} | \Psi_{T,0} \rangle$, where

$$\Psi_S = (\tfrac{1}{2})^{1/2}[\varphi_r(1)\varphi_s(2) + \varphi_r(2)\varphi_s(1)](\tfrac{1}{2})^{1/2}(\alpha_1\beta_2 - \alpha_2\beta_1)$$

$$= \Phi_S\Theta_{00} \quad (11.19)$$

$$\Psi_{T,0} = (\tfrac{1}{2})^{1/2}[\varphi_r(1)\varphi_t(2) - \varphi_r(2)\varphi_t(1)](\tfrac{1}{2})^{1/2}(\alpha_1\beta_2 + \alpha_2\beta_1)$$

$$= \Phi_T\Theta_{10} \quad (11.20)$$

The definitions of Θ_{00} and Θ_{10} are given by Eq. (11.9). The singlet state, with two singly occupied MO's φ_r and φ_s is denoted S. The triplet state, with two singly occupied MO's φ_r and φ_t, is denoted T. The lower-lying closed shells of electrons are unspecified, for reasons of brevity. Since $\hat{H}_{so}$ is a sum of one-electron operators, the matrix element is zero if the two state functions differ individually in the occupancy of more than one spin orbital. Expanding the matrix element in question, we find

$$\langle \Psi_S | \hat{H}_{so} | \Psi_{T,0} \rangle = \langle \Phi_S\Theta_{00} | \sum_K \xi_K(r_1)\hat{l}_{1K}\cdot\hat{s}_1 + \sum_K \xi_K(r_2)\hat{l}_{2K}\cdot\hat{s}_2 | \Phi_T\Theta_{10} \rangle$$

$$= \langle \Phi_S | \sum_K \xi_K(r_1)\hat{l}_{1K} | \Phi_T \rangle\langle \Theta_{00} | \hat{s}_1 | \Theta_{10} \rangle$$

$$+ \langle \Phi_S | \sum_K \xi_K(r_2)\hat{l}_{2K} | \Phi_T \rangle\langle \Theta_{00} | \hat{s}_2 | \Theta_{10} \rangle \quad (11.21)$$

Let us now consider the z component of one such term; the spin part (in units of $\hbar$) is evaluated as

$$\langle \Theta_{00} | \hat{s}_{1z} | \Theta_{10} \rangle = \tfrac{1}{2}\langle \alpha_1\beta_2 - \alpha_2\beta_1 | \tfrac{1}{2}(\alpha_1\beta_2 - \alpha_2\beta_1) \rangle = \tfrac{1}{2} \quad (11.22)$$

The orbital part of the same term becomes, in an obvious notation [see

Eqs. (11.19) and (11.20)]

$$\langle \Phi_S \mid \sum_K \xi_K(r_1)\hat{l}_{1K,z} \mid \Phi_T \rangle = \tfrac{1}{2}\langle r_1 s_2 + r_2 s_1 \mid \sum_K \xi_K(r_1)\hat{l}_{1K,z} \mid r_1 t_2 - r_2 t_1 \rangle \quad (11.23)$$

The operator of Eq. (11.23) contains the coordinates of electron 1 only; thus, since the three MO's r, s, and t are orthogonal, Eq. (11.23) reduces to

$$\langle \Phi_S \mid \sum_K \xi_K(r_1)\hat{l}_{1K,z} \mid \Phi_T \rangle = -\tfrac{1}{2}\langle r_2 \mid r_2 \rangle\langle s_1 \mid \sum_K \xi_K(r_1)\hat{l}_{1K,z} \mid t_1 \rangle \quad (11.24)$$

Expansion in terms of atomic orbitals and introduction of the one-center approximation, yields

$$\langle \Phi_S \mid \sum_K \xi_K(r_1)\hat{l}_{1K,z} \mid \Phi_T \rangle = -\tfrac{1}{2}\langle \sum_\mu c_{s\mu}\chi_\mu \mid \sum_K \xi_K \hat{l}_{K,z} \mid \sum_\nu c_{t\nu}\chi_\nu \rangle$$

$$= -\tfrac{1}{2} \sum{}' c_{s\mu}{}^* c_{t\nu} \zeta_{nl,K} \langle \chi_\mu' \mid \hat{l}_{K,z} \mid \chi_\nu' \rangle \quad (11.25)$$

where χ' represents the angular part of χ and $\sum'$ indicates summation over all one-center terms only.[5]

It may be shown similarly that identical results are found for the z component of the operator containing the coordinates of electron 2. All x and y components of the operator yield null matrix elements; therefore,

$$\langle \Psi_S \mid \hat{H}_{so} \mid \Psi_T \rangle = -\tfrac{1}{2} \sum{}' c_{s\mu}{}^* c_{t\nu} \zeta_{nl,K} \langle \chi_\mu' \mid \hat{l}_{K,z} \mid \chi_\nu' \rangle \quad (11.26)$$

A somewhat more convenient, though less transparent notation can be obtained if we do not expand the matrix elements over the AO's as in Eq. (11.25) but if, instead, we generalize Eq. (11.24) by setting

$$\tfrac{1}{2} \sum_K \xi_K \hat{l}_{K,z} \equiv \hat{H}_{l,z} \quad (11.27)$$

Then, Eq. (11.26) becomes

$$\langle \Psi_S \mid \hat{H}_{so} \mid \Psi_{T,0} \rangle = -\langle s \mid \hat{H}_{l,z} \mid t \rangle \quad (11.28)$$

Using this notation, we can express the matrix elements between the singlet and the triplet wave functions in a general way. If we denote the two remaining components of the triplet state by $\Psi_{T,1}$ and $\Psi_{T,-1}$, so that

$$\Psi_{T,1} = \Phi_T\Theta_{1,1}; \qquad \Psi_{T,-1} = \Phi_T\Theta_{1,-1} \quad (11.29)$$

the results found are summarized in Table 11.5.

---

[5] Note that $\zeta_{nl,K}$ contains $\hbar^2$. Therefore, as of now, $\hat{l}$ and $\hat{s}$ operators are expressed in units of $\hbar$.

TABLE 11.5. *Matrix elements of* $\hat{H}_{so}$ *between singlet and triplet states.*[3],[4]
[States are defined in Eqs. (11.19) and (11.20).]

	$\langle \Psi_s \mid \hat{H}_{so} \mid \Psi_{T,1} \rangle$	$\langle \Psi_s \mid \hat{H}_{so} \mid \Psi_{T,0} \rangle$	$\langle \Psi_s \mid \hat{H}_{so} \mid \Psi_{T,-1} \rangle$
$\hat{H}_{so,x}$	$(\frac{1}{2})^{1/2} \langle s \mid \hat{H}_{l,x} \mid t \rangle$	0	$(\frac{1}{2})^{1/2} \langle s \mid \hat{H}_{l,x} \mid t \rangle$
$\hat{H}_{so,y}$	$(\frac{1}{2})^{1/2} i \langle s \mid \hat{H}_{l,y} \mid t \rangle$	0	$(\frac{1}{2})^{1/2} i \langle s \mid \hat{H}_{l,y} \mid t \rangle$
$\hat{H}_{so,z}$	0	$- \langle s \mid \hat{H}_{l,z} \mid t \rangle$	0

(J) Multiplicity Mixing

Equation (11.26) and Table 11.5 illustrate the origin of *multiplicity mixing*: *pure singlet* and *pure triplet* states are no longer eigenfunctions of the total Hamiltonian. The spin-orbit interaction between states of different multiplicity generates perturbed wave functions which may dominantly relate to some *nominal* multiplicity, but which assuredly also contain admixtures of states of both higher and lower nominal multiplicity. This point is discussed in detail in Sec. 7.

(K) A Note on Group Theory

Since $\hat{H}_{so}$ is an additive part of the Hamiltonian, it follows that $\hat{H}_{so}$ transforms as the totally symmetric representation. Thus, the group-theoretical requirement for a nonzero matrix element $\langle \Psi_1 \mid \hat{H}_{so} \mid \Psi_2 \rangle$ is, simply, that $\Gamma(\Psi_1) = \Gamma(\Psi_2)$.

The manner in which spin functions behave when subjected to various symmetry operations of the group has been discussed by several authors.[4]-[6] For an *odd* number of electrons, symmetry considerations lead to the concept of double groups. For an *even* number of electrons, the spin wave functions may be labeled using the formalism of the usual single group—double-group symbolism is unnecessary. Especially simple and useful results are obtained for the two-electron problem: The singlet spin function is always totally symmetric; the three triplet spin functions always transform as do the three rotations around the Cartesian axes.

2. FIRST-ORDER ENERGY SPLITTING: AN EXAMPLE CALCULATION

We suppose that we have available the eigenfunctions and eigen-energies of a zeroth-order spin-free Hamiltonian—for example, the MWH eigenvectors and eigenenergies of chapter 4. We now use these wave functions as the starting point for a perturbation treatment based on $\hat{H}_{so}$.

We consider orbitally degenerate zero-order states only. Zero-order states which are not orbitally degenerate have zero expectation values for the operators $\hat{l}$ and/or $\hat{L}$ [see Sec. 1(G)]: For example, the interaction matrix of the three orbitally nondegenerate spin-degenerate functions $\Psi_{T,1}$, $\Psi_{T,0}$, and $\Psi_{T,-1}$ of Table 11.5 is

$\hat{L}\cdot\hat{S}$	$\lvert \Psi_{T,1} \rangle$	$\lvert \Psi_{T,0} \rangle$	$\lvert \Psi_{T,-1} \rangle$
$\langle \Psi_{T,1} \rvert$	0	0	0
$\langle \Psi_{T,0} \rvert$	0	0	0
$\langle \Psi_{T,-1} \rvert$	0	0	0

$$(11.30)$$

and is a null matrix. This is due to the fact that a spatially nondegenerate wave function is always real and a real wave function is characterized by zero angular momentum. Thus, the spin degeneracy of orbitally nondegenerate states cannot be removed by first-order spin-orbit coupling.

Heilman and Ballhausen[7] have calculated the first-order energy splitting of the orbitally degenerate $^2\Pi$ ground state of NO within the framework of MO theory; they have also correlated experiment and theory for this molecule. We do not discuss any one specific molecule; we direct attention, instead to a general one-electron case. A two-electron problem is discussed in Sec. 3.

The many-electron problem introduces nothing new physically; it does, however, necessitate an inordinate increase in labor.[6] We do not discuss the many-electron problem.

(A) Specification of Problem

Spin-orbit energy splittings are not always observable in the band spectra of molecules. However, if one or more of the atomic spin-orbit coupling constants ζ_{nl} is large, the splittings may become detectable. It follows from Eq. (11.4) that ζ_{nl} is proportional to Z^4 and that ζ_{nl} should be

---

[6] If the zero-order set of eigenfunctions consists of one-electron MO's, these MO's must be combined into properly antisymmetrized configurational wave functions and these latter diagonalized with respect to the perturbation operators $\Sigma_{i<j}e^2/r_{ij}$ and $\hat{H}_{so}$. If one of these perturbation influences is much larger than the other, it is usual to effect exact diagonalization with respect to that component and to treat the smaller component by a perturbation approach imposed on the results of this prior diagonalization. A number of very nice examples relevant to this point are available in Ballhausen (Ref. [5]). Since this author adopts a crystal-field stance, he can make extensive use of the more sophisticated methods developed in the theory of atomic spectra.

larger for heavier atoms. The shielding due to inner electrons is quite incomplete and, as a result, the effective nuclear field seen by the valence electrons of a heavy atom can be quite large. It seems more relevant, therefore, to consider a transition-metal complex, rather than a typical organic containing C, H, O, or N atoms, as our computational example.

Consider a transition-metal complex of D_{4h} symmetry, the central ion of which contains *one* d electron outside a closed shell configuration (e.g., Ti^{3+}, V^{4+}, Zr^{3+}, Nb^{4+}, Re^{6+},...). Suppose that the result of the zero-order (i.e., spin-independent) calculation is that shown in Fig. 11.2 and Table 11.6. The spin-orbital wave functions

$$\{\varphi_{xz}\alpha, \varphi_{xz}\beta, \varphi_{yz}\alpha, \varphi_{yz}\beta, \varphi_{xy}\alpha, \varphi_{xy}\beta, \varphi_{z^2}\alpha, \varphi_{z^2}\beta, \varphi_{x^2-y^2}\alpha, \varphi_{x^2-y^2}\beta\} \quad (11.31)$$

form an orthonormal set of eigenfunctions of the spin-free Hamiltonian $\hat{H}_0$. The straightforward procedure involves calculation of the $10 \times 10 = 100$ matrix elements of $\hat{H}_0 + \hat{H}_{so}$. A number of the off-diagonal matrix elements may be expected to be nonzero. Diagonalization of this energy matrix yields the required eigenvalues and eigenvectors. Such a calculation is *exact* in the

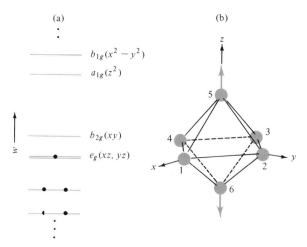

FIG. 11.2.   (a) MO energy-level diagram for a transition-metal complex in a D_{4h} environment (an octahedron stretched along the z dimension). (b) Ligand numbering. The only excited configurations considered are those obtained by promoting the e_g electron to the b_{2g}, a_{1g}, and b_{1g} MO's. In view of this, the space parts of all state wavefunctions transform the same as the singly occupied orbital. The coordinate systems on the ligands and on the central metal ion (not shown) are all parallel to the depicted set.

sense that it does not introduce approximations other than those inherent in the one-electron one-center picture. However, this can be a rather tedious affair and in most instances, much work can be evaded by using a group-theoretic partitioning of the matrix.

(B) Interaction Within Degenerate Ground Configuration

The highest-energy occupied MO of the ground configuration is the degenerate e_g level. Since the complex is supposed to contain an odd number of electrons and since the e_g level (see Fig. 11.2) is occupied by only one electron, the ground state is of species ^2E_g. Thus, the subset

$$\{\varphi_{xz}\alpha, \ \varphi_{xz}\beta, \ \varphi_{yz}\alpha, \ \varphi_{yz}\beta\} \tag{11.32}$$

contains four wave functions which, in the approximation of Table 11.6, are degenerate. We might expect that $\hat{H}_{so}$ causes a comparatively strong inter-action among these four wave functions. Indeed, the subset of states based on those MO's which are energetically distant from e_g are likely to have only secondary effects on the ground energy levels. Therefore, as a first approx-imation, we neglect all the nonoccupied molecular orbitals and study the effects of the operator $\hat{H}_{so}$ on the subset of Eq. (11.32).

TABLE 11.6. *Molecular orbitals of a complex of D_{4h} symmetry.*
{The atomic orbitals on the different ligands are identified by subscript [see Fig. 11.2 (b)]. The metal orbital is combined only with the most important ligand group orbital of the same symmetry; other group orbitals of the proper symmetry are excluded because consideration of them adds little that is new physically. Coefficients are unspecified in order to make the discussion more generally applicable to any one-electron problem.} (For an *elongated octahedral* molecule, $c_3 < c_2$; to a first approximation, we can also set $c_3 = 0$. For a *compressed octa-hedron*, such as is considered in Sec. 4 and depicted in Fig. 11.7, we have $c_3 > c_2$; in this case, it is more reasonable to set $c_2 = 0$.)

$$\varphi_{x^2-y^2} = c_1'''(3d_{x^2-y^2}) + c_2'''(-2p_{x1} + 2p_{y2} + 2p_{x3} - 2p_{y4})$$

$$\varphi_{z^2} = c_1''(3d_{z^2}) + c_2''(2p_{x1} + 2p_{y2} - 2p_{x3} - 2p_{y4}) + c_3''(2p_{z6} - 2p_{z5})$$

$$\varphi_{xy} = c_1'(3d_{xy}) + c_2'(2p_{y1} + 2p_{x2} - 2p_{y3} - 2p_{x4})$$

$$\varphi_{xz} = c_1(3d_{xz}) + c_2(2p_{z1} - 2p_{z3}) + c_3(2p_{x5} - 2p_{x6})$$

$$\varphi_{yz} = c_1(3d_{yz}) + c_2(2p_{z2} - 2p_{z4}) + c_3(2p_{y5} - 2p_{y6})$$

The interaction matrix of $\hat{H} = \hat{H}_0 + \hat{H}_{so}$ is

$\hat{H}$	$\|\varphi_{xz}\alpha\rangle$	$\|\varphi_{yz}\alpha\rangle$	$\|\varphi_{xz}\beta\rangle$	$\|\varphi_{yz}\beta\rangle$
$\langle\varphi_{xz}\alpha\|$	$W^0(E_g)$	$-c_1{}^2\dfrac{i\zeta_{3d}}{2}$	0	0
$\langle\varphi_{yz}\alpha\|$	$c_1{}^2\dfrac{i\zeta_{3d}}{2}$	$W^0(E_g)$	0	0
$\langle\varphi_{xz}\beta\|$	0	0	$W^0(E_g)$	$c_1{}^2\dfrac{i\zeta_{3d}}{2}$
$\langle\varphi_{yz}\beta\|$	0	0	$-c_1{}^2\dfrac{i\zeta_{3d}}{2}$	$W^0(E_g)$

$$(11.33)$$

All diagonal elements of $\hat{H}_{so}$, for instance, $\langle\varphi_{xz}\alpha \mid \sum_K \xi_K \hat{l}_K \cdot \hat{s} \mid \varphi_{xz}\alpha\rangle$, are zero, as could be anticipated [see Sec. 1(G)] from the real nature of the starting wave functions. As an example of an off-diagonal element, consider

$$\langle\varphi_{xz}\alpha \mid \sum_K \xi_K \hat{l}_K \cdot \hat{s} \mid \varphi_{yz}\alpha\rangle$$

$$= \tfrac{1}{2}\hbar\langle c_1(3d_{xz}) + c_2(2p_{x1} - 2p_{x3}) \mid \sum_K \xi_K \hat{l}_{K,z} \mid c_1(3d_{yz}) + c_2(2p_{x2} - 2p_{x4})\rangle$$

$$(11.34)$$

In the one-center approximation, this integral reduces[7] to $-i(c_1{}^2/2)\zeta_{3d}$ (see Table H.1).

The (4×4) matrix of Eq. (11.33) is reducible to two (2×2) matrices. The first of these

$$\begin{pmatrix} W^0(E_g) & -ic_1{}^2\dfrac{\zeta_{3d}}{2} \\[2ex] ic_1{}^2\dfrac{\zeta_{3d}}{2} & W^0(E_g) \end{pmatrix}$$

is diagonalized for the energy values

$$W_1 = W^0(E_g) + \tfrac{1}{2}c_1{}^2\zeta_{3d} \qquad (11.35)$$

---

[7] This reduction requires, in accord with Table 11.6, that c_3 equal to zero in Eq. (11.34).

with corresponding wave function[8]

$$\Psi_1 = \varphi_1 \alpha = (\tfrac{1}{2})^{1/2} (\varphi_{xz} + i\varphi_{yz}) \alpha \qquad (11.36)$$

and

$$W_2 = W^0(E_g) - \tfrac{1}{2} c_1{}^2 \zeta_{3d} \qquad (11.37)$$

with corresponding wave function

$$\Psi_2 = \varphi_2 \alpha = (\tfrac{1}{2})^{1/2} (\varphi_{xz} - i\varphi_{yz}) \alpha \qquad (11.38)$$

The eigenvalues and eigenfunctions of the second matrix are

$$W_3 = W^0(E_g) + \tfrac{1}{2} c_1{}^2 \zeta_{3d}; \qquad \Psi_3 = \varphi_2 \beta = (\tfrac{1}{2})^{1/2} (\varphi_{xz} - i\varphi_{yz}) \beta \quad (11.39)$$

$$W_4 = W^0(E_g) - \tfrac{1}{2} c_1{}^2 \zeta_{3d}; \qquad \Psi_4 = \varphi_1 \beta = (\tfrac{1}{2})^{1/2} (\varphi_{xz} + i\varphi_{yz}) \beta \quad (11.40)$$

Thus, the fourfold degenerate ^2E_g state is split by spin-orbit coupling into two twofold degenerate levels[9] as shown in Fig. 11.3. The ground state transforms as Γ_6, as may be verified by detailed analysis of the behavior of the ground-state wave functions under the relevant symmetry operations.

The level splitting of Fig. 11.3 is also predictable using group theory. In the double group D_{4h}' or, for that matter, in D_4', the orbital set $(\varphi_{xz}, \varphi_{yz})$ transforms as E_1' or Γ_5, whereas the spin functions transform as E_2' or Γ_6. The subset of Eq. (11.32) transforms as $\Gamma_5 \times \Gamma_6$ which is a four-dimensional *reducible* representation. This reducible representation can be resolved into the two *irreducible* representations $\Gamma_6 + \Gamma_7$, in accordance with Fig. 11.3.

(C) Numerical Evaluation of ζ_{nl}

The energy splitting, $c_1{}^2 \zeta_{nl}$, varies between a few hundred and a few thousand wave numbers. For any given metal ion, ζ_{nl} is a function of the

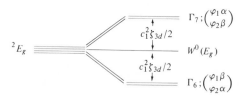

FIG. 11.3. Energy splitting of the ^2E_g level under the influence of spin-orbit coupling.

---

[8] Ψ_1 is a state wave function; it is a function of the coordinates of electrons—inner shell and valence. We neglect the role played by closed shells; there is but one valence electron. Therefore, we describe the many-electron state function approximately by a one-electron spin orbital. Hence, the mixed state and orbital notation of Eq. (11.35)ff.

[9] The remaining twofold degeneracy is termed a *Kramers degeneracy*; it can be removed by an external magnetic field. We return to this point in Sec. 3.

atomic charge and the electronic configuration of the atom. This conclusion is borne out by tables[8] of *experimental* values of ζ (i.e., values extracted from atomic spectroscopic data).

A Mulliken population analysis leads to an estimate of effective charge and to the electronic configuration of the metal ion. Such data, used interpolatively in conjunction with Dunn's tables,[8] should yield a reasonable semiempirical estimate of the spin-orbit coupling constant in question. For example, the calculation on the $TiF_6{}^{3-}$ complex of Chap. 4 yields the configuration $3d^{2.61}4s^{0.38}4p^{0.50}$ and the charge $+0.51$ a.u. for the titanium center. Assuming—somewhat arbitrarily, it must be admitted—that $4s$ and $4p$ occupation influence ζ_{3d} to the same extent, Dunn's tables yield $\zeta_{3d} = 94.5$ cm^{-1} for titanium in $TiF_6{}^{3-}$.

Some typical values for ζ_{nd} in the configuration $\ldots nd^1$ are[8]:

$$Ti^{3+}:\ 155\ cm^{-1}; \qquad V^{4+}:\ 250\ cm^{-1}; \qquad Zr^{3+}:\ 500\ cm^{-1};$$

$$Nb^{4+}:\ 750\ cm^{-1}; \qquad Re^{6+}:\ 3500\ cm^{-1}$$

(D)   Interaction With Excited Configurations

The spin functions transform as Γ_6. Thus, the excited configurations are classifiable as

$$\Gamma(x^2 - y^2) = B_{1g} = \Gamma_3; \qquad \Gamma_3 \times \Gamma_6 = \Gamma_7$$

$$\Gamma(z^2) = A_{1g} = \Gamma_1; \qquad \Gamma_1 \times \Gamma_6 = \Gamma_6$$

$$\Gamma(xy) = B_{2g} = \Gamma_4; \qquad \Gamma_4 \times \Gamma_6 = \Gamma_7 \qquad (11.41)$$

Therefore, the only excited state which can mix with the ground state $\{\varphi_1\beta,\ \varphi_2\alpha;\ \Gamma_6\}$ is $\{\varphi_{z^2}\alpha,\ \varphi_{z^2}\beta;\ \Gamma_6\}$.

The interaction matrix for the Γ_6 case is found to be

$\hat{H}$	$\lvert\varphi_1\beta\rangle$	$\lvert\varphi_{z^2}\alpha\rangle$	$\lvert\varphi_2\alpha\rangle$	$\lvert\varphi_{z^2}\beta\rangle$
$\langle\varphi_1\beta\rvert$	$W^0(E_g) - \frac{1}{2}c_1^2\zeta_{3d}$	A	0	0
$\langle\varphi_{z^2}\alpha\rvert$	A	$W^0(A_{1g})$	0	0
$\langle\varphi_2\alpha\rvert$	0	0	$W^0(E_g) - \frac{1}{2}c_1^2\zeta_{3d}$	$-A$
$\langle\varphi_{z^2}\beta\rvert$	0	0	$-A$	$W^0(A_{1g})$ (11.42)

where

$$A \equiv (\tfrac{1}{2})^{1/2}[-(3)^{1/2}c_1c_1{}''\zeta_{3d} + 2c_2c_2{}''\zeta_{2p}] \qquad (11.43)$$

The evaluation of the off-diagonal element A proceeds as follows:

$$\langle\varphi_{z^2}\alpha \mid \hat{H}_0 + \hat{H}_{so} \mid \varphi_1\beta\rangle = \langle\varphi_{z^2}\alpha \mid \hat{H}_{so} \mid \varphi_1\beta\rangle$$

$$= \langle\varphi_{z^2}\alpha \mid \sum_K \xi_K \hat{l}_{Kx}\hat{s}_x + \sum_K \xi_K \hat{l}_{Ky}\hat{s}_y \mid \varphi_1\beta\rangle \qquad (11.44)$$

In the one-center approximation, the x component reduces to

$$\langle \varphi_{z^2}\alpha \mid \sum_K \xi_K \hat{l}_{Kx}\hat{s}_x \mid \varphi_1\beta \rangle$$

$$= \tfrac{1}{2}\hbar \langle c_1''(3d_{z^2}) + c_2''(2p_{x1} + 2p_{y2} - 2p_{z3} - 2p_{y4} \mid \sum_K \xi_K \hat{l}_{Kx} \mid$$

$$\times \mid (\tfrac{1}{2})^{1/2}[c_1(3d_{xz}) + c_2(2p_{z1} - 2p_{z3}) + ic_1(3d_{yz}) + ic_2(2p_{z2} - 2p_{z4})]\rangle$$

$$= \tfrac{1}{2}(\tfrac{1}{2})^{1/2}[-c_1c_1''\zeta_{3d}(3)^{1/2} + 2c_2c_2''\zeta_{2p}] \tag{11.45}$$

The components of all other matrix elements, including the y component of A, can be evaluated similarly to yield the matrix of Eq. (11.42). This matrix is reducible to two (2×2) matrices, M and M', which possess identical eigenvalues and similar eigenvectors.

The solutions of symmetric 2×2 determinants of the type occurring in the partitioning of Eq. (11.42) are most conveniently expressed[5] in trigonometric form. If the secular equation is written

$$\begin{vmatrix} H_{AA} - W & H_{AB} \\ H_{AB} & H_{BB} - W \end{vmatrix} = 0 \tag{11.46}$$

and if we define

$$\tan 2\theta \equiv \frac{2H_{AB}}{H_{AA} - H_{BB}} \tag{11.47}$$

the solutions of Eq. (11.46) are given by

$$W_- = H_{AA} - H_{AB}\cot \theta; \qquad \mid \Psi_- \rangle = \sin \theta \mid \Psi_A \rangle - \cos \theta \mid \Psi_B \rangle$$

$$W_+ = H_{BB} + H_{AB}\cot \theta; \qquad \mid \Psi_+ \rangle = \cos \theta \mid \Psi_A \rangle + \sin \theta \mid \Psi_B \rangle \tag{11.48}$$

where[10]

$$\cot \theta = \frac{\tan 2\theta}{-1 - (1 + \tan^2 2\theta)^{1/2}} \tag{11.49}$$

[10] For each value of $\tan 2\theta$, there are *two* corresponding values of $\cot \theta$ which are given by

$$\cot \theta = \frac{\tan 2\theta}{-1 \pm (1 + \tan^2 2\theta)^{1/2}}$$

In the derivation of Eqs. (11.48), however, one must reject the solution with the positive sign. Hence, in this particular problem, $\cot \theta$ and $\tan 2\theta$ always have opposite sign. Therefore, $\tfrac{1}{2}\pi < 2\theta < \tfrac{3}{2}\pi$. The same point arises in taking $\sin \theta$ and $\cos \theta$. While it is generally true that

$$\sin^2 \theta - \cos^2 \theta = \pm (1 + \tan^2 2\theta)^{-1/2}$$

only the $+$ sign may be used. The relative signs of $\sin \theta$ and $\cos \theta$ are determined by

$$\cos \theta = \sin \theta \cot \theta$$

Applying Eqs. (11.47)–(11.49) to the interaction matrices M and M', we find for M:

$$\tan 2\theta = \frac{2A}{W^0(E_g) - \frac{1}{2}c_1^2\zeta_{3d} - W^0(A_{1g})} \qquad (11.50)$$

$$W_- = W^0(E_g) - \tfrac{1}{2}c_1^2\zeta_{3d} - A\cot\theta; \qquad |\Psi_-\rangle = \sin\theta\,|\varphi_1\beta\rangle - \cos\theta\,|\varphi_{z^2}\alpha\rangle$$

$$W_+ = W^0(A_{1g}) + A\cot\theta; \qquad |\Psi_+\rangle = \cos\theta\,|\varphi_1\beta\rangle + \sin\theta\,|\varphi_{z^2}\alpha\rangle$$
$$(11.51)$$

and, similarly, for M'

$$\tan 2\theta' = -\tan 2\theta \qquad (11.52)$$

$$W_-' = W^0(E_g) - \tfrac{1}{2}c_1^2\zeta_{3d} - A\cot\theta; \qquad |\Psi_-'\rangle = \sin\theta\,|\varphi_2\alpha\rangle + \cos\theta\,|\varphi_{z^2}\beta\rangle$$

$$W_+' = W^0(A_{1g}) + A\cot\theta; \qquad |\Psi_+'\rangle = -\cos\theta\,|\varphi_2\alpha\rangle + \sin\theta\,|\varphi_{z^2}\beta\rangle$$
$$(11.53)$$

If the energy difference between the interacting levels is much larger than the spin-orbit coupling interaction energy, it follows from Eq. (11.47) that $\tan 2\theta$ is a very small quantity. In the limiting case of $\tan 2\theta \to 0$, and $2\theta \to \pi$ (*not* $2\theta \to 0$—a possibility eliminated in footnote 10) it follows that $\sin\theta \to 1$ and $\cos\theta \to 0$ and that mixing of the excited Γ_6 levels into the ground state is very small. In any case, the result of the interaction between the two Γ_6 levels is that the upper level is pushed up by an amount

$$A\cot\theta = \tfrac{1}{2}\Delta W\{[1 + (2A/\Delta W)^2]^{1/2} - 1\} \qquad (11.54)$$

and that the lower level is stabilized by exactly the same amount. The quantity ΔW is the energy difference between the two interacting levels and is given by

$$\Delta W = W^0(A_{1g}) - W^0(E_g) + \tfrac{1}{2}c_1^2\zeta_{3d} \qquad (11.55)$$

The resulting energy splitting is diagrammed in Fig. 11.4.

The reader should note that the wave functions of Eqs. (11.51) and (11.53) are no longer eigenfunctions of $\hat{s}_z$: They possess mixed α and β character. The *goodness* of the spin quantization has been destroyed by spin-orbit coupling.

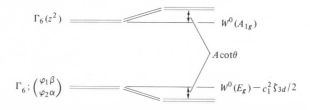

FIG. 11.4.   Schematic of the interaction of the Γ_6 levels.

If we adopt the typical values[11]: $\zeta_{3d} = 500$ cm^{-1}; $\zeta_{2p} = 50$ cm^{-1}; $\Delta W = 20\,000$ cm^{-1}; and $c_1 = c_1'' = 0.9$, Eq. (11.54) predicts an energy correction of about 10 cm^{-1}—which is practically negligible. Therefore, it is reasonable to expect that excited state-ground state mixing under the influence of $\hat{H}_{so}$ is important only when the metal atom is heavy (that is, large ζ) and the perturbation gap is small (i.e., $\Delta W \to 0$).

(E) Perturbation Theory

Since the effects predicted in Sec. 2(D) are oftentimes so small, it is of interest to compare the results of Eqs. (11.51)–(11.53) with those of first- and second-order perturbation theory. In the first order, the mixing of $\Gamma_6(z^2)$ into $\Gamma_6(\varphi_1\beta, \varphi_2\alpha)$ is given by

$$| \Psi_- \rangle = | \varphi_1\beta \rangle - \frac{A}{\Delta W} | \varphi_{z^2}\alpha \rangle$$

$$| \Psi_-' \rangle = | \varphi_2\alpha \rangle + \frac{A}{\Delta W} | \varphi_{z^2}\beta \rangle \qquad (11.56)$$

The mixing coefficient $-A/\Delta W$ should be compared to the ratio [see Eqs. (11.51) and (11.53)]

$$-\frac{\cos\theta}{\sin\theta} = -\cot\theta = \frac{-2A}{\Delta W\{1 + [1 + (2A/\Delta W)^2]^{1/2}\}} \qquad (11.57)$$

If

$$(2A/\Delta W)^2 \ll 1 \qquad (11.58)$$

Eq. (11.57) reduces to $-A/\Delta W$.

The first-order energy correction is zero. In the second order, the energy correction to $\Gamma_6(\varphi_1\beta, \varphi_2\alpha)$ due to the interaction with $\Gamma(z^2)$ is

$$W_{corr} = -\frac{A^2}{\Delta W} \qquad (11.59)$$

It is seen that Eq. (11.54) reduces to Eq. (11.59) if the condition of Eq. (11.58) is fulfilled:

$$-\frac{\Delta W}{2}\left\{\left[1 + \left(\frac{2A}{\Delta W}\right)^2\right]^{1/2} - 1\right\} \simeq -\frac{\Delta W}{2}\left[1 + \frac{1}{2}\left(\frac{2A}{\Delta W}\right)^2 - 1\right] = -\frac{A^2}{\Delta W}$$

$$(11.60)$$

---

[11] Usually in a complex, $\zeta_{metal} \gg \zeta_{ligands}$; oftentimes in other molecules, one of the atoms is much heavier than any of the others, so that one can neglect all but one of the ζ's.

Since Eq. (11.58) is oftentimes a very reasonable assumption, we conclude that perturbation theory reproduces the correct solutions quite well.

3. ZERO-FIELD SPLITTING:
AN EXAMPLE CALCULATION

It has been shown in Sec. 2 that an orbitally nondegenerate triplet state exhibits no first-order energy splitting. Nonetheless, triplet states are usually split into two or three separate levels, a few wave numbers apart. There are two obvious physical mechanisms which can lead to this removal of degeneracy. The first is a spin-spin magnetic dipole interaction between the two unpaired electrons of the triplet state. The second is a second-order spin-orbit effect which mixes various excited states into the three ground-state levels and thereby infuses them with some orbital angular momentum; if this mixing affects the three triplet wave functions to different extents, a small energy splitting should result. Spin-spin coupling is believed to be the dominant effect leading to the removal of spin degeneracy in triplet states of organic molecules; it is discussed in chapter 12. The discussion of the present section is restricted to zero-field splitting caused by second-order spin-orbit effects. These latter effects are principally responsible for the removal of spin degeneracies in transition-metal complexes which contain relatively heavy metal ions.

We have introduced the concept of zero-field splitting on the basis of a two-electron example—specifically, for a triplet state. One might suppose that a one-electron system, giving rise to doublet spin states, should also exhibit a zero-field splitting. That no splitting occurs in this latter case follows from an important theorem due to Kramers[6],[9]: *The energy levels of any odd-electron system are 2n-fold degenerate, where n is an integer.* It follows that a doublet state cannot be split by any internal effect, be it spin-spin coupling or spin-orbit coupling.

The order of magnitude of the zero-field splitting (~ 1 cm^{-1}) is such that it can most easily be studied by means of electron-paramagnetic-resonance (EPR) spectroscopy. We do not intend to present a full analysis of such spectra; indeed, we ignore nuclear effects completely. We intend no more than to illustrate the relationship between zero-field splitting and optical spectroscopic data and to show how both are affected and determined by spin-orbit coupling.

(A) Specification of Problem

Consider the D_{4h} system of Fig. 11.2 and Table 11.6; let there be *two* electrons in the highest-energy occupied e_g MO. In metal ions of the first transition series, electron repulsion is far more important than spin-orbit coupling. Thus, it is best to introduce first the repulsion interaction between the two electrons and then to discuss spin-orbit coupling as a slight extra perturbation. The $\ldots e_g{}^2$ configuration gives rise to A_{1g}, A_{2g}, B_{1g}, and B_{2g} states in D_{4h}. It is readily shown that the wave function with orbital symmetry A_{2g} is a spin triplet, whereas the other three states are spin singlets. The wave functions corresponding to the different states are

$$\Psi(^3A_{2g,i}) = (\tfrac{1}{2})^{1/2}[\varphi_{xz}(1)\varphi_{yz}(2) - \varphi_{xz}(2)\varphi_{yz}(1)]\Theta_{1i}; \qquad i = 1, 0, -1$$

$$\Psi(^1A_{1g}) = (\tfrac{1}{2})^{1/2}[\varphi_{xz}(1)\varphi_{xz}(2) + \varphi_{yz}(1)\varphi_{yz}(2)]\Theta_{00}$$

$$\Psi(^1B_{1g}) = (\tfrac{1}{2})^{1/2}[\varphi_{xz}(1)\varphi_{xz}(2) - \varphi_{yz}(1)\varphi_{yz}(2)]\Theta_{00}$$

$$\Psi(^1B_{2g}) = (\tfrac{1}{2})^{1/2}[\varphi_{xz}(1)\varphi_{yz}(2) + \varphi_{xz}(2)\varphi_{yz}(1)]\Theta_{00} \qquad (11.61)$$

The ground state is $^3A_{2g}$; it is an orbitally nondegenerate spin triplet. The previous considerations indicate that this state can exhibit a zero-field splitting. We now evaluate this splitting.

(B) Symmetry Considerations

The $M_S = 0$ component of $^3A_{2g}$, denoted $^3A_{2g,0}$ henceforth, transforms as the direct product $A_{2g} \times \Gamma(\hat{R}_z) = A_{2g} \times A_{2g} = A_{1g}$. The $^3A_{2g,\pm 1}$ components transform as $A_{2g} \times \Gamma(\hat{R}_x, \hat{R}_y) = A_{2g} \times E_g = E_g$. Hence, the result of spin-orbit perturbation is a decomposition of the initially threefold degenerate ground state into one nondegenerate A_{1g} state and one twofold degenerate E_g state.

Since the Hamiltonian $\hat{H}$ is totally symmetric, it can only induce mixing of states which possess identical transformation properties. Thus, the ground-state triplet can mix with all singlet states of species $^1A_{1g}$ and ^1E_g. Since the orbital symmetry of any *gerade* triplet state in D_{4h} is some one of A_{1g}, A_{2g}, B_{1g}, B_{2g}, or E_g, and since the spin symmetry of any triplet is $A_{2g} + E_g$, it follows that the direct-product representation always contains one or both of the A_{1g} and E_g representations. Therefore, the ground triplet state can mix with all excited *gerade* triplet states.

The lowest excited triplet state arises from the configuration[12] $\ldots e_g{}^1b_{2g}{}^1$; it is an orbitally degenerate state given by

$$\Psi(^3E_{g,i})_a = (\tfrac{1}{2})^{1/2}[\varphi_{xz}(1)\varphi_{xy}(2) - \varphi_{xz}(2)\varphi_{xy}(1)]\Theta_{1i}$$

$$\Psi(^3E_{g,i})_b = (\tfrac{1}{2})^{1/2}[\varphi_{yz}(1)\varphi_{xy}(2) - \varphi_{yz}(2)\varphi_{xy}(1)]\Theta_{1i} \qquad (11.62)$$

It may be shown that the ^3E_g state is not very distant energetically from the $^3A_{2g}$ ground state: The D_{4h} symmetry can be viewed as either an elongated or compressed O_h symmetry. In O_h, ^3E_g and $^3A_{2g}$ are degenerate, merging into a single $^3T_{1g}$ level.[13] If the tetragonal distortion of the octa-

---

[12] See Fig. 11.2.

[13] This may be understood using Eqs. (11.61) and (11.62): The spatial parts of the three wave functions $\Psi(^3A_{2g})$, $\Psi(^3E_g)_a$, and $\Psi(^3E_g)_b$ are related by a cyclic permutation of φ_{xz}, φ_{yz}, and φ_{xy} and these three MO's are degenerate in O_h.

hedron is not too large, the energy gap between ^3E_g and $^3A_{2g}$ should remain rather small. Hence, we may conclude that ^3E_g is the most important perturbing excited state.

The $^1A_{1g}$ state of Eq. (11.61) is of higher energy than ^3E_g and contributes less to the zero-field splitting.

We now restrict attention to the mixing of the ^3E_g and $^1A_{1g}$ excited states into the ground state.

(C)   Mixing of Ground State and ^3E_g State

Second-order perturbation theory yields

$$W(^3A_{2g},i) = W^0(^3A_{2g},i)$$

$$- \sum_j \frac{\langle \Psi(^3A_{2g},i) \mid \hat{H}_{so} \mid \Psi(^3E_g,j)_a \rangle \langle \Psi(^3E_g,j)_a \mid \hat{H}_{so} \mid \Psi(^3A_{2g},i) \rangle}{\Delta W}$$

$$- \sum_j \frac{\langle \Psi(^3A_{2g},i) \mid \hat{H}_{so} \mid \Psi(^3E_g,j)_b \rangle \langle \Psi(^3E_g,j)_b \mid \hat{H}_{so} \mid \Psi(^3A_{2g},i) \rangle}{\Delta W}$$

$$(11.63)$$

or, for short,

$$W(i) = W^0(i) - \sum_j \frac{\langle i \mid \hat{H}_{so} \mid j, a \rangle \langle j, a \mid \hat{H}_{so} \mid i \rangle}{\Delta W}$$

$$- \sum_j \frac{\langle i \mid \hat{H}_{so} \mid j, b \rangle \langle j, b \mid \hat{H}_{so} \mid i \rangle}{\Delta W} \qquad (11.64)$$

where

$$\Delta W = W^0(^3E_g) - W^0(^3A_{2g}) \qquad (11.65)$$

One now calculates the effects of the perturbation on each of the three i components ($i = 1, 0, -1$) separately.[14] It is useful, however, to note that certain (i,j) mixings are necessarily zero. The total symmetry of the relevant wave functions is given in Table 11.7. Thence, it may be seen that if $i = 0$, j must equal ± 1 in order that the matrix element be nonzero; similarly, if $i = \pm 1$, j must equal zero.

Let us consider, as an example, the matrix element

$$\langle 1, b \mid \hat{H}_{so} \mid 0 \rangle = \langle \Psi(^3E_g,1)_b \mid \sum_K \xi_K(1)\hat{\mathbf{l}}_{1K} \cdot \hat{\mathbf{s}}_1 + \xi_K(2)\hat{\mathbf{l}}_{2K} \cdot \hat{\mathbf{s}}_2 \mid \Psi(^3A_{2g},0) \rangle$$

$$(11.66)$$

---

[14] Equations (11.63) and (11.64) presuppose that second-order perturbation effects do not mix the $^3A_{2g,i}$ states with each other. This is correct in D_{4h} symmetry, but not in some lower symmetries.

We consider only the x and y components of the $\hat{\mathbf{l}}\cdot\hat{\mathbf{s}}$ operators because $(\hat{l}_{1z}\hat{s}_{1z} + \hat{l}_{2z}\hat{s}_{2z})$ cannot mix $M_S = \pm 1$ with $M_S = 0$. Using the expressions of Table 11.6 one finds

$$\langle 1, b \mid \hat{H}_{\mathrm{so}} \mid 0 \rangle = \frac{i}{2}\left(\frac{1}{2}\right)^{1/2} (c_1 c_1' \zeta_{3d} + 2c_2 c_2' \zeta_{2p}) \equiv \left(\frac{1}{2}\right)^{1/2} \frac{ik}{2} \quad (11.67)$$

Similarly, one obtains

$$\langle 1, a \mid \hat{H}_{\mathrm{so}} \mid 0 \rangle = -\langle -1, a \mid \hat{H}_{\mathrm{so}} \mid 0 \rangle = -\langle 0, a \mid \hat{H}_{\mathrm{so}} \mid 1 \rangle$$

$$= \langle 0, a \mid \hat{H}_{\mathrm{so}} \mid -1 \rangle = \left(\frac{1}{2}\right)^{1/2} \frac{k}{2} \quad (11.68)$$

$$\langle 1, b \mid \hat{H}_{\mathrm{so}} \mid 0 \rangle = \langle -1, b \mid \hat{H}_{\mathrm{so}} \mid 0 \rangle = \langle 0, b \mid \hat{H}_{\mathrm{so}} \mid 1 \rangle$$

$$= \langle 0, b \mid \hat{H}_{\mathrm{so}} \mid -1 \rangle = \left(\frac{1}{2}\right)^{1/2} \frac{ik}{2} \quad (11.69)$$

Elements of the type $\langle i \mid \hat{H}_{\mathrm{so}} \mid j, (a \text{ or } b) \rangle$ are complex conjugates of the corresponding elements $\langle j, (a \text{ or } b) \mid \hat{H}_{\mathrm{so}} \mid i \rangle$. Consequently, we have

$$W(^3A_{2g,1}) = W^0(^3A_{2g,1}) - \frac{1}{4}\frac{k^2}{\Delta W}$$

$$W(^3A_{2g,-1}) = W^0(^3A_{2g,-1}) - \frac{1}{4}\frac{k^2}{\Delta W}$$

$$W(^3A_{2g,0}) = W^0(^3A_{2g,0}) - \frac{1}{2}\frac{k^2}{\Delta W} \quad (11.70)$$

Since $W^0(^3A_{2g,1}) = W^0(^3A_{2g,-1}) = W^0(^3A_{2g,0})$, the splitting pattern is that

TABLE 11.7. *Transformation properties of* $\Psi(^3A_{2g})$ *and* $\Psi(^3E_g)$ *in spin-orbital space.*

Wave function	Total Symmetry
$\Psi(^3A_{2g,0})$	A_{1g}
$\Psi(^3A_{2g,\pm 1})$	E_g
$\Psi(^3E_{g,0})$	E_g
$\Psi(^3E_{g,\pm 1})$	$A_{1g} + A_{2g} + B_{1g} + B_{2g}$

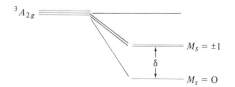

FIG. 11.5. Zero-field splitting of the $^3A_{2g}$ ground state of a D_{4h} molecule caused by mixing with the ^3E_g state.

shown in Fig. 11.5. The total energy splitting of the ground state caused by mixing with ^3E_g is given by

$$\delta = \frac{1}{2}\frac{k^2}{\Delta W} - \frac{1}{4}\frac{k^2}{\Delta W} = \frac{1}{4}\frac{k^2}{\Delta W} = \frac{1}{4}\frac{(c_1c_1'\zeta_{3d} + 2c_2c_2'\zeta_{2p})^2}{\Delta W} \quad (11.71)$$

For a small tetragonal perturbation (say, $\Delta W = 1000$ cm^{-1}) and for ζ_{3d} of the order of a few hundred cm^{-1}, one estimates δ to be approximately 10 cm^{-1}.

(D) Mixing of Ground State and $^1A_{1g}$ State

The $^1A_{1g}$ state, being of quite high energy, contributes only slightly to zero-field splitting.

The explicit expression for $\Psi(^1A_{1g})$ is given[15] in Eq. (11.61); it mixes only with the $M_S = 0$ component of $^3A_{2g}$. Thus, reduction occurs to

$$\langle \Psi(^3A_{2g,0}) \mid \hat{H}_{so} \mid \Psi(^1A_{1g}) \rangle$$

$$= \tfrac{1}{4}\langle [\varphi_{xz}(1)\varphi_{yz}(2) - \varphi_{yz}(1)\varphi_{xz}(2)]\Theta_{10} \mid \sum_K \xi_K(1)\hat{l}_{1z,K}\hat{s}_{1z} + \sum_K \xi_K(2)\hat{l}_{2z,K}\hat{s}_{2z} \mid$$

$$\times \mid [\varphi_{xz}(1)\varphi_{xz}(2) + \varphi_{yz}(1)\varphi_{yz}(2)]\Theta_{00}\rangle$$

$$= -\tfrac{1}{2}ic_1^2\zeta_{3d} \quad (11.72)$$

Therefore, the second-order energy correction due to the $^1A_{1g}$, $^3A_{2g,0}$ mixing is given by

$$W(^3A_{2g,0}) = W^0(^3A_{2g,0}) - \frac{c_1^4\zeta_{3d}^2}{4[W^0(^1A_{1g}) - W^0(^3A_{2g})]} \quad (11.73)$$

Thus, the $M_S = 0$ component is lowered relative to the $M_S = \pm 1$ doublet and the zero-field splitting increases. The effect, however, is comparatively small—for an energy gap of 20 000 cm^{-1} and $\zeta_{3d} = 300$ cm^{-1}, we find $\Delta\delta$ of approximately 0.5 cm^{-1}. This should be compared to the value of $\delta \sim$ 10 cm^{-1} caused by $^3A_{2g}$, ^3E_g mixing.

[15] See also Table 11.6.

4. *g* FACTOR IN ORBITALLY NONDEGENERATE STATES

When a molecule is subjected to a homogeneous magnetic field **H** its energy levels are modified and most of the degeneracies present at zero field are removed. This splitting, termed *Zeeman splitting*, is of the order of 1 cm^{-1} per 10-kG applied field. We now direct attention to the behavior of one- and two-electron systems and to the energy splitting exhibited by the degenerate states of such systems in externally applied magnetic fields.

Why treat the calculation of Zeeman effects in a chapter devoted to spin-orbit coupling? The interaction of an external magnetic field with a molecule does not appear, at least at first sight, to be dependent on internal spin-orbit interactions in the molecule. Indeed, the energy operator associated with the application of a magnetic field does *not* contain a spin-orbit term; it is given simply by

$$\hat{H}_{ze} = \beta_e \mathbf{H}(\hat{\mathbf{L}} + g_f \hat{\mathbf{S}}) \qquad (11.74)$$

where β_e is the Bohr magneton[16,17] and g_f is a dimensionless number reflecting the rather peculiar magnetic properties of a free spinning electron.[18] Despite this, spin-orbit coupling is relevant to calculations of Zeeman effects. The $\hat{\mathbf{L}}$ part of the operator of Eq. (11.74) elicits the orbital angular momentum associated with any wave function upon which it operates. Since spatially nondegenerate wave functions have no angular momentum (i.e., $L = 0$ in the zero-order approximation of neglect of spin-orbit coupling), the magnetic field produces energy splittings of magnitude $g_f \beta_e \mid \mathbf{H} \mid$ for such states. However, the introduction of spin-orbit interactions leads to the mixing of some orbital angular momentum into the zero-order wave functions. Application of Eq. (11.74) to these spin-orbit perturbed wave functions now yields a somewhat different splitting, namely, $g_f \beta_e \mid \mathbf{H} \mid$ *plus some orbital contribution.* The net result may be described in terms of an *effective g* factor—one which incorporates both spin and orbital effects—such that the resulting energy splitting is given by $g \beta_e \mid \mathbf{H} \mid$. It is clear that this *g* factor depends on the details of spin-orbit coupling in the molecule.

The magnitude of the energy splitting $g \beta_e \mid \mathbf{H} \mid$ is also a function of the orientation of the magnetic field **H** with respect to the molecular axes. Indeed, each relative orientation is characterized by its own *g* factor. This phenomenon is usually referred to as the *g-factor anisotropy*; it is revealed

---

[16] $\beta_e = e\hbar/2mc = (0.92731)10^{-20}$ erg/G.

[17] Note that $\hbar$ is now incorporated into β_e. Hence, this factor no longer appears in either $\hat{\mathbf{L}}$ or $\hat{\mathbf{S}}$—although both of these operators, as originally defined, did contain the $\hbar$ factor.

[18] $g_f = 2.002322$.

by accurate EPR measurements on monocrystals. The variation of the g factor with direction is partly determined by molecular symmetry.

It should be clear from the discussion of Sec. 3 that the measurement of zero-field splitting is closely connected to the observation of Zeeman splittings. For one thing, both are of the same order of magnitude. Thus, if a given triplet state shows no zero-field splitting, only one resonance field may be exhibited in the EPR experiment; if the triplet does show zero-field splitting, there may be two observable resonance fields—these statements are clarified in Fig. 11.6. The experimental tie between these two effects—zero-field splitting and Zeeman effects—is mathematically paralleled by the concept of a *spin Hamiltonian* which we introduce in Sec. 5.

A large amount of EPR data has evolved in the last ten years; it provides us with valuable information about the ground states of many molecules. For that reason, we concentrate attention on the behavior of ground states. Moreover, we consider paramagnetic transition-metal complexes only; these are usually stable and they can be studied relatively easily.

(A) Specification of Problem

The procedure for evaluating g factors is slightly different depending on whether the state in question is or is not orbitally degenerate. If the state is orbitally degenerate, it exhibits a first-order energy splitting under the influence of spin-orbit coupling. Per Sec. 2, this splitting may be of the order of several hundred wave numbers, whereas splittings produced by a 10-kG magnetic field are of the order of 1 cm^{-1}. Hence, $\hat{H}_{so}$ has to be introduced first; afterwards, one may investigate the influence of $\hat{H}_{ze}$ on the spin-orbit

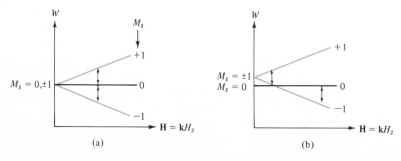

(a)   (b)

FIG. 11.6. Paramagnetic resonance effects in the triplet state of a molecule which (a) exhibits no zero-field splitting and (b) which exhibits zero-field splitting. Transitions are induced by means of a radio frequency (rf) radiation field of constant frequency, ν_{rf}, whereas the magnetic field is varied continuously. The selection rule $\Delta M_S = \pm 1$ applies in the absence of spin-orbit coupling. The resonance field positions are shown by double-headed arrows whose length corresponds to $h\nu_{rf}$.

perturbed wave functions. If, on the other hand, the ground state is orbitally nondegenerate, it is reasonable to introduce $\hat{H}_{ze}$ first—$\hat{H}_{so}$ being introduced later as a perturbation. However, the methods of calculation are very similar in both instances. We limit ourselves here to the case of an orbitally nondegenerate ground state.

The molecule we consider has D_{4h} symmetry. It is diagrammed in Fig. 11.7. It corresponds to an octahedron which has experienced compression of its z dimension. The converse case of the elongated octahedron is diagrammed in Fig. 11.2; the MO energy-level diagram of Figs. 11.2 and 11.7 are identical except for two MO interchanges: The e_g and b_{2g} MO's interchange location, as do the a_{1g} and b_{1g} MO's. The formal MO expressions of Table 11.6 pertain to both elongation and compression of the octahedron—Figs. 11.2 and 11.7, respectively—and are used here.

The highest-energy populated MO is b_{2g}; this MO is supposed to contain one electron. Hence, the zero-order ground state is $^2B_{2g}\{\varphi_{xy}\alpha, \varphi_{xy}\beta\}$.

(B) Neglect of Spin-Orbit Coupling

We first evaluate the g gactors using zero-order wave functions (that is, no spin-orbit coupling). If the magnetic field is applied along the z direction, then $H_z = |\mathbf{H}|$ and $H_x = H_y = 0$. Furthermore, the energies of the two

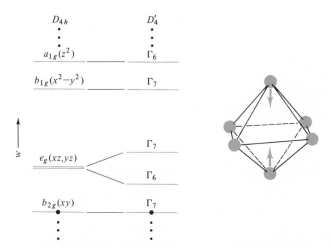

FIG. 11.7. Partial energy-level diagram of a compressed octahedron of D_{4h} symmetry. The zero-order ground state of a d^1 system is $^2B_{2g}$. The transformation properties cited under D_{4h} refer to the MO's. The transformation properties cited under D_4' refer to spin-orbital wave functions descriptive of the states produced by excitation of the b_{2g} electron to the various higher-energy MO's.

ground-state functions, $\varphi_{xy}\alpha$ and $\varphi_{xy}\beta$, are[19]

$$\langle \varphi_{xy}\alpha \mid \beta_e H_z(\hat{l}_z + 2\hat{s}_z) \mid \varphi_{xy}\alpha \rangle = \beta_e H_z \qquad (11.75)$$

and

$$\langle \varphi_{xy}\beta \mid \beta_e H_z(\hat{l}_z + 2\hat{s}_z) \mid \varphi_{xy}\beta \rangle = -\beta_e H_z \qquad (11.76)$$

The off-diagonal matrix elements are zero. Consequently, the resulting energy splitting equals $2\beta_e H_z$ and $g_{\parallel}$ equals g_f. Thus, the g factor relevant to application of a magnetic field parallel to the main symmetry axis (that is, $g_{\parallel}$) is identical to that for a free-electron spin (i.e., g_f).

If the magnetic field is directed along the x axis, only the off-diagonal matrix elements are nonzero. We find

$$\langle \varphi_{xy}\alpha \mid \beta_e H_x(\hat{l}_x + 2\hat{s}_x) \mid \varphi_{xy}\alpha \rangle = 0 \qquad (11.77)$$

$$\langle \varphi_{xy}\alpha \mid \beta_e H_x(\hat{l}_x + 2\hat{s}_x) \mid \varphi_{xy}\beta \rangle = \beta_e H_x \qquad (11.78)$$

The perturbation matrix is

$$\begin{array}{c} \langle \varphi_{xy}\alpha \mid \\ \\ \langle \varphi_{xy}\beta \mid \end{array} \begin{vmatrix} 0 & \beta_e H_x \\ \\ \beta_e H_x & 0 \end{vmatrix} \qquad (11.79)$$

This matrix is diagonal for $W = \pm\beta_e H_x$. The energy splitting is again $2\beta_e H_x$ and $g_{\perp}$ equals g_f. Because of symmetry reasons, a magnetic field in the y direction generates exactly the same energy splitting and can be described by the same $g_{\perp}$ factor. In fact, as long as the magnetic field remains in the (x, y) plane, the energy splitting is constant and given by $g_{\perp}\beta_e H$. Thus, the use of the symbol $\perp$, meaning the magnetic field is perpendicular to the main symmetry axis, is justified.

The neglect of spin-orbit coupling leads to the prediction that the complex behaves like an isolated single spin system. This is often expressed by stating that *the orbital angular momentum is completely quenched.*

(C) Mixing with Lowest-Energy Excited Γ_7 State

In the absence of a magnetic field, but considering spin-orbit effects, we know that the ground state is not adequately specified by $\{\varphi_{xy}\alpha, \varphi_{xy}\beta\}$; in fact, the ground state contains some admixture of all zero-order excited states of the same total symmetry. The total symmetry of the different wave functions in the spin-orbital space is provided in Fig. 11.7. The lowest-energy excited state[20] which can interact with $\Gamma_7(^2B_{2g})$ is $\Gamma_7\{\varphi_1\alpha, \varphi_2\beta\}$.

---

[19] For simplicity, we replace $g_f = 2.002322$ by $g_f = 2$ in Eq. (11.75). We continue this practice henceforth.

[20] For the analytic form of $\Gamma_7\{\varphi_1\alpha, \varphi_2\beta\}$, see Eqs. (11.36) and (11.39).

The interaction of these two states is a problem of the same nature as those discussed in Sec. 2. For example, we find

$$\langle \varphi_1 \alpha \mid \hat{H}_{so} \mid \varphi_{xy} \alpha \rangle = \langle \varphi_1 \alpha \mid \sum_K \xi_K \hat{l}_{K,z} \hat{s}_z \mid \varphi_{xy} \alpha \rangle = 0 \tag{11.80}$$

$$\langle \varphi_1 \alpha \mid \hat{H}_{so} \mid \varphi_{xy} \beta \rangle = \langle \varphi_1 \alpha \mid \sum_K \xi_K \hat{l}_{K,x} \hat{s}_x + \sum_K \xi_K \hat{l}_{K,y} \hat{s}_y \mid \varphi_{xy} \beta \rangle$$

$$= (\tfrac{1}{2})^{1/2} i (c_1 c_1' \zeta_{3d} + 2 c_2 c_2' \zeta_{2p}) \equiv iB \tag{11.81}$$

The interaction matrix is given by[21]

$\hat{H}_0 + \hat{H}_{so}$	$\mid \varphi_{xy} \alpha \rangle$	$\mid \varphi_2 \beta \rangle$	$\mid \varphi_{xy} \beta \rangle$	$\mid \varphi_1 \alpha \rangle$
$\langle \varphi_{xy} \alpha \mid$	0	$-iB$	0	0
$\langle \varphi_2 \beta \mid$	iB	$W^0(^2E_g) + \tfrac{1}{2} c_1^2 \zeta_{3d} + c_3^2 \zeta_{2p}$	0	0
$\langle \varphi_{xy} \beta \mid$	0	0	0	$-iB$
$\langle \varphi_1 \alpha \mid$	0	0	iB	$W^0(^2E_g) + \tfrac{1}{2} c_1^2 \zeta_{3d} + c_3^2 \zeta_{2p}$

$$\tag{11.82}$$

The interaction energy of Eq. (11.81) is of the order of magnitude of ζ_{3d} and, if the tetragonal distortion is not too large, $W^0(^2E_g) - W^0(^2B_{2g}) = W^0(^2E_g)$ might be of the same order of magnitude. Therefore, it is not proper to use perturbation theory—Eq. (11.82) should be diagonalized exactly. Since the 4×4 matrix reduces to two identical submatrices, we can use Eqs. (11.46)–(11.51). If

$$\tan 2\theta \equiv - \frac{2B}{W^0(^2E_g) + \tfrac{1}{2} c_1^2 \zeta_{3d} + c_3^2 \zeta_{2p}} \tag{11.83}$$

the perturbed ground state becomes

$$\mid + \rangle = \sin \theta \mid \varphi_{xy} \alpha \rangle - i \cos \theta \mid \varphi_2 \beta \rangle$$

$$\mid - \rangle = \sin \theta \mid \varphi_{xy} \beta \rangle - i \cos \theta \mid \varphi_1 \alpha \rangle \tag{11.84}$$

where $\mid + \rangle$ is the perturbed $M_S = \tfrac{1}{2}$ function $(\varphi_{xy} \alpha)$, and $\mid - \rangle$ is the perturbed $M_S = -\tfrac{1}{2}$ function $(\varphi_{xy} \beta)$.

Let us now investigate the manner in which the set $\{+, -\}$ behaves when a magnetic field is impressed along the z axis. One of the diagonal elements of $\hat{H}_{so}$ is

$$\langle - \mid \beta_e H_z (\hat{l}_z + 2 \hat{s}_z) \mid - \rangle = \beta_e H_z [\sin^2 \theta \langle \varphi_{xy} \beta \mid \hat{l}_z + 2 \hat{s}_z \mid \varphi_{xy} \beta \rangle$$

$$+ \cos^2 \theta \langle \varphi_1 \alpha \mid \hat{l}_z + 2 \hat{s}_z \mid \varphi_1 \alpha \rangle] \tag{11.85}$$

---

[21] The geometry considered here, being a compressed octahedron, does not allow neglect of the coefficients c_3 of Table 11.6.

Now, it is readily seen that

$$\langle \varphi_{xy}\beta \mid \hat{l}_z + 2\hat{s}_z \mid \varphi_{xy}\beta \rangle = \langle \varphi_{xy}\beta \mid 2\hat{s}_z \mid \varphi_{xy}\beta \rangle = -1 \qquad (11.86)$$

Using Eqs. (11.36) and (11.39) and the MO's of Table 11.6, it also follows that

$$\langle \varphi_1\alpha \mid \hat{l}_z + 2\hat{s}_z \mid \varphi_1\alpha \rangle$$

$$= \tfrac{1}{2}\langle \varphi_{xz} + i\varphi_{yz} \mid \hat{l}_z \mid \varphi_{xz} + i\varphi_{yz} \rangle + 1$$

$$= 1 + \tfrac{1}{2}c_1^2 \langle 3d_{xz} + i3d_{yz} \mid \hat{l}_z \mid 3d_{xz} + i3d_{yz} \rangle$$

$$+ \tfrac{1}{2}c_1c_2 \langle 3d_{xz} + i3d_{yz} \mid \hat{l}_z \mid 2p_{z1} - 2p_{z3} + i2p_{z2} - i2p_{z4} \rangle$$

$$+ \tfrac{1}{2}c_2c_1 \langle 2p_{z1} - 2p_{z3} + i2p_{z2} - i2p_{z4} \mid \hat{l}_z \mid 3d_{xz} + i3d_{yz} \rangle$$

$$+ \tfrac{1}{2}c_2^2 \langle 2p_{z1} - 2p_{z3} + i2p_{z2} - i2p_{z4} \mid \hat{l}_z \mid 2p_{z1} - 2p_{z3} + i2p_{z2} - i2p_{z4} \rangle$$

$$+ \tfrac{1}{2}c_3^2 \langle 2p_{x5} - 2p_{x6} + ip_{y5} - i2p_{y6} \mid \hat{l}_z \mid 2p_{x5} - 2p_{x6} + i2p_{y5} - i2p_{y6} \rangle$$

$$+ \tfrac{1}{2}c_1c_3 \langle 3d_{xz} + i3d_{yz} \mid \hat{l}_z \mid 2p_{x5} - 2p_{x6} + i2p_{y5} - i2p_{y6} \rangle$$

$$+ \tfrac{1}{2}c_3c_1 \langle 2p_{x5} - 2p_{x6} + i2p_{y5} - i2p_{y6} \mid \hat{l}_z \mid 3d_{xz} + i3d_{yz} \rangle$$

$$+ \tfrac{1}{2}c_2c_3 \langle 2p_{z1} - 2p_{z3} + i2p_{z2} - i2p_{z4} \mid \hat{l}_z \mid 2p_{x5} - 2p_{x6} + i2p_{y5} - i2p_{y6} \rangle$$

$$+ \tfrac{1}{2}c_3c_2 \langle 2p_{x5} - 2p_{x6} + i2p_{y5} - i2p_{y6} \mid \hat{l}_z \mid 2p_{z1} - 2p_{z3} + i2p_{z2} - i2p_{z4} \rangle$$

$$\qquad (11.87)$$

Calculating each term separately, one obtains

$$\langle 3d_{xz} + i3d_{yz} \mid \hat{l}_z \mid 3d_{xz} + i3d_{yz} \rangle = 2$$

$$\langle 3d_{xz} + i3d_{yz} \mid \hat{l}_z \mid 2p_{z1} - 2p_{z3} + i2p_{z2} - i2p_{z4} \rangle$$

$$= \langle 2p_{z1} - 2p_{z3} - i2p_{z2} + i2p_{z4} \mid -\hat{l}_z \mid 3d_{xz} - i3d_{yz} \rangle$$

$$= 4S(p\pi, d\pi) \qquad (11.88)$$

where $S(p\pi, d\pi)$ or, for short S_π, is the π-overlap integral between a $3d_{xz}$ AO on the metal and a $2p\pi$ AO on ligand 1. Similarly, one finds

$$\langle 2p_{z1} - 2p_{z3} + i2p_{z2} - i2p_{z4} \mid \hat{l}_z \mid 3d_{xz} + i3d_{yz} \rangle = 4S_\pi \qquad (11.89)$$

$$\langle 2p_{z1} - 2p_{z3} + i2p_{z2} - i2p_{z4} \mid \hat{l}_z \mid 2p_{z1} - 2p_{z3} + i2p_{z2} - i2p_{z4} \rangle$$

$$= 2i\langle 2p_{z1} - 2p_{z3} \mid \hat{l}_z \mid 2p_{z2} - 2p_{z4} \rangle \qquad (11.90)$$

Equation (11.90) reduces to zero if ligand-ligand (LL) overlap can be neglected. Treating the other terms of Eq. (11.87) similarly, and combining these with Eqs. (11.85)–(11.90), one finds

$$\langle - \mid \beta_e H_z(\hat{l}_z + 2\hat{s}_z) \mid - \rangle = \beta_e H_z[-\sin^2\theta + \cos^2\theta$$

$$\times (1 + c_1^2 + 4c_1c_2 S_\pi + 2c_3^2 + 4c_1c_3 S_\pi')] \qquad (11.91)$$

The quantity S_π' of Eq. (11.91) is the π overlap between a $3d_{xz}$ AO of metal and a $2p_x$ AO of ligand 5. In a compressed octahedron S_π' is greater than S_π.

Two important remarks apply to the integral evaluations in Eqs. (11.88) and (11.89). First, the two-center terms of the operator of Eq. (11.74) are obviously not negligible—in distinct contrast to the behavior of the spin-orbit operator, per Sec. 1(C), where two-center terms were neglected throughout. Second, the $\hat{l}$ operator should *not* be decomposed into a sum $\sum_K \hat{l}_K$ over all nuclei as in the case of spin-orbit interaction. Instead, the $\hat{l}$ operator has to be defined with respect to one fixed, but arbitrary origin. The angular momentum of the electron and its interaction with the magnetic field are calculated with respect to this origin as in Eqs. (11.88) and (11.89).

In a manner identical to that used in generating Eq. (11.91), one finds

$$\langle + \mid \beta_e H_z(\hat{l}_z + 2\hat{s}_z) \mid + \rangle = \beta_e H_z[\sin^2\theta - \cos^2\theta$$

$$\times (1 + c_1^2 + 4c_1c_2 S_\pi + 2c_3^2 + 4c_1c_3 S_\pi')] \quad (11.92)$$

The $g_{||}$ factor is now obtained as the difference of energies of Eqs. (11.92) and (11.91). It is given by

$$g_{||} = 2[\sin^2\theta - \cos^2\theta(1 + c_1^2 + 4c_1c_2 S_\pi + 2c_3^2 + 4c_1c_3 S_\pi')] \quad (11.93)$$

Making use of Eq. (11.83) and the equations which are footnoted to Eq. (11.49), one finds the equivalent alternative expression

$$g_{||} = \frac{(2 + c_1^2 + 2c_3^2 + 4c_1c_2 S_\pi + 4c_1c_3 S_\pi')(W_E^0 + \frac{1}{2}c_1^2\zeta_{3d} + c_3^2\zeta_{2p})}{[(W_E^0 + \frac{1}{2}c_1^2\zeta_{3d} + c_3^2\zeta_{2p})^2 + 2(c_1c_1'\zeta_{3d} + 2c_2c_2'\zeta_{2p})^2]^{1/2}}$$

$$- c_1^2 - 2c_3 - 4c_1c_2 S_\pi - 4c_1c_3 S_\pi' \quad (11.94)$$

where for brevity, $W^0(^2E_g) \equiv W_E^0$. For some typical values of the different parameters, say, $c_1 = 0.85$; $c_2 = -0.3$; $c_3 = -0.4$; $c_1' = 0.9$; $c_2' = -0.3$; $S_\pi = 0.08$; $S_\pi' = 0.1$; $W^0(^2E_g) = 1000$ cm^{-1}; $\zeta_{3d} = 500$ cm^{-1}; $\zeta_{2p} = 100$ cm^{-1}, one obtains $g_{||} = 1.72$. This should be compared to the value $g_{||} = 2$ found in Sec. 4(B), where spin-orbit coupling was neglected.

The calculation of $g_\perp$ proceeds along similar lines. In this instance, however, the two diagonal matrix elements are zero. The off-diagonal elements are given by

$$\langle - \mid \beta_e H_z(\hat{l}_x + 2\hat{s}_x) \mid + \rangle = \langle + \mid \beta_e H_z(\hat{l}_x + 2\hat{s}_x) \mid - \rangle$$

$$= \beta_e H_z[\sin^2\theta - (2)^{1/2}\sin\theta\cos\theta$$

$$\times (c_1c_1' + 2c_1'c_2 S_\pi + 4c_1c_2' S_\pi + 2c_2c_2' + 2c_1'c_3 S_\pi')]$$

$$(11.95)$$

The magnetic field scrambles the two original functions $\langle + |$ and $\langle - |$ to yield the linear combinations $(\frac{1}{2})^{1/2}\{\langle + | \pm \langle - |\}$. The resulting energy gap is described by the $g_\perp$ factor

$$g_\perp = 2[\sin^2 \theta - (2)^{1/2} \sin \theta \cos \theta$$

$$\times (c_1 c_1' + 2c_1'c_2 S_\pi + 4c_1 c_2' S_\pi + 2c_2 c_2' + 2c_1'c_3 S_\pi')] \quad (11.96)$$

Substituting for trigonometric functions yields

$$g_\perp = 1 + \cfrac{(W_E^0 + \frac{1}{2}c_1^2 \zeta_{3d} + c_3^2 \zeta_{2p}) - 2(c_1 c_1' \zeta_{3d} + 2c_2 c_2' \zeta_{2p})}{[(W_E^0 + \frac{1}{2}c_1^2 \zeta_{3d} + c_3^2 \zeta_{2p})^2 + 2(c_1 c_1' \zeta_{3d} + 2c_2 c_2' \zeta_{2p})^2]^{1/2}}$$

$$\qquad\qquad \times (c_1 c_1' + 2c_1'c_2 S_\pi + 4c_1 c_2' S_\pi + 2c_2 c_2' + 2c_1'c_3 S_\pi')$$

$$(11.97)$$

Adopting the numerical values quoted in the previous paragraph one evaluates $g_\perp = 1.43$; the value appropriate to the limiting case of no spin-orbit coupling is $g_\perp = 2$.

It is seen that the effects of spin-orbit coupling are twofold: First, the production of a marked change in the value of the g factors and, second, the introduction of considerable anisotropy.

(D) Influence of Delocalization

The molecular orbitals which we are using are concentrated mainly on the central metal ion (that is, $| c_1 |$ is much larger than $| c_2 |$, etc.). The crystal-field theory describes the electronic structure of transition-metal complexes by pushing this attitude to the extreme and by supposing that the orbitals in question are situated *entirely* on the metal; in other words, it is supposed that

$$c_1 = c_1' = c_1'' = c_1''' = 1; \qquad c_2 = c_2' = c_2'' = c_2''' = c_3'' = 0 \quad (11.98)$$

Substituting Eq. (11.98) into Eqs. (11.93) and (11.96), one obtains

$$g_{||} = 2(\sin^2 \theta - 2 \cos^2 \theta) \qquad\qquad (11.99)$$

$$g_\perp = 2[\sin^2 \theta - (2)^{1/2} \sin \theta \cos \theta] \qquad\qquad (11.100)$$

Alternatively, from Eqs. (11.94) and (11.97), one finds

$$g_{||} = \frac{3(2W_E^0 + \zeta_{3d})}{[(2W_E^0 + \zeta_{3d})^2 + 8\zeta_{3d}^2]^{1/2}} - 1 \qquad\qquad (11.101)$$

$$g_\perp = \frac{2W_E^0 - 3\zeta_{3d}}{[(2W_E^0 + \zeta_{3d})^2 + 8\zeta_{3d}^2]^{1/2}} + 1 \qquad\qquad (11.102)$$

For the same numerical example used earlier (that is, $W_E{}^0 = 1000 \text{ cm}^{-1}$ and $\zeta_{3d} = 500 \text{ cm}^{-1}$) we now find $g_{||} = 1.61$ and $g_\perp = 1.17$. There is serious discrepancy between these values and the MO results of the previous subsection. It is clear that delocalization effects should not be neglected in the calculation of g factors.

Often, there is also a rather serious discrepancy between the experimental g factors and the ones calculated by means of the crystal-field formulas. In order to improve the agreement, one can introduce ζ_{3d}', an *effective* spin-orbit coupling constant, which incorporates deviations from the crystal-field picture in an empirical way. The most obvious of these deviations is, of course, the delocalization effect. From a large number of experimental data, it is usually found that $\zeta_{3d}' \sim (0.7 \text{ or } 0.8)\zeta_{3d}$.

It is useful to apply the same attitude to the numerical discrepancies between crystal-field and the MO results. For instance, Eq. (11.101) yields the MO result $g_{||} = 1.72$ for $\zeta_{3d}' = 394 \text{ cm}^{-1}$ and Eq. (11.102) yields the MO result $g_\perp = 1.43$ for $\zeta_{3d}' = 315 \text{ cm}^{-1}$. An average value for ζ_{3d}' of about 350 cm^{-1}, compared to the original 500 cm^{-1}, is immediately suggested. This means that the crystal-field formulas, in our chosen example, yield the MO result if one uses an effective constant of $350/500 = 0.7$ times the free-ion value. This is about the same factor one had to introduce empirically to correlate experiment with the crystal-field results. This result seems to indicate that the MO treatment of the previous sections accounts more or less adequately for delocalization effects.

(E) Mixing with Higher-Energy Excited Γ_7 State

The remaining excited Γ_7 state, namely, $\{\varphi_{x^2-y^2}\alpha, \varphi_{x^2-y^2}\beta\}$, possesses orbital symmetry B_{1g} and lies at higher energy; it might be expected, therefore, that this state would perturb the ground state much less significantly than does $\Gamma_7\{\varphi_1\alpha, \varphi_2\beta\}$. Consequently, this mixing may be evaluated using perturbation theory.

It is found that $|-\rangle$ mixes with $|\varphi_{x^2-y^2}\beta\rangle$ only, whereas $|+\rangle$ mixes with $|\varphi_{x^2-y^2}\alpha\rangle$ only. The interaction elements are

$$\langle\varphi_{x^2-y^2}\beta \,|\, \hat{H}_{\text{so}} \,|\, -\rangle = -\langle\varphi_{x^2-y^2}\alpha \,|\, \hat{H}_{\text{so}} \,|\, +\rangle$$

$$= i \sin\theta(c_1'c_1'''\zeta_{3d} - 2c_2'c_2'''\zeta_{2p})$$

$$+ i \cos\theta(\tfrac{1}{2})^{1/2}(c_1c_1'''\zeta_{3d} - 2c_2c_2'''\zeta_{2p})$$

$$\equiv iC \tag{11.103}$$

Therefore, first-order perturbation theory yields

$$|\tfrac{1}{2}\rangle = |+\rangle + \frac{iC}{W^0(^2B_{1g})}\,|\varphi_{x^2-y^2}\alpha\rangle$$

$$= \sin\theta\,|\varphi_{xy}\alpha\rangle - i\cos\theta\,|\varphi_2\beta\rangle + \frac{iC}{W^0(^2B_{1g})}\,|\varphi_{x^2-y^2}\alpha\rangle$$

$$|-\tfrac{1}{2}\rangle = |-\rangle - \frac{iC}{W^0(^2B_{1g})}\,|\varphi_{x^2-y^2}\beta\rangle$$

$$= \sin\theta\,|\varphi_{xy}\beta\rangle - i\cos\theta\,|\varphi_1\alpha\rangle - \frac{iC}{W^0(^2B_{1g})}\,|\varphi_{x^2-y^2}\beta\rangle \quad (11.104)$$

where $|\tfrac{1}{2}\rangle$ and $|-\tfrac{1}{2}\rangle$ are the corrected functions. In general, it is convenient to label the perturbed functions by means of the M_S quantum number of the parent unperturbed functions.

The behavior of these two corrected functions under the perturbation $\beta_e H_z(\hat{l}_z + 2\hat{s}_z)$ is determined by the matrix elements

$$\langle -\tfrac{1}{2}\,|\,\hat{l}_z + 2\hat{s}_z\,|\,\tfrac{1}{2}\rangle = 0$$

$$\langle -\tfrac{1}{2}\,|\,\hat{l}_z + 2\hat{s}_z\,|\,-\tfrac{1}{2}\rangle = \langle -\,|\,\hat{l}_z + 2\hat{s}_z\,|\,-\rangle$$

$$+ \frac{2C\sin\theta}{W^0(^2B_{1g})}\,[2c_1'c_1''' + 4(3)^{1/2}c_1'c_2'''S_\sigma + 8c_2'c_1'''S_\pi]$$

$$- \frac{C^2}{[W^0(^2B_{1g})]^2} \quad (11.105)$$

where S_σ is the σ-overlap integral, $S(2p\sigma, d_{z^2}\sigma)$, at the appropriate metal-ligand distance. Therefore, a correction term, $\Delta g_{\parallel}$, should be added to the value of the $g_{\parallel}$ factor obtained in Sec. 4(B). This correction term, per Eq. (11.105), is

$$\Delta g_{\parallel} = -2\Bigg\{\frac{2C\sin\theta}{W^0(^2B_{1g})}\,[2c_1'c_1''' + 4(3)^{1/2}c_1'c_2'''S_\sigma + 8c_1'''c_2'S_\pi]$$

$$- \frac{C^2}{[W^0(^2B_{1g})]^2}\Bigg\} \quad (11.106)$$

Let us now check the importance of this correction *via* numerical example. We again set all coefficients for the metal orbitals, namely, c_1, c_1', etc., equal to 0.9; we set $c_2' = -0.33$ and $c_2''' = -0.45$. Further, we let $S_\pi = 0.1$; $S_\sigma = 0.13$; $W^0(^2E_g) = 1000$ cm^{-1}; $W^0(^2B_{1g}) = 20\,000$ cm^{-1};

$\zeta_{3d} = 500$ cm^{-1}; and $\zeta_{2p} = 100$ cm^{-1}. It is found that $B = 305$ cm^{-1}; $\sin \theta = 0.973$; $\cos \theta = 0.232$; and $C = 425$ cm^{-1}. Hence, we find $\Delta g_{||} = -0.08$. It is already obvious that the mixing of a state at an energy separation of $\sim$20 000 cm^{-1} cannot be neglected even when there is another interacting level only 1000 cm^{-1} above the ground state. The final value of the $g_{||}$ factor is $1.72 - 0.08 = 1.64$.

In the crystal-field approximation, Eq. (11.106) reduces to

$$\Delta g_{||} = -2 \left[\frac{4C \sin \theta}{W^0(^2B_{1g})} - \frac{C^2}{[W^0(^2B_{1g})]^2} \right] \tag{11.107}$$

If the small second-order term is neglected and if C is replaced by[22] $\zeta_{3d} \sin \theta + (\frac{1}{2})^{1/2} \zeta_{3d} \cos \theta$, Eq. (11.107) simplifies to

$$\Delta g_{||} \simeq \frac{8\zeta_{3d}[\sin \theta + (\frac{1}{2})^{1/2} \cos \theta] \sin \theta}{W^0(^2B_{1g})} \tag{11.108}$$

The previous numerical example now yields $\Delta g_{||} = -0.20$.

The conclusions which result from this discussion are quite simple:

(i)   It is not proper to exclude mixing with $\Gamma_7(^2B_{1g})$ within the context of either the delocalized or crystal-field pictures; the effects of its inclusion are simply too large.

(ii)   On the other hand, it is not very useful to include mixing with $\Gamma_7(^2B_{1g})$ in the crystal-field picture: The effects of its inclusion are just as wrong relative to the -0.08 value (that is, the value with delocalization) as are the effects of its neglect.

(iii)   It is difficult—at least, for the specific case of $\Gamma_7(^2B_{1g})$ mixing—to reconcile the MO and the crystal-field results by means of any sort of effective spin-orbital coupling constant. For example, calculation using Eq. (11.108) with $\zeta_{3d}' = 350$ cm^{-1} yields $\Delta g_{||} = -0.16$; this, of course, is still a gross overestimate of the molecular-orbital value of $\Delta g_{||}$. In order to reproduce the MO value of $\Delta g_{||}$, namely, -0.08, a value $\zeta_{3d}'' = 190$ cm^{-1} is required. Since $\zeta_{3d}'' = 190$ cm^{-1} is only 40% of $\zeta_{3d} = 500$ cm^{-1}, it seems that empiricism coupled with crystal-field theory has over-stepped reasonable bounds of propriety.

5. SPIN HAMILTONIAN

We do not introduce any new phenomenology in this section; we merely introduce a *different formalism* which renders discussion of zero-field splitting and Zeeman splitting more tractable.

---

[22] The expression $C = \zeta_{3d} \sin \theta + (\frac{1}{2})^{1/2} \zeta_{3d} \cos \theta$ follows from Eq. (11.103) in the context of the crystal-field approximation.

Consider the orbitally nondegenerate ground state, $^2B_{2g}\{\varphi_{xy}\alpha, \varphi_{xy}\beta\}$ discussed in Sec. 4(B). Application of the Zeeman operator $\beta_e H(\hat{1} + 2\hat{s})$ leads to the prediction of an isotropic g factor $g_{||} = g_{\perp} = g_f$. Spin-orbit mixing of the ground state and the excited states contaminates $|\varphi_{xy}\alpha\rangle$ with some $|\varphi_2\beta\rangle$ and some $|\varphi_{x^2-y^2}\alpha\rangle$ character; similar mixing contaminates $|\varphi_{xy}\beta\rangle$ with some $|\varphi_1\alpha\rangle$ and some $|\varphi_{x^2-y^2}\beta\rangle$ character. The resulting g factor is now anisotropic; it equals g_f *plus* an orbital contribution which depends on the relative orientation of the magnetic field vector and the molecular symmetry axes.

We now ask if it is possible to describe this specified anisotropy on the assumption that the magnetic moment of the molecule is entirely and solely due to electron spin. In other words, we seek a *fictitious* spin—the properties of which are drastically different, of course, from those of the *true* electron spin—in terms of which we can discuss g-factor anisotropy. Since the magnetic properties of this fictitious spin must also incorporate the anisotropic orbital effects, its behavior can only be described by means of a g tensor. The specification of a fictitious spin immediately entails the introduction of fictitious spin operators which we define[10] as

$$\hat{s}_{fz} \left| \tfrac{1}{2} \right\rangle = \tfrac{1}{2} \left| \tfrac{1}{2} \right\rangle \qquad \text{(analogous to } \hat{s}_z | \alpha\rangle = \tfrac{1}{2} | \alpha\rangle)$$

$$\hat{s}_{fz} \left| -\tfrac{1}{2} \right\rangle = -\tfrac{1}{2} \left| -\tfrac{1}{2} \right\rangle \qquad \text{(analogous to } \hat{s}_z | \beta\rangle = -\tfrac{1}{2} | \beta\rangle)$$

$$\hat{s}_{fx} \left| \tfrac{1}{2} \right\rangle = \tfrac{1}{2} \left| -\tfrac{1}{2} \right\rangle \qquad \text{(analogous to } \hat{s}_x | \alpha\rangle = \tfrac{1}{2} | \beta\rangle)$$

$$\hat{s}_{fx} \left| -\tfrac{1}{2} \right\rangle = \tfrac{1}{2} \left| \tfrac{1}{2} \right\rangle \qquad \text{(analogous to } \hat{s}_x | \beta\rangle = \tfrac{1}{2} | \alpha\rangle)$$

$$\hat{s}_{fy} \left| \tfrac{1}{2} \right\rangle = \tfrac{1}{2}i \left| -\tfrac{1}{2} \right\rangle \qquad \text{(analogous to } \hat{s}_y | \alpha\rangle = \tfrac{1}{2}i | \beta\rangle)$$

$$\hat{s}_{fy} \left| -\tfrac{1}{2} \right\rangle = -\tfrac{1}{2}i \left| \tfrac{1}{2} \right\rangle \qquad \text{(analogous to } \hat{s}_y | \beta\rangle = -\tfrac{1}{2}i | \alpha\rangle) \quad (11.109)$$

The subscript f, as in $\hat{s}_f$, distinguishes a fictitious spin operator, $\hat{s}_f$ or $\hat{S}_f$, from a true spin operator, $\hat{s}$ or $\hat{S}$. The symbols $|\tfrac{1}{2}\rangle$ and $|-\tfrac{1}{2}\rangle$ represent the perturbed $|\varphi_{xy}\alpha\rangle$ and $|\varphi_{xy}\beta\rangle$ states, respectively; these symbols are more precisely defined in Eqs. (11.104).

The operator $\beta_e H(\hat{L} + 2\hat{S})$ is now replaced by an effective spin-only operator given by

$$\hat{H}_{ze} = \beta \mathbf{H} \cdot g \cdot \hat{\mathbf{S}}_f = (H_x \, H_y \, H_z) \begin{pmatrix} g_{xx} & g_{xy} & g_{xz} \\ g_{yx} & g_{yy} & g_{yz} \\ g_{zx} & g_{zy} & g_{zz} \end{pmatrix} \begin{pmatrix} \hat{S}_{fx} \\ \hat{S}_{fy} \\ \hat{S}_{fz} \end{pmatrix} \quad (11.110)$$

If the coordinate axes coincide with the symmetry axes of the molecule, the g-tensor will be diagonal and Eq. (11.110) simplifies to

$$\hat{H}_{ze} = \beta_e (H_x g_{xx} \hat{S}_{fx} + H_y g_{yy} \hat{S}_{fy} + H_z g_{zz} \hat{S}_{fz}) \quad (11.111)$$

where g_{xx}, g_{yy} and g_{zz} are parameters which remain to be determined.

A description of zero-field splitting can also be attempted in terms of the spin Hamiltonian concept. Because of our point of view and because H in this instance equals zero, it follows that zero-field splitting must be supposed to arise from fictitious spin-fictitious spin interactions. The Hamiltonian, therefore, will contain terms in $\hat{S}_{fx}\hat{S}_{fx}$, $\hat{S}_{fy}\hat{S}_{fy}$ and $\hat{S}_{fz}\hat{S}_{fz}$. Indeed, it may be shown that the zero-field splitting of a two-electron system can be described adequately by (see Sec. 11.8 and chapter 12)

$$\hat{H}_{ss} = D(\hat{S}_{fz}^2 - \tfrac{2}{3}) + E(\hat{S}_{fx}^2 - \hat{S}_{fy}^2) \qquad (11.112)$$

where D and E are parameters which are to be determined. Both D and E depend on spin-orbit interactions (see Sec. 11.3) as well as on true spin-true spin coupling (see chapter 12). The true spin-true spin contribution is usually negligible in the case of transition metal complexes.

(A) Utility of the Spin Hamiltonian Concept

The spin Hamiltonian concept is of wide applicability. It is readily extended to systems containing more than two electrons; it can be used to describe nuclear spin effects; in all instances, it specifies the number of parameters which are required in order to account for a given set of experimental results; and it also prescribes the manner in which these parameters relate to both experiment and theory.

A given problem, therefore, consists of two parts:

(i)   One fits, or attempts to fit, the experimental results to a suitable spin Hamiltonian. In this way, one obtains empirical values for the parameters $g_{xx}, \ldots, D$, and E;

(ii)   One now calculates the parameters on the basis of some theoretical model. Such a calculation, of course, is equivalent to ones already outlined in Secs. 3 and 4. However, the spin Hamiltonian concept provides a particularly convenient junction between theory on the one hand and experiment on the other.

We attempt illustration of this utility in the next two paragraphs.

Suppose one wishes to investigate the manner in which energy levels are affected when a magnetic field is applied along some arbitrary direction in a molecule. It is very easy to use the spin Hamiltonian of Eq. (11.111) to calculate the energy splitting as a function of the parameters g_{xx}, g_{yy}, and g_{zz}. For instance, in D_{4h}, the general interaction matrix for a one-electron problem is given by

$\beta_e(H_x g_{xx}\hat{s}_{fx} + H_y g_{yy}\hat{s}_{fy} + H_z g_{zz}\hat{s}_{fz})$	$\lvert \tfrac{1}{2} \rangle$	$\lvert -\tfrac{1}{2} \rangle$
$\langle \tfrac{1}{2} \rvert$	$\tfrac{1}{2}g_{\parallel}\beta_e H_z$	$\tfrac{1}{2}g_{\perp}(H_x - iH_y)$
$\langle -\tfrac{1}{2} \rvert$	$\tfrac{1}{2}g_{\perp}(H_x + iH_y)$	$-\tfrac{1}{2}g_{\parallel}\beta_e H_z$

$$(11.113)$$

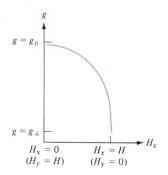

FIG. 11.8. Variation of the g factor as a function of orientation of magnetic field to molecular symmetry axes. The field is of constant magnitude and is situated in the (x, z) plane.

where $g_{||} = g_{zz}$ and $g_\perp = g_{xx} = g_{yy}$. When the magnetic field is in the (x, y) plane—namely, when $H_z = 0$—the resulting energy splitting is given by $g_\perp \beta_e (H_x^2 + H_y^2)^{1/2}$. Since $(H_x^2 + H_y^2)^{1/2}$ is the absolute value of the applied magnetic field, we find the previously specified result: The g factor is constant as long as the magnetic field remains in the (x, y) plane. If **H** is in the (x, z) plane, the energy splitting is given by $\beta_e(g_{||}^2 H_z^2 + g_\perp^2 H_x^2)^{1/2}$. The predicted variation of g as a function of direction is given in Fig. 11.8. All results emerge simply—no detailed g-factor calculations of the sort outlined in Sec. 4 are necessary.

The advantages of the spin Hamiltonian become even more apparent when the molecule exhibits a zero-field splitting. The Hamiltonian for a two-electron system is given by

$$\hat{H}_{ss} = \beta_e(H_x g_{xx}\hat{S}_{tx} + H_y g_{yy}\hat{S}_{ty} + H_z g_{zz}\hat{S}_{tz}) + D(\hat{S}_{tz}^2 - \tfrac{2}{3}) + E(\hat{S}_{tx}^2 - \hat{S}_{ty}^2)$$

$$(11.114)$$

The interaction matrix is

	$\lvert 1 \rangle$	$\lvert 0 \rangle$	$\lvert -1 \rangle$
$\langle 1 \rvert$	$g_{zz}\beta_e H_z + \tfrac{1}{3}D$	$(2)^{-1/2}(g_{xx}\beta_e H_z - ig_{yy}\beta_e H_y)$	E
$\langle 0 \rvert$	$(2)^{-1/2}(g_{xx}\beta_e H_x + ig_{yy}\beta_e H_y)$	$-\tfrac{2}{3}D$	$(2)^{-1/2}(g_{xx}\beta_e H_x - ig_{yy}\beta_e H_y)$
$\langle -1 \rvert$	E	$(2)^{-1/2}(g_{xx}\beta_e H_x + ig_{yy}\beta_e H_y)$	$-g_{zz}\beta_e H_z + \tfrac{1}{3}D$

$$(11.115)$$

where $|1\rangle$, $|0\rangle$, and $|-1\rangle$ are the spin-orbit corrected $M_S = 1, 0, -1$ functions.[23] The energy splittings produced by simultaneous second-order spin-orbit interaction and external magnetic field interaction can be quite complex. However, Eq. (11.115)—or the solutions to it—indicate that such splittings can always be expressed as some function of the same parametric set.

(B) A One-Electron Problem

Consider the molecule discussed in Sec. 4. Let the magnetic field be oriented along the z axis. The interaction matrix of Eq. (11.113) becomes

$g_{zz}\beta_e H_z \hat{s}_{fz}$	$\lvert \tfrac{1}{2} \rangle$	$\lvert -\tfrac{1}{2} \rangle$
$\langle \tfrac{1}{2} \rvert$	$\tfrac{1}{2} g_{zz}\beta_e H_z$	0
$\langle -\tfrac{1}{2} \rvert$	0	$-\tfrac{1}{2} g_{zz}\beta_e H_z$

$$(11.116)$$

Equations (11.113) and (11.116) are based on the simplified Hamiltonian of Eq. (11.111). We now verify that application of the more general operator of Eq. (11.110) leads also to $g_{pq} = \delta_{pq}$ in the particular coordinate system we have chosen. The matrix of Eq. (11.116) is now replaced by

$\beta_e H_z (g_{zx}\hat{s}_{fx} + g_{zy}\hat{s}_{fy} + g_{zz}\hat{s}_{fz})$	$\lvert \tfrac{1}{2} \rangle$	$\lvert -\tfrac{1}{2} \rangle$
$\langle \tfrac{1}{2} \rvert$	$\tfrac{1}{2} g_{zz}\beta_e H_z$	$\tfrac{1}{2}\beta_e H_z (g_{zx} - ig_{zy})$
$\langle -\tfrac{1}{2} \rvert$	$\tfrac{1}{2}\beta_e H_z (g_{zx} + ig_{zy})$	$-\tfrac{1}{2} g_{zz}\beta_e H_z$

$$(11.117)$$

It is required that the spin Hamiltonian matrix of Eq. (11.117) should be identical to the matrix of the true Hamiltonian $\beta_e H_z(\hat{l}_z + 2\hat{s}_z)$. Consequently, we must find

$$\langle \tfrac{1}{2} \lvert \hat{l}_z + 2\hat{s}_z \rvert \tfrac{1}{2} \rangle = \tfrac{1}{2} g_{zz}$$
$$\langle \tfrac{1}{2} \lvert \hat{l}_z + 2\hat{s}_z \rvert -\tfrac{1}{2} \rangle = \tfrac{1}{2}(g_{zx} - ig_{zy}) \qquad (11.118)$$

Using the analytic forms for $|\tfrac{1}{2}\rangle$ and $|-\tfrac{1}{2}\rangle$ given by Eq. (11.104), it is readily shown that $g_{zx} = g_{zy} = 0$; the same expression for g_{zz} as was obtained earlier in Sec. 4 is also found. Thus, Eqs. (11.116) and (11.117) are identical and g_{zz} is evaluable.

A number of convenient formulas for the g factor have been derived by Pryce[11] and Abragam and Pryce.[12] They used the spin Hamiltonian

---

[23] By analogy with the convention used in the one-electron problem of the previous paragraph, where the perturbed wave functions were genealogically labeled $|\tfrac{1}{2}\rangle$ and $|-\tfrac{1}{2}\rangle$ [see Eq. (11.104) also].

formalism in conjunction with perturbation theory; their treatment, the so-called Λ formalism, offers a very handy way of solving both one- and many-electron problems. As the derivation of these formulas is rather involved, and as the formalism is restricted to a crystal-field point of view,[24] we simply refer the reader to the original papers.

6. SPIN HAMILTONIAN APPROACH TO ZERO-FIELD SPLITTING IN A TRIPLET STATE

The purpose of this section is to redo most of the computations of Sec. 3 in the spin Hamiltonian approach. This accomplishes two objectives:

(i)   The nature of the spin Hamiltonian concept is clarified;
(ii)   understanding of symmetry effects on zero-field splitting is amplified.

The general spin Hamiltonian interaction matrix for a triplet state is given in Eq. (11.115). Setting $|\mathbf{H}| = H_x = H_y = H_z = 0$, Eq. (11.115) reduces to

	$\|1\rangle$	$\|0\rangle$	$\|-1\rangle$
$\langle 1\|$	$\frac{1}{3}D$	0	E
$\langle 0\|$	0	$-\frac{2}{3}D$	0
$\langle -1\|$	E	0	$\frac{1}{3}D$

$$(11.119)$$

This is the interaction matrix for zero-field splitting caused by fictitious spin-fictitious spin coupling. The eigenvalues of this matrix are

$$W_1 = \tfrac{1}{3}D - E; \qquad W_2 = \tfrac{1}{3}D + E; \qquad W_3 = -\tfrac{2}{3}D \quad (11.120)$$

The zero-order ground-state functions for the $^3A_{2g}$ state are given, as in Sec. 3, by

$$\Psi(^3A_{2g,i}) = (\tfrac{1}{2})^{1/2}[\varphi_{xz}(1)\varphi_{yz}(2) - \varphi_{xz}(2)\varphi_{yz}(1)]\Theta_{1i} \quad (11.61)$$

The functions $|1\rangle$, $|0\rangle$, and $|-1\rangle$ are precisely these same three functions perturbed by spin-orbit mixing with the excited ^3E_g state and, to a lesser extent, with the excited $^1A_{1g}$ state. The wave functions of the ^3E_g state are

$$\Psi(^3E_{g,i})_a = (\tfrac{1}{2})^{1/2}[\varphi_{xz}(1)\varphi_{xy}(2) - \varphi_{xz}(2)\varphi_{xy}(1)]\Theta_{1i}$$

$$\Psi(^3E_{g,i})_b = (\tfrac{1}{2})^{1/2}[\varphi_{yz}(1)\varphi_{xy}(2) - \varphi_{yz}(2)\varphi_{xy}(1)]\Theta_{1i} \quad (11.62)$$

We abbreviate $\Psi(^3A_{2g,i})$ to $|A, i\rangle$, $\Psi(^3E_{g,i})_a$ to $|a, i\rangle$, and $\Psi(^3E_{g,j})_b$ to $|b, j\rangle$.

---

[24] The Λ formalism, however, is susceptible to MO extension.[13]-[15]

For simplicity, consider mixing with the excited ^3E_g state only. Application of second-order perturbation theory leads to

$$H_{ij} = -\sum_k \frac{\langle A,i|\hat{H}_{so}|a,k\rangle\langle a,k|\hat{H}_{so}|A,j\rangle + \langle A,i|\hat{H}_{so}|b,k\rangle\langle b,k|\hat{H}_{so}|A,j\rangle}{\Delta W}$$

(11.121)

where ΔW is defined in Eq. (11.65) and the index k runs over the values $-1, 0, 1$ in Θ_{1k}. Rather than evaluate these matrix elements directly—as was done in Sec. 3—let us demonstrate, instead, the manner in which they are transformed into matrix elements of the operator of Eq. (11.112). Expanding $\hat{H}_{so}$ as

$$\hat{H}_{so} = \sum_K \xi_K(1)\hat{\mathbf{l}}_{1k}\cdot\hat{\mathbf{s}}_1 + \sum_K \xi_K(2)\hat{\mathbf{l}}_{2K}\cdot\hat{\mathbf{s}}_2$$

$$= \tfrac{1}{2}\sum_p \{[\sum_K \xi_K(1)\hat{l}_{1p} + \sum_K \xi_K(2)\hat{l}_{2p}](\hat{s}_{1p} + \hat{s}_{2p})$$

$$+ [\sum_K \xi_K(1)\hat{l}_{1p} - \sum_K \xi_K(2)\hat{l}_{2p}](\hat{s}_{1p} - \hat{s}_{2p})\}$$

(11.122)

where p denotes any one of x, y, and z. Using the explicit expressions for the wave functions, one obtains[25]

$$H_{ij} = -\sum_k \frac{\sum_p \langle\varphi_{yz}|\sum_K \xi_K\hat{l}_p|\varphi_{xy}\rangle\langle\varphi_{xy}|\sum_K \xi_K\hat{l}_p|\varphi_{yz}\rangle\langle\Theta_{1i}|\hat{S}_p|\Theta_{1k}\rangle\langle\Theta_{1k}|\hat{S}_p|\Theta_{1j}\rangle}{4\Delta W}$$

$$-\sum_k \frac{\sum_p \langle\varphi_{xz}|\sum_K \xi_K\hat{l}_p|\varphi_{xy}\rangle\langle\varphi_{xy}|\sum_K \xi_K\hat{l}_p|\varphi_{xz}\rangle\langle\Theta_{1i}|\hat{S}_p|\Theta_{1k}\rangle\langle\Theta_{1k}|\hat{S}_p|\Theta_{1j}\rangle}{4\Delta W}$$

(11.123)

We now note the equivalence

$$\sum_k \langle\Theta_{1i}|\hat{S}_p|\Theta_{1k}\rangle\langle\Theta_{1k}|\hat{S}_p|\Theta_{1j}\rangle = \langle\Theta_{1i}|\hat{S}_p^2|\Theta_{1j}\rangle$$

(11.124)

We also introduce a new set of symbols X, Y, and Z where X is defined as

$$X = \frac{\langle\varphi_{xz}|\sum_K \xi_K\hat{l}_x|\varphi_{xy}\rangle\langle\varphi_{xy}|\sum_K \xi_K\hat{l}_x|\varphi_{xz}\rangle + \langle\varphi_{yz}|\sum_K \xi_K\hat{l}_x|\varphi_{xy}\rangle\langle\varphi_{xy}|\sum_K \xi_K\hat{l}_x|\varphi_{yz}\rangle}{\Delta W}$$

(11.125)

with similar expressions for Y and Z. Use of Eqs. (11.124) and (11.125)

---

[25] In general, Eq. (11.123) should also contain cross terms of the type

$$\langle|\sum\hat{l}_p|\rangle\langle|\sum\hat{l}_q|\rangle\langle|\hat{S}_p|\rangle\langle|\hat{S}_q|\rangle.$$

All such terms are zero in the present instance because we have purposely chosen a particular coordinate system—one whose axes coincide with the symmetry axes of the molecule—in order to make the calculations as simple as possible.

reduces Eq. (11.123) to

$$H_{ij} = -\tfrac{1}{4}\langle \Theta_{1i} \,|\, X\hat{S}_x{}^2 + Y\hat{S}_y{}^2 + Z\hat{S}_z{}^2 \,|\, \Theta_{1j}\rangle \qquad (11.126)$$

or, alternatively, to

$$H_{ij} = -\tfrac{1}{4}\langle \Theta_{1i} \,|\, D(\hat{S}_z{}^2 - \tfrac{2}{3}) + E(\hat{S}_x{}^2 - \hat{S}_y{}^2) \,|\, \Theta_{1j}\rangle \qquad (11.127)$$

where D and E are defined as

$$D \equiv \tfrac{1}{4}[\tfrac{1}{2}(X + Y) - Z]; \qquad E \equiv \tfrac{1}{8}(Y - X) \qquad (11.128)$$

The quantity H_{ij} is the matrix element of a true spin operator acting on regular spin function operands. The use of the spin Hamiltonian formalism amounts to a replacement of the true spin functions $\Theta_{1,1}$, $\Theta_{1,0}$, and $\Theta_{1,-1}$ by perturbed spin-orbital functions $|-1\rangle$, $|0\rangle$, and $|-1\rangle$, and a simultaneous replacement of the $\hat{S}_p$ operators by $\hat{S}_{fp}$ operators. The formal results are identical but Eq. (11.128) has the advantage of permitting ready quantitative calculation of the parameters D and E. The derivation of Eqs. (11.121)–(11.128) shows quite clearly how second-order spin-orbit effects generate an effective spin-spin coupling mechanism.

The elaboration of Eq. (11.128) leads to

$$X = Y = \frac{(c_1 c_1' \zeta_{3d} + 2c_2 c_2' \zeta_{2p})^2}{\Delta W}; \qquad Z = 0 \qquad (11.129)$$

and

$$D = \frac{1}{4} \frac{(c_1 c_1' \zeta_{3d} + 2c_2 c_2' \zeta_{2p})^2}{\Delta W} = \delta; \qquad E = 0 \qquad (11.130)$$

This last result is identical with that obtained in Sec. 3 [see Eq. (11.71)].

Equation (11.120) indicates that when $E = 0$, the triplet level splits into one nondegenerate and one doubly degenerate set of levels. This always is the case in a molecule of axial symmetry where $X = Y$. When the molecular symmetry is lower than axial (i.e., D_{2h}, $C_{2v}, \ldots$), so that $X \neq Y$, the triplet state splits into three nondegenerate levels. When the molecular symmetry is higher than axial (i.e., O_h, $T_d, \ldots$), so that $X = Y = Z$ (and $D = E = 0$), the three triplet levels remain degenerate and these molecules do not exhibit any zero-field splitting.

7. SPIN-FORBIDDEN TRANSITIONS

The electric-dipole transition probability between two states of different spin multiplicity is zero—see Theorem 1, chapter 9. The electric-dipole selection rules are $\Delta S = 0$, $\Delta M_S = 0$; however, these rules are strictly correct only in the absence of spin-orbital coupling.

The operator $\hat{H}_{so}$ mixes states of different multiplicity: A zero-order singlet acquires some triplet contamination, and *vice versa*; doublets acquire some quartet character, and so on.

Multiplicity mixing provides a mechanism which can be held accountable for the occurrence of *spin-forbidden* transitions. Such transitions are well known; they are observed both in absorption and in emission. The transition probability is usually very low: For most organic molecules, it is of the order of 10^5 times smaller than the more intense spin-allowed electronic transitions. The explanation which we provide runs as follows: The two states under consideration are both *very nearly* pure multiplets for which $\Delta S = 1$; however, the spin-orbit perturbation negates *absolute* purity of these spin states. Thus, a spin-forbidden transition acquires allowedness to the extent that it is contaminated with spin-allowed transitions. An attempt to illustrate this mechanism is made in Fig. 11.9.

(A) An Example Calculation: Formaldehyde

The ground state of formaldehyde possesses C_{2v} geometry. A partial energy-level scheme for H_2CO is shown in Fig. 11.10; the corresponding molecular orbitals and energies[16] are listed in Table 11.8.

The ground state of H_2CO, having the electron configuration $\ldots a_1{}^2 b_1{}^2 b_2{}^2$, is a $^1A_1$ state. The first-excited electron configuration, $\ldots a_1{}^2 b_1{}^2 b_2{}^2 b_1$, results from a $b_1 \leftarrow b_2$ ($\pi^* \leftarrow n$) transition; it gives rise to a $^3A_2$ and a $^1A_2$ state.

(B) Group-Theoretic Considerations

A triplet state with orbital symmetry $\Gamma(U)$ can be contaminated with a singlet state, ^1V, if $\Gamma(V) = \Gamma(U) \times A_2$, B_1, or B_2. This follows from the fact that, in C_{2v}, the three rotation operators, which transform identically to the three triplet spin functions, generate the representations $A_2(\hat{R}_z)$,

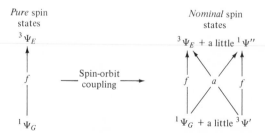

FIG. 11.9. Spin-forbidden transition $^3\Psi_E \leftrightarrow {}^1\Psi_G$ becomes partly allowed through admixture of the spin-allowed transitions $^3\Psi_E \leftrightarrow {}^3\Psi'$ and $^1\Psi'' \leftrightarrow {}^1\Psi_G$. The contamination of $^3\Psi_E$ by $^1\Psi''$ and of $^1\Psi_G$ by $^3\Psi'$ is due to spin-orbit coupling. The symbol f denotes spin forbidden and the symbol a denotes spin allowed.

TABLE 11.8.   *MO functions and energies for* H_2CO.

{An atomic orbital on oxygen has the subscript O; on carbon, it has the subscript C. The only hydrogen AO's that are included are $1s$ orbitals; they are denoted $1s_1$ and $1s_2$ where the subscripts 1 and 2 denote the hydrogen center in question. [After Carroll *et al.* Ref. [16].}

Representation	MO Wave function	Energy (eV)
a_1	$\varphi(\sigma^*) = 0.123\,(2s)_C + 0.029\,(2p_z)_C$ $- 1.394\,(3s)_C + 0.509\,(3s)_O$ $- 0.072\,(2p_z)_O - 0.090\,(1s_1 + 1s_2)$	-5.141
b_1	$\varphi(\pi^*) = 0.876\,(2p_x)_C - 0.809\,(2p_x)_O$ $- 0.124\,(3p_x)_C + 0.111\,(3p_x)_O$	-7.851
b_2	$\varphi(n) = -0.127\,(2p_y)_C + 0.976\,(2p_y)_O$ $+ 0.239\,(1s_1 - 1s_2)$	-11.029
b_1	$\varphi(\pi) = 0.569\,(2p_x)_C + 0.665\,(2p_x)_O$ $+ 0.016\,(3p_x)_C - 0.016\,(3p_x)_O$	-13.237
a_1	$\varphi(\sigma) = -0.198\,(2s)_C + 0.346\,(2p_z)_C$ $+ 0.060\,(3s)_C + 0.259\,(2s)_O$ $- 0.774\,(2p_z)_O - 0.047\,(3s)_O$ $+ 0.184\,(1s_1 + 1s_2)$	-14.518

$B_1(\hat{R}_y)$, and $B_2(\hat{R}_x)$. This singlet admixture makes the transition $^3U \leftrightarrow\, ^1A$ partly allowed if $^1V \leftrightarrow\, ^1A_1$ is itself an allowed transition. Now, the three dipole components, which transform identically to the translations along the three Cartesian axes x, y, and z, generate the representations A_1, B_1, and B_2. We can conclude, therefore, that allowedness is conferred on the transition $^3U \leftarrow\, ^1A_1$ if $\Gamma(V) = A_1$, B_1, or B_2.

Consequently, if the ground state is a $^1A_1$ state, the group-theoretic requirement for the mixing of a *useful* singlet[26] into ^3U can be formulated as follows: If one of the translations—say, $\hat{T}_p$, where $p = x$, y, z—transforms as $\Gamma(U) \times \Gamma(\hat{R}_q)$, where $\hat{R}_q$ is a rotation, then any singlet state ^1V with symmetry $\Gamma(\hat{T}_p)$ is a useful singlet. The polarization of the transition is the same as that of the $^1V \leftrightarrow\, ^1A_1$ transition (that is, it is p polarization).

This group-theoretic requirement must be supplemented by two other requirements which stem from the one-electron nature of both the spin-orbit coupling operator and the electric-dipole moment operator. These

---

[26] By *useful* singlet mixing into ^3U we mean the admixture of singlets which are connected to the ground state through nonzero matrix elements of the electric-dipole operator $e\mathbf{r}$.

requirements are (see chapter 8):

(i) ^1V can differ from the ground state in the occupation of at most one molecular orbital;

(ii) ^1V can differ from ^3U in the occupation of at most one molecular orbital.

These rules can be extended when the group contains two- or three-dimensional representations—that is, when degeneracies are involved.

The ground state with orbital symmetry $^1A_1$ can be contaminated with a triplet state ^3V if $\Gamma(V) = A_2(\hat{R}_z)$, $B_1(\hat{R}_y)$, or $B_2(\hat{R}_x)$. The transition $^3U \leftrightarrow {}^1A_1$ becomes possible when $\Gamma(U) \times \Gamma(V) = A_1(x)$, $B_1(y)$, or $B_2(z)$. In other words, if one of the rotations—say, $\hat{R}_p$—transforms like $\Gamma(U) \times \Gamma(\hat{T}_q)$ where $\hat{T}_q$ is one of the translations, then any triplet state, ^3V, which transforms like $\hat{R}_p$ is group theoretically a useful triplet. The resulting polarization is identical to that of the transition $^3V \leftrightarrow {}^3U$ (i.e., it is q polarization). Thus, all useful triplets transform as rotations whereas all useful singlets transform as translations. Hence, if a particular triplet state—say, ^3V—is useful, this does *not* necessarily entail that the singlet arising from the corresponding configuration, ^1V, is a useful singlet. However, the polarization conferred on the $^3U \leftrightarrow {}^1A_1$ transition through admixture of the

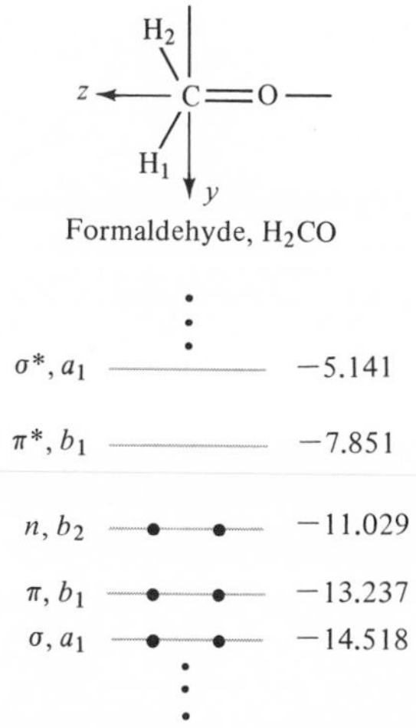

FIG. 11.10. Molecular geometry and partial MO energy-level scheme (Ref. [16]) of formaldehyde. The n MO is *nonbonding;* the asterisk denotes *antibonding;* σ and π have the usual meaning. Energies on the right are quoted in eV units. The x axis is perpendicular to the molecular plane.

singlet manifold into ^3U is the same as that conferred by admixture of the triplet manifold into $^1A_1$.

If the states under consideration possess permanent dipole moments, mixing of ^3U into $^1A_1$ and *vice versa* can also contribute to the allowedness of the $^3U \leftrightarrow {}^1A_1$ transition.

(C) Calculation of $^3A_2 \leftrightarrow {}^1A_1$ Transition Probability

The spin-orbit coupling constant for any of the atoms of H_2CO is small; hence, we can use perturbation theory. The $^1A_1(\pi^* \leftarrow \pi)$ state is closer energetically to $^3A_2$ than any other useful singlet state. Indeed, it may be shown that the $^1A_1(\pi^* \leftarrow \pi)$ state is by far the most important perturber of the $^3A_2$ state: In fact, mixing with $^1A_1$ alone can account for the observed transition probability of $^3A_2 \leftarrow {}^1A_1$. Therefore, we consider this mixing only.

Approximate expressions for the relevant wave functions are

$$^1A_1(\pi^* \leftarrow \pi) = (\tfrac{1}{2})^{1/2}(\pi_1^*\pi_2 + \pi_2^*\pi_1)\Theta_{00}$$

$$^3A_2(\pi^* \leftarrow n) = (\tfrac{1}{2})^{1/2}(\pi_1^*n_2 - \pi_2^*n_1)\Theta_{1i} \qquad (11.131)$$

where π_1 represents $\varphi(\pi)(1)$, and so forth. Substituting π for s and n for t, we now apply the results of Table 11.5. Because of symmetry reasons, the only $\hat{l}$ component which connects π and n transforms as $b_2 \times b_1 = a_2$. Hence, the operative component is $\hat{l}_z$ or, more precisely $\hat{H}_{l,z} = \tfrac{1}{2}\sum_K \xi_K \hat{l}_{z,K}$. Substituting the expressions of Table 11.8 for π and n, we obtain

$$\langle {}^1A_1(\pi^* \leftarrow \pi) \mid \hat{H}_{so} \mid {}^3A_2 \rangle = -\langle \pi \mid \hat{H}_{l,z} \mid n \rangle$$

$$= \tfrac{1}{2}i(-0.072\zeta_{2pc} + 0.649\zeta_{2po}) \qquad (11.132)$$

Therefore, the perturbed triplet wave function is

$$\Psi'(^3A_2) = \Psi(^3A_2) + \frac{i}{2}\left(\frac{0.072\zeta_{2pc} - 0.649\zeta_{2po}}{W^0(^1A_1, \pi^* \leftarrow \pi) - W^0(^3A_2)}\right)\Psi(^1A_1, \pi^* \leftarrow \pi)$$

$$(11.133)$$

where the prime denotes the perturbed state.

The experimental values of the spin-orbit coupling constants for the $2p$ orbitals on carbon and oxygen are[17]: $\zeta_{2pc} = 28$ cm^{-1} and $\zeta_{2po} = 152$ cm^{-1}. We set $W^0(^1A_1, \pi^* \leftarrow \pi) - W^0(^3A_2)$ equal to the experimental value of 4.5 eV. Substituting these values into Eq. (11.133), we find

$$\Psi'(^3A_2) = \Psi(^3A_2) + 0.0014i\Psi(^1A_1, \pi^* \leftarrow \pi) \qquad (11.134)$$

The resulting singlet character of $\Psi'(^3A_2)$ is very small indeed.

The transition moment connecting $\Psi'(^3A_2)$ with the ground state is

$$M_z = \langle 0.0014i\Psi(^1A_1, \pi^* \leftarrow \pi) \mid ez \mid \Psi(^1A_1, \pi^2) \rangle$$

$$= -0.0014ei(2)^{1/2}\langle \pi^* \mid z \mid \pi \rangle \qquad (11.135)$$

The integral $\langle \pi^* \,|\, z \,|\, \pi \rangle$ can be evaluated in the manner of chapter 9; its value is found to be -0.62 Å. The intrinsic lifetime τ_0 of the $^3A_2 \to {}^1A_1(\pi^2)$ phosphorescence is the reciprocal of the transition probability; it is given by

$$\frac{1}{\tau_0} = \frac{64\pi^4 \bar{\nu}_t^3}{3h} \,|\, \mathbf{M} \,|^2 \tag{11.136}$$

where $\bar{\nu}_t$ is the wave number corresponding to the singlet-triplet transition: $\bar{\nu}_t \sim 26\,000$ cm^{-1}. Numerically, Eq. (11.136) yields $\tau_0 \sim 5 \times 10^{-3}$ sec. This result correlates well with the available experimental data: The lifetime of the $\pi^* \to n$ phosphorescence is reported[17] to range from $\sim 8 \times 10^{-2}$ to $\sim 6 \times 10^{-4}$ sec in different molecules which contain the carbonyl group. Moreover, the predicted z polarization of the $\pi^* \to n$ triplet $\to$ singlet transition has been confirmed experimentally by several authors.[15]−[20]

EXERCISES

1. Using the definitions of the spin operators given in Eq. (11.8), derive the effects of $\hat{s}_+ = \hat{s}_x + i\hat{s}_y$, and $\hat{s}_- = \hat{s}_x - i\hat{s}_y$ on the spin functions α and β.

Answer: The results are given in the bottom part of Table 11.1.

2. Calculate the matrix elements of the operators $\hat{S}_y$, $\hat{S}_z$, $\hat{S}_y^2$, $\hat{S}_z^2$ in the basis Θ_{1-1}, Θ_{10}, and Θ_{11}. The functions Θ_{1i} are defined in Eq. (11.9) and $\hat{S}_x = \hat{s}_{x1} + \hat{s}_{x2}$, etc.

Answer:

$\hat{S}_y$	$|\Theta_{11}\rangle$	$|\Theta_{10}\rangle$	$|\Theta_{1-1}\rangle$
$\langle\Theta_{11}|$	0	$-\dfrac{i}{2^{1/2}}$	0
$\langle\Theta_{10}|$	$\dfrac{i}{2^{1/2}}$	0	$-\dfrac{i}{2^{1/2}}$
$\langle\Theta_{1-1}|$	0	$\dfrac{i}{2^{1/2}}$	0

$\hat{S}_z$	$|\Theta_{11}\rangle$	$|\Theta_{10}\rangle$	$|\Theta_{1-1}\rangle$
$\langle\Theta_{11}|$	1	0	0
$\langle\Theta_{10}|$	0	0	0
$\langle\Theta_{1-1}|$	0	0	-1

$\hat{S}_y^2$	$|\Theta_{11}\rangle$	$|\Theta_{10}\rangle$	$|\Theta_{1-1}\rangle$
$\langle\Theta_{11}|$	$\frac{1}{2}$	0	$-\frac{1}{2}$
$\langle\Theta_{10}|$	0	1	0
$\langle\Theta_{1-1}|$	$-\frac{1}{2}$	0	$\frac{1}{2}$

$\hat{S}_z^2$	$|\Theta_{11}\rangle$	$|\Theta_{10}\rangle$	$|\Theta_{1-1}\rangle$
$\langle\Theta_{11}|$	1	0	0
$\langle\Theta_{10}|$	0	0	0
$\langle\Theta_{1-1}|$	0	0	1

3. Calculate the effect of the three $\hat{l}_p$ operators on the atomic orbitals: (a) $2p_x$; (b) $3d_{xy}$, $3d_{z^2}$; (c) $5f_{z^3}$, $5f_{xyz}$.

Hints: Use the polar coordinate representation of Eq. (11.11) for the $\hat{l}_p$ operators. The radial parts of the wave functions are unaffected by the $\hat{l}_p$ operators; therefore, the principal quantum number is left unchanged.

The answers are available in Table 11.3 of the text and in Appendix H.

4. Apply the crystal-field approximation to the example treated in Sec. 2 and evaluate the first-order energy splittings. Derive the modified expressions for the energies and the wave functions. Find the effective ζ_{3d}' which reproduces the energy splitting predicted by MO theory when $c_1 = 0.85$.

Answer: The modified expressions can be found by setting $c_1 = 1$ in Eqs. (11.36)–(11.40). The energy splitting predicted by MO theory can be reproduced in crystal-field theory by using an effective ζ_{3d}' such that $\zeta_{3d}' = (0.85)^2 \zeta_{3d}$ (or $\zeta_{3d}' \sim 0.72 \zeta_{3d}$).

The orbital wave functions become pure metal d orbitals. Γ_7 is characterized by $j = \frac{3}{2}$ and Γ_6 by $j = \frac{1}{2}$.

5. Consider a linear diatomic molecule AB (symmetry $C_{\infty v}$; z axis along the internuclear axis). If the highest-energy occupied molecular orbital of the ground-state configuration has orbital symmetry π (twofold degenerate) and if it contains one electron, the zero-order ground state is $^2\Pi$. Calculate the first-order energy splitting due to spin-orbit coupling on the assumption that the orbital parts of the zero-order $^2\Pi$ state are given by

$$\varphi_1 = (\tfrac{1}{2})^{1/2}[(2p_x)_A + (2p_x)_B]; \qquad \varphi_2 = (\tfrac{1}{2})^{1/2}[(2p_y)_A + (2p_y)_B]$$

The functions φ_1 and φ_2 represent the two components, x and y, of the π MO.

Answer: It is possible to conclude on symmetry grounds that spin-orbit coupling removes the degeneracy. Indeed, in $C_{\infty v}$, $\{\alpha, \beta\}$ form a basis for the irreducible representation $E_{1/2}$. Since $E_{1/2} \times \Pi = E_{1/2} + E_{3/2}$, two different spin-orbital states can evolve.

It is possible to calculate the energy splitting by the procedure used in Sec. 2. The result is

$$\Delta W = \tfrac{1}{2}(\zeta_{2p})_A + \tfrac{1}{2}(\zeta_{2p})_B$$

The $E_{1/2}$ level drops energetically by an amount $\tfrac{1}{2}\Delta W$ whereas $E_{3/2}$ increases by an equal amount. This latter conclusion has been derived by Herzberg on the basis of qualitative arguments.[21] For the more specific example of the NO molecule, see Ref. [7] of text.

6. In Sec. 3 we assumed that electron repulsion was much more important than spin-orbit coupling in determining the state energy-level diagram. Investigate how the energy-level diagram is affected as spin-orbit coupling becomes increasingly important.

Answer: Figure 11.11 provides the answer.

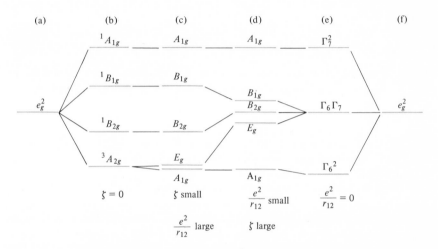

FIG. 11.11. Starting from the . . . e_g^2 configuration [(a) or (f)], we can introduce electron repulsion and ignore spin-orbit coupling (b), or we can introduce spin-orbit coupling while ignoring electron repulsion (e). Both extreme situations can be corrected by allowing the missing effect to enter as a perturbation; thus, we obtain (c) and (d). The group-theoretical symbols relate to the symmetry of the total (spin + orbital) wave functions except in cases (a), (b), and (f).

7. Derive Eq. (11.105).

Hint: Use the procedure outlined in Sec. 4(C). Note, especially, Eqs. (11.88) and (11.89).

8. Within the framework of crystal-field theory, Eq. (11.107) gives $\Delta g_{\parallel}$ caused by the mixing of $3d_{x^2-y^2}$ into the ground state. Putting $\sin \theta = 1$ amounts to neglecting the effects of the lowest excited Γ_7 state $\{\varphi_1\alpha, \varphi_2\beta\}$ [see Eq. (11.84)]. It corresponds to

the assumption that there are only two interacting states: $\Gamma_7\{d_{xy}\alpha, d_{xy}\beta\}$ and $\Gamma_7\{d_{x^2-y^2}\alpha, d_{x^2-y^2}\beta\}$. Calculate the g factor directly for this situation and check whether or not the same result is obtained by setting $\sin\theta = 1$ in Eq. (11.107).

Answer: $$g_{||} = 2 - \frac{8\zeta}{W^0(^2B_{1g})} + \frac{2\zeta^2}{W^0(^2B_{1g})^2} \tag{11.137}$$

9. Can the spin Hamiltonian of Eq. (11.112) be used to calculate a first-order energy splitting?

Answer: No!

10. Establish selection rules for singlet-triplet transitions in a molecule with C_{4v} symmetry and totally symmetric ground state.

Hint: Use the same procedure as in Sec. 7. Note that the two translations $\hat{T}_x$ and $\hat{T}_y$ are degenerate, as also the two rotations $\hat{R}_x$ and $\hat{R}_y$.

BIBLIOGRAPHY

[1] E. U. Condon and G. H. Shortley, *The Theory of Atomic Spectra* (Cambridge University Press, London, 1964).

[2] W. Marshall and R. Stuart, *Phys. Rev.* **123,** 2048 (1961).

[3] D. S. McClure, *J. Chem. Phys.* **20,** 682 (1952).

[4] S. P. McGlynn, T. Azumi, and M. Kinoshita, *Molecular Spectroscopy of the Triplet State* (Prentice-Hall, Inc., Englewood Cliffs, N.J., 1969).

[5] C. J. Ballhausen, *Introduction to Ligand Field Theory* (McGraw-Hill Book Co., New York, 1962).

[6] V. Heine, *Group Theory in Quantum Mechanics* (Pergamon Press, London, 1960).

[7] O. J. Heilmann and C. J. Ballhausen, *Theoret. Chim. Acta* **3,** 159 (1965).

[8] T. M. Dunn, *Trans. Faraday Soc.,* **57,** 1441 (1961).

[9] H. A. Kramers, *Quantum Mechanics* (North-Holland Publishing Co., Amsterdam, 1957).

[10] A. Carrington and A. D. McLachlan, *Introduction to Magnetic Resonance* (Harper and Row Publishers, New York, 1967).

[11] M. H. L. Pryce, *Proc. Phys. Soc. (London)* **A63,** 25 (1950).

[12] A. Abragam and M. H. L. Pryce, *Proc. Roy. Soc. (London)* **A205,** 135 (1951).

[13] A. H. Maki and B. R. McGarvey, *J. Chem. Phys.* **29,** 31 (1958).

[14] D. Kivelson and R. Neiman, *J. Chem. Phys.* **35,** 149 (1961).

[15] B. R. McGarvey, in *Transition Metal Chemistry,* edited by R. L. Carlin (Marcel Dekker, New York, 1966), Vol. 3 pp. 89–201.

[16] D. G. Carroll, L. G. Vanquickenborne, and S. P. McGlynn, *J. Chem. Phys.* **45,** 2777 (1966).

[17] E. H. Gilmore, G. E. Gibson, and D. S. McClure, *J. Chem. Phys.* **20,** 829 (1952).

[18] R. Shimada and L. Goodman, *J. Chem. Phys.* **43,** 2027 (1965).

[19] J. W. Sidman, *J. Am. Chem. Soc.* **78,** 2363 (1956).

[20] J. W. Sidman, *J. Chem. Phys.* **27,** 820 (1957).

[21] G. Herzberg, *Molecular Spectra and Molecular Structure. I. Spectra of Diatomic Molecules* (Van Nostrand, New York 1950), Chap. V.

General References

1. S. P. McGlynn, T. Azumi, and M. Kinoshita, *Molecular Spectroscopy of the Triplet State* (Prentice-Hall, Inc., Englewood Cliffs, N.J., 1969).

2. C. J. Ballhausen, *Introduction to Ligand Field Theory* (McGraw-Hill Book Co., New York, 1962).

3. J. S. Griffith, *The Theory of Transition-Metal Ions* (Cambridge University Press, Cambridge, England, 1961).

4. A. Carrington and H. C. Longuet-Higgins, *Quart. Rev.* **14,** 427 (1960).

5. M. H. L. Pryce, *Nuovo Cimento Suppl.* **6,** 817 (1957).

6. A. Carrington and A. D. McLachlan, *Introduction to Magnetic Resonance* (Harper and Row, New York, 1967).

7. B. R. McGarvey, in *Transition Metal Chemistry*, edited by R. L. Carlin (Marcel Dekker, New York, 1966), Vol. 3, pp. 89–201.

8. V. Heine, *Group Theory in Quantum Mechanics* (Pergamon Press, London, 1960).

9. G. E. Pake, *Paramagnetic Resonance* (W. A. Benjamin, Inc., New York, 1962).

10. E. König, in *Physical Methods in Advanced Inorganic Chemistry*, edited by H. A. O. Hill and P. Day (Interscience Publishers, Inc., New York, 1968).

CHAPTER 12

Spin-Spin Coupling[1]

The spin degeneracy of any electronic state (doublet, triplet, etc) may be removed by an external magnetic field. The resultant splitting—the Zeeman spectrum—is a very important tool for the characterization of electronic energy levels. However, even at zero applied magnetic field strength, the spin degeneracy is oftentimes only approximate; its absence can be detected by delicate-measurement techniques.

Our sole purpose here is to discuss the absence of exact spin degeneracy at zero applied magnetic field strength. Spin-orbit perturbation and its contribution to zfs[2] is discussed in chapter 11.

The greatest relevance of this chapter pertains to triplet states. The recent abundance of EPR studies of such triplet states[1]−[3] has aroused

---

[1] The term *spin-spin interactions* (or *spin-spin coupling*), as we use it in this chapter, refers to *intramolecular electron-spin electron-spin magnetic dipolar interactions*. As commonly used, however, this same term serves as a catch-all for various types of interactions: nuclear-spin nuclear-spin dipolar and exchange interactions; nuclear-spin electron-spin dipolar interactions; electron-spin electron-spin dipolar and exchange interactions; etc. Furthermore, the listed interactions can exist *intermolecularly* as well as *intramolecularly*.

[2] We abbreviate zero-field splitting to zfs.

393

considerable interest[4]–[24] in zfs, with the result that this field is now very large.

The zfs of electronic states of molecules which consist of low atomic-weight elements is known to arise predominantly from the internal magnetic fields produced by unpaired electrons; the spin-orbit interactions do contribute, but only in a minor way.[25]–[27] As examples, we might list the ground states of O_2, NH, CH_2 and most organic biradicals; and the excited triplet states[3] of molecules such as ethylene, benzene, and formaldehyde.

The zfs of compounds of the heavier elements, as exemplified by rare-earth and transition-metal complexes, is almost wholly attributable to spin-orbit effects; the spin-spin interactions are implicated—but only as slight extra perturbations which can be handled, at least in part, using the attitudes of this chapter.

1. ZERO-ORDER WAVE FUNCTIONS

The molecular Hamiltonian $\hat{H}$ is

$$\hat{H} = \hat{H}_0 + \hat{H}_d \tag{12.1}$$

where $\hat{H}_0$ is the spin-free part of $\hat{H}$, and $\hat{H}_d$ is the spin-spin dipolar interaction Hamiltonian. An explicit form for $\hat{H}_d$ is given in Eq. (12.15). We regard $\hat{H}_0$ as the zero*th*-order Hamiltonian and $\hat{H}_d$ as the perturbing Hamiltonian. If the eigenfunctions of $\hat{H}_0$ be antisymmetric with respect to electron interchange and if they be eigenfunctions of the total spin operators $\hat{S}^2$ and $\hat{S}_z$, we obtain a set of singlet eigenfunctions[4] $^1\Psi_S(1, 2, \ldots, n)$ with energies ^1W_S, a set of triplet eigenfunctions $^3\Psi_T(1, 2, \ldots, n)$ with energies ^3W_T, and so forth; thus,

$$\hat{H}_0 \, {}^{2S+1}\Psi_R(1, 2, \ldots, n) = {}^{2S+1}W_R \, {}^{2S+1}\Psi_R(1, 2, \ldots, n) \tag{12.2}$$

$$(S = 0, 1, 2, \ldots \text{ when } n \text{ is even})$$

Since we are concerned only with the lowest-energy triplet state, we disregard the subscript T and write

$$\hat{H}_0 \, {}^3\Psi(1, 2, \ldots, n) = {}^3W \, {}^3\Psi(1, 2, \ldots, n) \tag{12.3}$$

(A) Two-Electron Wave Functions

It may be shown that the zfs caused by spin-spin interaction is dependent only on those two electrons for which the wave function is antisym-

---

[3] Doublet states of monoradicals are excluded because they contain only one spin system; quartet states of organic molecules are unknown; quintet states of such molecules have only been detected recently.

[4] The electron labeling $(1, 2, \cdots, n)$ represents both space and spin coordinates, for example, $i \equiv (x_i, y_i, z_i; \omega_i) \equiv (r_i; \omega_i)$.

metric in the *space variables*.[28] In other words, in the case of the lowest-energy singly excited triplet state, the electrons in filled molecular orbitals do not contribute to the zfs. Therefore, it is convenient to reduce the many-electron wave function $^3\Psi(1, 2,\ldots, n)$ to a simple two-electron wave function $^3\Psi(1, 2)$, where

$$^3\Psi(1, 2,\ldots, n) \supset {}^3\Psi(1, 2) \tag{12.4}$$

The spin eigenfunctions of $\hat{S}^2$ and $\hat{S}_z$ are obtained by the methods of chapter 7. They are

$$\Theta_{11}(\omega_1,\omega_2) \equiv \Theta_{11}(1, 2) \equiv \alpha(\omega_1)\alpha(\omega_2); \qquad S=1; \quad M_S=1 \tag{12.5}$$

$$\Theta_{10}(1, 2) = (\tfrac{1}{2})^{1/2}[\alpha(\omega_1)\beta(\omega_2) + \beta(\omega_1)\alpha(\omega_2)]; \quad S=1; \quad M_S=0 \tag{12.6}$$

$$\Theta_{1-1}(1,2) = \beta(\omega_1)\beta(\omega_2); \qquad S=1; \quad M_S=-1 \tag{12.7}$$

$$\Theta_{00}(1, 2) = (\tfrac{1}{2})^{1/2}[\alpha(\omega_1)\beta(\omega_2) - \beta(\omega_1)\alpha(\omega_2)]; \quad S=0; \quad M_S=0 \tag{12.8}$$

Since the spin functions with $S = 1$ are symmetric in the spin variables, the coupled space function $\Phi(1, 2)$ must be antisymmetric in the space variables. Such a space function can be constructed from two different singly occupied molecular orbitals, φ_r and φ_s, as

$$\Phi(\mathbf{r}_1, \mathbf{r}_2) \equiv \Phi(1, 2) = (\tfrac{1}{2})^{1/2}[\varphi_r(1)\varphi_s(2) - \varphi_s(1)\varphi_r(2)]$$
$$\equiv | \varphi_r(1)\varphi_s(2) | \tag{12.9}$$

Thus, the triplet state wave functions are

$$^3\Psi_{M_S}(1, 2) = \Phi(1, 2)\Theta_{1M_S}(1, 2) \qquad (M_S = \pm1, 0) \tag{12.10}$$

where the subscript M_S on $^3\Psi(1, 2)$ is attached in order to distinguish the different values of M_S.

In the zero*th*-order approximation, the three functions $^3\Psi_{M_S}(1, 2)$ are degenerate because the operator $\hat{H}_0$ of Eq. (12.1) does not contain any spin-dependent parts. Therefore, any linear combination of the wave functions $^3\Psi_{M_S}$ is also an eigenfunction of $\hat{H}_0$ with the same eigenvalue ^3W. In other words,

$$\hat{H}_0\Big[\sum_{M_S=0,\pm1} C_{M_S}\, {}^3\Psi_{M_S}(1, 2)\Big] = {}^3W\Big[\sum_{M_S=0,\pm1} C_{M_S}\, {}^3\Psi_{M_S}(1, 2)\Big] \tag{12.11}$$

Such a linear combination need not be an eigenfunction of $\hat{S}_z$. Perturbation theory based on the zero*th*-order set of Eq. (12.10) and the spin-spin interaction operator scrambles this basis set and generates a new set of three orthogonal functions having the more general form of Eq. (12.11).

2. DIPOLAR PERTURBATION

We now treat the effect of $\hat{H}_d$ on the wave function of Eq. (12.11); we use first-order perturbation theory. As justification, we note that the energy associated with $\hat{H}_d$ is usually very small[5] relative to that associated with $\hat{H}_0$. Before doing so, however, it is convenient to transform $\hat{H}_d$ into an appropriate form.

The interaction energy of two magnetic dipoles μ_1 and μ_2 is

$$[r_{12}{}^2(\mu_1 \cdot \mu_2) - 3(\mathbf{r}_{12} \cdot \mu_1)(\mathbf{r}_{12} \cdot \mu_2)]r_{12}{}^{-5} \tag{12.12}$$

where $\mathbf{r}_{12}$ is the vector joining the centers of the two dipoles. In the case of a magnetic dipole associated with an electron spin $\hat{\mathbf{s}}$, the magnetic dipole μ is defined as

$$\mu = -g\beta_e \hat{\mathbf{s}} \tag{12.13}$$

where g is a spectroscopic splitting factor[6] and β_e is the Bohr magneton:

$$\beta_e = |e|\hbar/2mc = 0.9273 \times 10^{-21} \text{ erg G}^{-1} \tag{12.14}$$

where $|e|$ is the absolute value of an electronic charge. Therefore, $\hat{H}_d$ can be written as

$$\hat{H}_d = g^2\beta_e{}^2[r_{12}{}^2(\hat{\mathbf{s}}_1 \cdot \hat{\mathbf{s}}_2) - 3(\mathbf{r}_{12} \cdot \hat{\mathbf{s}}_1)(\mathbf{r}_{12} \cdot \hat{\mathbf{s}}_2)]r_{12}{}^{-5} \tag{12.15}$$

Expansion of Eq. (12.15) into its components leads to

$$\hat{H}_d = g^2\beta_e{}^2[\hat{s}_{1x}\hat{s}_{2x}(r_{12}{}^2 - 3x_{12}{}^2) + \hat{s}_{1y}\hat{s}_{2y}(r_{12}{}^2 - 3y_{12}{}^2)$$
$$+ \hat{s}_{1z}\hat{s}_{2z}(r_{12}{}^2 - 3z_{12}{}^2) - 3(\hat{s}_{1x}\hat{s}_{2y} + \hat{s}_{1y}\hat{s}_{2x})x_{12}y_{12}$$
$$- 3(\hat{s}_{1y}\hat{s}_{2z} + \hat{s}_{1z}\hat{s}_{2y})y_{12}z_{12} - 3(\hat{s}_{1z}\hat{s}_{2x} + \hat{s}_{1x}\hat{s}_{2z})x_{12}z_{12}]r_{12}{}^{-5} \tag{12.16}$$

Since the two electron spins are correlated, it is convenient to use the total spin operators instead of the individual spin operators. By use of the relations[7]:

$$\hat{s}_{1x}\hat{s}_{2x} = \tfrac{1}{2}\hat{S}_x{}^2 - \tfrac{1}{4}, \text{ etc.} \tag{12.17a}$$

$$\hat{s}_{1x}\hat{s}_{2y} + \hat{s}_{1y}\hat{s}_{2x} = \tfrac{1}{2}(\hat{S}_x\hat{S}_y + \hat{S}_y\hat{S}_x), \text{ etc.} \tag{12.17b}$$

---

[5] A factor of 10^4 times smaller is by no means unusual.
[6] The term *spectroscopic splitting factor* is commonly abbreviated to: *g factor* or *g value*.
[7] See Exercise 1.

we obtain

$$\hat{H}_d = \tfrac{1}{2}g^2\beta_e^2[\hat{S}_x^2(r_{12}^2 - 3x_{12}^2) + \hat{S}_y^2(r_{12}^2 - 3y_{12}^2) + \hat{S}_z^2(r_{12}^2 - 3z_{12}^2)$$
$$- 3(\hat{S}_x\hat{S}_y + \hat{S}_y\hat{S}_x)x_{12}y_{12} - 3(\hat{S}_y\hat{S}_z + \hat{S}_z\hat{S}_y)y_{12}z_{12}$$
$$- 3(\hat{S}_x\hat{S}_z + \hat{S}_z\hat{S}_x)x_{12}z_{12}]r_{12}^{-5} \tag{12.18}$$

(A) Spin Hamiltonian

In order to obtain the coefficients C_{M_S} of Eq. (12.11), we now apply perturbation theory for a degenerate state. The secular determinant is

$$|\, H_{M_S,M_{S'}} - W\delta_{M_S,M_{S'}} \,| = 0 \tag{12.19}$$

$$H_{M_S,M_{S'}} = \langle\Phi(1,2)\Theta_{1M_S}(1,2) \,|\, \hat{H}_d \,|\, \Phi(1,2)\Theta_{1M_{S'}}(1,2)\rangle \tag{12.20}$$

If we introduce the notation

$$D_{xx} = \tfrac{1}{2}g^2\beta_e^2\langle\Phi(1,2) \,|\, (r_{12}^2 - 3x_{12}^2)/r_{12}^5 \,|\, \Phi(1,2)\rangle, \text{ etc.} \tag{12.21a}$$

$$D_{xy} = \tfrac{1}{2}g^2\beta_e^2\langle\Phi(1,2) \,|\, -3x_{12}y_{12}/r_{12}^5 \,|\, \Phi(1,2)\rangle, \text{ etc.} \tag{12.21b}$$

we may rewrite Eq. (12.20) as

$$H_{M_S,M_{S'}} = \langle\Theta_{1M_S}(1,2) \,|\, \sum_p \sum_q D_{pq}\hat{S}_p\hat{S}_q \,|\, \Theta_{1M_{S'}}(1,2)\rangle, \qquad (p,q = x,y,z)$$

$$\equiv \langle\Theta_{1M_S}(1,2) \,|\, \hat{H}_s \,|\, \Theta_{1M_{S'}}(1,2)\rangle \tag{12.22}$$

Equation (12.22) defines a subsidiary Hamiltonian $\hat{H}_s$ given by

$$\hat{H}_s \equiv \hat{\mathbf{S}}\cdot\tilde{D}\cdot\hat{\mathbf{S}} \tag{12.23}$$

where $\tilde{D}$ is a symmetric tensor conventionally referred to as the zfs tensor; its components are given by averages over the triplet-state orbital wave function $\Phi(1,2)$ as shown in Eq. (12.21).

(B) Principal Axes

Since the total Hamiltonian is invariant under all symmetry operations of the molecular point group, the spin Hamiltonian is also invariant with respect to the same transformations. In the case of orthorhombic or higher symmetry,[8] one may choose a rectangular coordinate system—axes x, y, and z—such that the Hamiltonian is invariant with respect to rotations by $180°$ about the x, y, and z axes.[29] Therefore, a choice of such principal axes reduces the spin Hamiltonian to a diagonal form

$$\hat{H}_s = D_{xx}\hat{S}_x^2 + D_{yy}\hat{S}_y^2 + D_{zz}\hat{S}_z^2 \tag{12.24}$$

---

[8] Even for a lower molecular symmetry, the zfs tensor, being symmetric, may be diagonalized by a proper unitary transformation.

In view of the relations

$$\hat{S}_x{}^2 + \hat{S}_y{}^2 + \hat{S}_z{}^2 = \hat{S}^2$$

$$D_{xx} + D_{yy} + D_{zz} = 0 \tag{12.25}$$

we may reduce Eq. (12.24) to the alternative form[9]

$$\hat{H}_s = D(\hat{S}_z{}^2 - \tfrac{1}{3}\hat{S}^2) + E(\hat{S}_x{}^2 - \hat{S}_y{}^2) \tag{12.26}$$

The quantities D and E of Eq. (12.26) are the zfs parameters defined as

$$D \equiv \tfrac{3}{2}D_{zz} = \langle \Phi(1,2) \mid (3g^2\beta_e{}^2/4)(r_{12}{}^2 - 3z_{12}{}^2)/r_{12}{}^5 \mid \Phi(1,2) \rangle \tag{12.27}$$

$$E \equiv \tfrac{1}{2}(D_{xx} - D_{yy}) = \langle \Phi(1,2) \mid (3g^2\beta_e{}^2/4)(y_{12}{}^2 - x_{12}{}^2)/r_{12}{}^5 \mid \Phi(1,2) \rangle \tag{12.28}$$

(C) zfs

The zfs is given by the solution of Eq. (12.19). Using the spin functions of Eqs. (12.5)–(12.7) and the spin Hamiltonian of Eq. (12.26), we now calculate the matrix elements of Eq. (12.22). Substitution of these matrix elements into Eq. (12.19), yields

$$\begin{vmatrix} \tfrac{1}{3}D - W & 0 & E \\ 0 & -\tfrac{2}{3}D - W & 0 \\ E & E & \tfrac{1}{3}D - W \end{vmatrix} = 0 \tag{12.29}$$

The roots are

$$W_x = \tfrac{1}{3}D - E = -D_{xx}$$

$$W_y = \tfrac{1}{3}D + E = -D_{yy}$$

$$W_z = -\tfrac{2}{3}D = -D_{zz} \tag{12.30}$$

Therefore, the first-order energies are $^3W + W_x$, $^3W + W_y$, and $^3W + W_z$. In other words, the triplet degeneracy is removed by the small perturbation due to the spin-spin interaction. The energy-level splittings are described by the zfs parameters D and E in the manner shown in Fig. 12.1.

(D) Perturbed Wave Functions

Following standard procedure, we obtain those linear combinations of the zero-order wave functions which diagonalize the secular determinant. In other words, we obtain the proper set of coefficients C_{M_S} of Eq. (12.11).

---

[9] See Exercise 2.

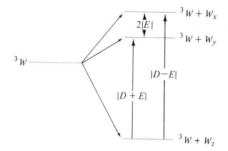

FIG. 12.1. zfs of a triplet state.

The results are

$$^3\Psi_x(1, 2) = (\tfrac{1}{2})^{1/2}[^3\Psi_{-1}(1, 2) - {}^3\Psi_1(1, 2)] \equiv \Phi(1, 2)\Theta_x(1, 2)$$

$$^3\Psi_y(1, 2) = (-\tfrac{1}{2})^{1/2}[^3\Psi_{-1}(1, 2) + {}^3\Psi_1(1, 2)] \equiv \Phi(1, 2)\Theta_y(1, 2)$$

$$^3\Psi_z(1, 2) = {}^3\Psi_0(1, 2) \equiv \Phi(1, 2)\Theta_z(1, 2) \qquad (12.31)$$

The new set of spin functions which has been introduced is

$$\Theta_x(1, 2) \equiv (\tfrac{1}{2})^{1/2}[\Theta_{1-1}(1, 2) - \Theta_{11}(1, 2)]$$

$$\Theta_y(1, 2) \equiv (-\tfrac{1}{2})^{1/2}[\Theta_{1-1}(1, 2) + \Theta_{11}(1, 2)]$$

$$\Theta_z(1, 2) \equiv \Theta_{10}(1, 2) \qquad (12.32)$$

These new functions are not eigenfunctions of $\hat{S}_z$. However, they do have their spins polarized in the yz, zx, and xy planes, respectively. This polarization is seen in Table 12.1, where the mode of action of the total spin operators is detailed. Thus, the secular determinant of Eq. (12.29) is now

TABLE 12.1. *Mode of action of spin operators on Θ_x, Θ_y, and Θ_z.*

	$\Theta_x(1, 2)$	$\Theta_y(1, 2)$	$\Theta_z(1, 2)$
$\hat{S}_x$	0	$i\Theta_z(1, 2)$	$-i\Theta_y(1, 2)$
$\hat{S}_y$	$-i\Theta_z(1, 2)$	0	$i\Theta_x(1, 2)$
S_z	$i\Theta_y(1, 2)$	$-i\Theta_x(1, 2)$	0
$\hat{S}_x^2$	0	$\Theta_y(1, 2)$	$\Theta_z(1, 2)$
$\hat{S}_y^2$	$\Theta_x(1, 2)$	0	$\Theta_z(1, 2)$
$\hat{S}_z^2$	$\Theta_x(1, 2)$	$\Theta_y(1, 2)$	0

diagonal:

$$\begin{vmatrix} \tfrac{1}{3}D - E - W & 0 & 0 \\ \\ 0 & \tfrac{1}{3}D + E - W & 0 \\ \\ 0 & 0 & -\tfrac{2}{3}D - W \end{vmatrix}$$

$$= \begin{vmatrix} -D_{xx} - W & 0 & 0 \\ \\ 0 & -D_{yy} - W & 0 \\ \\ 0 & 0 & -D_{zz} - W \end{vmatrix} = 0 \quad (12.33)$$

Since the latter form is particularly convenient mathematically, the spin Hamiltonian of Eq. (12.24) is frequently used for the interpretation of experimental results.[10]

3. ZERO-FIELD MAGNETIC RESONANCE

The magnetic dipole transition between any pair of the three triplet components is allowed. This conclusion follows from a simple magnetic dipole transition probability computation:

$$| \langle {}^3\Psi_x(1, 2) \mid g\beta_e\mathbf{H}_{rf}\cdot\hat{\mathbf{S}} \mid {}^3\Psi_y(1, 2) \rangle |^2 = (g\beta_e H_{rf}\gamma_z)^2$$
$$| \langle {}^3\Psi_y(1, 2) \mid g\beta_e\mathbf{H}_{rf}\cdot\hat{\mathbf{S}} \mid {}^3\Psi_z(1, 2) \rangle |^2 = (g\beta_e H_{rf}\gamma_x)^2$$
$$| \langle {}^3\Psi_z(1, 2) \mid g\beta_e\mathbf{H}_{rf}\cdot\hat{\mathbf{S}} \mid {}^3\Psi_x(1, 2) \rangle |^2 = (g\beta_e H_{rf}\gamma_y)^2 \quad (12.34)$$

where $\mathbf{H}_{rf}$ is the magnetic field associated with the radio-frequency electromagnetic wave and γ_x, γ_y, and γ_z are the direction cosines of $\mathbf{H}_{rf}$ with respect to the x, y, and z axes, respectively. The corresponding transition energies are

$$h\nu_z = | W_x - W_y | = | 2E | = | D_{xx} - D_{yy} |$$
$$h\nu_x = | W_y - W_z | = | D + E | = | D_{yy} - D_{zz} |$$
$$h\nu_y = | W_z - W_x | = | D - E | = | D_{zz} - D_{xx} | \quad (12.35)$$

Thus, the magnitudes of the zfs parameters are determinable at zero applied magnetic field strength.[11] This type of experiment is often referred to as the *zero-field magnetic resonance* method—it is a difficult experimental technique.

---

[10] A further replacement of D_{xx}, D_{yy}, and D_{zz} by $-X$, $-Y$, and $-Z$, respectively, is usually adopted, namely,

$$\hat{H}_s \equiv - X\hat{S}_x{}^2 - Y\hat{S}_y{}^2 - Z\hat{S}_z{}^2$$

[11] Of course, the zfs parameters are also determinable using standard electron-paramagnetic-resonance techniques.

4. zfs PARAMETERS

The physical meaning of the parameters D and E (or D_{xx}, D_{yy}, and D_{zz}) has been investigated in Sec. 2. The two sets of parameters which we have used are correlated by Eq. (12.30). We now proceed to the calculation of D and E as specified in Eqs. (12.27) and (12.28). For simplicity, we introduce the following notations:

$$\hat{D} \equiv (3g^2\beta_e{}^2/4)\,(r_{12}{}^2 - 3z_{12}{}^2)\,r_{12}{}^{-5} \tag{12.36}$$

$$\hat{E} \equiv (3g^2\beta_e{}^2/4)\,(y_{12}{}^2 - x_{12}{}^2)\,r_{12}{}^{-5} \tag{12.37}$$

Since the spatial wave function $\Phi(1, 2)$ is given by Eq. (12.9), the expressions for D and E become

$$
\begin{aligned}
D &= \langle |\,\varphi_r(1)\varphi_s(2) \,||\, \hat{D} \,||\, \varphi_r(1)\varphi_s(2)\,| \rangle \\
&= \langle \varphi_r(1)\varphi_s(2) - \varphi_s(1)\varphi_r(2) \,|\, \hat{D} \,|\, \varphi_r(1)\varphi_s(2) \rangle
\end{aligned}
\tag{12.38}
$$

$$
\begin{aligned}
E &= \langle |\,\varphi_r(1)\varphi_s(2) \,||\, \hat{E} \,||\, \varphi_r(1)\varphi_s(2)\,| \rangle \\
&= \langle \varphi_r(1)\varphi_s(2) - \varphi_s(1)\varphi_r(2) \,|\, \hat{E} \,|\, \varphi_r(1)\varphi_s(2) \rangle
\end{aligned}
\tag{12.39}
$$

The evaluation of the integrals of Eqs. (12.38) and (12.39) requires wave functions for φ_r and φ_s; we adopt the LCAO-MO formalism.

(A) Orbital Wave Functions

The wave functions are

$$\varphi_r(1) = \sum_{\mu} c_{r\mu}\chi_{\mu}(1) \tag{12.40}$$

$$\varphi_s(2) = \sum_{\nu} c_{s\nu}\chi_{\nu}(2) \tag{12.41}$$

where χ_{μ} and χ_{ν} denote atomic orbitals on atoms μ and ν, respectively. Substitution of Eqs. (12.40) and (12.41) into Eqs. (12.38) and (12.39) yields

$$D = \sum_{\mu}\sum_{\nu}\sum_{\rho}\sum_{\sigma} (c_{r\mu}c_{s\nu} - c_{s\mu}c_{r\nu})\,c_{r\rho}c_{s\sigma} \times I(\mu\nu\rho\sigma)_D \tag{12.42}$$

$$I(\mu\nu\rho\sigma)_D = \langle \chi_{\mu}(1)\chi_{\nu}(2) \,|\, \hat{D} \,|\, \chi_{\rho}(1)\chi_{\sigma}(2) \rangle \tag{12.43}$$

$$E = \sum_{\mu}\sum_{\nu}\sum_{\rho}\sum_{\sigma} (c_{r\mu}c_{s\nu} - c_{s\mu}c_{r\nu})\,c_{r\rho}c_{s\sigma} \times I(\mu\nu\rho\sigma)_E \tag{12.44}$$

$$I(\mu\nu\rho\sigma)_E = \langle \chi_{\mu}(1)\chi_{\nu}(2) \,|\, \hat{E} \,|\, \chi_{\rho}(1)\chi_{\sigma}(2) \rangle \tag{12.45}$$

Therefore, our problem has been reduced to the evaluation of the atomic integrals: $I(\mu\nu\rho\sigma)_{D,E}$. These integrals may be of one-, two-, three-, and four-center type.

In order to proceed further, we must assume some functional form for the atomic orbitals χ_μ. Indeed, we may classify computational methods into two groupings on the basis of the form chosen for the AO's. One grouping utilizes Slater-type functions,[4]–[18] either singly or in a multi-ζ approximation to the SCF AO's; the other grouping utilizes Gaussian-type functions.[19]–[24] The use of Gaussian-type functions markedly simplifies the calculation of atomic integrals.

(B)   Slater-Type Functions

If we adopt a Slater basis set, some special two-center integrals,[12] as well as all types of one-center integrals,[13] are obtainable in closed analytic forms by a method developed by Geller.[9],[10],[12],[13]

Approximations must be used for two-center exchange-type integrals[14] (that is, $\mu = \sigma$ and $\nu = \rho$) and multicenter integrals. Approximations usually adopted are

(i)   Overlap integrals of nonadjacent atoms are neglected.

(ii)   The point-charge approximation is used for integrals where the centers are widely separated (that is, R is large).

(iii)   The Mulliken approximation is adopted for multicenter integrals:

$$I(\mu\nu\rho\sigma) \cong (S^2/4)\left[I(\mu\nu\mu\nu) + I(\mu\sigma\mu\sigma) + I(\rho\nu\rho\nu) + I(\rho\sigma\rho\sigma)\right] \quad (12.46)$$

where the pairs of AO's μ and ρ, and ν and σ are on adjacent centers, where the origins of μ and ν may not be coincident, and where the overlap S is

$$S \equiv \langle \chi_\mu(1) \mid \chi_\rho(1) \rangle = \langle \chi_\nu(2) \mid \chi_\sigma(2) \rangle \quad (12.47)$$

(iv)   The operators are expanded in spherical harmonics.

(C)   Gaussian-Type Functions[19],[20]

The atomic orbitals $\chi_\mu(\mathbf{r})$, and so forth, in the integrals of Eqs. (12.43) and (12.45) may be expanded in terms of a finite set of Gaussian functions.

---

[12] $\langle \chi_\mu(1)\chi_\nu(2) \mid D$ or $\hat{E} \mid \chi_\mu{}'(1)\chi_\nu{}'(2) \rangle$, where χ_μ and $\chi_\mu{}'$ refer to two different Slater AO's on the same center.

[13] $\langle \chi_\mu(1)\chi_\mu{}'(2) \mid \hat{D}$ or $\hat{E} \mid \chi_\mu(1)\chi_\mu{}'(2) \rangle$ and $\langle \chi_\mu(1)\chi_\mu{}'(2) \mid \hat{D}$ or $\hat{E} \mid \chi_\mu{}'(1)\chi_\mu(2) \rangle$ where χ_μ and $\chi_\mu{}'$ refer to two different AO's on the same center.

[14] $\langle \chi_\mu(1)\chi_\nu(2) \mid \hat{D}$ or $\hat{E} \mid \chi_\nu(1)\chi_\mu(2) \rangle$.

In the present case[15] of $2p_z$ AO's, this expansion is

$$\chi_\mu(\mathbf{r}_1) = \sum_{i=1}^{N} c_{\mu i} G_z(\alpha_{\mu i}, \mathbf{r}_{\mu 1}) \tag{12.48}$$

where N is the number of Gaussian functions with different orbital exponents $\alpha_{\mu i}$, where the coefficients $c_{\mu i}$ are chosen to maximize[16] the overlap of the approximate function of Eq. (12.48) with the exact AO function $\chi_\mu(\mathbf{r}_1)_{\text{exact}}$, and $G_z(\alpha_{\mu i}, \mathbf{r}_{\mu 1})$ is the Gaussian $2p_z$ AO defined as

$$G_z(\alpha_{\mu i}, \mathbf{r}_{\mu 1}) \equiv (128\alpha_{\mu i}{}^5/\pi^3)^{1/4} z_{\mu 1} \exp(-\alpha_{\mu i} r_{\mu 1}{}^2) \tag{12.49}$$

The integrals $I(\mu\nu\rho\sigma)_{D,E}$ are now rewritten in terms of Gaussian functions

$$I(\mu\nu\rho\sigma)_{D,E} = \sum_i \sum_j \sum_k \sum_l c_{\mu i} c_{\nu j} c_{\rho k} c_{\sigma l} \times I'(\mu\nu\rho\sigma)_{D,E} \tag{12.50}$$

$$I'(\mu\nu\rho\sigma)_D = \langle G_z(\alpha_{\mu i}, \mathbf{r}_{\mu 1}) G_z(\alpha_{\nu j}, \mathbf{r}_{\nu 2}) \,|\, \hat{D} \,|\, G_z(\alpha_{\rho k}, \mathbf{r}_{\rho 1}) G_z(\alpha_{\sigma l}, \mathbf{r}_{\sigma 2}) \rangle \tag{12.51}$$

with a similar equation for $I'(\mu\nu\rho\sigma)_E$. Boorstein and Gouterman[20,21] have derived the following equations[17]:

$$I'(\mu\nu\rho\sigma)_{D,E}$$

$$= \langle G_z(\alpha_\mu, \mathbf{r}_{\mu 1}) G_z(\alpha_\nu, \mathbf{r}_{\nu 2}) \,\left|\, \begin{matrix} \hat{D} \\ \hat{E} \end{matrix} \,\right|\, G_z(\alpha_\rho, \mathbf{r}_{\rho 1}) G_z(\alpha_\sigma, \mathbf{r}_{\sigma 2}) \rangle$$

$$= (-3g^2\beta_e{}^2/4hca_0{}^3R_{\eta\xi}{}^3)[\alpha_\mu\alpha_\nu\alpha_\rho\alpha_\sigma/\{[(\alpha_\mu + \alpha_\rho)/2]^2[(\alpha_\nu + \alpha_\sigma)/2]^2\}]^{5/4}$$

$$\times \exp\left[-\{\alpha_\mu\alpha_\rho/(\alpha_\mu + \alpha_\rho)\}r_{\mu\rho}{}^2 - \{\alpha_\nu\alpha_\sigma/(\alpha_\nu + \alpha_\sigma)\}r_{\nu\sigma}{}^2\right]$$

$$\times \begin{pmatrix} 1 \\ \cos 2\phi_{\eta\xi} \end{pmatrix} \times \left\{ P(r_0) \begin{bmatrix} -1 + 9/r_0{}^2 - 225\omega/4r_0{}^4 \\ -1 + 5/r_0{}^2 - 105\omega/4r_0{}^4 \end{bmatrix} \right.$$

$$\left. + Q(r_0) \begin{bmatrix} 225\omega/2r_0{}^3 + (75\omega/2 - 18)/r_0 + (15\omega/2 - 4)r_0 + \omega r_0{}^3/2 \\ 105\omega/2r_0{}^3 + (35\omega/2 - 10)/r_0 + (7\omega/2 - 4/3)r_0 + \omega r_0{}^3/2 \end{bmatrix} \right\} \tag{12.52}$$

[15] The zfs is being investigated for the specific case of a triplet state of $\pi^* \leftarrow \pi$ molecular-orbital excitation nature.

[16] In other words, $\chi_\mu(\mathbf{r}_1)$ is the best possible shape mimic of $\chi_\mu(\mathbf{r}_1)_{\text{exact}}$.

[17] In Eqs. (12.52)–(12.54), the subscripts i, j, k, and l are dropped for simplicity.

where

$$P(r_0) = (1/2\pi)^{1/2} \int_{-r_0}^{r_0} \exp\left(-t^2/2\right) dt$$

$$Q(r_0) = (1/2\pi)^{1/2} \exp\left(-r_0^2/2\right)$$

$$r_0^2 = 2(\alpha_\mu + \alpha_\rho)(\alpha_\nu + \alpha_\sigma) R_{\eta\xi}^2 / (\alpha_\mu + \alpha_\nu + \alpha_\rho + \alpha_\sigma)$$

$$R_{\eta\xi} = |\mathbf{r}_\eta - \mathbf{r}_\xi| = |\mathbf{R}_{\eta\xi}|$$

$$\mathbf{R}_{\eta\xi} = (X_{\eta\xi}, Y_{\eta\xi}, 0)$$

$$\mathbf{r}_\eta = (\alpha_\mu \mathbf{r}_\mu + \alpha_\rho \mathbf{r}_\rho)/(\alpha_\mu + \alpha_\rho)$$

$$\mathbf{r}_\xi = (\alpha_\nu \mathbf{r}_\nu + \alpha_\sigma \mathbf{r}_\sigma)/(\alpha_\nu + \alpha_\sigma)$$

$$\omega = 4(\alpha_\mu + \alpha_\rho)(\alpha_\nu + \alpha_\sigma)/(\alpha_\mu + \alpha_\nu + \alpha_\rho + \alpha_\sigma)^2$$

$$\cos 2\phi_{\eta\xi} = (Y_{\eta\xi}^2 - X_{\eta\xi}^2)/(Y_{\eta\xi}^2 + X_{\eta\xi}^2) \tag{12.53}$$

We are now able to evaluate the integrals $I'(\mu\nu\rho\sigma)_{D,E}$ and, consequently, the zfs parameters in all instances except for $R_{\eta\xi} = 0$. For this exception, the integrals $I'(\mu\nu\rho\sigma)_{D,E}$ reduce to one-center two-electron integrals, and yield

$$I'(\mu\nu\rho\sigma)_D = 4(3g^2\beta_e^2/hca_0^3)\{\alpha_\mu\alpha_\nu\alpha_\rho\alpha_\sigma/[(\alpha_\mu + \alpha_\nu + \alpha_\rho + \alpha_\sigma)/4]^2\}^{5/4}(\pi)^{-1/2}$$

$$\times \exp\{-[\alpha_\mu\alpha_\rho/(\alpha_\mu + \alpha_\rho)]r_{\mu\rho}^2 - [\alpha_\nu\alpha_\sigma/(\alpha_\nu + \alpha_\sigma)]r_{\nu\sigma}^2\}$$

$$\times\left[-\frac{1}{15}\left(\frac{1}{\alpha_\mu + \alpha_\rho} + \frac{1}{\alpha_\nu + \alpha_\sigma}\right) + \frac{2}{7}\frac{1}{\alpha_\mu + \alpha_\nu + \alpha_\rho + \alpha_\sigma}\right]$$

$$\tag{12.54}$$

$$I'(\mu\nu\rho\sigma)_E = 0 \tag{12.55}$$

Derivation of these equations for Gaussian AO's has been thoroughly reviewed.[30],[31]

5. ETHYLENE: AN EXAMPLE CALCULATION

We discuss the zfs parameters D and E for the lowest triplet state of ethylene. We assume that ethylene is planar in this state and that we can neglect all AO's on the four hydrogen atoms as well as all σ AO's on the two carbons. We perform the calculation in a Gaussian basis; for reasons of simplicity, we assume that the $2p_z$ AO on each carbon center can be represented by a single Gaussian function. This last approximation is quite drastic; nonetheless, it serves to illustrate the method and the reader should experience no great difficulty in extending the procedure to the case in which the AO representation is given by a linear combination of Gaussian functions. Thus, in the single Gaussian basis, the AO on carbon 1 is given by

$$\chi_\mu(\mathbf{r}_1) \equiv G_z(\alpha, \mathbf{r}_{\mu 1}) \equiv (128\alpha^5/\pi^3)^{1/4} z_{\mu 1} \exp\left[-\alpha r_{\mu 1}^2\right] \qquad (12.56)$$

The lowest-energy triplet state of ethylene is represented exactly by a single-configuration wave function

$$\Phi(1, 2) = (\tfrac{1}{2})^{1/2}[\varphi_1(1)\varphi_2(2) - \varphi_2(1)\varphi_1(2)] \equiv |\varphi_1(1)\varphi_2(2)| \qquad (12.57)$$

where the MO's φ_1 and φ_2 are the simple Hückel type

$$\varphi_1(i) = c_{11}\chi_1(i) + c_{12}\chi_2(i) \qquad (i = 1, 2)$$

$$\varphi_2(j) = c_{21}\chi_1(j) + c_{22}\chi_2(j) \qquad (j = 1, 2)$$

$$c_{11} = c_{12} = c_{21} = -c_{22} = (\tfrac{1}{2})^{1/2} \qquad (12.58)$$

Using Eqs. (12.38), (12.39), (12.42), and (12.44), the elements D and E are

$$\binom{\hat{D}}{\hat{E}} = \left\langle \varphi_1(1)\varphi_2(2) \,\middle|\, \begin{matrix} \hat{D} \\ \hat{E} \end{matrix} \,\middle|\, \varphi_1(1)\varphi_2(2) \right\rangle$$

$$= \sum_{\mu=1}^{2}\sum_{\nu=1}^{2}\sum_{\rho=1}^{2}\sum_{\sigma=1}^{2} (c_{1\mu}c_{2\nu} - c_{2\mu}c_{1\nu})c_{1\rho}c_{2\sigma}\left\langle \chi_\mu(1)\chi_\nu(2) \,\middle|\, \begin{matrix} \hat{D} \\ \hat{E} \end{matrix} \,\middle|\, \chi_\rho(1)\chi_\sigma(2) \right\rangle$$

$$= (c_{11}c_{22} - c_{21}c_{12})^2 \left\langle \chi_1(1)\chi_2(2) - \chi_2(1)\chi_1(2) \,\middle|\, \begin{matrix} \hat{D} \\ \hat{E} \end{matrix} \,\middle|\, \chi_1(1)\chi_2(2) \right\rangle$$

$$= \left\langle \chi_1(1)\chi_2(2) \,\middle|\, \begin{matrix} \hat{D} \\ \hat{E} \end{matrix} \,\middle|\, \chi_1(1)\chi_2(2) \right\rangle - \left\langle \chi_2(1)\chi_1(2) \,\middle|\, \begin{matrix} \hat{D} \\ \hat{E} \end{matrix} \,\middle|\, \chi_1(1)\chi_2(2) \right\rangle$$

$$= I'(1212) - I'(2112) \qquad (12.59)$$

The integral $I'(1212)$ can be evaluated using Eq. (12.52) with $R_{\eta\xi} = R$, $r_0^2 = 2\alpha R^2$, $\omega = 1$, and $\cos 2\phi_{\eta\xi} = 1$ (that is, the y axis is taken parallel to the C=C bond); R is the C=C internuclear distance. The second term of Eq. (12.59) can be reduced to a one-center integral, and can be evaluated from Eq. (12.54). These integrals, as a whole, are written

$$\binom{\hat{D}}{\hat{E}} = \left(\frac{3g^2\beta_e^2}{4hca_0^3}\right) \times \left(\frac{1}{R^3}\right) \times \left\{ P(r_0) \begin{bmatrix} 225/4r_0^4 - 9/r_0^2 + 1 \\ 105/4r_0^4 - 5/r_0^2 + 1 \end{bmatrix} \right.$$

$$\left. - Q(r_0) \begin{bmatrix} 225/2r_0^3 + 39/2r_0 + (7/2)r_0 + (113/210)r_0^3 \\ 105/2r_0^3 + 15/2r_0 + (13/6)r_0 + r_0^3/2 \end{bmatrix} \right\} \qquad (12.60)$$

Taking $R = 2.64$ a.u. and $\alpha = 0.403$ a.u.$^{-2}$, we obtain

$$D = 0.185 \text{ cm}^{-1}; \qquad E = 0.225 \text{ cm}^{-1} \qquad (12.61)$$

EXERCISES

1. Verify the relation of Eq. (12.17) and derive Eq. (12.18) from Eq. (12.15).

Hint: $\hat{S}_x^2 = (\hat{s}_{1x} + \hat{s}_{2x})^2 = 2\hat{s}_{1x}\hat{s}_{2x} + \hat{s}_{1x}^2 + \hat{s}_{2x}^2 = 2\hat{s}_{1x}\hat{s}_{2x} + \frac{1}{2}$ \qquad (12.62)

because

$$\hat{s}_{1x}^2 \begin{pmatrix} \alpha \\ \beta \end{pmatrix} = \hat{s}_{2x}^2 \begin{pmatrix} \alpha \\ \beta \end{pmatrix} = \frac{1}{4} \begin{pmatrix} \alpha \\ \beta \end{pmatrix} \qquad (12.63a)$$

$$\begin{aligned}
\hat{S}_x \hat{S}_y &= (\hat{s}_{1x} + \hat{s}_{2x})(\hat{s}_{1y} + \hat{s}_{2y}) \\
&= (\hat{s}_{1x}\hat{s}_{2y} + \hat{s}_{2x}\hat{s}_{1y}) + (\hat{s}_{1x}\hat{s}_{1y} + \hat{s}_{2x}\hat{s}_{2y}) \\
&= (\hat{s}_{1x}\hat{s}_{2y} + \hat{s}_{2x}\hat{s}_{1y}) + i(\hat{s}_{1z} + \hat{s}_{2z})/2 \qquad (12.64)
\end{aligned}$$

$$\hat{S}_y \hat{S}_x = (\hat{s}_{1x}\hat{s}_{2y} + \hat{s}_{2x}\hat{s}_{1y}) - i(\hat{s}_{1z} + \hat{s}_{2z})/2 \qquad (12.65)$$

because

$$\hat{s}_{1x}\hat{s}_{1y} \begin{pmatrix} \alpha \\ \beta \end{pmatrix} = -\hat{s}_{1y}\hat{s}_{1x} \begin{pmatrix} \alpha \\ \beta \end{pmatrix} = (i\hat{s}_{1z}/2) \begin{pmatrix} \alpha \\ \beta \end{pmatrix} \qquad (12.63b)$$

2. Derive Eq. (12.26) from Eq. (12.24).

Answer: $D_{xx}\hat{S}_x^2 + D_{yy}\hat{S}_y^2 + D_{zz}\hat{S}_z^2 = \frac{1}{2}(D_{xx} - D_{yy})(\hat{S}_x^2 - \hat{S}_y^2)$

$$\begin{aligned}
&\qquad + \tfrac{1}{2}(D_{xx} + D_{yy})(\hat{S}_x^2 + \hat{S}_y^2) + D_{zz}\hat{S}_z^2 \\
&= \tfrac{1}{2}(D_{xx} - D_{yy})(\hat{S}_x^2 - \hat{S}_y^2) \\
&\qquad + \tfrac{1}{2}(-D_{zz})(\hat{S}^2 - \hat{S}_z^2) + D_{zz}\hat{S}_z^2 \\
&= \tfrac{1}{2}(D_{xx} - D_{yy})(\hat{S}_x^2 - \hat{S}_y^2) \\
&\qquad + \tfrac{3}{2}D_{zz}(\hat{S}_z^2 - \hat{S}^2/3) \qquad (12.66)
\end{aligned}$$

3. What is the appearance of the EPR spectrum of the triplet state of naphthalene when an external magnetic field is applied along the long axis (x axis) of the naphthalene molecule ($D = +0.1003$ cm^{-1}, $E = -0.0134$ cm^{-1}, and $\nu = 9400$ MHz)?

4. Show that $\hat{S}_u | \Theta_u \rangle = 0 (u = x, y, z)$ in Table 12.1 when $\mathbf{H} = 0$. How would the eigenfunctions Θ_u be modified when a small magnetic field is applied along the z axis?

Answer: The perturbing spin Hamiltonian is now

$$\hat{H}' = g\beta_e H_z \hat{S}_z + D_x \hat{S}_x^2 + D_y \hat{S}_y^2 + D_z \hat{S}_z^2 \tag{12.67}$$

and the secular determinant becomes

$$\begin{vmatrix} -D_x & -ig\beta_e H_z & 0 \\ ig\beta_e H_z & -D_y & 0 \\ 0 & 0 & -D_z \end{vmatrix} = \begin{vmatrix} \frac{1}{3}D - E & -ig\beta_e H_z & 0 \\ ig\beta_e H_z & \frac{1}{3}D + E & 0 \\ 0 & 0 & -\frac{2}{3}D \end{vmatrix} \tag{12.68}$$

Therefore, Θ_z is still an eigenfunction, but Θ_x and Θ_y are now mixed with each other. The eigenfunctions are as follows:

$$(\Theta_x - i\lambda\Theta_y)/(1 + \lambda^2)^{1/2}; \quad (\Theta_y - i\lambda\Theta_x)/(1 + \lambda^2)^{1/2}; \quad \Theta_z \tag{12.69}$$

where

$$\lambda \equiv [(\delta - E)/(\delta + E)]^{1/2}; \qquad \delta^2 \equiv E^2 + (g\beta_e H_z)^2 \tag{12.70}$$

5. The two highest-energy occupied and the two lowest-energy unoccupied molecular orbitals of naphthalene are tabulated below.

φ_r	1	2	3	4	5	c_{ri} 6	7	8	9	10
$\varphi_{4'}$	0	$-c$	c	0	$-c$	0	c	$-c$	0	c
$\varphi_{5'}$	a	$-b$	$-b$	a	0	$-a$	b	b	$-a$	0
φ_5	a	b	$-b$	$-a$	0	a	b	$-b$	$-a$	0
φ_4	0	c	c	0	$-c$	0	c	c	0	$-c$

$$a \equiv 0.42533, \qquad b \equiv 0.26287, \qquad c \equiv 0.40825 \tag{12.71}$$

Out of these, two L_a triplet-state configurations are obtained as follows:

$$\Phi_1(1, 2) = |\varphi_5(1)\varphi_{5'}(2)|; \qquad \Phi_2(1, 2) = |\varphi_4(1)\varphi_{4'}(2)| \tag{12.72}$$

The approximate spinless state function can now be constructed as

$$\Phi(1, 2) = (\sin\phi)\Phi_1(1, 2) + (\cos\phi)\Phi_2(1, 2) \tag{12.73}$$

Now, using a single Gaussian function representation [i.e., Eq. (12.56)] for each carbon $2p_z$ AO, calculate the zfs parameters

D and E for this state. (The parametric angle ϕ is to be determined by comparison with the experimental data.) For simplicity, take account of only two-center nearest-neighbor terms (C—C bondlength $= 1.40$ Å).

Answer:

$$D = \langle\, |\, \varphi_5\varphi_{5'}\, ||\, \hat{D}\, ||\, \varphi_5\varphi_{5'}\, |\rangle \sin^2\phi + \langle\, |\, \varphi_4\varphi_{4'}\, ||\, \hat{D}\, ||\, \varphi_4\varphi_{4'}\, |\rangle \cos^2\phi$$

$$+ 2\langle\, |\, \varphi_5\varphi_{4'}\, ||\, \hat{D}\, ||\, \varphi_4\varphi_{5'}\, |\rangle \sin\phi \cos\phi$$

$$= 0.0440 \sin^2\phi + 0.0616 \cos^2\phi$$

$$+ 2[0.0446 \sin\phi \cos\phi]\ (\text{cm}^{-1}) \tag{12.74}$$

$$E = -0.0139 \sin^2\phi + 0.0749 \cos^2\phi$$

$$- 2[0.0271 \sin\phi \cos\phi]\ (\text{cm}^{-1}) \tag{12.75}$$

BIBLIOGRAPHY

[1]   C. A. Hutchison, Jr., and B. W. Mangum, *J. Chem. Phys.* **29,** 952 (1958); **34,** 908 (1961).

[2]   J. H. van der Waals and M. S. deGroot, *Mol. Phys.* **2,** 333 (1959); **3,** 190 (1960).

[3]   W. A. Yager, E. Wasserman, and R. M. R. Cramer, *J. Chem. Phys.* **37,** 1148 (1962).

[4]   H. F. Hameka, *J. Chem. Phys.* **31,** 315 (1959).

[5]   R. M. Pitzer and H. F. Hameka, *J. Chem. Phys.* **37,** 2725 (1963).

[6]   J. Higuchi, *J. Chem. Phys.* **38,** 1237 (1963); **39,** 1339 (1963); **39,** 1847 (1963).

[7]   J. H. van der Waals and G. ter Maten, *Mol. Phys.* **8,** 301 (1964).

[8]   R. D. Sharma, *J. Chem. Phys.* **38,** 2350 (1963); **41,** 3259 (1964).

[9]   M. Geller, *J. Chem. Phys.* **39,** 853 (1963).

[10] M. Geller and R. W. Griffith, *J. Chem. Phys.* **40,** 2309 (1964).

[11] H. Sternlicht, *J. Chem. Phys.* **38,** 2316 (1963).

[12] J. B. Lounsbury, *J. Chem. Phys.* **42,** 1549 (1965); **46,** 2193 (1967); **47,** 1566 (1967).

[13] J. B. Lounsbury and G. W. Barry, *J. Chem. Phys.* **44,** 4367 (1966).

[14] R. McWeeny, *J. Chem. Phys.* **34,** 399 (1961); **34,** 1065 (1961).

[15] R. McWeeny and Y. Mizuno, *Proc. Roy. Soc. (London)* **259A,** 554 (1961).

[16] A. D. McLachlan, *Mol. Phys.* **5,** 51 (1962); **6,** 441 (1963).

[17] G. A. Peterson and A. D. McLachlan, *J. Chem. Phys.* **45,** 628 (1966).

[18] J. A. R. Coope, J. B. Farmer, C. L. Gardner, and C. A. McDowell, *J. Chem. Phys.* **42,** 54 (1965).

[19] M. Gouterman and W. Moffitt, *J. Chem. Phys.* **30,** 1107 (1959).

[20] M. Gouterman, *J. Chem. Phys.* **30,** 1369 (1959).

[21] S. A. Boorstein and M. Gouterman, *J. Chem. Phys.* **39,** 2443 (1963); **41,** 2776 (1964); **42,** 3070 (1965).

[22] Y.-N. Chiu, *J. Chem. Phys.* **39,** 2736 (1963); **39,** 2749 (1963).

[23] C. Thomson, *Mol. Phys.* **10,** 309 (1966); **11,** 197 (1966).

[24] M. Godfrey, C. W. Kern, and M. Karplus, *J. Chem. Phys.* **44,** 4459 (1966).

[25] S. J. Fogel and H. F. Hameka, *J. Chem. Phys.* **42,** 132 (1965).

[26] J. W. McIver, Jr., and H. F. Hameka, *J. Chem. Phys.* **45,** 767 (1966); **46,** 825 (1967).

[27] H. Lefebvre-Brion and C. M. Moser, *J. Chem. Phys.* **46,** 819 (1967).

[28] H. M. McConnell, *Proc. Nat. Acad. Sci. U.S.* **45,** 172 (1959).

[29] K. W. H. Stevens, *Proc. Roy. Soc.* (*London*) **214A,** 237 (1952).

[30] I. Shavit and M. Karplus, *J. Chem. Phys.* **36,** 55 (1962); **43,** 398 (1965).

[31] C. W. Kern and M. Karplus, *J. Chem. Phys.* **43,** 415 (1965).

General References

1. A. Carrington and A. D. McLachlan, *Introduction to Magnetic Resonance* (Harper and Row, Publishers, Inc., New York, 1967).

2. *The Triplet State*, edited by A. Zahlan (Cambridge University Press, Cambridge, England, 1967).

3. S. P. McGlynn, T. Azumi, and M. Kinoshita, *Molecular Spectroscopy of the Triplet State* (Prentice-Hall, Inc., Englewood Cliffs, N.J., 1969).

APPENDIX A

Atomic Orbitals for Hydrogenlike Atoms*

1. FUNCTIONS $\Phi_m(\phi)$

$$(m = 0, \pm 1, \pm 2, \ldots)$$

$\Phi_m(\phi) = (1/2\pi)^{1/2}e^{im\phi}$ General Expression

$\Phi_0 = (1/2\pi)^{1/2}$

$\Phi_1 = (1/2\pi)^{1/2}e^{i\phi}$

$\Phi_{-1} = (1/2\pi)^{1/2}e^{-i\phi}$

etc.

2. FUNCTIONS $\Theta_{lm}(\theta)$

$(l \geq |m|;\ l$ is a non-negative integer.$)$

$$\Theta_{lm}(\theta) = \left(\frac{(2l+1)(l-|m|)!}{2(l+|m|)!}\right)^{1/2} P_l^{|m|}(\cos\theta) \qquad \text{General Expression}$$

* L. Pauling and E. B. Wilson, *Introduction to Quantum Mechanics* (McGraw-Hill Book Co., Inc., New York, 1935), Chap. 5.

where

$$P_l^{|m|}(\cos \theta) = (1 - \cos^2 \theta)^{|m|/2} \frac{d^{|m|}}{dz^{|m|}} P_l(\cos \theta)$$

where

$$P_l(\cos \theta) = \frac{1}{2^l l!} \frac{d^l (\cos^2 \theta - 1)^l}{dz^l}; \quad \text{for } l = 1, 2, 3, \ldots$$

and

$$P_0(\cos \theta) = 1; \quad \text{for } l = 0 \text{ only}$$

$$\Theta_{00} = (\tfrac{1}{2})^{1/2}$$

$$\Theta_{10} = (\tfrac{3}{2})^{1/2} \cos \theta$$

$$\Theta_{1\pm 1} = \mp (\tfrac{3}{4})^{1/2} \sin \theta$$

$$\Theta_{20} = (5/8)^{1/2} (3 \cos^2 \theta - 1)$$

$$\Theta_{2\pm 1} = \mp (15/4)^{1/2} \sin \theta \cos \theta$$

$$\Theta_{2\pm 2} = (15/16) \sin^2 \theta$$

$$\Theta_{30} = (7/8)^{1/2} (5 \cos^3 \theta - 3 \cos \theta)$$

$$\Theta_{3\pm 1} = \mp (21/32)^{1/2} \sin \theta (5 \cos^2 \theta - 1)$$

$$\Theta_{3\pm 2} = (105/16)^{1/2} \sin^2 \theta \cos \theta$$

$$\Theta_{3\pm 3} = \mp (35/32)^{1/2} \sin^3 \theta$$

3. FUNCTIONS Y_{lm}

(The Spherical Harmonics in Real and Imaginary Form.)

l	$\lvert m\rvert$	Type	Real Form Function	l	m	Type	Imaginary Form Function
0	0	s	$(1/4\pi)^{1/2}$	0	0	s	$(1/4\pi)^{1/2}$
1	0	p_z	$(3/4\pi)^{1/2}\cos\theta$	1	0	$(p_z)p_0$	$(3/4\pi)^{1/2}\cos\theta$
1	1	p_x	$(3/4\pi)^{1/2}\sin\theta\cos\phi$	1	1	p_{+1}	$(3/8\pi)^{1/2}\sin\theta\, e^{i\phi}$
1	1	p_y	$(3/4\pi)^{1/2}\sin\theta\sin\phi$	1	-1	p_{-1}	$-(3/8\pi)^{1/2}\sin\theta\, e^{-i\phi}$
2	0	d_{z^2}	$(5/16\pi)^{1/2}(3\cos^2\theta-1)$	2	0	$(d_{z^2})d_0$	$(5/16\pi)^{1/2}(3\cos^2\theta-1)$
2	1	d_{xz}	$(15/4\pi)^{1/2}\cos\theta\sin\theta\cos\phi$	2	1	d_{+1}	$-(15/8\pi)^{1/2}\cos\theta\sin\theta\, e^{i\phi}$
2	1	d_{yz}	$(15/4\pi)^{1/2}\cos\theta\sin\theta\sin\phi$	2	-1	d_{-1}	$(15/8\pi)^{1/2}\cos\theta\sin\theta\, e^{-i\phi}$
2	2	d_{xy}	$(15/8\pi)^{1/2}\sin^2\theta(2\sin\phi\cos\phi)$	2	$+2$	d_{+2}	$(15/32\pi)^{1/2}\sin^2\theta\, e^{2i\phi}$
2	2	$d_{x^2-y^2}$	$(15/8\pi)^{1/2}\sin^2\theta(\cos^2\phi-\sin^2\phi)$	2	-2	d_{-2}	$(15/32\pi)^{1/2}\sin^2\theta\, e^{-2i\phi}$
3	0	f_{z^3}	$(7/16\pi)^{1/2}(5\cos^3\theta-3\cos\theta)$	3	0	$(f_{z^3})f_0$	$(7/16\pi)^{1/2}(5\cos^3\theta-3\cos\theta)$
3	1	f_{xz^2}	$(21/16\pi)^{1/2}\sin\theta(5\cos^2\theta-1)\cos\phi$	3	1	f_{+1}	$-(21/64\pi)^{1/2}\sin\theta(5\cos^2\theta-1)e^{i\phi}$
3	1	f_{yz^2}	$(21/16\pi)^{1/2}\sin\theta(5\cos^2\theta-1)\sin\phi$	3	-1	f_{-1}	$(21/64\pi)^{1/2}\sin\theta(5\cos^2\theta-1)e^{-i\phi}$
3	2	$f_{z(x^2-y^2)}$	$(105/8\pi)^{1/2}\sin^2\theta\cos\theta(\cos^2\theta-\sin^2\theta)$	3	2	f_{+2}	$(105/32\pi)^{1/2}\sin^2\theta\cos\theta\, e^{2i\phi}$
3	2	f_{xyz}	$(105/2\pi)^{1/2}\sin^2\theta\cos\theta\sin\phi\cos\phi$	3	-2	f_{-2}	$(105/32\pi)^{1/2}\sin^2\theta\cos\theta\, e^{-2i\phi}$
3	3	$f_{x(x^2-y^2)}$	$(35/16\pi)^{1/2}\sin^3\theta(\cos^3\phi-3\sin^2\phi\cos\phi)$	3	3	f_{+3}	$-(35/64\pi)^{1/2}\sin^3\theta\, e^{3i\phi}$
3	3	$f_{y(x^2-y^2)}$	$(35/16\pi)^{1/2}\sin^3\theta(3\sin\phi\cos^2\phi-\sin^3\phi)$	3	-3	f_{-3}	$(35/64\pi)^{1/2}\sin^3\theta\, e^{-3i\phi}$

4. FUNCTIONS $R_{nl}'(r) = rR_{nl}$

$$(n = 1, 2, 3, \ldots; l = n - 1, n - 2, \ldots 0.)$$

$$R_{nl}(r) = -\left[\left(\frac{2Z}{na_0}\right)^3 \frac{(n - l - 1)!}{2n[(n + l)!]^3}\right]^{1/2} e^{-\rho/2}\rho^l L_{n+l}{}^{2l+1}(\rho)$$

where

$$\rho \equiv \frac{2Z}{na_0}r$$

and where

$$L_{n+l}{}^{2l+1}(\rho) = \sum_{k=0}^{n-l-1} \frac{[(n + l)!]^2(-1)^{k+1}}{(n - l - 1 - k)!(2l + 1 + k)!k!}\rho^k$$

$$R_{10}' = R'(1s) = -2re^{-r}$$

$$R_{20}' = R'(2s) = -(1/2)^{1/2}r(1 - r/2)e^{-r/2}$$

$$R_{21}' = R'(2p) = -(1/24)^{1/2}r^2 e^{-r/2}$$

$$R_{30}' = R'(3s) = -(4/27)^{1/2}r(1 - 2r/3 + 2r^2/27)e^{-r/3}$$

$$R_{31}' = R'(3p) = -(64/4374)^{1/2}r^2(1 - r/6)e^{-r/3}$$

$$R_{32}' = R'(3d) = -(16/196\,830)^{1/2}r^3 e^{-r/3}$$

$$R_{40}' = R'(4s) = -(1/4)r(1 - 3r/4 + r^2/8 - r^3/192)e^{-r/4}$$

$$R_{41}' = R'(4p) = -(5/768)^{1/2}r^2(1 - r/4 + r^2/80)e^{-r/4}$$

$$R_{42}' = R'(4d) = -(1/20\,480)^{1/2}r^3(1 - r/12)e^{-r/4}$$

$$R_{43}' = R'(4f) = -(1/20\,643\,840)^{1/2}r^4 e^{-r/4}$$

NOTES

(i)   The functions R_{nl}' are given here because separate normalization of the radial eigenfunctions requires the identity

$$\int_0^\infty r^2 R_{nl}{}^2 dr = \int_0^\infty (R_{nl}')^2 dr = 1$$

(ii) The length r is measured in atomic units (i.e., in bohrs).

(iii) The specific eigenfunctions quoted (as opposed to the general eigenfunction) are for $Z = 1$; eigenfunctions R_{nl}' for any Z and arbitrary length units are obtained by multiplying the specific functions quoted by $(Z/a_0)^{1/2}$ and replacing r by Zr/a_0, where a_0 is the Bohr radius in those same length units.

(iv) The radial eigenfunctions R_{nl} for any Z and arbitrary length units are obtained by dividing the R_{nl}' of item (iii) by r. Thus, for example,

$$R_{10} = R(1s) = 2(Z/a_0)^{3/2}e^{-(Z/a_0)r}$$

$$R_{20} = R(2s) = (1/2)^{1/2}(Z/a_0)^{3/2}[1 - (Z/2a_0)r]e^{-(Z/2a_0)r}$$

$$R_{32} = R(3d) = (16/196\ 830)^{1/2}(Z/a_0)^{7/2}r^2e^{-(Z/3a_0)r}$$

APPENDIX B

Atomic Orbitals for Multielectron Atoms and Ions

1. SLATER'S RULES FOR Z

As far as we are concerned, a Slater-type orbital is given by

$$\chi = A r^{n-1} e^{-\zeta r} Y_{lm}(\theta, \phi) \tag{B1}$$

where n is the principal quantum number and A is a normalization constant. Slater[1] devised a set of simple rules for the purposes of best representation of SCF energies and atom sizes by a single radial function of the form $r^b e^{-\zeta r}$. The quantities b and ζ are given by

$$b = n^* - 1 \tag{B2}$$

$$\zeta = (Z - \sigma)/n^* = Z/n^* \tag{B3}$$

where n^* is an effective principal quantum number and where σ is a screening constant. A discussion of this type of representation has been given in Sec. 6(A) of chapter 1. The rules for determining σ and n^* are as follows:

(A) *Value of n^*:* The value of n^* is related to the principal quantum number n according to the following tabulation:

$n =$	1	2	3	4	5	6
$n^* =$	1	2	3	3.7	4	4.2

(*B*) *Value of* σ: We first arrange the atomic orbitals into the following groupings:

$$(1s)\,(2s2p)\,(3s3p)\,(3d)\,(4s4p)\,(4d)\,(4f)\,(5s5p)\,(5d)\ \text{etc.}$$

The value of σ for a given orbital is then determined as the sum of the following contributions:

(i) Zero from electrons in groups outside that of the orbital being considered.

(ii) A contribution of 0.35 from each other electron in the same grouping if the electron being considered is p, d, f, etc., or ns where $n \neq 1$. A contribution of 0.30 from each other electron in the same grouping if the electron being considered is $1s$.

(iii) A contribution of 0.85 from each electron with principal quantum number $n - 1$ and 1.00 from each electron with principal quantum number less than $n - 1$ if the electron being considered is ns or np.

(iv) A contribution of 1.00 from each electron in inner groupings if the electron being considered is d or f.

2. A COMPILATION OF SLATER ORBITALS

The atomic orbitals $\chi(1s)$, $\chi(2p_x)$, etc., are abbreviated to $1s$, $2p_x$, etc. The distances r, x, y, z are in atomic units.

$n = 1$ $\qquad 1s = (\mathbf{Z}^3/\pi)^{1/2}e^{-\mathbf{Z}r}$

$n = 2$ $\qquad 2s = (\mathbf{Z}^5/96\pi)^{1/2}re^{-\mathbf{Z}r/2}$

$\qquad\qquad 2p_x = (\mathbf{Z}^5/32\pi)^{1/2}xe^{-\mathbf{Z}r/2}$

$\qquad\qquad 2p_y = (\mathbf{Z}^5/32\pi)^{1/2}ye^{-\mathbf{Z}r/2}$

$\qquad\qquad 2p_z = (\mathbf{Z}^5/32\pi)ze^{-\mathbf{Z}r/2}$

$n = 3$ $\qquad 3s = (2\mathbf{Z}^7/5\pi 3^9)^{1/2}r^2e^{-\mathbf{Z}r/3}$

$\qquad\qquad 3p_x = (2\mathbf{Z}^7/5\pi 3^8)^{1/2}xre^{-\mathbf{Z}r/3}$

$\qquad\qquad 3d_{xy} = (\mathbf{Z}^7/2\pi 3^8)^{1/2}2xye^{-\mathbf{Z}r/3}$

$\qquad 3d_{x^2-y^2} = (\mathbf{Z}^7/2\pi 3^8)^{1/2}(x^2 - y^2)e^{-\mathbf{Z}r/3}$

$\qquad\qquad 3d_{xz} = (\mathbf{Z}^7/2\pi 3^8)^{1/2}2xze^{-\mathbf{Z}r/3}$

$\qquad\qquad 3d_{yz} = (\mathbf{Z}^7/2\pi 3^8)^{1/2}2yze^{-\mathbf{Z}r/3}$

$\qquad\qquad 3d_{z^2} = (\mathbf{Z}^7/2\pi 3^8)^{1/2}\dfrac{3z^2 - r^2}{(3)^{1/2}}e^{-\mathbf{Z}r/3}$

$$n = 4 \qquad 4s = \left(\frac{2^{6.4}Z^{8.4}}{(3.7)^{8.4}\pi\Gamma(8.4)}\right)^{1/2} r^{2.7}e^{-Zr/3.7} \qquad \text{see footnote 1}$$

$$4p_x = \left(\frac{(3)\,2^{6.4}Z^{8.4}}{(3.7)^{8.4}\pi\Gamma(8.4)}\right)^{1/2} xr^{1.7}e^{-Zr/3.7}$$

$$4d_{xy} = \left(\frac{(15)\,2^{4.4}Z^{8.4}}{(3.7)^{8.4}\pi\Gamma(8.4)}\right)^{1/2} 2xyr^{0.7}e^{-Zr/3.7}$$

$$4f_{yz^2} = \left(\frac{(21)\,2^{3.4}Z^{8.4}}{(3.7)^{8.4}\pi\Gamma(8.4)}\right)^{1/2} (5z^2 - r^2)yr^{-0.3}e^{-Zr/3.7}$$

see footnote 2

$$n = 5 \qquad 5s = (Z^9/35\pi)^{1/2}(1/1536)\,r^3e^{-Zr/4}$$

$$5p_x = (3Z^9/35\pi)^{1/2}(1/1536)\,xr^2e^{-Zr/4}$$

$$5d_{xy} = (6Z^9/7\pi)^{1/2}(1/3072)\,2xyre^{-Zr/4}$$

$$5f_{yz^2} = (6Z^9/5\pi)^{1/2}(1/6144)\,(5z^2 - r^2)ye^{-Zr/4}$$

$$5f_{xz^2} = (6Z^9/5\pi)^{1/2}(1/6144)\,(5z^2 - r^2)xe^{-Zr/4}$$

$$5f_{z^3} = (6Z^9/5\pi)^{1/2}(1/6144)\,(2/3)^{1/2}(5z^2 - 3r^2)ze^{-Zr/4}$$

$$5f_{xyz} = (6Z^9/5\pi)^{1/2}(1/6144)\,2(10)^{1/2}xyze^{-Zr/4}$$

$$5f_{z(x^2-y^2)} = (6Z^9/5\pi)^{1/2}(1/6144)\,(10)^{1/2}(x^2 - y^2)ze^{-Zr/4}$$

$$5f_{y(x^2-y^2)} = (6Z^9/5\pi)^{1/2}(1/6144)\,(5/3)^{1/2}(3x^2 - y^2)ye^{-Zr/4}$$

$$5f_{x(x^2-y^2)} = (6Z^9/5\pi)^{1/2}(1/6144)\,(5/3)^{1/2}(x^2 - 3y^2)xe^{-Zr/4}$$

3. SIMPLE ORBITAL REPRESENTATIONS

[1] J. C. Slater, *Phys. Rev.* **36,** 57 (1930). Slater's rules as tabulated above [In this regard, see also C. Zener, *Phys. Rev.* **36,** 51 (1930).]

[2] W. E. Duncanson and C. A. Coulson, *Proc. Roy. Soc. (Edinburgh)* **62,** 37 (1944). Improved orbitals of the Slater variety for atoms in the first row of the Periodic Table.

---

[1] Γ is the standard notation for a gamma function; its numerical value can be found in standard mathematical tables.

[2] As a matter of convenience, the complete set of f orbitals is listed under $n = 5$, rather than $n = 4$; the fractional powers for $n = 4$ ($n^* = 3.7$) require a rather awkward notation involving gamma functions.

[3] L. Pauling, *Proc. Roy. Soc. (London)* **A114,** 181 (1927); L. Pauling and J. Sherman, *Z. Krist.* **81,** 1 (1932). Screening constants for all atoms for use in hydrogenlike wave functions. These constants were evolved on the basis of empirical considerations relating, for example, to x-ray term values, molecular refraction values, etc.

[4] E. Clementi and D. L. Raimondi, *J. Chem. Phys.* **38,** 2686 (1963). Present a modified set of Slater rules for best single ζ's and $n^* = n$. Neutral atoms with 2 to 36 electrons. The set of rules presented is based on energy minimization considerations. [See also C. C. J. Roothaan, Technical Report, Laboratory of Molecular Structure and Spectra, The University of Chicago, Chicago, Illinois, 1955 (unpublished).]

[5] Document No. 7545, ADI Auxiliary Publications Project, Photoduplication Service, Library of Congress, Washington 25, D.C. ($27.50: photoprints; $7.75: microfilm). As [4] above, but for excited states of atoms and ions containing 2 to 36 electrons.

[6] G. Burns, *J. Chem. Phys.* **41,** 1521 (1964). Present modified Slater rules for both n^* and σ for *ns, np, nd,* and *nf* electrons in atoms up to and including rare earths. Based on comparison of moments $\langle r^q \rangle$ to previously evaluated Hartree–Fock AO's.

[7] L. C. Cusachs, D. G. Carroll, B. Trus, and S. P. McGlynn, *Int. J. Quantum Chem.,* Slater Symposium Issue, 1967. Single STF's ($n^* = n$ or $n - 1$) for first- and second-row atoms and a general method for obtaining such AO's for any orbital for which a Hartree–Fock AO is available. Based on mimicry of interatom SCF-AO overlap integrals.

4. COLLECTIONS OF ANALYTIC MULTI-STF REPRESENTATIONS

[8] E. Clementi, C. C. J. Roothaan, and M. Yoshimine, *Phys. Rev.* **127,** 1618 (1962). First-row atoms.

[9] E. Clementi, *J. Chem. Phys.* **38,** 996 (1962). Ground and excited states of isoelectronic series with 2 to 10 electrons.

[10] E. Clementi and A. D. McLean, *Phys. Rev.* **133,** A419 (1964). Li⁻, B⁻, C⁻, N⁻, O⁻, F⁻.

[11] E. Clementi, *J. Chem. Phys.* **38,** 1001 (1964). 11 to 18 electrons.

[12] E. Clementi, A. D. McLean, D. L. Raimondi, and M. Yoshimine, *Phys. Rev.* **133,** A1274 (1964). Na⁻, Al⁻, Si⁻, P⁻, S⁻, Cl⁻.

[13] E. Clementi, *J. Chem. Phys.* **41,** 295 (1964). 19 to 30 electrons.

[14] E. Clementi, *J. Chem. Phys.* **41,** 303 (1964). 31 to 36 electrons.

[15] E. Clementi, IBM, *J. Res. Develop. Suppl.* **9,** 2 (1965).

[16] E. Clementi, R. Matcha, and A. Veillard, *J. Chem. Phys.* **47,** 1865 (1967). Ground states of atoms of third period of atomic table: 19 to 36 electrons.

[17] P. S. Bagus and T. L. Gilbert. These AO's are found in A. D. McLean and M. Yoshimine, *IBM J. Res. Develop. Suppl.* **12,** 206 (1967).

[18] J. W. Richardson, W. C. Nieuwpoort, R. R. Powell, and W. F. Edgell, *J. Chem. Phys.* **36,** 1057 (1962). $3d$ and $4s$ AO's of Ti through Cu.

[19] J. W. Richardson, R. R. Powell, and W. C. Nieuwpoort, *J. Chem. Phys.* **38,** 796 (1963). $4p$ and $4d$ AO's of Ti through Cu.

[20] H. Basch and H. B. Gray, *Teoret. Chim. Acta (Berl.)* **4,** 367 (1967). Analytical representation of the numerical Hartree–Fock–Slater SCF AO's for the second- and third-row transition metals. It is difficult to say how good the analytic representation is. The numerical AO's which were fitted were taken from F. Herman and S. Skillman, *Atomic Structure Calculations* (Prentice-Hall, Inc., Englewood Cliffs, N.J., 1963).

5. OTHER COLLECTIONS OF SELF-CONSISTENT FIELD ATOMIC ORBITALS

[21] J. C. Slater, *Quantum Theory of Atomic Structure* (McGraw-Hill Book Co., Inc., New York, 1960), Vol. 1, Appendix 16. A collection of references through 1956.

[22] F. Herman and S. Skillman, *Atomic Structure Calculations* (Prentice-Hall, Inc., Englewood Cliffs, N.J., 1963). Hartree–Fock–Slater numeric SCF AO's for all the elements.

[23] M. Cohen, Preprints No. 42, Quantum Theory Project, University of Florida, Gainesville, Fla., 1963 (unpublished). Elements containing 1 to 10 elements. These AO's are hydrogenlike.

APPENDIX C

Sources of Overlap Integrals

1. TABLES OF OVERLAP INTEGRALS

1. R. S. Mulliken, C. A. Rieke, D. Orloff, and H. Orloff, *J. Chem. Phys.* **17,** 1248 (1949); $1s$, $2s$, $2p$, $3s$, $3p$, $5s$, $5p$.
2. H. H. Jaffé and G. O. Doak, *J. Chem. Phys.* **21,** 196 (1953); $2s$, $2p$, $5s$, $5p$, $3d$.
3. H. H. Jaffé, *J. Chem. Phys.* **21,** 258 (1953); $2p$, $3d$, $5d$.
4. J. L. Roberts and H. H. Jaffé, *J. Chem. Phys.* **27,** 883 (1957); $2s$, $2p$, $3d$, $5d$.
5. L. Leifer, F. A. Cotton, and J. R. Leto, *J. Chem. Phys.* **28,** 364 (1958); **28,** 1258 (1958); $2s$, $2p$, $3s$, $4s$, $4p$, $4d$.
6. D. A. Brown, *J. Chem. Phys.* **29,** 1086 (1958); $2p$, $4p$.
7. D. P. Craig, A. Maccoll, R. S. Nyholm, L. E. Orgel, and L. E. Sutton, *J. Chem. Soc.* 354 (1954); $3s$, $3p$, $3d$.
8. D. A. Brown and N. J. Fitzpatrick, *J. Chem. Phys.* **46,** 2005 (1967); useful tables including f orbitals.
9. E. A. Boudreaux, L. Chopin Cusachs and L. Dureaux, *Numerical Tables of Two-Center Overlap Integrals* (W. A. Benjamin, Inc., New York, 1970).

2. TABLES OF *A* AND *B* INTEGRALS AND/OR EXPANSION OF *S* IN TERMS OF *A* AND *B* FUNCTIONS

10. A. Lofthus, *Mol. Phys.* **5,** 105 (1962); gives a wide selection of master formulas for *s*, *p*, *d*, *f*, *g* and *h* orbitals.
11. M. Kotani, A. Amemiya, E. Ishiguro, and T. Kimura, *Tables of Molecular Integrals* (Maruzen Co., Ltd., Tokyo, 1963), 2nd ed.
12. H. Preuss, *Integraltafeln zur Quantenchemie* (Springer-Verlag, Berlin, 1956–60), Vols. I–IV.
13. J. Miller, J. M. Gerhauser, and F. A. Matsen, *Quantum Chemistry Integrals and Tables* (University of Texas Press, Austin, Tex., 1959).
14. S. Flodmark, *Table of Molecular A and B Functions* (Institute of Theoretical Physics, Stockholm, Sweden, 1957).
15. E. A. Magnusson, *Rev. Pure Appl. Chem.* **14,** 57 (1964); provides a full list of references which deal with integrals of all kinds useful in quantum chemistry computations.

3. OVERLAP INTEGRAL COMPUTER PROGRAMS

The following programs present overlap integrals between all dual combinations of orbitals listed. These tables usually cover the ranges of *p* and *t* variables which produce non-negligible *S*.

16. W. A. Yeranos, QCPE Program 82 (Quantum Chemistry Program Exchange, Indiana University, Bloomington, Ind. 47401). FORTRAN II, IBM 1620.
17. Peter O'Donnell Offenhartz, *J. Chem. Ed.* **44,** 604 (1967); overlap integral program for *s*, *p*, and *d* orbitals—general *n*—FORTRAN II and IV.
18. L. C. Cusachs, Department of Chemistry, Tulane University— FORTRAN II and IV—*s*, *p*, *d*, and *f* orbitals—*n* equals 1 to 7.
19. There are several overlap integral programs included as part of larger programs (e.g., QCPE program 30).

APPENDIX D

Valence Orbital Ionization Energies

Valence orbital ionization energies are discussed in chapter 4. A reasonable approximation to these VOIE's is[1]

$$VOIE \propto (Z - \sigma)^2/(n^*)^2 \propto (Z^2 - 2\sigma Z + \sigma^2)/(n^*)^2 \qquad (D1)$$

where Z is the atomic number, σ is the shielding constant, and n^* is the effective principal quantum number. Thus, the VOIE's of an isoelectronic sequence should follow an equation[2]−[4]

$$VOIE = \alpha Z^2 + \beta Z + \gamma \qquad (D2)$$

Relativistic as well as other effects introduce terms in Z^{-1}, Z^3, and Z^4. We assume these terms to be small.

The effective nuclear charge may also be written as

$$Z - \sigma(\mu) = q + \sum_{\nu} [1 - n_\nu \sigma_\nu(\mu)] \qquad (D3)$$

where $\sigma(\mu)$ is the extent to which an electron in χ_μ is shielded from the nuclear charge Z; q is the excess charge on the atom; $\sigma_\nu(\mu)$ is the extent to which an electron in χ_ν shields one in the χ_μ AO; n_ν is the number of electrons in the χ_ν AO; and the summation is over all AO's. Insertion of Eq. (D3)

423

TABLE D.1. *Valence orbital ionization energies in the form* $VOIE = Aq^2 + Bq + C$ *(hydrogen through potassium).*[a]

Number of Electrons	Configuration	Valence Orbital	Standard Deviation	A (kK)	B (kK)	C (kK)
1	$1s$	$1s$	0.0	109.84	219.2	109.7
2	$1s^2$	$1s$	0.1	109.82	301.7	198.4
2	$1s2p$	$1s$	0.1	109.96	386.8	357.85
3	$1s^22p$	$1s$	0.1	110.15	467.2	524.8
1	$2s$	$2s$	0.0	27.48	54.8	27.4
3	$1s^22s$	$2s$	0.1	27.62	76.0	43.4
4	$(He)2s^2$	$2s$	0.1	27.64	100.3	75.1
4	$(He)2s2p$	$2s$	0.3	27.76	81.3	47.9
5	$(He)2s^22p$	$2s$	0.1	27.82	120.6	113.4
5	$(He)2s2p^2$	$2s$	0.3	27.91	119.1	122.25
6	$(He)2s^22p^2$	$2s$	0.2	27.95	141.6	156.6
6	$(He)2s2p^3$	$2s$	0.4	28.00	141.2	171.0
7	$(He)2s^22p^3$	$2s$	0.2	28.16	162.2	206.2
7	$(He)2s2p^4$	$2s$	0.1	28.05	163.3	226.0
8	$(He)2s^22p^4$	$2s$	0.3	27.95	184.6	260.8
9	$(He)2s^22p^5$	$2s$	0.5	28.07	205.7	323.6
10	$(He)2s^22p^6$	$2s$	0.1	28.29	227.0	390.9
1	$2p$	$2p$	0.0	27.48	54.8	27.4
2	$1s2p$	$2p$	0.1	27.52	57.8	28.6
3	$1s^22p$	$2p$	0.2	27.74	59.1	28.4
4	$(He)2s2p$	$2p$	0.2	27.72	97.6	79.8
4	$(He)2p^2$	$2p$	0.4	27.57	76.1	45.35
5	$(He)2s^22p$	$2p$	0.2	27.78	102.4	66.75
5	$(He)2s2p^2$	$2p$	0.3	28.02	96.1	67.0
5	$(He)2p^3$	$2p$	0.1	27.25	94.0	61.4
6	$(He)2s^22p^2$	$2p$	0.3	27.95	118.2	85.8
6	$(He)2s2p^3$	$2p$	0.4	28.03	111.95	86.9
6	$(He)2p^4$	$2p$	0.1	28.06	105.4	88.1
7	$(He)2s^22p^3$	$2p$	0.2	28.16	133.2	106.4
7	$(He)2s2p^4$	$2p$	2.1	30.01	114.0	129.4
8	$(He)2s^22p^4$	$2p$	0.4	27.94	149.75	127.4
8	$(He)2s2p^5$	$2p$	0.1	27.76	145.2	126.4
9	$(He)2s^22p^5$	$2p$	0.4	27.93	165.5	150.4
9	$(He)2s2p^6$	$2p$	0.8	28.22	157.7	155.1
10	$(He)2s^22p^6$	$2p$	0.3	28.25	180.2	173.9
11	$(Ne)3s$	$3s$	0.6	13.18	68.0	41.0
12	$(Ne)3s^2$	$3s$	0.5	13.13	78.2	61.25
12	$(Ne)3s3p$	$3s$	0.5	13.14	78.6	71.7

TABLE D.1.   (*Continued*).

Number of Electrons	Configuration	Valence Orbital	Standard Deviation	A (kK)	B (kK)	C (kK)
13	(Ne)$3s^2 3p$	$3s$	0.8	13.15	89.0	90.8
13	(Ne)$3s 3p^2$	$3s$	0.0	9.50	103.6	89.4
14	(Ne)$3s^2 3p^2$	$3s$	1.3	13.08	99.9	119.6
14	(Ne)$3s 3p^3$	$3s$	2.0	11.12	118.2	111.2
15	(Ne)$3s^2 3p^3$	$3s$	0.3	14.27	106.7	151.4
16	(Ne)$3s^2 3p^4$	$3s$	0.3	12.23	124.0	166.7
17	(Ne)$3s^2 3p^5$	$3s$	0.2	13.70	126.7	203.8
18	(Ne)$3s^2 3p^6$	$3s$	0.3	13.24	138.6	235.6
11	(Ne)$3p$	$3p$	0.7	13.33	49.4	23.9
12	(Ne)$3s 3p$	$3p$	0.6	13.21	60.5	36.0
12	(Ne)$3p^2$	$3p$	0.6	12.05	61.9	41.3
13	(Ne)$3s^2 3p$	$3p$	0.6	13.29	71.1	47.85
13	(Ne)$3s 3p^2$	$3p$	0.6	12.44	75.65	42.8
14	(Ne)$3s^2 3p^2$	$3p$	0.3	13.02	81.7	62.5
14	(Ne)$3s 3p^3$	$3p$	2.8	8.82	110.7	19.4
14	(Ne)$3s^2 3p 4s$	$3p$	0.9	13.36	86.3	90.5
15	(Ne)$3s 3p^3$	$3p$	0.0	15.25	83.9	81.6
15	(Ne)$3s 3p^4$	$3p$	1.3	14.25	83.85	100.1
15	(Ne)$3s^2 3p^2 4s$	$3p$	0.7	14.63	91.3	114.8
16	(Ne)$3s^2 3p^4$	$3p$	0.7	13.17	98.5	93.4
16	(Ne)$3s 3p^5$	$3p$	0.7	13.88	94.9	99.7
16	(Ne)$3s^2 3p^3 4s$	$3p$	0.4	13.57	102.4	131.6
17	(Ne)$3s^2 3p^5$	$3p$	0.4	13.49	106.3	110.4
17	(Ne)$3s 3p^6$	$3p$	0.0	13.40	106.4	116.0
17	(Ne)$3s^2 3p^4 4s$	$3p$	0.2	13.36	112.0	153.3
18	(Ne)$3s^2 3p^6$	$3p$	0.1	13.36	116.6	127.5
18	(Ne)$3s^2 3p^5 4s$	$3p$	0.6	13.39	121.5	175.3
19	(Ar)$3d$	$3d$	0.1	13.10	24.5	12.2
11	(Ne)$4s$	$4s$	0.3	7.47	29.0	15.5
12	(Ne)$3s 4s$	$4s$	0.1	7.67	32.2	19.8
13	(Ne)$3s^2 4s$	$4s$	0.2	7.65	36.1	22.85
14	(Ne)$3s^2 3p 4s$	$4s$	0.8	7.79	39.7	25.15
15	(Ne)$3s^2 3p^2 4s$	$4s$	1.0	9.30	38.5	31.8
16	(Ne)$3s^2 3p^3 4s$	$4s$	0.2	7.82	45.35	30.1
17	(Ne)$3s^2 3p^4 4s$	$4s$	0.3	8.00	48.0	32.1
18	(Ne)$3s^2 3p^5 4s$	$4s$	0.4	7.99	51.9	33.6
19	(Ar)$4s$	$4s$	0.4	8.09	53.5	34.7

[a] From Basch, Viste, and Gray, Ref. [5] of text. Reprinted by permission.

TABLE D.2. *Valence orbital ionization energies in the form* $VOIE = Aq^2 + Bq + C$ *(in kK)* *(titanium through nickel).*[a-d]

Parameter	Configuration	Valence Orbital	VOIE Curve Number	$v = 4$ Ti	$v = 5$ V	$v = 6$ Cr	$v = 7$ Mn	$v = 8$ Fe	$v = 9$ Co	$v = 10$ Ni
A	d^v	d	1	17.15	15.8	14.75	14.1	13.8	13.85	14.2
A	$d^{v-1}s$	d	2	18.45	14.0	9.75	5.5	13.8	13.85	14.2
A	$d^{v-1}p$	d	3	18.45	14.0	9.75	5.5	13.8	13.85	14.2
A	$d^{v-1}s$	s	4	9.3	8.55	8.05	7.6	7.35	7.25	7.35
A	$d^{v-2}s^2$	s	5	9.3	8.55	8.05	7.6	7.35	7.25	7.35
A	$d^{v-2}sp$	s	6	9.3	8.55	8.05	7.6	7.35	7.25	7.35
A	$d^{v-1}p$	p	7	7.8	7.45	7.25	7.2	7.3	7.55	7.95
A	$d^{v-2}p^2$	p	8	7.8	7.45	7.25	7.2	7.3	7.55	7.95
A	$d^{v-2}sp$	p	9	7.8	7.45	7.25	7.2	7.3	7.55	7.95
B	d^v	d	1	60.85	68.0	74.75	80.8	86.2	91.15	95.5
B	$d^{v-1}s$	d	2	77.85	87.0	95.95	105.0	101.5	106.25	110.7
B	$d^{v-1}p$	d	3	76.75	87.3	96.95	106.0	101.9	105.55	108.2
B	$d^{v-1}s$	s	4	50.4	54.15	57.55	60.9	63.85	66.65	69.05
B	$d^{v-2}s^2$	s	5	58.5	62.95	66.85	70.3	73.05	75.25	77.05
B	$d^{v-2}sp$	s	6	55.0	57.55	60.45	63.8	67.35	71.35	75.65
B	$d^{v-1}p$	p	7	35.6	45.45	47.55	49.3	50.8	51.95	52.85
B	$d^{v-2}p^2$	p	8	48.9	50.85	52.85	55.2	57.8	60.65	63.75
B	$d^{v-2}sp$	p	9	48.9	50.85	52.85	55.2	57.8	60.65	63.75

C										
C	d^v	d	1	27.4	31.4	35.1	38.6	41.9	44.8	47.6
C	$d^{v-1}s$	d	2	44.6	51.4	57.9	64.1	70.0	75.6	80.9
C	$d^{v-1}p$	d	3	55.4	61.4	67.7	74.3	81.2	88.4	95.9
C	$d^{v-1}s$	s	4	48.6	51.0	53.2	55.3	57.3	59.1	60.8
C	$d^{v-2}s^2$	s	5	57.2	60.4	63.3	65.9	68.3	70.5	72.3
C	$d^{v-2}sp$	s	6	66.0	70.6	74.7	78.3	81.4	84.0	86.0
C	$d^{v-1}p$	p	7	26.9	27.7	28.4	29.2	29.9	30.7	31.4
C	$d^{v-2}p^2$	p	8	35.9	36.8	37.8	38.8	39.7	40.7	41.6
C	$d^{v-2}sp$	p	9	34.4	36.4	38.1	39.4	40.3	40.8	40.9

[a] From Basch, Viste, and Gray, Ref. [5] of text. Reprinted by permission.

[b] The number v across the top of the table is the total number of electrons in the $3d$, $4s$, and $4p$ atomic orbitals of the neutral species.

[c] A given curve applies to configurations which, for successive q's, differ only in the number of d electrons.

[d] As an example, the $3d$ VOIE of Mn^{++} in the configuration d^4p is obtained from curve 3 at $q = 2$ as

$$3d\ \text{VOIE}\,(Mn^{++};\ldots d^4p) = 5.5(2)^2 + 106.0(2) + 74.3 = 308.3\ \text{kK}.$$

into Eq. (D1) yields[5] an equation [see Eq. (4.22)]

$$\chi_\mu \, \text{VOIE} = Aq^2 + Bq + C \qquad (D4)$$

for an isoelectronic sequence. Compilations of VOIE's in the format of Eq. (D4) are given in Tables D.1 and D.2. Some valence orbital ionization energies for gallium through zinc are given in Table D.3. VOIE's for the elements which are missing in Tables D.1–D.3 (namely, Ca, Sc, Cu, and Zn) are available in Ballhausen and Gray.[6]

Valence orbital ionization energies in the form [see Eq. (4.47)]

$$-\chi_\mu \, \text{VOIE} = A''(pop) + B'' + C''q = H_{\mu\mu} \qquad (D5)$$

are given in Table D.4. The 1s VOIE of hydrogen has been given[8] as

$$-1s \, \text{VOIE} = 0.121q^3 - 13.97q^2 - 26.93q - 13.6 \qquad (D6)$$

TABLE D.3. *Valence orbital ionization energies for 4s, 4p, and 5s electrons (in kK)*[a–d] *(gallium through krypton).*

Ionization Configuration	4s (0→1) $4s^2 4p^v$	4s (1→2) $4s^2 4p^{v-1}$	4p (0→1) $4s^2 4p^v$	4p (1→2) $4s^2 4p^{v-1}$	5s (0→1) $4s^2 4p^{v-1} 5s$	5s (1→2) $4s^2 4p^{v-2} 5s$
Ga (1)	103.2	(164.7)[e]	47.4	···	23.5	···
Ge (2)	122.8	201.2	60.9	127.5	25.7	66.3
As (3)	144.8	234.9	74.1	(147.3)[e]	27.7	74.1
Se (4)	168.1	265.7	87.0	166.7	29.6	79.7
Br (5)	(193.8)[e]	(243.7)[e]	99.6	185.5	31.4	83.0
Kr (6)	221.7	319.0	111.8	204.0	32.9	84.0
Standard Deviation	2.4	6.8	0.8	1.5	0.5	1.1

[a] From Basch, Viste, and Gray, Ref. [5] of text. Reprinted by permission.

[b] The superscript v, as in $4s^2 4p^v$, is given by the number in parentheses which follows the element in the first column of the table.

[c] The symbols in parentheses across the top row of the table, as in $4s (0 \rightarrow 1)$, refer to the value of q before and after removal of an electron.

[d] As an example, the 4s VOIE of As$^+$ in the configuration $\cdots 4s^2 4p^2$ is 234.9 kK.

[e] The numbers in parentheses are extrapolated values.

TABLE D.4. *Valence orbital ionization energies in the form[a,b]*

$$-\chi_\mu \; VOIE = A''(pop) + B + Cq \; (in \; eV) = H_{\mu\mu}.$$

| Quantum Numbers | | Atom | Atomic Number | | | |
n	l		Z	A''	B''	C''
2	0	Li	3	1.1	−6.5	−4.3
2	1	Li	3	0.3	−3.9	−4.3
2	0	Be	4	1.3	−11.5	−7.2
2	1	Be	4	0.7	−6.7	−7.2
2	0	B	5	1.5	−17.0	−9.8
2	1	B	5	1.0	−9.6	−9.8
2	0	C	6	1.7	−22.9	−11.9
2	1	C	6	1.5	−12.9	−11.9
3	0	C	6	0.0[c]	−3.66[c]	−4.25[c]
3	1	C	6	0.0[c]	−2.51[c]	−3.90[c]
2	0	N	7	1.9	−29.3	−13.7
2	1	N	7	1.9	−16.3	−13.7
2	0	O	8	2.1	−36.2	−15.2
2	1	O	8	2.3	−19.9	−12.0[d]
3	0	O	8	0.0	−4.31	−8.01
2	0	F	9	2.3	−45.9	−16.2
2	1	F	9	2.8	−23.8	−16.2
3	0	Na	11	1.33	−6.55	−4.85
3	1	Na	11	1.22	−4.26	−4.85
3	0	Mg	12	1.33	−10.25	−6.04
3	1	Mg	12	1.12	−6.06	−6.09
3	0	Al	13	1.44	−14.06	−7.12
3	1	Al	13	1.13	−7.96	−7.12
3	0	Si	14	1.66	−17.98	−8.09
3	1	Si	14	1.12	−9.84	−8.09
3	0	P	15	1.99	−22.01	−8.95
3	1	P	15	1.13	−11.74	−8.95
3	0	S	16	2.43	−26.15	−9.70
3	1	S	16	1.12	−13.62	−9.70
3	2	S	16	0.0	−3.67	−6.20
4	0	S	16	0.0	−3.76	−6.40
4	1	S	16	0.0	−3.023	−4.42
3	0	Cl	17	2.98	−30.40	−10.66
3	1	Cl	17	1.13	−15.52	−10.66
4	0	Cl	17	0.0	−4.01	−6.28
4	1	Cl	17	0.0	−2.605	−5.111

[a] For general reference, see Refs. [7] and [8] of text.

[b] This table was prepared by A. T. Armstrong and B. Bertus.

[c] Based on the VOIE data of Hosoya, Ref. [9] of text.

[d] This value for C'' was used by Carroll *et al.*, Ref. [8] of text. For general usage, this C'' should be replaced by −15.2.

TABLE D.5.   *Other sources of VOIE data.*

Type of Data	Author(s)	Reference Number
VOIE's of valence-configuration type (H to Cl)	Hinze–Jaffé	[16]
VOIE's of valence-configuration type (H to F and ions thereof; Cl, Cl⁻, Br, Br⁻, I, I⁻)	Mulliken	[17]
VOIE's of valence-configuration type (H to Cl and ions thereof)	Skinner–Pritchard	[18]
VOIE's of valence-configuration type (Be to O and ions thereof)	Pilcher–Skinner	[19]
s and p VOIE's of valence configurations $ns^snp_x^xnp_y^ynp_z^z$ where the occupancy numbers s, x, y, z are 1 or 2 (H to A; Sc to Zn; Y to Cd; Hf to Hg)	Pritchard–Skinner	[20]
$2s$ and $2p$ VOIE's of valence configurationss s^pp^p where occupancy numbers are $s = 0, 1, 2$; $p = 0, 1 \ldots, 6$ (Li to F; S)	Cusachs–Reynolds	[7]
$3s$, $3p$, $3d$ VOIE's (Li to F)	Hosoya	[9]
$4s$, $4p$, $3d$ VOIE's (Si to Cl)	Palmieri–Zauli	[10]
$4s$, $4p$, $3d$ VOIE's of valence-configuration type (Na to Cl and ions thereof)	Cusachs–Linn	[11]
VOIE's of valence AO's of valence configurations $d^ds^sp^p$ (Ca to Cu and ions thereof)	Hinze–Jaffé	[21]
VOIE's of valence-configuration type (positive ions of Li to A; neutral atoms K, Ca, Ga to Kr; Rb, Sr, In to Xe)	Hinze–Jaffé	[22]
s, p, and d VOIE's of valence-configuration type (Li to Br)	Cusachs–Corrington	[12]
Atomic energy levels	Moore; Striganov–Sventitskii	[13]
Atomic ionization energies	Moore; Striganov–Sventitskii	[13]
Electron Affinity of H (I)	Pekeris	[14]
Electron Affinities [He (I) through Cl (I)]	Edlén	[15]

VOIE data are also available in a number of other sources. These are tabulated in Table D.5. VOIE data which are required, but which are not available, may be evaluated from tables of atomic energy levels[13] or from accurate HF-SCF energies such as are discussed in Appendix B. Atomic ionization energies are available in the same source[13]; electron affinity values for H(I) have been given by Pekeris[14] and for He(I) through Cl(I) by Edlén.[15]

BIBLIOGRAPHY

[1]   J. C. Slater, *Phys. Rev.* **36**, 57 (1930).

[2]   E. C. Baughan, *Trans. Faraday Soc.* **57**, 1863 (1961).

[3]   G. Glockler, *Phys. Rev.* **46**, 111 (1934).

[4]   H. O. Pritchard and H. A. Skinner, *J. Chem. Phys.* **22**, 1963 (1954).

[5]   H. Basch, A. Viste, and H. B. Gray, *Theoret. Chim. Acta (Berl.)* **3**, 458 (1965).

[6]   C. J. Ballhausen and H. B. Gray, *Molecular Orbital Theory* (W. Benjamin, Inc., New York, 1964), pp. 120–122. [See also: H. Basch, A. Viste, and H. B. Gray, *J. Chem. Phys* **44**, 10 (1966).]

[7]   L. C. Cusachs and J. W. Reynolds, *J. Chem. Phys.*, **43**, 160S (1965).

[8]   D. G. Carroll, A. T. Armstrong, and S. P. McGlynn, *J. Chem. Phys.* **44**, 1865 (1966).

[9]   H. Hosoya, *J. Chem. Phys.* **48**, 1380 (1968).

[10] P. Palmieri and C. Zauli, *J. Chem. Soc.* **A**813 (1967).

[11] L. C. Cusachs and J. R. Linn, Jr., *J. Chem. Phys.* **46**, 2919 (1967).

[12] L. C. Cusachs and J. H. Corrington, *Sigma Molecular Orbital Theory*, O. Sinanoğlu and K. Wiberg, Eds., Yale University Press, New Haven, 1970, p. 256.

[13] C. E. Moore, *Atomic Energy Levels* (National Bureau of Standards Circular 467, Washington, D.C.), Vol. I (1949), II (1952), and III (1958); A. R. Striganov and N. S. Sventitskii, *Tables of Spectral Lines of Neutral and Ionized Atoms* (Plenum Press, New York, 1968).

[14] C. L. Pekeris, *Phys. Rev.* **112**, 1649 (1948).

[15] B. Edlén, *J. Chem. Phys.* **33**, 98 (1960).

[16] J. Hinze and H. H. Jaffé, *J. Am. Chem. Soc.* **84**, 540 (1962).

[17] R. S. Mulliken, *J. Chem. Phys.* **2**, 782 (1934).

[18] H. A. Skinner and H. O. Pritchard, *Trans. Faraday Soc.* **49**, 1254 (1955).

[19] G. Pilcher and H. A. Skinner, *J. Inorg. Nucl. Chem.* **24**, 937 (1962).

[20] H. O. Pritchard and H. A. Skinner, *Chem. Revs.* **55**, 745 (1955).

[21] J. Hinze and H. H. Jaffé, *Can. J. Chem.* **41**, 1315 (1963).

[22] J. Hinze and H. H. Jaffé, *J. Phys. Chem.* **67**, 1501 (1963).

APPENDIX E

Atomic Integrals
$(\mu \zeta \mid \nu \eta)$ and $h_{\mu\nu}$

(A) One- and Two-Center Integrals: Evaluation in a Slater basis is possible. Reference should be made to C. Sandorfy, *Electronic Spectra and Quantum Chemistry* (Prentice-Hall, Inc., Englewood Cliffs, N.J., 1964), see pp. 272–281 and reference listed therein.

(B) Two- and Three-Center Integrals: Evaluation in a Slater basis is difficult. A number of methods of computation are available:

(i) The ζ-function approach is based on an expansion of all AO functions about some one single center. This method is discussed by M. P. Barnett and C. A. Coulson, *Phil. Trans. Roy. Soc.* (*London*) **A243,** 221 (1951); M. P. Barnett, in *Methods in Computational Physics* edited by B. J. Alder (Academic Press Inc., New York, 1963), Vol. 2, p. 95ff.

(ii) The axial expansion approach is based on an elaboration of all charge distributions in a series of exponentials whose centers are all colinear. This method is discussed by S. F. Boys and I. Shavitt, Technical Report WIS-AF-13, University of Wisconsin Naval Research Laboratory, Madison, Wis. (unpublished).

(iii) Numerical integration methods have been discussed by E. A. Magnusson and C. Zauli, *Proc. Phys. Soc.* (*London*) **78,** 53 (1961).

(iv) A Gaussian expansion approach, in which each AO function is expanded into a series of Gaussians to least-squares limitations, is discussed by Shavitt [Technical Report WIS-AF-13, University of Wisconsin Research Laboratory, Madison, Wis. (unpublished), p. 1ff].

(v) A Gaussian transform method is also discussed by Shavitt [see (iv) above]. The Gaussian approach is probably the easiest to use. As such, it has gained considerably in stature in the last four years.

(vi) Approximate methods are often used to express three- and four-center integrals in terms of one- and two-center integrals. The most common of these are:

(a) Sklar–London Approximation: The integral $(\mu\zeta \mid \nu\eta)$ is written as

$$(\mu\zeta \mid \nu\eta) = S_{\mu\zeta} S_{\nu\eta} (\overline{\mu\zeta}\ \overline{\mu\zeta} \mid \overline{\nu\eta}\ \overline{\nu\eta}) \tag{E1}$$

where $\overline{\mu\zeta}$ is an atomic orbital situated at the geometric midpoint of the μ and ζ centers. This procedure is quite difficult to use and not very accurate.

(b) The Mulliken Approximation: This approximation is widely used. It is given by

$$(\mu\zeta \mid \nu\eta) = \tfrac{1}{4} S_{\mu\zeta} S_{\nu\eta} [(\mu\mu \mid \nu\nu) + (\mu\mu \mid \eta\eta) + (\zeta\zeta \mid \nu\nu) + (\zeta\zeta \mid \eta\eta)] \tag{E2}$$

It is a simple procedure and it is fairly accurate.

(c) Zero-Differential Overlap Approximation: This procedure is under intensive study in the present literature. It is assumed that

$$\langle \chi_\mu \mid \chi_\nu \rangle = \delta_{\mu,\nu} \tag{E3}$$

whence, from the Mulliken approximation of Eq. (E2), we find

$$(\mu\zeta \mid \nu\eta) = \delta_{\mu,\zeta} \delta_{\nu,\eta} (\mu\mu \mid \nu\nu) \tag{E4}$$

This is indeed a drastic procedure; surprisingly, it appears to yield good results in practice. This approach for π-electron theories is discussed by I. Fischer-Hjalmars, *J. Chem. Phys.* **42,** 1962 (1965); *Modern Quantum Chemistry*, edited by O. Sinanoğlu (Academic Press Inc., New York, 1965), Vol. 1, p. 185ff.

(*C*) *Three- and Four-Center Integrals:* See F. E. Harris and H. H. Michels, Advances in Chemical Physics, edited by I. Prigogine (Interscience, New York, 1967), Vol. 13, p. 205.

APPENDIX F

The Transformed Products $\langle \chi \mid \hat{\mathbf{r}}$ in a Slater Basis

The products $\langle \chi \mid \hat{\mathbf{r}}$ are atomic orbitals, $\langle \chi' \mid$, whose principal quantum number is larger than that of $\langle \chi \mid$. The primed orbitals possess the same orbital exponent, $\zeta = Z/n^*$, as the original unprimed orbital. The primed orbitals, possessing a higher value of n (and, therefore, of n^*—see Appendix B), possess a higher value of the effective charge, Z'. The prime stresses the fact that these atomic orbitals are *not ordinary atomic orbitals of the particular atom under consideration*. A specific example of the transformation $\langle \chi \mid \hat{\mathbf{r}} = a\langle \chi' \mid$ is found in Sec. 4(D) of chapter 9. Atomic orbitals with $n = 4$ ($n^* = 3.7$) are not included: $\langle \chi \mid \hat{\mathbf{r}}$, in this case, leads to radial wave functions containing noninteger powers of r (i.e., $r^{4.7}$).

$$\langle 1s \mid x = \frac{1}{Z} \langle 2p_x' \mid$$

$$\langle 2s \mid x = \frac{(10)^{1/2}}{Z} \langle 3p_x' \mid$$

$$\langle 3s \mid x = \frac{(42)^{1/2}}{Z} \langle 5p_x' \mid$$

$$\langle 2p_x \mid x = \frac{(2)^{1/2}}{Z} \{(3)^{1/2}\langle 3d_{x^2-y^2}' \mid - \langle 3d_{z^2}' \mid + (5)^{1/2}\langle 3s' \mid\}$$

$$\langle 2p_x \mid y = \frac{(6)^{1/2}}{Z} \langle 3d_{xy}' \mid$$

$$\langle 2p_x \mid z = \frac{(6)^{1/2}}{Z} \langle 3d_{xz}' \mid$$

$$\langle 2p_y \mid y = \frac{(2)^{1/2}}{Z} \{-(3)^{1/2}\langle 3d_{x^2-y^2}' \mid - \langle 3d_{z^2}' \mid + (5)^{1/2}\langle 3s' \mid\}$$

$$\langle 2p_z \mid z = \frac{(2)^{1/2}}{Z} \{2\langle 3d_{z^2}' \mid + (5)^{1/2}\langle 3s' \mid\}$$

$$\langle 3p_x \mid x = \frac{1}{Z} (\tfrac{42}{5})^{1/2}\{(3)^{1/2}\langle 5d_{x^2-y^2}' \mid - \langle 5d_{z^2}' \mid + (5)^{1/2}\langle 5s' \mid\}$$

$$\langle 3p_x \mid y = \frac{3}{Z} (\tfrac{14}{5})^{1/2}\langle 5d_{xy}' \mid$$

$$\langle 3d_{xy} \mid x = \frac{3}{Z} \{(3)^{1/2}\langle 5f_{y^3}' \mid - (\tfrac{1}{5})^{1/2}\langle 5f_{yz^2}' \mid + 2(\tfrac{7}{10})^{1/2}\langle 5p_y' \mid\}$$

$$\langle 3d_{xy} \mid y = \frac{3}{Z} \{-(3)^{1/2}\langle 5f_{x^3}' \mid - (\tfrac{1}{5})^{1/2}\langle 5f_{xz^2}' \mid + 2(\tfrac{7}{10})^{1/2}\langle 5p_x' \mid\}$$

$$\langle 3d_{xy} \mid z = \frac{3}{Z} (2)^{1/2}\langle 5f_{xyz}' \mid$$

$$\langle 3d_{x^2-y^2} \mid x = \frac{3}{Z} \{(3)^{1/2}\langle 5f_{x^3}' \mid - (\tfrac{1}{5})^{1/2}\langle 5f_{xz^2}' \mid + 2(\tfrac{7}{10})^{1/2}\langle 5p_x' \mid\}$$

$$\langle 3d_{x^2-y^2} \mid y = \frac{3}{Z} \{(3)^{1/2}\langle 5f_{y^3}' \mid + (\tfrac{1}{5})^{1/2}\langle 5f_{yz^2}' \mid - 2(\tfrac{7}{10})^{1/2}\langle 5p_y' \mid\}$$

$$\langle 3d_{x^2-y^2} \mid z = \frac{3}{Z} (2)^{1/2}\langle 5f_{z(x^2-y^2)}' \mid$$

$$\langle 3d_{xz} \mid x = \frac{6}{Z} \{(\tfrac{1}{2})^{1/2}\langle 5f_{z(x^2-y^2)}' \mid - (\tfrac{3}{10})^{1/2}\langle f_{z^3}' \mid + (\tfrac{7}{10})^{1/2}\langle 5p_z' \mid\}$$

$$\langle 3d_{xz} \mid y = \frac{3}{Z} (2)^{1/2} \langle 5f_{xyz}' \mid$$

$$\langle 3d_{xz} \mid z = \frac{6}{Z} \{ (\tfrac{4}{5})^{1/2} \langle 5f_{xz^2}' \mid + (\tfrac{7}{10})^{1/2} \langle 5p_x' \mid \}$$

$$\langle 3d_{yz} \mid x = \frac{3}{Z} (2)^{1/2} \langle 5f_{xyz}' \mid$$

$$\langle 3d_{yz} \mid y = \frac{6}{Z} \{ -(\tfrac{1}{2})^{1/2} \langle 5f_{z(x^2-y^2)}' \mid - (\tfrac{3}{10})^{1/2} \langle 5f_{z^3}' \mid + (\tfrac{7}{10})^{1/2} \langle 5p_z' \mid \}$$

$$\langle 3d_{yz} \mid z = \frac{6}{Z} \{ (\tfrac{4}{5})^{1/2} \langle 5f_{yz^2}' \mid + (\tfrac{7}{10})^{1/2} \langle 5p_y' \mid \}$$

$$\langle 3d_{z^2} \mid x = \frac{6}{Z} \{ (\tfrac{3}{5})^{1/2} \langle 5f_{xz^2}' \mid - (\tfrac{7}{30})^{1/2} \langle 5p_x' \mid \}$$

$$\langle 3d_{z^2} \mid y = \frac{6}{Z} \{ (\tfrac{3}{5})^{1/2} \langle 5f_{yz^2}' \mid - (\tfrac{7}{30})^{1/2} \langle 5p_y' \mid \}$$

$$\langle 3d_{z^2} \mid z = \frac{6}{Z} \{ (\tfrac{9}{10})^{1/2} \langle 5f_{z^3}' \mid + 2(\tfrac{7}{30})^{1/2} \langle 5p_z' \mid \}$$

APPENDIX G

One-Center Matrix Elements of $\hat{\mathbf{r}}$ in a Slater Basis

TABLE G.1. *One-center matrix elements of $\hat{\mathbf{r}}$ between s and p orbitals.*

(Units depend on principal quantum numbers and are given immediately below Table G.1.)

	$\langle p_x \mid$	$\langle p_y \mid$	$\langle p_z \mid$
$x \mid s \rangle$	1	0	0
$y \mid s \rangle$	0	1	0
$x \mid s \rangle$	0	0	1

The units are atomic units of length. Z_{ns} and Z_{np} are the *effective nuclear charges* (i.e., **Z**) felt by an electron in an ns or an np orbital

$$\langle 1s \mid x \mid 2p_x \rangle = \left(\frac{Z_{1s}{}^3 Z_{2p}{}^5}{2}\right)^{1/2} \frac{2^8}{(2Z_{1s} + Z_{2p})^5}$$

$$\langle 1s \mid x \mid 3p_x \rangle = (10 Z_{1s}{}^3 Z_{3p}{}^7)^{1/2} \frac{(3^2)(2^5)}{(3Z_{1s} + Z_{3p})^6}$$

437

$$\langle 1s \mid x \mid 4p_x \rangle = \left(\frac{Z_{1s}{}^3 Z_{4p}{}^{8.4}}{3} \right)^{1/2} \frac{\Gamma(6.7)}{[\Gamma(8.4)]^{1/2}} \frac{(2)^{5.2}(3.7)^{2.5}}{[(3.7)Z_{1s} + Z_{4p}]^{6.7}}$$

$$\langle 1s \mid x \mid 5p_x \rangle = \left(\frac{15 Z_{1s}{}^3 Z_{5p}{}^9}{7} \right)^{1/2} \frac{(2)^{11}}{(4Z_{1s} + Z_{5p})^7}$$

$$\langle 2s \mid x \mid 2p_x \rangle = \left(\frac{Z_{2s}{}^5 Z_{2p}{}^5}{3} \right)^{1/2} \frac{5(2)^6}{(Z_{2s} + Z_{2p})^6}$$

$$\langle 2s \mid x \mid 3p_x \rangle = (15 Z_{2s}{}^5 Z_{3p}{}^7)^{1/2} \frac{(3)^3(2)^{11}}{(3Z_{2s} + 2Z_{3p})^7}$$

$$\langle 2s \mid x \mid 4p_x \rangle = \left(\frac{Z_{2s}{}^5 Z_{4p}{}^{8.4}}{18} \right)^{1/2} \frac{\Gamma(7.7)}{[\Gamma(8.4)]^{1/2}} \frac{(2)^{10.9}(3.7)^{3.5}}{[(3.7)Z_{2s} + 2Z_{4p}]^{7.7}}$$

$$\langle 2s \mid x \mid 5p_x \rangle = \left(\frac{35 Z_{2s}{}^5 Z_{5p}{}^9}{2} \right)^{1/2} \frac{(2)^{11}}{(2Z_{2s} + Z_{5p})^8}$$

$$\langle 3s \mid x \mid 2p_x \rangle = (15 Z_{2p}{}^5 Z_{3s}{}^7)^{1/2} \frac{(3)^3(2)^{11}}{(3Z_{2p} + 2Z_{3s})^7}$$

$$\langle 3s \mid x \mid 3p_x \rangle = (3 Z_{3s}{}^7 Z_{3p}{}^7)^{1/2} \frac{(7)(2)^7}{(Z_{3s} + Z_{3p})^8}$$

$$\langle 3s \mid x \mid 4p_x \rangle = \left(\frac{Z_{3s}{}^7 Z_{4p}{}^{8.4}}{5} \right)^{1/2} \frac{\Gamma(8.7)}{[\Gamma(8.4)]^{1/2}} \frac{(2)^{5.7}(3)^{3.7}(3.7)^{4.5}}{[(3.7)Z_{3s} + 3Z_{4p}]^{8.7}}$$

$$\langle 3s \mid x \mid 5p_x \rangle = (14 Z_{3s}{}^7 Z_{5p}{}^9)^{1/2} \frac{(3)^5(2)^{18}}{(3Z_{5p} + 4Z_{3s})^9}$$

$$\langle 4s \mid x \mid 2p_x \rangle = \left(\frac{Z_{2p}{}^5 Z_{4s}{}^{8.4}}{18} \right)^{1/2} \frac{\Gamma(7.7)}{[\Gamma(8.4)]^{1/2}} \frac{(2)^{10.9}(3.7)^{3.5}}{[2Z_{4s} + (3.7)Z_{2p}]^{7.7}}$$

$$\langle 4s \mid x \mid 3p_x \rangle = \left(\frac{Z_{3p}{}^7 Z_{4s}{}^{8.4}}{5} \right)^{1/2} \frac{\Gamma(8.7)}{[\Gamma(8.4)]^{1/2}} \frac{(2)^{5.7}(3)^{3.7}(3.7)^{4.5}}{[3Z_{4s} + (3.7)Z_{3p}]^{8.7}}$$

$$\langle 4s \mid x \mid 4p_x \rangle = \left(\frac{Z_{4s}^{8.4} Z_{4p}^{8.4}}{3}\right)^{1/2} \frac{(8.4)(3.7)(2)^{8.4}}{[Z_{4s} + Z_{4p}]^{9.4}}$$

$$\langle 4s \mid x \mid 5p_x \rangle = \left(\frac{Z_{4s}^{8.4} Z_{5p}^{9}}{945}\right)^{1/2} \frac{\Gamma(9.7)}{[\Gamma(8.4)]^{1/2}} \frac{(2)^{15.6}(3.7)^{5.5}}{[4Z_{4s} + (3.7)Z_{5p}]^{9.7}}$$

$$\langle 5s \mid x \mid 2p_x \rangle = \left(\frac{35 Z_{3s}^{9} Z_{2p}^{5}}{2}\right)^{1/2} \frac{(2)^{11}}{(Z_{5s} + 2Z_{2p})^{8}}$$

$$\langle 5s \mid x \mid 3p_x \rangle = (14 Z_{5s}^{9} Z_{3p}^{7})^{1/2} \frac{(3)^{5}(2)^{18}}{(3Z_{5s} + 4Z_{3p})^{9}}$$

$$\langle 5s \mid x \mid 4p_x \rangle = \left(\frac{Z_{4p}^{8.4} Z_{5s}^{9}}{945}\right)^{1/2} \frac{\Gamma(9.7)}{[\Gamma(8.4)]^{1/2}} \frac{(2)^{15.6}(3.7)^{5.5}}{[4Z_{5s} + (3.7)Z_{4p}]^{9.7}}$$

$$\langle 5s \mid x \mid 5p_x \rangle = (Z_{5s}^{9} Z_{5p}^{9})^{1/2} \frac{(3)(2)^{11}}{(Z_{5s} + Z_{5p})^{10}}$$

TABLE G.2. *One-center matrix elements of the transition moment operator $\hat{\mathbf{r}}$ between p and d orbitals.*

(Units depend on principal quantum numbers and are given immediately below Table G.2.)

	$\langle d_{x^2-y^2} \mid$	$\langle d_{z^2} \mid$	$\langle d_{xy} \mid$	$\langle d_{xz} \mid$	$\langle d_{yz} \mid$
$x \mid p_x \rangle$	1	$-(\frac{1}{3})^{1/2}$	0	0	0
$y \mid p_x \rangle$	0	0	1	0	0
$z \mid p_x \rangle$	0	0	0	1	0
$x \mid p_y \rangle$	0	0	1	0	0
$y \mid p_y \rangle$	-1	$-(\frac{1}{3})^{1/2}$	0	0	0
$z \mid p_y \rangle$	0	0	0	0	1
$x \mid p_z \rangle$	0	0	0	1	0
$y \mid p_z \rangle$	0	0	0	0	1
$z \mid p_z \rangle$	0	$(\frac{4}{3})^{1/2}$	0	0	0

The units are atomic units of length. Z_{np} and Z_{nd} are the *effective nuclear charges* (i.e., **Z**) felt by an electron in an *np* or an *nd* orbital.

$$\langle 2p_x \mid x \mid 3d_{x^2-y^2}\rangle = (Z_{2p}{}^5 Z_{3d}{}^7)^{1/2} \frac{(3)^4(2)^{11}}{(3Z_{2p}+2Z_{3d})^7}$$

$$\langle 2p_x \mid x \mid 4d_{x^2-y^2}\rangle = \left(\frac{Z_{2p}{}^5 Z_{4d}{}^{8.4}}{15}\right)^{1/2} \frac{\Gamma(7.7)}{[\Gamma(8.4)]^{1/2}} \frac{(2)^{10.4}(3.7)^{3.5}}{[(3.7)Z_{2p}+2Z_{3d}]^{7.7}}$$

$$\langle 2p_x \mid x \mid 5d_{x^2-y^2}\rangle = \left(\frac{21Z_{2p}{}^5 Z_{5d}{}^9}{2}\right)^{1/2} \frac{(2)^{11}}{(2Z_{2p}+Z_{5d})^8}$$

$$\langle 3p_x \mid x \mid 3d_{x^2-y^2}\rangle = \left(\frac{Z_{3p}{}^7 Z_{3d}{}^7}{5}\right)^{1/2} \frac{(21)(2)^7}{(Z_{3p}+Z_{3d})^8}$$

$$\langle 3p_x \mid x \mid 4d_{x^2-y^2}\rangle = \left(\frac{Z_{3p}{}^7 Z_{4d}{}^{8.4}}{25}\right)^{1/2} \frac{\Gamma(8.7)}{[\Gamma(8.4)]^{1/2}} \frac{(2)^{5.7}(3)^{4.2}(3.7)^{4.5}}{[(3.7)Z_{3p}+3Z_{3d}]^{8.7}}$$

$$\langle 3p_x \mid x \mid 5d_{x^2-y^2}\rangle = \left(\frac{21Z_{3p}{}^7 Z_{5d}{}^9}{10}\right)^{1/2} \frac{(3)^5(2)^{19}}{(4Z_{3p}+3Z_{5p})^9}$$

$$\langle 4p_x \mid x \mid 3d_{x^2-y^2}\rangle = \left(\frac{Z_{3d}{}^7 Z_{4p}{}^{8.4}}{25}\right)^{1/2} \frac{\Gamma(8.7)}{[\Gamma(8.4)]^{1/2}} \frac{(2)^{5.7}(3)^{4.2}(3.7)^{4.5}}{[3Z_{3p}+(3.7)Z_{3d}]^{8.7}}$$

$$\langle 4p_x \mid x \mid 4d_{x^2-y^2}\rangle = \left(\frac{Z_{4p}{}^{8.4} Z_{4d}{}^{8.4}}{5}\right)^{1/2} \frac{(3.7)(8.4)(2)^{8.4}}{[Z_{4p}+Z_{4d}]^{9.4}}$$

$$\langle 4p_x \mid x \mid 5d_{x^2-y^2}\rangle = \left(\frac{Z_{4p}{}^{8.4} Z_{5d}{}^9}{1575}\right)^{1/2} \frac{\Gamma(9.7)}{[\Gamma(8.4)]^{1/2}} \frac{(2)^{15.6}(3.7)^{5.5}}{[(3.7)Z_{5d}+4Z_{4p}]^{9.7}}$$

$$\langle 5p_x \mid x \mid 3d_{x^2-y^2}\rangle = \left(\frac{21Z_{5p}{}^9 Z_{3d}{}^7}{10}\right)^{1/2} \frac{(3)^5(2)^{19}}{(3Z_{5p}+4Z_{3d})^9}$$

$$\langle 5p_x \mid x \mid 4d_{x^2-y^2}\rangle = \left(\frac{Z_{5p}{}^9 Z_{4d}{}^{8.4}}{1575}\right)^{1/2} \frac{\Gamma(9.7)}{[\Gamma(8.4)]^{1/2}} \frac{(2)^{15.6}(3.7)^{5.5}}{[(3.7)Z_{5p}+4Z_{4d}]^{9.7}}$$

$$\langle 5p_x \mid x \mid 5d_{x^2-y^2}\rangle = \left(\frac{Z_{5p}{}^9 Z_{5d}{}^9}{5}\right)^{1/2} \frac{(3)^2(2)^{11}}{(Z_{5p}+Z_{5d})^{10}}$$

TABLE G.3. *One-center matrix elements of the transition moment operator $\hat{\mathbf{r}}$ between d and f orbitals.*

(Units depend on principal quantum numbers and are given immediately following Table G.3.)

	$\langle f_{z^3}\vert$	$\langle f_{yz^2}\vert$	$\langle f_{xz^2}\vert$	$\langle f_{xyz}\vert$	$\langle f_{z(x^2-y^2)}\vert$	$\langle f_{y^3}\vert$	$\langle f_{x^3}\vert$
$x\,\vert\,d_{xy}\rangle$	0	-1	0	0	0	$(15)^{1/2}$	0
$y\,\vert\,d_{xy}\rangle$	0	0	-1	0	0	0	$-(15)^{1/2}$
$z\,\vert\,d_{xy}\rangle$	0	0	0	$(10)^{1/2}$	0	0	0
$x\,\vert\,d_{xz}\rangle$	$-(6)^{1/2}$	0	0	0	$(10)^{1/2}$	0	0
$y\,\vert\,d_{xz}\rangle$	0	0	0	$(10)^{1/2}$	0	0	0
$z\,\vert\,d_{xz}\rangle$	0	0	4	0	0	0	0
$x\,\vert\,d_{yz}\rangle$	0	0	0	$(10)^{1/2}$	0	0	0
$y\,\vert\,d_{yz}\rangle$	$-(6)^{1/2}$	0	0	0	$-(10)^{1/2}$	0	0
$z\,\vert\,d_{yz}$	0	4	0	0	0	0	0
$x\,\vert\,d_{x^2-y^2}\rangle$	0	0	-1	0	0	0	$-(15)^{1/2}$
$y\,\vert\,d_{x^2-y^2}\rangle$	0	1	0	0	0	$(15)^{1/2}$	0
$z\,\vert\,d_{x^2-y^2}\rangle$	0	0	0	0	$(10)^{1/2}$	0	0
$x\,\vert\,d_{z^2}\rangle$	0	0	$(12)^{1/2}$	0	0	0	0
$y\,\vert\,d_{z^2}\rangle$	0	$(12)^{1/2}$	0	0	0	0	0
$z\,\vert\,d_{z^2}\rangle$	$(18)^{1/2}$	0	0	0	0	0	0

The units of Table G.3 are quoted in terms of the atomic unit of length (a.u.). Z_{nd} and Z_{nf} are the *effective nuclear charges* (i.e., Z) felt by an electron in an nd or an nf orbital.

$$\langle 3d_{x^2-y^2}\,\vert\,y\,\vert\,4f_{yz^2}\rangle = \left(\frac{Z_{3d}{}^7 Z_{4f}{}^{8.4}}{175}\right)^{1/2}\frac{\Gamma(8.7)}{[\Gamma(8.4)]^{1/2}}\frac{(2)^{5.2}(3)^{4.2}(3.7)^{4.5}}{[(3.7)Z_d + 3Z_f]^{8.7}}$$

$$\langle 3d_{x^2-y^2}\,\vert\,y\,\vert\,5f_{yz^2}\rangle = \left(\frac{3Z_{3d}{}^7 Z_{5f}{}^9}{5}\right)^{1/2}\frac{(3)^5(2)^{18}}{(4Z_{3d}+3Z_{5f})^9}$$

$$\langle 4d_{x^2-y^2}\,\vert\,y\,\vert\,4f_{yz^2}\rangle = \left(\frac{Z_{4d}{}^{8.4}Z_{4f}{}^{8.4}}{35}\right)^{1/2}\frac{(3.7)(8.4)(2)^{7.9}}{[Z_{4f}+Z_{4d}]^{9.4}}$$

$$\langle 4d_{x^2-y^2}\,\vert\,y\,\vert\,5f_{yz^2}\rangle = (Z_{4d}{}^{8.4}Z_{5f}{}^9)^{1/2}\frac{\Gamma(9.7)}{[\Gamma(8.4)]^{1/2}}\frac{(2)^{15.1}(3.7)^{5.5}}{(105)[4Z_{4d}+(3.7)Z_{5f}]^{9.7}}$$

$$\langle 5d_{x^2-y^2}\,\vert\,y\,\vert\,4f_{yz^2}\rangle = (Z_{4f}{}^{8.4}Z_{5d}{}^9)^{1/2}\frac{\Gamma(9.7)}{[\Gamma(8.4)]^{1/2}}\frac{(2)^{15.1}(3.7)^{5.5}}{(105)[4Z_{4f}+(3.7)Z_{5d}]^{9.7}}$$

$$\langle 5d_{x^2-y^2}\,\vert\,y\,\vert\,5f_{yz^2}\rangle = \left(\frac{2Z_{5d}{}^9 Z_{5f}{}^9}{35}\right)^{1/2}\frac{(3)^2(2)^{10}}{(Z_{5f}+Z_{5d})^{10}}$$

APPENDIX H

Effect of the Orbital Operators $\hat{\ell}_p$, where $p = x, y, z$, on Real d and f Atomic Orbitals (Units of $\hbar$)

[The effects of the $\hat{l}_p$ operators on p atomic orbitals are given in Table 11.4. The $\hat{l}_p$ operators are discussed in Secs. 1(F), 1(G), and 1(H), chapter 11.]

TABLE H.1.

$\hat{l}_x$	$\lvert d_{xy} \rangle$	$\lvert d_{xz} \rangle$	$\lvert d_{yz} \rangle$	$\lvert d_{x^2-y^2} \rangle$	$\lvert d_{z^2} \rangle$
$\langle d_{xy} \rvert$	0	$-i$	0	0	0
$\langle d_{xz} \rvert$	i	0	0	0	0
$\langle d_{yz} \rvert$	0	0	0	$-i$	$-i\,(3)^{1/2}$
$\langle d_{x^2-y^2} \rvert$	0	0	i	0	0
$\langle d_{z^2} \rvert$	0	0	$-i\,(3)^{1/2}$	0	0

TABLE H.2.

$\hat{l}_y$	$\lvert d_{xy}\rangle$	$\lvert d_{xz}\rangle$	$\lvert d_{yz}\rangle$	$\lvert d_{x^2-y^2}\rangle$	$\lvert d_{z^2}\rangle$
$\langle d_{xy}\rvert$	0	0	i	0	0
$\langle d_{xz}\rvert$	0	0	0	$-i$	$i(3)^{1/2}$
$\langle d_{yz}\rvert$	$-i$	0	0	0	0
$\langle d_{x^2-y^2}\rvert$	0	i	0	0	0
$\langle d_{z^2}\rvert$	0	$-i(3)^{1/2}$	0	0	0

TABLE H.3.

$\hat{l}_z$	$\lvert d_{xy}\rangle$	$\lvert d_{xz}\rangle$	$\lvert d_{yz}\rangle$	$\lvert d_{x^2-y^2}\rangle$	$\lvert d_{z^2}\rangle$
$\langle d_{xy}\rvert$	0	0	0	$2i$	0
$\langle d_{xz}\rvert$	0	0	$-i$	0	0
$\langle d_{yz}\rvert$	0	i	0	0	0
$\langle d_{x^2-y^2}\rvert$	$-2i$	0	0	0	0
$\langle d_{z^2}\rvert$	0	0	0	0	0

TABLE H.4.

$\hat{l}_x$	$\lvert f_{z^3}\rangle$	$\lvert f_{xz^2}\rangle$	$\lvert f_{yz^2}\rangle$	$\lvert f_{xyz}\rangle$	$\lvert f_{z(x^2-y^2)}\rangle$	$\lvert f_{x^3}\rangle$	$\lvert f_{y^3}\rangle$
$\langle f_{z^3}\rvert$	0	0	$i(6)^{1/2}$	0	0	0	0
$\langle f_{xz^2}\rvert$	0	0	0	$i(\tfrac{5}{2})^{1/2}$	0	0	0
$\langle f_{yz^2}\rvert$	$-i(6)^{1/2}$	0	0	0	$-i(\tfrac{5}{2})^{1/2}$	0	0
$\langle f_{xyz}\rvert$	0	$-i(\tfrac{5}{2})^{1/2}$	0	0	0	$-i(\tfrac{3}{2})^{1/2}$	0
$\langle f_{z(x^2-y^2)}\rvert$	0	0	$i(\tfrac{5}{2})^{1/2}$	0	0	0	$i(\tfrac{3}{2})^{1/2}$
$\langle f_{x^3}\rvert$	0	0	0	$i(\tfrac{3}{2})^{1/2}$	0	0	0
$\langle f_{y^3}\rvert$	0	0	0	0	$-i(\tfrac{3}{2})^{1/2}$	0	0

TABLE H.5.

$\hat{l}_y$	$\lvert f_{z^3}\rangle$	$\lvert f_{xz^2}\rangle$	$\lvert f_{yz^2}\rangle$	$\lvert f_{xyz}\rangle$	$\lvert f_{z(x^2-y^2)}\rangle$	$\lvert f_{x^3}\rangle$	$\lvert f_{y^3}\rangle$
$\langle f_{z^3}\rvert$	0	$-i(6)^{1/2}$	0	0	0	0	0
$\langle f_{xz^2}\rvert$	$i(6)^{1/2}$	0	0	0	$-i(\tfrac{5}{2})^{1/2}$	0	0
$\langle f_{yz^2}\rvert$	0	0	0	$-i(\tfrac{5}{2})^{1/2}$	0	0	0
$\langle f_{xyz}\rvert$	0	0	$i(\tfrac{5}{2})^{1/2}$	0	0	0	$-i(\tfrac{3}{2})^{1/2}$
$\langle f_{z(x^2-y^2)}\rvert$	0	$i(\tfrac{5}{2})^{1/2}$	0	0	0	$-i(\tfrac{3}{2})^{1/2}$	0
$\langle f_{x^3}\rvert$	0	0	0	0	$i(\tfrac{3}{2})^{1/2}$	0	0
$\langle f_{y^3}\rvert$	0	0	0	$i(\tfrac{3}{2})^{1/2}$	0	0	0

TABLE H.6.

$\hat{l}_z$	$\|f_{z^3}\rangle$	$\|f_{xz^2}\rangle$	$\|f_{yz^2}\rangle$	$\|f_{xyz}\rangle$	$\|f_{z(x^2-y^2)}\rangle$	$\|f_{x^3}\rangle$	$\|f_{y^3}\rangle$
$\langle f_{z^3}\|$	0	0	0	0	0	0	0
$\langle f_{xz^2}\|$	0	0	$-i$	0	0	0	0
$\langle f_{yz^2}\|$	0	i	0	0	0	0	0
$\langle f_{xyz}\|$	0	0	0	0	$2i$	0	0
$\langle f_{z(x^2-y^2)}\|$	0	0	0	$-2i$	0	0	0
$\langle f_{x^3}\|$	0	0	0	0	0	0	$-3i$
$\langle f_{y^3}\|$	0	0	0	0	0	$3i$	0

APPENDIX I

Character Tables for Symmetry Groups Used in this Text

$$C_i$$

C_i	E	i	
A_g	1	1	$\hat{R}_x, \hat{R}_y, \hat{R}_z$
A_u	1	-1	x, y, z

(This point group is also reproduced in Table 7.3. However, in an effort to maintain a relationship with the $D_{\infty h}$ group, the representations A_g and A_u have been designated as Σ_g and Σ_u, respectively, in Table 7.3.)

445

$$C_{2v}$$

C_{2v}	E	C_2	$\sigma_v(xz)$	$\sigma_v'(yz)$	
A_1	1	1	1	1	z
A_2	1	1	-1	-1	$\hat{R}_z$
B_1	1	-1	1	-1	$x, \hat{R}_y$
B_2	1	-1	-1	1	$y, \hat{R}_x$

(This point group is also reproduced in Table 7.1.)

$$C_{4v}$$

C_{4v}	E	$2C_4$	C_2	$2\sigma_v$	$2\sigma_d$	
A_1	1	1	1	1	1	z
A_2	1	1	1	-1	-1	$\hat{R}_z$
B_1	1	-1	1	1	-1	
B_2	1	-1	1	-1	1	
E	2	0	-2	0	0	$(x, y), (\hat{R}_x, \hat{R}_y)$

$$C_{\infty v}$$

$C_{\infty v}$	E	$2C_\infty^\phi$	$\cdots$	$\infty\,\sigma_v$	
Σ^+	1	1	$\cdots$	1	z
Σ^-	1	1	$\cdots$	-1	$\hat{R}_z$
Π	2	$2\cos\phi$	$\cdots$	0	$(x, y), (\hat{R}_x, \hat{R}_y)$
Δ	2	$2\cos 2\phi$	$\cdots$	0	
Φ	2	$2\cos 3\phi$	$\cdots$	0	
$\cdot$	$\cdot$	$\cdot$	$\cdots$	$\cdot$	
$\cdot$	$\cdot$	$\cdot$	$\cdots$	$\cdot$	
$\cdot$	$\cdot$	$\cdot$	$\cdots$	$\cdot$	

$$D_6$$

D_6	E	$2C_6$	$2C_3$	C_2	$3C_2'$	$3C_2''$	
A_1	1	1	1	1	1	1	
A_2	1	1	1	1	-1	-1	$z, \hat{R}_z$
B_1	1	-1	1	-1	1	-1	
B_2	1	-1	1	-1	-1	1	
E_1	2	1	-1	-2	0	0	(x, y), $(\hat{R}_x, \hat{R}_y)$
E_2	2	-1	-1	2	0	0	

(This point group is also reproduced in Table 7.5.)

$$D_{2h}$$

D_{2h}	E	$C_2(z)$	$C_2(y)$	$C_2(x)$	i	$\sigma(xy)$	$\sigma(xz)$	$\sigma(yz)$	
A_g	1	1	1	1	1	1	1	1	
B_{1g}	1	1	-1	-1	1	1	-1	-1	$\hat{R}_z$
B_{2g}	1	-1	1	-1	1	-1	1	-1	$\hat{R}_y$
B_{3g}	1	-1	-1	1	1	-1	-1	1	$\hat{R}_x$
A_u	1	1	1	1	-1	-1	-1	-1	
B_{1u}	1	1	-1	-1	-1	-1	1	1	z
B_{2u}	1	-1	1	-1	-1	1	-1	1	y
B_{3u}	1	-1	-1	1	-1	1	1	-1	x

$$D_{4h}$$

D_{4h}	E	$2C_4$	C_2	$2C_2'$	$2C_2''$	i	$2S_4$	σ_h	$2\sigma_v$	$2\sigma_d$	
A_{1g}	1	1	1	1	1	1	1	1	1	1	
A_{2g}	1	1	1	−1	−1	1	1	1	−1	−1	$\hat{R}_z$
B_{1g}	1	−1	1	1	−1	1	−1	1	1	−1	
B_{2g}	1	−1	1	−1	1	1	−1	1	−1	1	
E_g	2	0	−2	0	0	2	0	−2	0	0	$(\hat{R}_x, \hat{R}_y)$
A_{1u}	1	1	1	1	1	−1	−1	−1	−1	−1	
A_{2u}	1	1	1	−1	−1	−1	−1	−1	1	1	z
B_{1u}	1	−1	1	1	−1	−1	1	−1	−1	1	
B_{2u}	1	−1	1	−1	1	−1	1	−1	1	−1	
E_u	2	0	−2	0	0	−2	0	2	0	0	(x, y)

$$D_{6h}$$

D_{6h}	E	$2C_6$	$2C_3$	C_2	$3C_2'$	$3C_2''$	i	$2S_3$	$2S_6$	σ_h	$3\sigma_d$	$3\sigma_v$	
A_{1g}	1	1	1	1	1	1	1	1	1	1	1	1	
A_{2g}	1	1	1	1	-1	-1	1	1	1	1	-1	-1	$\hat{R}_z$
B_{1g}	1	-1	1	-1	1	-1	1	-1	1	-1	1	-1	
B_{2g}	1	-1	1	-1	-1	1	1	-1	1	-1	-1	1	
E_{1g}	2	1	-1	-2	0	0	2	1	-1	-2	0	0	$(\hat{R}_x, \hat{R}_y)$
E_{2g}	2	-1	-1	2	0	0	2	-1	-1	2	0	0	
A_{1u}	1	1	1	1	1	1	-1	-1	-1	-1	-1	-1	
A_{2u}	1	1	1	1	-1	-1	-1	-1	-1	-1	1	1	z
B_{1u}	1	-1	1	-1	1	-1	-1	1	-1	1	-1	1	
B_{2u}	1	-1	1	-1	-1	1	-1	1	-1	1	1	-1	
E_{1u}	2	1	-1	-2	0	0	-2	-1	1	2	0	0	(x, y)
E_{2u}	2	-1	-1	2	0	0	-2	1	1	-2	0	0	

$$D_{\infty h}$$

$D_{\infty h}$	E	$2C_\infty{}^\phi$	$\cdots$	$\infty\,\sigma_v$	i	$2S_\infty{}^\phi$	$\cdots$	$\infty\,C_2$	
$\Sigma_g{}^+$	1	1	$\cdots$	1	1	1	$\cdots$	1	
$\Sigma_g{}^-$	1	1	$\cdots$	-1	1	1	$\cdots$	-1	$\hat{R}_z$
Π_g	2	$2\cos\phi$	$\cdots$	0	2	$-2\cos\phi$	$\cdots$	0	$(\hat{R}_x, \hat{R}_y)$
Δ_g	2	$2\cos 2\phi$	$\cdots$	0	2	$2\cos 2\phi$	$\cdots$	0	
$\cdot$	$\cdot$	$\cdot$	$\cdots$	$\cdot$	$\cdot$	$\cdot$	$\cdots$	$\cdot$	
$\cdot$	$\cdot$	$\cdot$	$\cdots$	$\cdot$	$\cdot$	$\cdot$	$\cdots$	$\cdot$	
$\cdot$	$\cdot$	$\cdot$	$\cdots$	$\cdot$	$\cdot$	$\cdot$	$\cdots$	$\cdot$	
$\Sigma_u{}^+$	1	1	$\cdots$	1	-1	-1	$\cdots$	-1	z
$\Sigma_u{}^-$	1	1	$\cdots$	-1	-1	-1	$\cdots$	1	
Π_u	2	$2\cos\phi$	$\cdots$	0	-2	$2\cos\phi$	$\cdots$	0	(x, y)
Δ_u	2	$2\cos 2\phi$	$\cdots$	0	-2	$-2\cos 2\phi$	$\cdots$	0	
$\cdot$	$\cdot$	$\cdot$	$\cdots$	$\cdot$	$\cdot$	$\cdot$	$\cdots$	$\cdot$	
$\cdot$	$\cdot$	$\cdot$	$\cdots$	$\cdot$	$\cdot$	$\cdot$	$\cdots$	$\cdot$	
$\cdot$	$\cdot$	$\cdot$	$\cdots$	$\cdot$	$\cdot$	$\cdot$	$\cdots$	$\cdot$	

$$T_d$$

T_d	E	$8C_3$	$3C_2$	$6S_4$	$6\sigma_d$	
A_1	1	1	1	1	1	
A_2	1	1	1	-1	-1	
E	2	-1	2	0	0	
T_1	3	0	-1	1	-1	$(\hat{R}_x, \hat{R}_y, \hat{R}_z)$
T_2	3	0	-1	-1	1	(x, y, z)

$$O_h$$

O_h	E	$8C_3$	$6C_2$	$6C_4$	$3C_2$	i	$6S_4$	$8S_6$	$3\sigma_h$	$6\sigma_d$	
A_{1g}	1	1	1	1	1	1	1	1	1	1	
A_{2g}	1	1	-1	-1	1	1	-1	1	1	-1	
E_g	2	-1	0	0	2	2	0	-1	2	0	
T_{1g}	3	0	-1	1	-1	3	1	0	-1	-1	$(\hat{R}_x, \hat{R}_y, \hat{R}_z)$
T_{2g}	3	0	1	-1	-1	3	-1	0	-1	1	
A_{1u}	1	1	1	1	1	-1	-1	-1	-1	-1	
A_{2u}	1	1	-1	-1	1	-1	1	-1	-1	1	
E_u	2	-1	0	0	2	-2	0	1	-2	0	
T_{1u}	3	0	-1	1	-1	-3	-1	0	1	1	(x, y, z)
T_{2u}	3	0	1	-1	-1	-3	1	0	1	-1	

Author Index

Subject Index